Mass Spectrometry for Microbial Proteomics

Mass Spectrometry for Microbial Proteomics

Edited by

HAROUN N. SHAH

Health Protection Agency Centre for Infections, London, UK

and

SAHEER E. GHARBIA

Health Protection Agency Centre for Infections, London, UK

A John Wiley & Sons, Ltd., Publication

Registered office

John Wiley & Sons Ltd, The Atrium, Southern Gate, Chichester, West Sussex, PO19 8SQ, United Kingdom

For details of our global editorial offices, for customer services and for information about how to apply for permission to reuse the copyright material in this book please see our website at www.wiley.com.

Library of Congress Cataloging-in-Publication Data

Mass spectrometry for microbial proteomics / edited by Haroun Shah and Saheer Gharbia.
     p. cm.
  Includes bibliographical references and index.
  ISBN 978-0-470-68199-2
  1. Microbial proteins–Spectra.   2. Mass spectrometry.   3. Proteomics–Methodology.   I. Shah, Haroun N.
II. Gharbia, Saheer.
  QR92.P75.M37 2010
  572'.36–dc22

                                          2010001891

A catalogue record for this book is available from the British Library.

ISBN 978-0-470-68199-2 (H/B)

Set in 10 on 12 pt Times by Toppan Best-set Premedia Limited
Printed and bound in Singapore by Markono Print Media Pte Ltd

For Laila, Louise and Camille.

Contents

Preface

Dictated by the genome, the architectural scaffold of the cell and its communication molecules are largely proteins. The impetus to study proteins predates the era of biochemistry when the latter was at its peak; classical enzymology, radiolabeling methods, chromatography approaches and electrophoresis were used as preparative methods to determine the immensely diverse range of cellular molecules. The reaction of the isothiocyanates with amines to produce fluorescent dansyl amino acid derivatives was ingeniously modified by Edman (referred to as Edman degradation) to deduce the amino acid sequence of proteins by a very tedious and technically demanding process. Mass spectrometry at that time was limited to the characterisation of molecules such as respiratory quinones, lipids, and peptides that were within the analytical range of these instruments, namely ~1000 Da. The simultaneous inventions of new forms of mass spectrometry – matrix-assisted laser desorption/ionisation time of flight mass spectrometry (MALDI-TOF-MS) which was pioneered by Franz Hillenkamp Michael Karas and their colleagues and Koichi Tanaka and his group and, Electrospray Ionisation Mass Spectrometry by John Bennet Fenn, ushered in a new era of biological sciences and its impact has been immense. Protein analysis became more accessible and for the first time could be analysed in real time. This saw a quantum leap in its applications to medicine and biosciences and the number of publications rose so steeply that several new journals were launched to meet this output.

Applications in microbiology have lagged behind and when published, they typically fail to appear in microbiological journals. As such, the general microbiological scientific community is still largely ill at ease with the technology and what can be achieved by using it. Proteome maps have been created for various microbial species and placed on websites of many universities and research institutes. However, whilst the aesthetics are emphasised, little insight is given on the pitfalls and limitations of the data. Unlike DNA where a single or limited range of techniques can be used to derive the sequence of a genome, at present the coverage of a bacterial proteome cannot be deduced by a single technique. Thus, a range of methods and combinations such as 1D-SDS-PAGE, 2D-GE, DIGE, LC and multidimensional LC, ProteinChip or Affinity capture methods are used with MS-MS analysis to derive data. It is generally stated that the correlation between methods is only about 30%; this emphasises the need to use several approaches in tandem. Therefore, one of the aims of this book is to incorporate a range of methods to provide the reader with several options to gain a more holistic view of the proteome of a bacterium.

While there are many books in the field of proteomics and mass spectrometry of higher organisms, there is a notable lack of reference books on lower order, unicellular life forms that make up the most abundant species on earth. Protein analysis of microbes is a burgeoning field but papers are scattered across a broad range of journals. This book aims to capture broad microbial applications of proteomics and mass spectrometry for complementing genome analysis which is now well advanced. Interestingly, mass spectrometry – long considered to be inappropriate for genomic analysis due largely to the poor ionisation of nucleotide bases – is now being actively explored. Accordingly, microbial applications are at the forefront of these applications and are covered in the final chapters of the book.

The chapters follow the general progression in which most microbiology laboratories are approaching the subject:

Transition → Tools → Preparation → Profiling by Patterns → Target Proteins → Data Analysis → DNA Resequencing.

This book is consequently divided into seven sections, as follows:

1. Microbial Characterisation; the Transition from Conventional Methods to Proteomics.
2. Proteomics Tools and Biomarker Discovery.
3. Protein Samples: Preparation Techniques
4. Characterisation of Microorganisms by Pattern Matching of Mass Spectral Profiles and Biomarker Approaches Requiring Minimal Sample Preparation.
5. Targeted Molecules and Analysis of Specific Microorganisms.
6. Statistical Analysis of 2D Gels and Analysis of Mass Spectral Data
7. DNA Resequencing by MALDI-TOF-Mass Spectrometry and its Application to Traditional Microbiological Problems.

At present, genomics and proteomics have developed as separate fields; few laboratories attempt to bridge both areas. Proteomics is almost synonymous with mass spectrometry; however, DNA sequencing by mass spectrometry is rapidly developing and providing valuable support as a high throughput method. The view taken here is that these two subjects are inseparable and while at present the technologies are different, we envisage that in the near future only one technology (mass spectrometry) will be needed to drive both areas.

List of Contributors

Ali Al-Shahib, Department for Bioanalysis and Horizon Technologies, Health Protection Agency Centre for Infections, 61 Colindale Avenue, London NW9 5EQ, UK

Haike Antelmann, Institute of Microbiology, Ernst-Moritz-Arndt-University of Greifswald, F.-L.-Jahn-Str. 15, D-17487 Greifswald, Germany

Catherine Arnold, Department for Bioanalysis and Horizon Technologies, Health Protection Agency Centre for Infections, 61 Colindale Avenue, London NW9 5EQ, UK

Hristo Atanassov, Pôle Biologie-Santé, Pavillon Médecine-Sud, Centre Hospitalier Universitaire – La Milétrie, 40, avenue du recteur Pineau, 86022 Poitiers Cedex, France

Graham Ball, School of Biomedical and Natural Sciences, Nottingham Trent University, Clifton Lane, Nottingham, NG11 8NS, UK

Lucie Balonova, Institute of Molecular Pathology, Faculty of Military Health Science UO, Trebesska 1575, 500 01 Hradec Kralove, Czech Republic

Giulia Bernardini, Department of Molecular Biology, University of Siena, via Fiorentina 1, 53100 Siena, Italy

Chloe Bishop, Department for Bioanalysis and Horizon Technologies, Health Protection Agency Centre for Infections, 61 Colindale Avenue, London NW9 5EQ, UK

Egisto Boschetti, Bio-Rad Laboratories, 92430 Marnes-la-Coquette, France

Daniela Braconi, Department of Molecular Biology, University of Siena, via Fiorentina 1, 53100 Siena, Italy

Elisabet Carlsohn, Proteomics Core Facility, Sahlgrenska Academy, University of Gothenburg, Gothenburg, Sweden

Eleonora Cerasoli, Biophysics and Biodiagnostics Quality of Life Division, National Physical Laboratory, Hampton Road, Teddington TW11 0LW, UK

Yong Chen, Informatics, Sequenom, Inc., 3595 John Hopkins Court, San Diego, CA 92121, USA

Caroline Chilton, Department for Bioanalysis and Horizon Technologies, Health Protection Agency Centre for Infections, 61 Colindale Avenue, London NW9 5EQ, UK

Darren Chooneea, Department for Bioanalysis and Horizon Technologies, Health Protection Agency Centre for Infections, 61 Colindale Avenue, London NW9 5EQ, UK

Malcolm R. Clench, Biomedical Research Centre, Sheffield Hallam University, Howard Street, Sheffield S1 1WB, UK

Jiri Dresler, Institute of Molecular Pathology, Faculty of Military Health Science UO, Trebesska 1575, 500 01 Hradec Kralove, Czech Republic

Vesela Encheva, Research and Technology Department, Laboratory of the Government Chemist, Queens Road, Teddington, Middlesex, TW11 OLY, UK

Marcel Erhard, AnagnosTec GmbH, Am Mühlenberg 11, D-14476 Potsdam OT Golm, Germany

Min Fang, Department for Bioanalysis and Horizon Technologies, Health Protection Agency Centre for Infections, 61 Colindale Avenue, London NW9 5EQ, UK

Simona Francese, Biomedical Research Centre, Sheffield Hallam University, Howard Street, Sheffield S1 1WB, UK

Alena Fucikova, Institute of Molecular Pathology, Faculty of Military Health Science UO, Trebesska 1575, 500 01 Hradec Kralove, Czech Republic

Tom Gaulton, Department for Bioanalysis and Horizon Technologies, Health Protection Agency Centre for Infections, 61 Colindale Avenue, London NW9 5EQ, UK

Saheer E. Gharbia, Department for Bioanalysis and Horizon Technologies, Health Protection Agency Centre for Infections, 61 Colindale Avenue, London NW9 5EQ, UK

Luc Guerrier, Bio-Rad Laboratories, 92430 Marnes-la-Coquette, France

Radhey S. Gupta, Department of Biochemistry, McMaster University, Hamilton, Ontario, Canada L8N 3Z5

Gillian Hallas, Department for Bioanalysis and Horizon Technologies, Health Protection Agency Centre for Infections, 61 Colindale Avenue, London NW9 5EQ, UK

Shea Hamilton, Department of Paediatrics, Imperial College London, St Mary's Campus, Norfolk Place, London W2 1PG, UK

Michael Hecker, Institute of Microbiology, Ernst-Moritz-Arndt-University of Greifswald, F.-L.-Jahn-Str. 15, D-17487 Greifswald, Germany

Lenka Hernychova, Institute of Molecular Pathology, Faculty of Military Health Science UO, Trebesska 1575, 500 01 Hradec Kralove, Czech Republic

Franz Hillenkamp, Institute for Medical Physics and Biophysics, University of Muenster, Robert-Koch-Str. 31, D-48149 Muenster, Germany

Christiane Honisch, Molecular Applications, Sequenom, Inc., 3595 John Hopkins Court, San Diego, CA 92121, USA

Wibke Kallow, AnagnosTec GmbH, Am Mühlenberg 11, D-14476 Potsdam OT Golm, Germany

Roger Karlsson, Department of Chemistry, University of Gothenburg, Kemivägen 10, SE-41296 Gothenburg, Sweden

Natasha A. Karp, Wellcome Trust Sanger Institute,Wellcome Trust Genome Campus, Hinxton, Cambridge, CB10 1SA, UK

Ghalia Khoder, Pôle Biologie-Santé, Pavillon Médecine-Sud, Centre Hospitalier Universitaire – La Milétrie, 40, avenue du recteur Pineau, 86022 Poitiers Cedex, France

Jana Klimentova, Institute of Molecular Pathology, Faculty of Military Health Science UO, Trebesska 1575, 500 01 Hradec Kralove, Czech Republic

J. Simon Kroll, Department of Paediatrics, Imperial College London, St Mary's Campus, Norfolk Place, London W2 1PG, UK

Paul R. Langford, Department of Paediatrics, Imperial College London, St Mary's Campus, Norfolk Place, London W2 1PG, UK

Juraj Lenco, Institute of Molecular Pathology, Faculty of Military Health Science UO, Trebesska 1575, 500 01 Hradec Kralove, Czech Republic

Michael Levin, Department of Paediatrics, Imperial College London, St Mary's Campus, Norfolk Place, London W2 1PG, UK

Marek Link, Institute of Molecular Pathology, Faculty of Military Health Science UO, Trebesska 1575, 500 01 Hradec Kralove, Czech Republic

Raju Misra, Department for Bioanalysis and Horizon Technologies, Health Protection Agency Centre for Infections, 61 Colindale Avenue, London NW9 5EQ, UK

Robert Parker, University of British Columbia, Center for High-Throughput Biology, Vancouver, Canada, V6T, 1Z4

Lakshani Rajakaruna, Department for Bioanalysis and Horizon Technologies, Health Protection Agency Centre for Infections, 61 Colindale Avenue, London NW9 5EQ, UK

Paulina D. Rakowska, Biophysics and Biodiagnostics Quality of Life Division, National Physical Laboratory, Hampton Road, Teddington TW11 0LW, UK

Emmanuel Raptakis, Shimadzu Biotech-Kratos Analytical, Wharfside, Trafford Wharf Road, Manchester M17 1GP, UK

Pier Giorgio Righetti, Department of Chemistry, Materials and Chemical Engineering 'Giulio Natta', Politecnico di Milano, via Mancinelli 7, Milano 20131, Italy

Annalisa Santucci, Department of Molecular Biology, University of Siena, via Fiorentina 1, 53100 Siena, Italy

Haroun N. Shah, Department for Bioanalysis and Horizon Technologies, Health Protection Agency Centre for Infections, 61 Colindale Avenue, London NW9 5EQ, UK

Josef Stegemann, Kühler Grund 50, 69126 Heidelberg, Germany

Jiri Stulik, Institute of Molecular Pathology, Faculty of Military Health Science UO, Trebesska 1575, 500 01 Hradec Kralove, Czech Republic

Nicola C. Thorne, Department for Bioanalysis and Horizon Technologies, Health Protection Agency Centre for Infections, 61 Colindale Avenue, London NW9 5EQ, UK

Ben van Baar, TNO Defence, Security and Safety, PO Box 45, NL-2280 AA Rijswijk, The Netherlands

Robert Ventzki, Am Neuberg 2, D-69221 Dossenheim, Germany

Martin Welker, AnagnosTec GmbH, Am Mühlenberg 11, D-14476 Potsdam OT Golm, Germany

Part I

Microbial Characterisation; the Transition from Conventional Methods to Proteomics

1

Changing Concepts in the Characterisation of Microbes and the Influence of Mass Spectrometry

Haroun N. Shah[1], Caroline Chilton[1], Lakshani Rajakaruna[1], Tom Gaulton[1], Gillian Hallas[1], Hristo Atanassov[2], Ghalia Khoder[2], Paulina D. Rakowska[3], Eleonora Cerasoli[3] and Saheer E. Gharbia[1]

[1] *Department for Bioanalysis and Horizon Technologies, Health Protection Agency Centre for Infections, London, UK*
[2] *Pôle Biologie-Santé, Pavillon Médecine-Sud Centre Hospitalier Universitaire – La Milétrie, Poitiers Cedex, France*
[3] *Biophysics and Biodiagnostics Quality of Life Division, National Physical Laboratory, Teddington, UK*

1.1 Background and Early Attempts to Use Mass Spectrometry on Microbes

The study of diversity and interrelationships between species has been central to the development of microbiology and the driving force behind classification and taxonomy. To link this to the field of infectious diseases, clear circumscription of taxa must be underpinned by robust criteria for accurate diagnosis, epidemiology and studies of microbial pathogenicity. Consequently, since its inception, key features of microbes have been used to provide characters for the description of species. This inventory of properties of bacterial species was reported in the first edition of *Bergey's Manual of Determinative Bacteriology* (1923) and up to the 8th edition (1974) relied almost exclusively on morphological and physiological properties.

Mass Spectrometry for Microbial Proteomics   Edited by Haroun N. Shah and Saheer E. Gharbia
© 2010 John Wiley & Sons, Ltd

The 1950s witnessed a wave of new technologies/instruments into the Life Sciences including high resolution spectrophotometers, gas chromatographs, basic centrifuges and various analytical instruments including early mass spectrometers. Methods to determine intermediate/end products of metabolism were deduced using gas chromatography and were immediately used to characterise anaerobic bacterial species while a number of reports detailed the isolation and characterisation of DNA. DNA base compositions of bacteria were being reported for the first time and attempts were made to quantitatively define taxa by imposing limits of 5 and 10 mol% G+C contents for a species and genus, respectively. DNA-DNA hybridisation set the limits of a species (>70%) while DNA-RNA hybridisation enabled close interspecies phylogenetic relatedness to be inferred for the first time in microbiology (Buchanan and Gibbons, 1974; De Ley *et al.*, 1970).

During this period, the value of chemical analysis of macromolecules of the bacterial cell began. Work and Dewey (1953) and subsequently Cummins and Harris (1956) reported the analysis of amino acids and sugars. Schleifer and Kandler (1972) surveyed the pepti-doglycan chemotypes within the bacterial kingdom using complex chromatographic methods and demonstrated their value in microbial systematics. This was further enhanced by the publication of methods to analyse amino acids in complex mixtures (Atfield and Morris, 1961). Futhermore, Ornstein (1964) and Davis (1964) pioneered the applications of polyacrylamide gel electrophoresis techniques (vertical and horizontal slab gel techniques) and isoelectric focusing of proteins (Shah *et al.*, 1982) which found applications as 'protein-finger-printing' methods in microbiology. This approach was mirrored following the early application in 1952 of pyrolysis mass spectrometry to pyrolyse albumin and pepin (Zemany, 1952) and by 1965 a report in *Nature* described a pyrolysis gas-liquid chromatography method for the identification of bacterial isolates (Reiner, 1965). This was reinforced by Meuzelaar and Kistemaker (1973) and many years later by Barshick *et al.* (1999), who also reported a technique for fast and reproducible fingerprinting of bacteria by pyrolysis mass spectrometry. This method, however, never gained the broad acceptance as a fingerprinting tool, unlike sodium dodecyl sulfate polyacrylamide gel electrophoresis (SDS-PAGE), largely because the technology was cumbersome, inter-laboratory reproducibility was poor and results did not parallel DNA-DNA hybridisation data.

The impetus to dissect the microbe further came with the analysis of lipids using various forms of chromatography with the analysis being very much driven by mass spectrometry. Polar lipids such as phospholipids and nonpolar lipids such as respiratory quinones, porphyrins and long chain cellular fatty acids provided new powerful reliable tests for characterisation of microorganisms (Shah and Collins, 1983). Figure 1.1 shows the value of mass spectrometry in unambiguously identifying the structural variation in chain length of the isoprene units of the 1-4 naphthoquinone ring of menaquinones from the *Bacteroides* and their impact in the restructuring of this large and complex group of microorganisms (Shah and Collins, 1983).

Between the 1960s and the 1990s taxa from nearly all areas of the microbial kingdom were intensely studied using a variety of (bio)chemical analytical methods. However, a major goal of systematic microbiology from its earliest days was to arrange taxa in a phylogentically coherent manner. This was explored initially using rRNA oligonucelotide cataloguing but as DNA sequencing became more accessible, comparative DNA sequencing of the small subunit ribosomal RNA molecule replaced it (Ludwig and Klenk, 2001). By the 1990s, the arrival of polymerase chain reaction (PCR) technologies enabled the

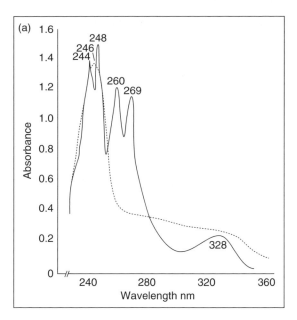

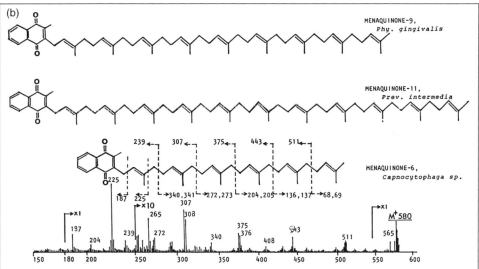

Figure 1.1 *(a) UV spectrum of menaquinoes in iso-octane solution and its reduction with potassium borohydride (dotted line). The characteristic spectrum and reduction shifts confirm the presence of a 1-4 naphthoquinone ring structure but requires mass spectrometry to elucidate its precise structure. (b) Mass spectra from different taxa within the Bacteroides showing variation in the polyprenyl side chain at position 3 of the ring. Menaquinone (MK)-6, MK-9 and MK-11 are shown. These structural variations were consistent among species and in accord with other phenotypic and genotypic properties and were used to restructure this genus*

most abundant and diverse range of taxa globally to be studied. Thus, the current edition of *Bergery's Manual of Systematic Bacteriology* has changed its format to reflect this pattern.

Presently, most of these tests have been superseded by comparative 16S rRNA sequencing and new species are often defined on very limited datasets. However, microbial taxonomists are apprehensive about using a single criterion for the description of taxonomic units and polyphasic approaches are generally regarded as more reliable. Proteomics offers a sound scientific basis to supersede traditional biochemical methods and the arrival of matrix assisted laser desorption/ionisation time-of-flight mass spectrometry (MALDI-TOF-MS) in particular is making this a realistic goal. Bacterial cells have been analysed by a variety of methods, including electrospray ionisation techniques, with considerable success.

1.2   Characterisation of Microorganisms by MALDI-TOF-MS; from Initial Ideas to the Development of the First Comprehensive Database

Mass Spectrometry has been utilised traditionally for chemical analysis and was limited to low molecular weight organic compounds such as respiratory quinones (Figure 1.1). The emergence of gentler ionisation processes led to more techniques for higher molecular mass determination of biological molecules. Plasma desorption, fast atom bombardment, laser desorption and electrospray ionisation have been utilised to analyse components of living organisms (Cotter, 1992). However, the discovery of MALDI perhaps represents the pinnacle of these studies as it permitted the analysis of biological molecules with no theoretical upper mass limit (Karas and Hillenkamp, 1988). In essence a matrix material is mixed with the biological sample and upon irradiation with a laser, molecules in the sample are ionised and desorbed to form a plume of gaseous ions. Nowadays the most common method used to detect this plume of ions is a TOF analyser, in which ions are separated and detected according to their molecular mass and charge. The resulting output of such analyses is a mass spectral profile representing molecular masses of ions in the original plume.

A number of studies have shown that MALDI-TOF-MS may be used for the rapid analysis of biological components of bacterial cells (Liang *et al.*, 1996; Chong *et al.*, 1997; Fenselau, 1997; Dai *et al.*, 1999; Nilsson, 1999). Furthermore these methods have been simplified to utilise intact bacterial cells, significantly reducing preparation time and biomass required for analysis (Claydon *et al.*, 1996; Holland *et al.,* 1996; Krishnamurthy *et al.*, 1996 Arnold *et al.*, 1998; Haag *et al.*, 1998; Wang *et al.*, 1998; Welham *et al.*, 1998; Domin *et al.*, 1999; Lynn *et al.*, 1999) (see Chapters 12 and 13). In its simplest form, intact bacteria are applied to a target plate, mixed with a matrix solution, dried and analysed in the mass spectrometer. The molecules analysed are generally surface components, which hitherto have not been systematically analysed. Since many of the interesting properties relating to microbial physiology (e.g. electron transport, signal transduction, etc.), virulence and pathogenicity (toxin assemble, haemagglutinins, ligands, binding receptors, etc.) are associated with the surface of cells, MALDI-TOF-MS theoretically offers the possibility for large scale comparative analysis of such molecules and provides a means of gaining insight into the diversity of such components among microorganisms.

A drawback of this approach is that surface-associated molecules of cells are notoriously affected by environmental parameters such as the composition of the growth medium, temperature, pH, etc., hence any attempt to utilise this method as a diagnostic technique demands that these parameters are standardised as far as possible and reproducibility studies be meticulously carried out. We reported the results of such a 4 year study (Keys *et al.*, 2004) in which several thousand strains (*iso*9001) were grown on a single medium (quality controlled, Columbia blood agar), and all processes, from the revival and subculture of each bacterial strain, sample preparation, application and MALDI-TOF-MS analysis were rigorously standardised to assemble a database. Considerable preliminary work was undertaken (Shah *et al.*, 2000, 2002) to set the parameters for the ensuing database development and validation. Once established, this database (~5000 spectra and 500 different species) was updated periodically as new species were analysed and, in multicentre studies, validated against strains from various laboratories (Dare, 2005).

Up to this point, the database was assembled and tested with a wide range of reference isolates. It was anticipated that the surface properties of each species would vary with its clinical counterpart but that there might be a core set of stable mass ions that was indicative of a particular species. It was not known which species might be stable and which would deviate, nor was the level of deviation known for a given species. To address this, a parallel study was undertaken where cultures recovered from patient specimens in a clinical laboratory, processed within the laboratory, were selected randomly for analysis. Of the 600 isolates collected, 18.4% of the total belonged to *Staphylococcus aureus*, and constituted the largest single group of isolates. A further set of isolates, obtained from the Staphylococcal Reference Unit, Health Protection Agency, were pooled and the entire collection of isolates analysed by MALDI-TOF-MS. The results obtained indicated that clinical isolates shared many common mass ions with type/reference strains which readily permitted their correct identification. The MicrobeLynx software successfully identified all but four isolates to the correct species. Those misidentified in the first instance were subsequently found to be mixed strains or their spectra showed low mass ion intensity. Once these were purified and re-analysed they were confirmed as *S. aureus* by both MALDI-TOF-MS and 16S rRNA sequence analysis. The high percentage of correct identifications coupled with the high speed and minimal sample preparation required, indicated that MALDI-TOF-MS has the potential to perform high throughput identification of clinical isolates of *S. aureus* despite the inherent diversity of this species.

This study was further extended to include another major nosocomical infectious agent, *Clostridium difficile*. Cultures were analysed from cells grown anaerobically on Columbia blood agar, Nutrient agar and Fastidious Anaerobic agar (FAA). However, unlike the resounding success achieved with *S. aureus*, the results obtained with these isolates were equivocal. Few strains were correctly identified to the species level and in most cases the resolution only reached the genus level when strains were cultured on FAA. Re-examination of the basic protocol eventually led to a change of the matrix solution from 5-chloro-2-mercaptobenzothiazole (CMBT) to 2,5-dihydroxy benzoic acid in acetonitrile:ethanol:water (1:1:1) with 0.3% trifluoroacetic acid (DBA). In collaboration with AnagnosTec (Potsdam, Germany) over 100 isolates of *C. difficile*, re-analysed by this method, yielded unequivocal results. Cells treated with DBA and held up to 1 h were shown by electron microscopy to gently disrupt the outer polymeric layers of the cell wall without disintegrating the cell (Figure 1.2). This enabled the DBA to reach the intracellular

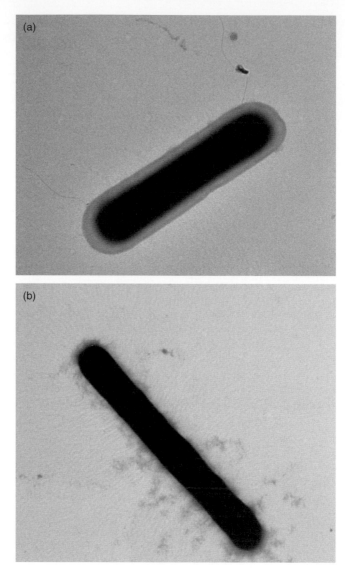

Figure 1.2   *(a) Electron micrograph of* Clostridium difficile *prior to treatment with 2,5-dihydroxy benzoic acid in acetonitrile:ethanol:water (1:1:1) with 0.3% trifluoroacetic acid (DBA). The external polymeric layers of the intact cell are clearly visible. (b) Electron micrograph of C. difficile following the addition of DBA. The disruption of the cell wall polymers is evident but the cell remains intact*

proteins and preferentially ionise the abundant ribosomal proteins (see Chapter 12). This results in a stable and reproducible mass spectral profile irrespective of the culture medium used to grow the cells or duration of the growth curve. This is in marked contrast to the initial method which required the procedure to be rigorously standardised. Thus, following years of development, it is apparent that a method now exists that can be used as a diagnostic platform for the rapid identification of microorganisms. This is backed up by continually expanding databases.

1.3 Characterisation of Microorganisms from Their Intracellular/ Membrane Bound Protein Profiles Using Affinity Capture with Particular Reference to Surface-Enhanced Laser Desorption/ Ionisation (SELDI)-TOF-MS

The diagnostic nature of the profiles obtained from TOF-MS experiments enables the development of extensive databases that may be used for pattern retrieval and profile matching. The capacity to identify individual proteins and clusters of related biomolecules provides a fundamental tool for identifying functions of virulence and epidemiology markers, as well as establishing a foundation for protein selection for vaccine development and antigen identification for studies of communicable diseases. Such information will become vital in understanding the emergence of new pathogens and for following phenotypic patterns among existing pathogens. Mass spectral profiles of peptides/proteins using affinity capture is a relatively new technology for microbiology. Of the readily available methods, SELDI, which utilises ProteinChip arrays with a MALDI-TOF-MS-based analytical platform, is the most readily available. The wells of the MS target plate for the SELDI-MS are manufactured with different surface chemistries, similar in principle to miniature affinity columns and are designed for capturing various classes of proteins prior to MS. The software enables direct conversion of mass intensities into 'gel-view' images making it analogous to an SDS-PAGE profile which is already familiar to microbiologists. To date there appears to be three broad applications of this technology emerging in microbiology which may be summarised as follows:

- A protein fingerprinting platform to replace SDS-PAGE.
- A species-specific diagnostic method.
- A biomarker search tool.

1.3.1 A Protein Fingerprinting Platform to Replace SDS-PAGE

Despite the plasticity of microbial genomes and the frequency of horizontal gene transfer, microbial species retain a large number of stable traits that enables the assignment of isolates to a given taxon. Amongst the most fluid genomes, full genome analysis indicates that changes are often superimposed upon a backbone of core genes that are indispensable to a particular species. Attention is currently focused on a number of stable genes (e.g. 16S rRNA, *rpoB*, etc.) as indicators of evolutionary history and consequently species diversity. The addition of proteomic data has not kept pace with genomics, yet in many cases, the backbone of several currently described diagnostic schemes have emerged from protein/peptide analysis such as multilocus enzymes electrophoresis (MLEE), SDS-PAGE and to a lesser extent isoelectric focusing (IEF)-protein profiles. These have been used for decades for studies on microbial population structure and SDS-PAGE, in particular, even up to the present time is used frequently as a protein fingerprinting method because the data corroborates so well with genome analysis (Heylen *et al.*, 2007; Zanoni *et al.*, 2009).

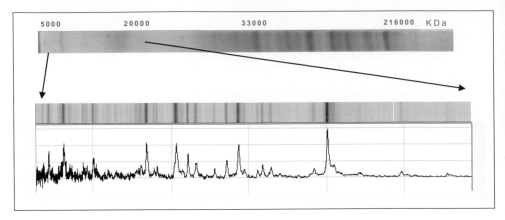

Figure 1.3 *Comparison of resolution between SDS-PAGE and SELDI spectral profiles. The SELDI 'gel view' representation of the mass spectrum between 5 kDa and 20 kDa highlights the greater sensitivity of this method using a bacterial extract. Courtesy of DHI Publishing, LLC. All rights reserved*

However, the technique is being phased out of microbiology, due possibly, to its cumbersome nature and the lack of a global database for inter-laboratory studies. The time is therefore ripe to build upon this foundation and supplant this established principle by more precise, sensitive and rapid mass spectral methods using, for example, affinity capture methods prior to MS analysis.

The SELDI ProteinChip method has drawbacks but because of its simplicity, versatility, sensitivity and reproducibility it is being used in a range of diverse microbiological applications. In practice, sample preparation is very simple; a minute volume (1–2 μl) of cell-free extract is added to several ProteinChips, the sample is overlaid with sinapinic acid, dried and analysed in tandem to give a comprehensive profile of an isolate. It is essential at the onset of a study to utilise several different arrays to optimise the method before seeking to undertake large scale spectral analyses. SELDI profiles are infinitely more sensitive than SDS-PAGE, as shown in Figure 1.3.

Unlike SDS-PAGE, SELDI enables different classes of proteins/peptides to be compared separately (Figure 1.4) or combined, making the technique more versatile and amenable for studying the huge diversity known to exist within the microbial kingdom. The SELDI mass ions are accurate molecular weight values which is in marked contrast to SDS-PAGE where the proteins/polypeptide patterns cannot be accurately sized and rely on calibration protein markers to approximate the molecular weight of various bands.

SELDI profiles of bacterial cells provide a comprehensive approach for high-throughput protein comparisons and, because of its 'gel-view' display, is similar in format to SDS-PAGE patterns (Figure 1.3). A global database based upon SEDLI profiles, if constructed, should overcome the inherent problems faced using SDS-PAGE since the method is rapid, robust, sensitive and the data can be transported electronically between laboratories. We have analysed the SELDI profiles from large numbers of strains of various microbial species using several ProteinChip arrays and, in general, characteristic signal mass ions

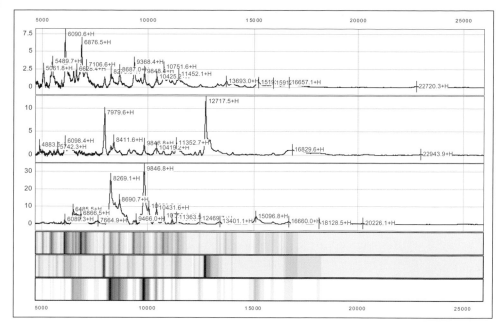

Figure 1.4   *SELDI on different ProteinChips; the top spectrum was acquired using a normal phase array (NP1), the middle spectrum using a strong anionic exchange array (SAX2) and the bottom spectrum using a weak cationic exchange array (WCX1). The gel view images are shown below the spectra*

of species were obtained, in addition to a variety of secondary biomarkers representing subclusters that are unique for a given taxon (Shah *et al.*, 2005). A multicentre study was carried out to demonstrate reproducibility between laboratories using a given set of standard operating procedures (unpublished) which lays the foundation for a public database similar to those used in genomics.

1.3.2   A Species-Specific Diagnostic Method

Extensive analysis over the last 5 years in our laboratory using several thousand isolates has led us to the conclusion, that the 'Biomarker' approach, which has been successfully employed in mammalian work, cannot be extended broadly to microbiology. This is due largely to the inherent diversity within a microbial population of a given species (Encheva *et al.*, 2005, 2006; Lancashire *et al.*, 2005; Shah *et al.*, 2005; Schmid *et al.*, 2006). For example, some species may be well defined and have a tight population structure where their genes are highly conserved and mutation rates are less frequent (e.g. *Mycobacterium tuberculosis*) while others, such as *Neisseria meningitidis,* possess such fluid genomes, that the diversity within the species may be represented as a very broad curve. Thus, the application of these technologies will be strongly influenced by the nature of the species and the problems to be addressed.

Table 1.1   Assignment of a cross-section of clinical strains of Neisseria spp. to N. gonorrhoeae based upon ANN values (<1.5) derived from SELDI-TOF-MS. A few exceptions are shown, namely, N. mucosa and Moraxella osloensis (1.4 and 1.31, respectively) and N. gonorrhoeae (1.65)

Organism	16S sequence identity (%)	Base pair ratio	SELDI ANN
N. gonorrhoeae	99.9	989/990	1.02
N. gonorrhoeae	100	1042/1042	1
N. gonorrhoeae	100	1034/1034	1.2
N. gonorrhoeae	99.90	1023/1024	1
N. gonorrhoeae	100	1030/1030	1
N. gonorrhoeae	100	1022/1022	1
N. gonorrhoeae	100	1033/1033	1
N. gonorrhoeae	99	756/760	1
N. gonorrhoeae	100	1024/1024	1
Kingella denitrificans	99.40	972/978	1.63
Kingella denitrificans	99.30	947/954	1.73
N. gonorrhoeae	100	1034/1034	1.26
N. gonorrhoeae	99.70	937/940	1.22
N. mucosa	99.80	981/983	1.88
N. mucosa	99.30	959/966	1.4
Moraxella osloensis	99.80	1016/1018	1.31
N. gonorrhoeae	99.90	1012/1013	1.02
N. gonorrhoeae	100	989/989	1
N. gonorrhoeae	100	1019/1019	1.65
N. gonorrhoeae	100	1016/1016	1
N. gonorrhoeae	100	1014/1014	1
N. gonorrhoeae	100	1016/1016	1.22

Knowledge of the diversity within a species and being able to assess the limits of acceptance is essential in designing a model for diagnosis of a pathogen with a high degree of confidence. Using a range of phenotypic and genetic markers, the diversity index of each taxon studied may be set. Consequently, it has been possible to select between 50 and 100 isolates that are bona fide members of Neisseria gonorrhoeae, S. aureus, C. difficile and other pathogens that reflect the diversity of each species and analyse them using SELDI-TOF-MS (Schmid et al., 2006). The mass profiles of each group of isolates is used initially as training data sets for artificial neural network (ANN) analysis, validated by a panel of strains and an index developed for acceptance or rejection of a test isolate (see Chapter 17). Table 1.1 shows examples of the data obtained when over 1000 clinical isolates were tested using this model which set a level of >1.5 for acceptance as a member of N. gonorrhoea. With the exception of a few aberrant strains (e.g. Neisseria mucosa = 1.4, Moraxella osloensis = 1.31) clinical isolates were confidently assigned to N. gonorrhoea (<1.5). Exceptions were <1% for misidentified N. gonorrhoea (see Table 1.1) (Schmid et al., 2006). The model was so successful than it was used as the initial screening test prior to 16S rRNA for identification.

In another application of this approach, a similar procedure was used to follow the microevolution of antibiotic resistance of S. aureus. For this study, ANNs were used to search for biomarker ions within the SELDI mass spectral profile. Analysis was carried

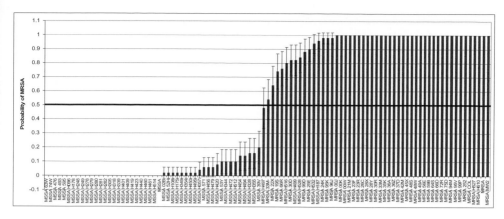

Figure 1.5 *Population distribution curve of methicillin-resistant and -sensitive* Staphylococcus aureus *(MRSA and MSSA, respectively) using ANN analysis. The curve shows that that most of the antibiotic resistant isolates were correctly predicted as MRSA with only two MSSA isolates being incorrectly predicted by the key ions*

out using a stepwise approach in order to rank the ions based on the ability to predict each strain as MRSA and MSSA. Unimportant and noisy values were removed and the remaining single ions (3–30 kDa) were fed into an ANN model (see Chapter 17). From the first set of inputs, ion 3081 Da was chosen as the best predictive ion because of its lower mean error compared with the rest of the ions and the process was repeated until the most important subset of ions was achieved which was in the range 3–19 kDa (3081, 5709, 5893, 7694, 9580, 15 308 and 18 896 Da). Having chosen the seven most predictive ions, each ion was re-analysed for predictive performance using 50 ANN models.

Figure 1.5 shows the population distribution curve of MRSA and MSSA. All methicillin-resistant isolates were correctly predicted as MRSA with only two MSSA (from 24 isolates) being incorrectly predicted by the key ions. As the MRSA isolates get closer to the predictive value of 1, most of these isolates were predicted to be 100 % as MRSA. This indicates that the seven ions chosen for the prediction are characteristic biomarkers for MRSA (Figure 1.5). A similar pattern was concluded for the MSSA isolates with the two misclassified isolates being atypical. Based on these results, an ANN-based model can be used as a rapid diagnostic tool when coupled with MS data and the model could be validated using a blind data set for further confirmation (see Chapter 17).

1.3.3   A Biomarker Search Tool

SELDI-TOF-MS has been promoted as a biomarker search tool (see Chapter 11) and some of its shortcomings have already been alluded to above. However, when confronted with an extremely diverse species and a background of numerous variable parameters that may predispose a pathogen to disease, SELDI because of its simplicity and high throughput nature is a useful platform for large scale screening. This is illustrated here in two studies to search for biomarkers, namely, a relatively small study (at the HPA Centre for Infections) involving *Enterococcus faecalis* and a far more extensive study undertaken at the

Université de Poitiers, France to search for proteomic biomarkers of the gastric pathogen, *Helicobacter pylori.*

1.3.3.1   E. faecalis

E. faecalis colonises the human colon but recent studies have shown that it has the capacity to leave its habitat and colonise other body sites where it is associated with disease. Consequently, they may be found in bacteraemia and a few have been reported in endocarditis (Huycke and Gilmore, 1995). Strains from the latter are often marked by the presence of a potent 'cytolysin' (Coburn *et al.,* 2004). One potential reservoir outside the intestine is the human dental root canal. To search for biomarkers that may help to elucidate the pathogenicity of this species, 60 strains of *E. faecalis* from the dental root canal were collected from Lithuanian and Finnish patients and analysed against strains from the intestinal tract and endocarditis using SELDI-TOF-MS (Reynaud af Geijersstam *et al.*, 2007). The mass spectral profiles and their corresponding 'gel view' images of 'cytolysin producing' strains against non-producers are shown in Figure 1.6(a). Distinct mass ions at 15 060, 15 350 and 16 250 Da were clearly evident among the cytolysin producers. A more comprehensive overview of the entire spectrum of representative strains is shown in the dendrogram in which the 'cytolysin' producing strains were recovered in a distinct cluster [Figure 1.6(b)]. Extensive studies with a wider range of strains are now required to assesss this approach to screen for the prevalence of these strains in patients with endocarditis to assist in evaluating their health risk prior to dental root canal treatment.

1.3.3.2   H. pylori

The gram negative, spiral-shaped and microaerophilic bacterium *H. pylori* is a highly successful human pathogen which colonizes the gastric mucosa and causes inflammation resulting in different clinical outcomes such as chronic and atrophic gastritis, gastro-duodenal ulcers, gastric MALT lymphoma and gastric cancer (Atherton, 2006). Although *H. pylori* has been extensively studied by a vast range of methods, from epidemiology of infected populations to structural biology and biophysics of specific virulence factors, some major issues have not been resolved that are both of clinical and fundamental interest. Clinically the question as to whether there exist combinations of *H. pylori* biomarkers that would be useful to differentiate and predict underlying gastric pathologies, and possibly, to contribute to the development of a successful vaccine remains unanswered. Any experimental strategy dealing with this issue is invariably confronted with the basic biological complexity of the pathogen which is not only characterized by one of the greatest genotypic and phenotypic diversities in the bacterial world, but also by its morphological plasticity. Depending on the environmental conditions and the cell growth phase, all *H. pylori* strains exist in two major forms, bacillary and coccoid, as well as in some intermediate 'U' forms. Another important issue to be taken into consideration is dealing with the changes in phenotype that occurs after multiple passages *in vitro* where the same strain may differ significantly from the phenotype of the initial culture. In addition, several *H. pylori* strains can coexist within the gastric mucosa of the same patient which further complicates studies on biomarker discovery (Hirschl *et al.*, 1994; Yakoob *et al.*, 2001).

The initial search for relevant *H. pylori* genomic biomarkers associated with a particular clinical outcome has proven to be of limited clinical relevance. For example, the

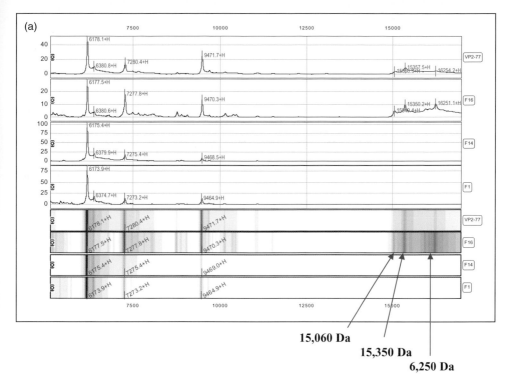

15,060 Da

15,350 Da

6,250 Da

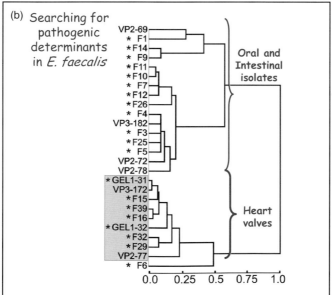

Figure 1.6 *(a) SELDI profiles of* E. faecalis *showing the 'cytolysin' producing strains (two above spectra) and the 'cytolysin' negative strains (two below). The red arrows indicate several distinct mass ions in this area of the spectrum that separates these isolates. Characteristic mass ions at 15060, 15350 and 16250 Da differentiate strains from both sites. (b) Dendrogram of SELDI profiles of* E. faecalis *showing the clustering of the 'cytolysin' producing strains (shaded) and the 'cytolysin' negative strains from the oral and intestinal sites. Reprinted from Reynaud af Geijersstam et al., Oral Microbiology and Immunology. Comparative analysis of virulence, determinants and mass spectral profiles of Finnish and Lithuanian endodontic Enterococcus faecalis isolates. (2007) 22,2 with permission from Wiley-Blackwell*

combination of the extensively studied factors of pathogenicity, *cagA, vacA* and *babA*, does not help to demarcate particular groups of *H. pylori* virulent strains because of the wide distribution of these factors over various biotypes (Hocker and Hohenberger, 2003). Proteomic technologies probably hold more promise for their classification because they represent the expression molecules of the cell. Comparative proteomic experimental protocols for studying *H. pylori* may be divided into two groups, indirect and direct methods. Using the 'indirect' immunoproteomics approach, *H. pylori* proteins were separated by two-dimensional gel electrophoresis (2D GE), transferred to membranes by Western blotting, and screened using a panel of sera collected from *H. pylori*-infected patients with different gastric pathologies (Haas *et al.*, 2002; Jungblut and Bumann, 2002; Utt *et al.*, 2002; Krah and Jungblut, 2004; Krah *et al.*, 2004; Lin *et al.*, 2006, 2007; Pereira *et al.*, 2006). The most frequently immuno-recognised spots were then used for precise localisation of protein antigens on the original gel or membrane used for Western transfer. By contrast, the 'direct' approach has relied on comparisons between 2D GE protein maps of *H. pylori* strains (Enroth *et al.*, 2000; Jungblut *et al.*, 2000; Cho *et al.*, 2002; Govorun *et al.*, 2003; Backert *et al.*, 2005; Park *et al.*, 2006) or SELDI protein/polypeptide profiles (Hynes *et al.*, 2003a,b; Das *et al.*, 2005; Ge *et al.*, 2008; Khoder *et al.*, 2009; Bernarde *et al.*, 2009). In both approaches, the proteins of interest were then identified by LC-MS/MS and matched against *H. pylori* strains with fully annotated genomes.

SELDI ProteinChip.   To date there are few reports of direct profiling *H. pylori* using SELDI ProteinChip technology (Hynes *et al.*, 2003a,b; Bernarde *et al.*, 2009; Khoder *et al.*, 2009). With regard to *H. pylori*, Das *et al.* (2005) and Ge *et al.* (2008) used SELDI to analyse human epithelial cells that had been infected by the pathogen. *H. pylori* strains from distinct geographic origins, Colombia, South Korea and France were studied by this method to search for distinct biomarkers (Bernarde *et al.*, 2009; Khoder *et al.*, 2009) (Figure 1.7). Strains were obtained from patients with gastric cancer (GC), low-grade gastric MALT lymphoma (LG-MALT) or duodenal ulcer (DU). The latter served as a control group, since it is well documented that patients with duodenal ulcer practically do not develop GC (Hansson *et al.*, 1996) or LG-MALT (Suzuki *et al.*, 2006). In an extensive study, 27 statistically significant biomarkers were identified from SELDI profiling, and five of them were shown to be highly correlated with disease. These were purified and eventually identified as a neutrophil-activating protein NapA, a RNA-binding protein, a DNA-binding histone-like protein HU, the 50S ribosomal protein L7/L12 and the urease A subunit. In these investigations strains were cultured several times prior to analysis to obtain a relatively stable phenotype. It is likely that after continuous passages under similar conditions *in vitro*, adaptation of the *H. pylori* strains resulted in stabilisation of the pattern of expressed proteins (Figure 1.8). This phenomenon has also been observed by Hynes *et al.*, (2003a) who conclude that common proteins are detected between initial clinical and culture collection strains of *H. pylori*, but greater variability occurs between the clinical and culture collections strains than among strains from the culture collection alone.

The SELDI ProteinChip method was not only useful for generation of the protein biomarkers defined by their masses, but also in purification of the most significant ones. Throughout the purification steps, all fractions obtained were systematically re-analysed by SELDI ProteinChip to ensure that there was no loss of the target protein (Figure 1.9). Nevertheless, the SELDI technology has limitations. First, irrespective of the nature of the

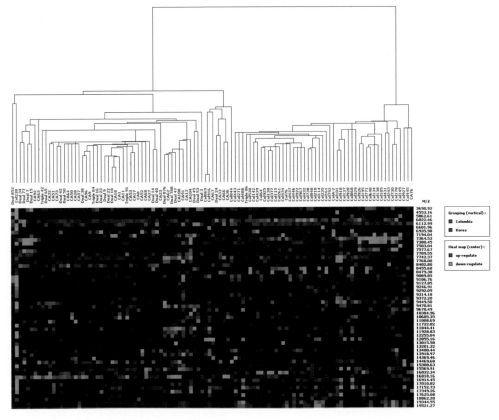

Figure 1.7   *Classification of Asian and South American H. pylori strains to illustrate the initial screening, by heat map with dendrogram, of H. pylori strains belonging to genotypically (and phenotypically) distinct groups. The 'outliers' (i.e. four Colombian strains within the Korean group and two Korean strains within the Colombian group) were excluded from further SELDI profiling thereby reducing the risk of error*

chromatography surface of the ProteinChip used, the most abundant protein patterns are obtained in the range 2.5–30 kDa. However, the number of proteins in the molecular weight range up to 30 kDa is only about 50% among the theoretically predicted proteomes of the sequenced *H. pylori* strains 26695, J99 and HPAG1. Another limitation of the SELDI ProteinChip method is that even in the range 2.5–30 kDa, the number of detectable proteins with signal-to-noise ratio (>5) was between 100 and 200. This may be extended by increasing the number of ProteinChip arrays used with different surface chemistries, and/or modifying the binding conditions (e.g. pH, ionic strength of buffers, etc.). In practice, increasing the initial conditions of protein binding in four of the most frequently used chromatographic ProteinChip arrays (ion-exchange CM10 and Q10, hydrophobic H50, and metal-affinity IMAC30 with chelated copper or zinc) increases the number of detectable proteins, but still remains well below those of the 2D GE (usually 1500–2000 spots per gel).

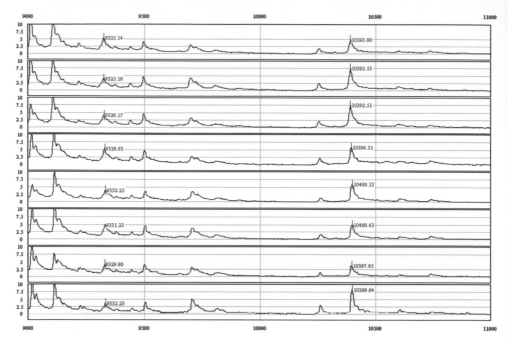

Figure 1.8   *Stable biomarker expression of* H. pylori *after eight passages of one and the same* H. pylori *strain. This enabled confirmation of the phenotypic stability of the two protein biomarkers (indicated in red) prior to further purification and sequencing. Reprinted from Journal of Chromatography B, 877/11-12, Khoder, Yamaoka, Fauchère, Burucoa, Atanassov, Proteomic Helicobacter pylori biomarkers discriminating between duodenal ulcer and gastric cancer. 1193–1199, Copyright 2009, with permission from Elsevier*

Despite these shortcomings, SELDI ProteinChip technology enables high-throughput comparative proteomics of hundreds of samples, which would otherwise be cumbersome and technically difficult by other methods. SELDI-TOF-MS analyses a cross-section of the proteome including some membrane bound proteins and, by comparison with existing methods, the technique is robust. The proteins analysed are generally low molecular weight (3–30 kDa) and of limited number, but the latter may be increased by use of different ProteinChip arrays. Also, only the most abundant proteins are likely to be captured; thus, other methods should also be employed, such as the use of nanoparticles (see below) or depletion methods (see Chapter 9).

1.4   Comparative Analysis of Proteomes of Diverse Strains within a Species; Use of 2D Fluorescence Difference Gel Electrophoresis (DIGE)

1.4.1   2D GE

2D GE is widely used for the analysis of complex protein mixtures extracted from cells. Proteins are separated in the first dimension by isoelectric focusing (IEF) and in the second by SDS-PAGESDS-PAGE according to their molecular weights. Complex mixtures of

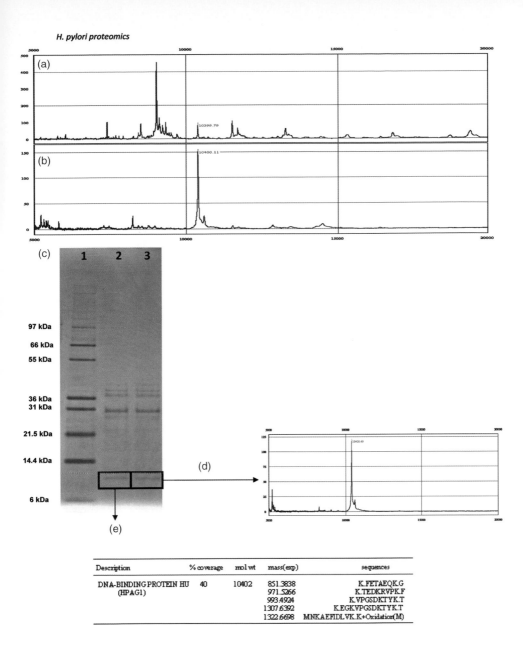

H. pylori proteomics

Description	% coverage	mol wt	mass(exp)	sequences
DNA-BINDING PROTEIN HU (HPAG1)	40	10402	851.3838	K.FETAEQK.G
			971.5266	K.TEDKRVPK.F
			993.4924	K.VPGSDKTYK.T
			1307.6392	K.EGKVPGSDKTYK.T
			1322.6698	MNKAEFIDLVK.K+Oxidation(M)

MNKAEFIDLV KKAGKYNSKREAEEAINAFTLAVETALSKGESVELVGFGKFETAEQKGKEGK
VPGSDKTY KTEDKRVPKFKPGKILKQKVEEGK

Figure 1.9 *Example of SELDI-assisted purification of an H. pylori protein biomarker with molecular mass of 10.4 kDa ('b10.4'). (a) Spectrum of the crude protein extract before fractionation. (b) Spectrum of the b10.4-enriched fraction obtained after RP-HPLC separation of the same crude protein extract in (a). (c) Separation by 1D SDS-PAGE of the b10.4-enriched HPLC fraction in (b). Lane 1, molecular weight markers; lanes 2 and 3, bands containing the same b10.4. (d) Quality control of purity of the passively eluted b10.4 from the 1D SDS-PAGE gel in (c), lane 3. (e) LC-MS/MS identification of b10.4 from the 1D SDS-PAGE gel in (c), lane 2. A DNA-binding protein (94 amino acids) was sequenced through four peptides marked in red, with 40% coverage. Reprinted from Journal of Chromatography B, 877/11-12, Khoder, Yamaoka, Fauchère, Burucoa, Atanassov, Proteomic Helicobacter pylori biomarkers discriminating between duodenal ulcer and gastric cancer. 1193–1199, Copyright 2009, with permission from Elsevier*

thousands of different proteins may be resolved and their identities determined by LC-MS/MS (see Chapter 14). This is often used in the first instance to obtain a map of the proteome of a particular species which can then be cross-referenced to its full genome to obtain an expression profile of a particular isolate under a given set of physiological and environmental parameters (see Chapter 15). This is now a well established procedure when searching for stable biomarkers, as in the case of vaccine targets (see Chapter 16). However, novel applications of 2D GE are now emerging as the landscape of proteomics changes. For example, investigations of antibiotic resistance have traditionally been based on phenotype assays such as disc testing, determinations of minimum inhibitory concentrations and molecular methods to probe the mechanisms involved. However, the increased usage of antibiotics has led to the evolution of novel resistance mechanisms, and a growing complexity of resistance phenotypes. Investigation of the underlying resistance mechanisms in such complex strains by conventional molecular methods is difficult, commonly requiring the screening of multiple known resistance genes, mutations or various cloning strategies. These techniques help to determine a resistance phenotype and allow mechanisms of resistance to be inferred through 'interpretative reading' (Livermore *et al.*, 2001) to monitor the prevalence and spread of resistance, However, due to the complexity of antibiotic resistance mechanisms, interpretations are likely to be incomplete, since many of the processes involved in cellular activities that confer resistance may not be immediately apparent, such as reduced cell permeability and upregulated efflux mechanisms (Szabo *et al.*, 2006).

Alternative approaches that better define these complex combinatorial resistance mechanisms may be fruitful. Proteomics may be an alternative method. This is demonstrated in the following example using a susceptible strain, *E. coli* J53 which was separated and compared with a resistant derivative, containing the multi-resistant plasmid pEK499, designated J499 (Woodford *et al.*, 2009). The plasmid contains a CTX-M enzyme, which hydrolyses extended spectrum β-lactam. The CTX-M enzymes provide their host with resistance against a wide variety of β-lactam antibiotics and their dissemination is global (Canton and Coque, 2006), bacteria were cultured to late log phase with $2\,\mu g\,ml^{-1}$ cefotaxime and cells harvested and mechanically homogenised. Samples of crude lysate from both strains were separated using 2D GE in the pH range 4–7. Gels were compared using gel imaging software and spots unique to each gel were identified. The antibiotic resistance proteins expressed only from the multi-resistant plasmid were excised and subjected to in-gel tryptic digestion, purified, and identified using MALDI-TOF-MS (Linear/Reflectron, Waters Ltd).

TEM-1 (a β-lactamase) and CTX-M-15 were both detected and distinguished by 2D GE and identified by MALDI-MS (Figure 1.10). Both these enzymes were produced in response to CTX, a β-lactam antibiotic. Additionally, the enzyme aminoglycoside acetyltransferase was detected highlighting the potential to gain further insight into bacterial resistance mechanisms. Additional proteins such as TEM-1 and AAC6, which also confer resistance, were also detected (Figure 1.10) – the latter was not screened for and was not thought to be expressed in response to CTX exposure.

In addition to identifying the proteins directly responsible for antibiotic resistance, the separated proteomes of resistant and susceptible organisms are used to determine the effect that acquisition of resistance mechanisms has on the host organism. Quantitative software highlights changes in protein expression by comparing the volume of the protein spots

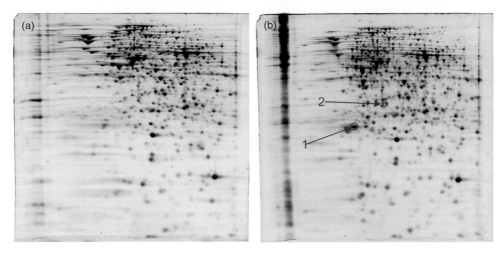

Figure 1.10   (a) Separated crude protein extract of susceptible J53-2 on a pH gradient of 4-7. (b) Separated protein extract of J499, a resistant derivative containing the plasmid pEK499. Circled are some of the novel proteins identified: 1, AAC-6, an aminoglycoside acetyltransferase; 2, TEM-1, a β-lactamase. CTX-M-15 was separated on a separate pH gradient

between gels. Such quantitative differentiation can provide insight into how the acquisition of antibiotic resistance genes or plasmids can affect the physiological processes of the host cell and may help to elucidate why certain resistances are retained long-term and why specific enzymes go on to successfully disseminate in the population.

1.4.2   The DIGE Technique

For comparative analysis such as differences in expression between strains grown under different physiological parameters (see Chapter 15), analysis between virulent and benign strains, antibiotic resistance and sensitive strains, etc., large numbers of isolates often need to be run and compared and this presents one of the major challenges for the technology. New high throughput electrophoretic systems are being developed (see Chapter 10) while others have been devised that can run 12 gels or more in parallel under standard conditions such as temperature, voltage, etc., to minimise variation between individual gels. There is much support for this approach, however, our own experience over several years does not support this view; even the same sample ran on different gels showed variation. Data analysis has improved with the introduction of a number of Software packages such as Delta2D, ImageMaster, Melanie, PDQuest, Progenesis and REDFIN, to match spots between gels of similar samples to search for differences in expression.

We have opted instead to undertake comparative studies of samples run only in the same gel using the DIGE technique which is currently the most convenient method for achieving this and involves the labelling of different proteins prior to 2D GE. The DIGE technology is based on the specific properties of the dyes, the CyDye™ DIGE Fluors. The two sets of dyes currently available, Cy™2, Cy3, and Cy5 minimal dyes, and Cy3 and

Cy5 saturation dyes have been designed to be both mass- and charge-matched. The DIGE chemistry uses *N*-hydroxysuccinimide ester reagents for low-stoichiometry labeling of ε-amino groups of the lysine side chains of proteins. Labeling reactions are optimised so that only 2–5% of the lysine residues are labeled. This is to ensure that quantitation is performed using protein molecules that have been labeled only once. In practice, two protein samples are labeled with a different fluorophor e.g. Cy 3 or Cy 5 and the samples combined and electrophoreised. Identical proteins labeled with each of the CyDye DIGE Fluors will migrate to the same position on a 2D gel and can then be visualised using a fluorescence imager and readily compared. The method has the added advantage that relative quantification of any given protein between two samples is more reliable. The technique enables samples to be multiplexed and compared with an internal standard, generally a mixture of the two samples to be compared. It has been noted that the Cy dyes may differentially label the same protein. To overcome this, reciprocal labeling of gels is recommended. The CyDyes have great sensitivity, detecting as little as 125 pg of protein and giving a linear response to protein concentration of up to four orders of magnitude (silver staining detects 1–60 ng of protein and less than a 100-fold dynamic range). Protein spots of interest may be excised from the gels and directly analysed by mass spectrometry. The latter is not affected by labelling because most peptides will not contain a label. An example involving studies on the pathogenicity of *C. difficile* is used here to illustrate the value of this approach.

1.4.2.1   C. difficile

C. difficile is an enteric pathogen responsible for a variety of gastrointestinal diseases including severe diarrhoea and pseudomembranous colitis (PMC) (Larson and Borriello, 1990; Kelly and Lamont, 1998). In the late 1980s, *C. difficile* emerged as a hospital acquired infectious agent, frequently associated with antibiotic treatment. In recent years, there has been an increase in number and severity of nosocomial infections, as well as an increase in community acquired infections, leading to this pathogen becoming an important focus for research. *C. difficile* is a rod-shaped, gram-positive, anaerobic, spore-forming bacterium. The two major toxins produced by many *C. difficile* strains are thought to be the main virulence factors, and mediate their effects through glucosylation of small GTP binding proteins such as Rho, interrupting cytoskeleton assembly (Borriello, 1998; Voth and Ballard, 2005). Although many of the more recent strains produce an additional binary toxin, the recent increase in virulence does not seem to be simply due to altered toxin expression, indicating that there are many interacting virulence factors playing a role in the severity of the emerging strains (Spigaglia and Mastrantonio, 2004).

The HPA Centre for Infections has completed the genome sequencing of three different *C. difficile* strains from different isolates spanning the last four decades. The target strains included a toxin producing strain from the 1970s (designated B), a low virulence strain from the 1980s (strain T), and a highly virulent recent strain. The latter belonged to the common ribotype (O27) associated with disease outbreaks, and is designated here as strain A. These genome sequences have provided only limited insight into the factors causing the increased virulence of more recently emerging strains, and cannot give any information on actual gene expression. Therefore, the proteomes of these three strains are now being investigated to complement the genomic data by highlighting differences in their protein expression. This may give indications of processes involved in making emerging strains more virulent, and aid diagnosis by pinpointing possible biomarkers. In order to compare the proteomes of the three strains, individual protein components of whole cell protein extracts from the three strains were identified using mass spectrometry, and compared to identify differentially expressed proteins. Mass spectrometry and the subsequent comparison of peptide mass fingerprints against a comprehensive database has become firmly established as a successful proteomics workflow for the identification of unknown proteins. However, in order to be used successfully for complex samples such as microbial cell protein extracts, the upstream protein separation techniques are key steps. Common methods of separation prior to mass spectrometry identification include chromatography and SDS-PAGE as described above.

Two separate approaches were used for this study, 1D SDS-PAGE coupled with LC-MS/MS using an LTQ-orbitrap (Thermo Fisher Scientific), and 2D SDS-PAGE and the spots analysed using a MALDI-TOF Linear/Reflectron (Waters Inc.). Using SDS-PAGE, the gel lanes were cut into a number of bands, each of which was subjected to in-gel trypsin digestion followed by LC-MS/MS (Figure 1.11).

This is a very sensitive method, and allows identification of a comprehensive list of proteins from a complex mixture. Protein extracts from different strains can be compared to determine the expressed profile for each protein extract. This method is relatively quick, and very sensitive, but is used mostly to determine the presence or absence of particular proteins, rather than for quantitative comparisons. A total of 397 *C. difficile* proteins were identified in this way; 52 distinct proteins were characterised in only one strain, which can then be further investigated as potential biomarkers (Figures 1.12 and 1.13).

Comparative analysis of the same strains were the undertaken using the DIGE technique discussed above. In order to compare the *C. difficile* strains, different biological and

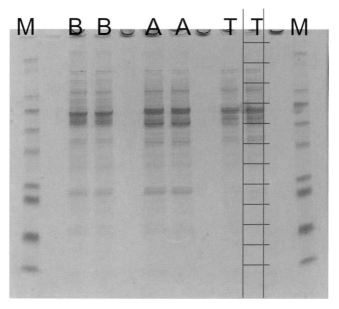

Figure 1.11 *Separation of polypeptides/proteins using SDS-PAGE and staining with Coomassie Brilliant Blue R (Sigma, London). 1D gel lane slices were cut into bands (see red ladder) for trypsin digestion and subjected to LC-MS/MS analysis*

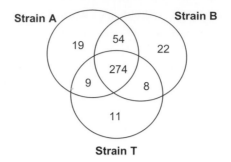

Figure 1.12 *Venn diagram to show the breakdown of proteins identified by LC-MS analysis according to the strains in which they were identified*

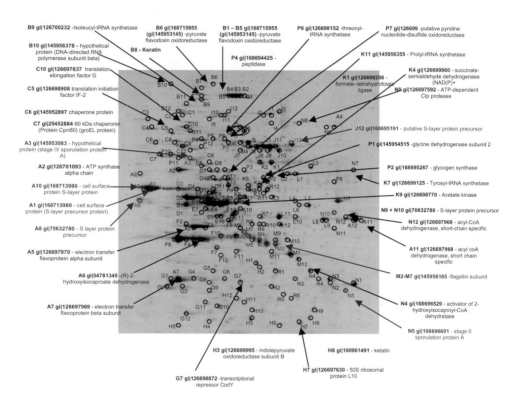

Figure 1.13 *An example of a protein reference map for a whole cell protein extract from* C. difficile *strain A, a hypervirulent O27 ribotype strain*

technical replicates were used and the differences in protein expression between the strains were statistically analysed using the Progenesis SameSpots software (Nonlinear Dynamics) (Figure 1.14). A total of 453 proteins were matched across the standards of all six gels. Proteins with an ANOVA value of $P < 0.05$ and a greater than ±2-fold difference were considered to differ significantly between the strains. Correlation analysis was used on the

Figure 1.14 *A DIGE gel image where protein extracts from two different* C. difficile *strains are run on the same gel. The Cy2 channel (blue) is the internal standard containing a mix of all protein extracts used in the experiment. The Cy3 channel (green) is a strain B protein extract while the Cy5 channel (red) is a strain A protein extract. Proteins appearing green or red are therefore different in the two strains*

112 spots which met these criteria in order compare expression profiles between strains. Twenty-eight proteins were shown to be up-regulated in the B strain, a further 28 were up-regulated in the A strain and 21 in the T strain. The DIGE images were then matched to the picking gels used for the reference maps to identify these proteins (Figure 1.15).

These analyses enabled the identification of proteins which are differentially regulated in virulent versus nonaggressive strains.

During the proteomic analysis of *C. difficile* strains, a number of proteins were identified by only one methodology. The SDS-PAGE-LC-MS/MS approach identified far more proteins and was a more sensitive method but there were also a number of proteins identified on the 2D reference maps that were not detected by other methods. The DIGE approach allowed relative quantification of protein expression between strains, identifying differentially expressed proteins. As this study highlights, the most comprehensive analysis of microbial proteomes is achieved by using a combination of different methodologies.

1.5   Nanoparticles as an Alternative Approach in the Analysis and Detection of Low Abundance and Low Molecular Weight Proteins Using MALDI-TOF-MS

The methods described above (e.g. 1D SDS-PAGE, 2D GE, IEF, SELDI-TOF-MS, etc.) are not selective and therefore any class of proteins that is over-represented (e.g. species

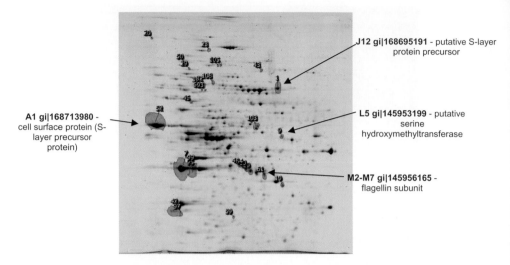

Figure 1.15   *Proteins identified as upregulated in strain A, with some identification from the protein reference map. The proteins identified as up-regulated in strain A by correlation analysis were matched to the 'picking gel' used to create the strain A reference map. The numbers indicate the rank of the protein, with protein 1 showing the greatest fold difference between the strains*

in high abundance) in a sample is likely to be preferentially analysed. In biological fluids, some of the most important analytes of interest are often present at very low levels and masked by high abundance biomolecules. Because of the over-representation of several proteins in biological fluids, e.g. albumin in serum (~60%), these are likely to be preferentially analysed. To date, improvements in mass spectroscopy-based detection has been made by pretreatment of complex samples by fractionation or removal of the most abundant components. Different depletion platforms are often applied, such as the use of multiple affinity removal system (MARS) top14 immuno-depletion spin columns (Agilent Technologies), which selectively removes 14 of the most abundant proteins, providing greater chances of detecting low abundance molecules. However, care should be taken when using such approaches for the analysis of low molecular weight analytes, as these species can be bound to the high abundance proteins and, therefore, removed in the depletion step. Microbial extracts, similarly, contain disproportionate classes of proteins, for example high levels of ribosomal proteins, but these vary between species. One novel depletion method has already been developed using combinatorial ligand libraries to increase the spectrum of proteins recovered (see Chapter 9). In all cases the depletion/concentration is followed by an elution step preceding analysis.

To specifically search for low abundant and low molecular weight (<1000 Da) analytes, considerable effort is presently being focused on applying nanoparticles as a capture medium prior to MALDI-MS analysis of complex samples. Their large surface area-to-volume ratio and exceptional ability to bind analytes of interest, when dispersed in biological fluids, are now used in many bioassays. Their tunable surface properties provide possibilities to create a range of highly selective and capturing species. Therefore, very low concentrations of the analytes can be separated and concentrated, for example by centrifugation, and directly introduced to MALDI-MS.

The choice of matrix plays an important role in the peptide and protein desorption process. In terms of proteomic analysis, organic matrices such as α-cyano-4-hydroxycinammic-acid (CHCA), 5-chloro-2-mercaptobenzothiazole (CMBT) and sinapinic acid (SA) are popular, due to their simple handling, ability to absorb UV radiation and ionize a diverse range of biomolecules such as proteins, peptides, lipids, sugars and DNA. However, they produce cluster ions that cause matrix related background in the low mass range, resulting in a decreased signal-to-noise ratio that obscures the analysis of small molecules (Hillenkamp and Peter-Katalinic, 2007). In 1988, Tanaka *et al.* (1988), using cobalt nanopowder suspended in glycerol, introduced the application of inorganic materials to replace conventional organic matrices. Since then, many different platforms have been developed using nanomaterials. The most promising are gold (Spencer *et al.*, 2008), silver (Hua *et al.*, 2007), manganese (Taira *et al.*, 2009) and magnetic (Fe_3O_4) nanoparticles (Lin *et al.*, 2007). The attractiveness of applying nanoparticles to MALDI-MS lies in their optical properties and tunable morphology. It has been reported that the use of nanoparticles can reduce the appearance of undesirable ions in the low molecular weight range, enhance ionisation capabilities, and strengthen the signal-to-noise ratio with little or no induced fragmentation of the analyte (Castellana and Russell, 2007; Chen *et al.*, 2007).

Nanoparticles offer new approaches to detect and concentrate analytes from cell extracts. The main focus of nanoparticle engineering is tailoring their surfaces to create selective, concentration probes for biological samples. The specific physicochemical properties of some nanomaterials enable their use for fractionation of complex samples by recognising and capturing specific classes of compounds. They may be designed for size dependent fractionation by coating them with a material containing size-selective pores (Cheng *et al.*, 2006). Some have been developed with surfaces covered by cationic or anionic functional groups to extract only negatively or positively charged species, respectively (Shrivas and Wu, 2008) (Figure 1.16).

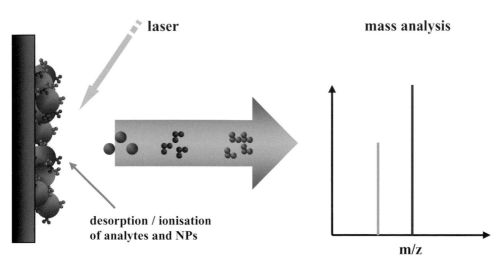

Figure 1.16   *Schematic representation of the desorption/ionisation process with the use of a nanoparticulate matrix*

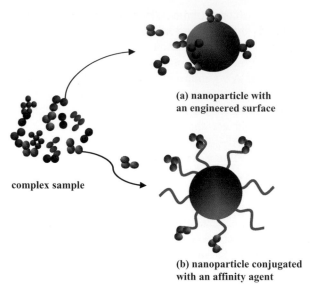

(a) nanoparticle with
an engineered surface

complex sample

(b) nanoparticle conjugated
with an affinity agent

Figure 1.17   *Schematic representation of nanoparticle-based capturing of analytes from a complex sample: (a) nanoparticle with selective surface properties; (b) nanoparticle conjugated with a capturing agent*

Following removal of the nanoparticles, the depleted fraction of the sample becomes easier to analyse. Alternatively, if the species bound to nanoparticles are subject to further analysis, they can be simply removed from the nanoparticle surface by treatment with an eluant. On the other hand, the elution step can be easily omitted and the nanoparticles, with the attached analytes, directly analysed by MALDI-MS. However, the addition of conventional matrices to enhance the desorption/ionisation process is still often required.

An interesting approach is to functionalise the nanoparticles with affinity agents to target particular analytes (Figure 1.17). An example has been reported by Chen *et al.*, where carbohydrate-encapsulated gold nanoparticles were used for capture and identification of the galactophilic *Pseudomonas aeruginosa* lectin I (Chen *et al.*, 2005).

Microbiological applications of this technology have been reported for a range of species (Lin *et al.*, 2005; Gültekin *et al.*, 2009). In such samples, the concentration of the pathogen is generally well below the threshold for MALDI-TOF-MS detection or analyses may be complicated by the interference of proteins and metabolites in the fluid. The challenge is therefore immense. In an early application of the technology, the antibiotic vancomycin, a potent inhibitor of peptidoglycan synthesis, which binds D-Ala-D-Ala moieties on the Gram-positive cell wall, was used to selectively trap these bacteria. In an elegant application of the method, vancomycin-modified magnetic nanoparticles were used as affinity capture probes to selectively trap such pathogens from biological fluids. The bacterial cells were then isolated from sample solutions by applying a magnetic field and characterised using MALDI-MS. This approach effectively reduced the interference of protein and metabolite signals in the mass spectra, because of the high specificity of vancomycin for the D-Ala-D-Ala units of the cell walls. Cell concentrations of $7-104 \, \text{cfu ml}^{-1}$ of *S. saprophyticus* and *S. aureus* were detected in the urine sample (Lin *et al.,* 2005).

In another study, dipicolinic acid, a characteristic residue of the bacterial spore, was used as a template to detect spores of *Bacillus* species. Here, a thiol ligand-capping method with polymerisable methacryloylamidocysteine was attached to gold–silver nanoclusters. It was designed as a reconstructed surface shell by synthetic host polymers based on a molecular imprinting method for recognition. Methacryloyl iminodiacetic acid-chrome Cr(III) was used as a new metal-chelating monomer via metal coordination–chelation interactions and dipicolinic acid. The latter simultaneously chelated to the Cr(III) metal ion and fitted into a shape-selective cavity. Thus, the interaction between the Cr(III) ion and free coordination spheres had an effect on the binding ability of the gold–silver nanoclusters nanosensor. The binding affinity of the dipicolinic acid imprinted nanoclusters and the determined affinity constants were found to be 18×10^6 and $9 \times 10^6 \, \text{mol l}^{-1}$, respectively, suggesting excellent use as an affinity template (Gültekin *et al.*, 2009).

A variety of covalent and noncovalent chemistries for derivatisation of nanoparticles with proteins and peptides have been reported (Aubin-Tam and Hamad-Schifferli, 2008) and the above examples serve to demonstrate the diverse applications of MALDI-TOF-MS across the fields of medicine and biology and the continuous improvements being made to increase the coverage of the proteome.

Though there are still technical challenges that need to be addressed, such as finding the most suitable nanoparticles to use as a matrix or to be able to minimise the interferences between the biomolecules and nanostructures, these new technologies are likely to have a major impact in elucidating the complex mechanism involved in host–bacterial interaction and lead to the discovery of new diagnostic targets and in biomarker discovery.

References

Arnold, R.J. and Reilly, J.P. (1998). Fingerprint matching of *E. coli* strains with matrix-assisted laser desorption/ionization time-of-flight mass spectrometry of whole cells using a modified correlation approach. Rapid Commun. Mass Spectrom. 12, 630–636.

Atfield, G.N. and Morris, C.J.O.R. (1961). Analytical separations by high voltage electrophoresis. Amino acids in protein hydrolysates. Biochem. J. 81, 606–614.

Atherton, J.C. (2006). The pathogenesis of *Helicobacter pylori*-induced gastro-duodenal diseases. Annu. Rev. Pathol. 1, 63–96.

Aubin-Tam, M.E. and Hamad-Schifferli, K. (2008). Structure and function of nanoparticle–protein conjugates. Biomed. Mater. 3, 034001.

Backert, S., Kwok, T., Schmid, M., Selbach, M., Moese, S., Peek Jr, R.M., Konig, W., Meyer, T.F. and Jungblut, P.R. (2005). Subproteomes of soluble and structure-bound *Helicobacter pylori* proteins analyzed by two-dimensional gel electrophoresis and mass spectrometry. Proteomics 5, 1331–1345.

Barshick, S.A., Wolf, D.A. and Vass, A.A. (1999). Differentiation of microorganisms based on pyrolysis-ion trap mass spectrometry using chemical ionization. Anal. Chem. 71, 633–641.

Bernarde, C., Khoder, G., Lehours, P., Burucoa, C., Fauchère, J.L., Delchier, J.C., Mégraud, F. and Atanassov, C. (2009). Proteomic *Helicobacter pylori* biomarkers discriminative of low-grade gastric MALT lymphoma and duodenal ulcer. Proteomics Clin. Appl. 3, 672–681.

Borriello, S.P. (1998). Pathogenesis of *Clostridium difficile* infection. J. Antimicrob. Chemother. 41, 13–19.

Buchanan, R.E. and Gibbons, N.E. (1974). *Bergey's Manual of Determinative Bacteriology*, 8th Edition, Williams & Wilkins, Baltimore.

Canton, R. and Coque, T.M. (2006). The CTX-M β-lactamase pandemic. Curr. Opin. Microbiol. 9, 466–475.

Castellana, E.T. and Russell, D.H. (2007). Tailoring nanoparticle surface chemistry to enhance laser desorption ionization of peptides and proteins. Nano. Lett. 7, 3023–3025.

Chen, Y., Luo, G., Diao, J., Chornoguz, O., Reeves, M. and Vertes, A. (2007). Laser desorption/ionization from nanostructured surfaces: nanowires, nanoparticle films and silicon microcolumn arrays. J. Phys. Conf. Ser. 59, 544–548.

Chen, Y.J., Chen, S.H., Chien, Y.Y., Chang, Y.W., Liao, H.K., Chang, C.Y., Jan, M.D., Wang, K.T. and Lin, C.C. (2005). Carbohydrate-encapsulated gold nanoparticles for rapid target-protein identification and binding-epitope mapping. ChemBioChem. 6, 1169–1173.

Cheng, M.M.C., Cuda, G., Bunimovich, Y.L., Gaspari, M., Heath, J.R., Hill, H.D., Mirkin, C.A., Nijdam, A.J., Terracciano, R., Thundat, T. and Ferrari, M. (2006). Nanotechnologies for biomolecular detection and medical diagnostics. Curr. Opin. Chem. Biol. 10, 11–19.

Cho, M.J., Jeon, B.S., Park, J.W., Jung, T.S., Song, J.Y., Lee, W.K., Choi, Y.J., Choi, S.H., Park, S.G., Park, J.U., Choe, M.Y., Jung, S.A., Byun, E.Y., Baik, S.C., Youn, H.S., Ko, G.H., Lim, D. and Rhee, K.H. (2002). Identifying the major proteome components of *Helicobacter pylori* strain 26695. Electrophoresis 23, 1161–1173.

Chong, B.E., Wall, D.B., Lubman, D.M. and Flynn, S.J. (1997). Rapid profiling of *E. coli* proteins up to 500 kDa from whole cell lysates using matrix-assisted laser desorption/ionization time-of-flight mass spectrometry. Rapid Commun. Mass Spectrom. 11, 1900–1908.

Claydon, M.A., Davey, S.N., Edwards-Jones, V. and Gordon, D.B. (1996). The rapid identification of intact microorganisms using mass spectrometry. Nat. Biotechnol. 14, 1584–1586.

Coburn, P.S., Pillar, C.M., Jett, B.D., Haas, W. and Gilmore, M.S. (2004). *Enterococcus faecalis* senses target cells and in response expresses cytolysin. Science 306, 2270–2272.

Cotter, R.J. (1992). Time-of-flight mass spectrometry for the structural analysis of biological molecules. Anal. Chem. 64, 1027A–1039A.

Cummins, C.S. and Harris, H. (1956). The chemical composition of the cell wall in some Gram-positive bacteria and its possible value as a taxonomic character. J. Gen. Microbiol. 14, 583–600.

Dai, Y., Li, L., Roser, D.C. and Long, S.R. (1999). Detection and identification of low-mass peptides and proteins from solvent suspensions of Escherichia coli by high performance liquid chromatography fractionation and matrix-assisted laser desorption/ionization mass spectrometry. Rapid Commun. Mass Spectrom. 13, 73–78.

Dare, D. (2005). Microbial Identification using MALDI-TOF-MS. In *Encyclopedia of Rapid Microbiological Methods*, Vol. 3, Ed. M. J. Miller, DHI Publishing, LLC, River Grove, IL, pp. 19–56.

Das, S., Sierra, J.C., Soman, K.V., Suarez, G., Mohammad, A.A., Dang, T.A., Luxon, B.A. and Reyes, V.E. (2005). Differential protein expression profiles of gastric epithelial cells following *Helicobacter pylori* infection using ProteinChips. J. Proteome Res. 4, 920–930.

Davis B.J. (1964). Disc electrophoresis -11. Methods and application to human serum proteins. Ann. N. Y. Acad. Sci. 121, 404–427.

De Ley, J., Cattoir, H. and Reynaerts, A. (1970). The quantitative measurement of DNA hybridization from renaturation rates. Eur. J. Biochem. 12, 133–142.

Domin, M.A., Welham, K.J. and Ashton, D.S. (1999). The effect of solvent and matrix combinations on the analysis of bacteria by matrix-assisted laser desorption/ionisation time-of-flight mass spectrometry. Rapid Commun. Mass Spectrom. 13, 222–226.

Encheva, V., Wait, R., Gharbia, S.E., Begum, S. and Shah, H.N. (2005). Proteome analysis of serovars Typhimurium and Pullorum of *Salmonella enterica* subspecies I. BMC Microbiology 5, 42.

Encheva, V., Wait, R., Gharbia, S.E., Begum, S. and Shah, H.N. (2006). Comparison of extraction procedures for proteome analysis of *Streptococcus pneumoniae* and a basic reference map. Proteomics 6, 3306–3317.

Enroth, H., Akerlund, T., Sillen, A. and Engstrand, L. (2000). Clustering of clinical strains of *Helicobacter pylori* analyzed by two-dimensional gel electrophoresis. Clin. Diagn. Lab. Immunol. 7, 301–306.

Fenselau, C. (1997). MALDI MS and strategies for protein analysis. Anal. Chem. 69, 661A–665A.

Ge, Z., Zhu, Y.L., Zhong, X., Yu, J.K., and Zheng, S. (2008). Discovering differential protein expression caused by CagA-induced ERK pathway activation in AGS cells using the SELDI-ProteinChip platform. World J. Gastroenterol. 14, 554–562.

Govorun, V.M., Moshkovskii, S.A., Tikhonova, O.V., Goufman, E.I., Serebryakova, M.V., Momynaliev, K.T., Lokhov, P.G., Khryapova, E.V., Kudryavtseva, L.V., Smirnova, O.V., Toropyguine, I.Y., Maksimov, B.I. and Archakov, A.I. (2003). Comparative analysis of proteome maps of *Helicobacter pylori* clinical isolates. Biochemistry (Mosc). 68, 42–49.

Gültekin, A., Diltemiz, S.E., Ersöz, A., Sarlözlü, N.Y., Denizli, A. and Say, R. (2009). Gold-silver nanoclusters having dipicolinic acid imprinted nanoshell for *Bacillus cereus* spores recognition. Talanta 78, 1332–1338.

Haas, G., Karaali, G., Ebermayer, K., Metzger, W.G., Lamer, S., Zimny-Arndt, U., Diescher, S., Goebel, U.B., Vogt, K., Roznowski, A.B., Wiedenmann, B.J., Meyer, T.F., Aebischer, T. and Jungblut, P.R. (2002). Immunoproteomics of *Helicobacter pylori* infection and relation to gastric disease. Proteomics 2, 313–324.

Haag, A.M., Taylor, M., S.N., Johnston, K.H. and Cole, R.B. (1998). Rapid identification and speciation of Haemophilus bacteria by matrix-assisted laser desorption/ionization time-of-flight mass spectrometry. J. Mass. Spectrom. 33, 750–756.

Hansson, L.E., Nyrén, O., Hsing, A.W., Bergström, R., Josefsson, S., Chow, W.H., Fraumeni Jr, J.F. and Adami, H.O. (1996). The risk of stomach cancer in patients with gastric or duodenal ulcer disease. New Engl. J. Med. 335, 242–249.

Heylen, K., Vanparys, B., Peirsegaele, F., Lebbe, L. and De Vos, P. (2007). *Stenotrophomonas terrae* sp. nov. and *Stenotrophomonas humi* sp. nov., two nitrate-reducing bacteria isolated from soil. Int. J. Syst. Evol. Microbiol. 57, 2056–2061.

Hillenkamp, F. and Peter-Katalinic, J. (2007). *MALDI MS: A Practical Guide to Instrumentation, Methods and Applications*, Wiley-VCH, Weinheim.

Hirschl, A.M., Richter, M., Makristathis, A., Pruckl, P.M., Willinger, B., Schutze, K. and Rotter, M.L. (1994). Single and multiple strain colonization in patients with *Helicobacter pylori*-associated gastritis: detection by macrorestriction DNA analysis. J. Infect. Dis. 170, 473–475.

Hocker, M. and Hohenberger, P. (2003). *Helicobacter pylori* virulence factors – one part of a big picture. Lancet. 362, 1231–1233.

Holland, R.D., Wilkes, J.G., Rafii, F., Sutherland, J.B., Persons, C.C., Voorhees, K.J. and Lay Jr, J.O. (1996). Rapid identification of intact whole bacteria based on spectral patterns using matrix-assisted laser desorption/ionization with time-of-flight mass spectrometry. Rapid Commun. Mass Spectrom. 10, 1227–1232.

Hua, L., Chen, J., Ge, L. and Tan, S.N. (2007). Silver nanoparticles as matrix for laser desorption/ionization mass spectrometry of peptides. J. Nanopart. Res. 9, 1133–1138.

Huycke, M.M. and Gilmore, M.S. (1995). Frequency of aggregation substance and cytolysin genes among *Enterococcal endocarditis* isolates. Plasmid 34, 152–156.

Hynes, S.O., McGuire, J., Falt, T. and Wadstrom, T. (2003a). The rapid detection of low molecular mass proteins differentially expressed under biological stress for four *Helicobacter spp.* using ProteinChip technology. Proteomics 3, 273–278.

Hynes, S.O., McGuire, J. and Wadstrom, T. (2003b). Potential for proteomic profiling of *Helicobacter pylori* and other *Helicobacter spp.* using a ProteinChip array. FEMS Immunol. Med. Microbiol. 36, 151–158.

Jungblut, P.R. and Bumann, D. (2002). Immunoproteome of *Helicobacter pylori*. Methods Enzymol. 358, 307–316.

Jungblut, P.R., Bumann D., Haas, G., Zimny-Arndt, U., Holland, P., Lamer, S., Siejak, F., Aebischer, A. and Meyer, T.F. (2000). Comparative proteome analysis of *Helicobacter pylori*. Mol. Microbiol. 36, 710–725.

Karas, M. and Hillenkamp, F. (1988). Laser desorption ionization of proteins with molecular masses exceeding 10,000 daltons. Anal. Chem. 60, 2299–2301.

Kelly, C.P. and Lamont, J.T. (1998). *Clostridium difficile* infection. Annu. Rev. Med. 49, 375–390.

Keys, C.J., Dare, D.J., Sutton, H., Wells, G., Lunt, M., McKenna, T., McDowall, M. and Shah, H.N. (2004). Compilation of a MALDI-TOF mass spectral database for the rapid screening and characterisation of bacteria implicated in human infectious diseases. Infect. Genet. Evol. 4, 221–242.

Khoder, G., Yamaoka, Y., Fauchère, J.L., Burucoa, C. and Atanassov, C. (2009). Proteomic *Helicobacter pylori* biomarkers discriminating between duodenal ulcer and gastric cancer. J. Chromatogr., B. 877, 1193–1939.

Krah, A. and Jungblut, P.R. (2004). Immunoproteomics. Methods Mol. Med. 94, 19–32.

Krah, A., Miehlke, S., Pleissner, K.P., Zimny-Arndt, U., Kirsch, C., Lehn, N., Meyer, T.F., Jungblut, P.R. and Aebischer, T. (2004). Identification of candidate antigens for serologic detection of *Helicobacter pylori*-infected patients with gastric carcinoma. Int. J. Cancer 108, 456–463.

Krishnamurthy, T. and Ross, P.L. (1996). Rapid identification of bacteria by direct matrix-assisted laser desorption/ionization mass spectrometric analysis of whole cells. Rapid Commun. Mass Spectrom. 10, 1992–1996.

Lancashire, L., Schmid, O., Shah, H.N. and Ball, G. (2005). Classification of bacterial species from proteomic data using combinatorial approaches incorporating artificial neural networks, cluster analysis and principal components analysis. Bioinformatics 21, 2191–2199.

Larson, H.E. and Borriello, S.P. (1990). Quantitative study of antibiotic-induced susceptibility to *Clostridium difficile* Enterocecitis in hamsters. Antimicrob. Agents Chemother. 34, 1348–1353.

Liang, X., Zheng, K., Qian, M.G. and Lubman, D.M. (1996). Determination of bacterial protein profiles by matrix-assisted laser desorption/ionization mass spectrometry with high-performance liquid chromatography. Rapid Commun. Mass Spectrom. 10, 1219–1226.

Lin, P.C., Tseng, M.C., Su, A.K., Chen, Y.J. and Lin, C.C. (2007). Functionalized magnetic nanoparticles for small-molecule isolation, identification, and quantification. Anal. Chem. 79, 3401–3408.

Lin, Y.F., Wu, M.S., Chang, C.C., Lin, S.W., Lin, J.T., Sun, Y.J., Chen, D.S. and Chow, L.P. (2006). Comparative immunoproteomics of identification and characterization of virulence factors from *Helicobacter pylori* related to gastric cancer. Mol. Cell. Proteomics 5, 1484–1496.

Lin, Y.S., Tsai, P.J., Weng, M.F. and Chen Y.C. (2005). Affinity capture using vancomycin-bound magnetic nanoparticles for the MALDI-MS analysis of bacteria. Anal. Chem. 77, 1753–1760.

Livermore, D.M., Winstanley, T.G. and Shannon, K.P. (2001). Interpretative reading: recognising the unusual and inferring resistance mechanisms from resistance phenotypes. J. Antimicrob. Chemother. 48, 87–102.

Ludwig, W. and Klenk, H.P. (2001). Overview: A phylogenetic backbone and taxonomic framework for prokaryotic systematics. In *Bergey's Manual of Systematic Bacteriology*, 2nd Edition, Eds D.R. Boone and R.W. Castenholz, Springer-Verlag.

Lynn, E.C., Chung, M.C., Tsai, W.C. and Han, C.C. (1999). Identification of Enterobacteriaceae bacteria by direct matrix-assisted laser desorptiom/ionization mass spectrometric analysis of whole cells. Rapid Commun. Mass Spectrom. 13, 2022–2027.

Meuzelaar, H.L.C. and Kistemaker, P.G. (1973). A technique for fast and reproducible fingerprinting of bacteria by pyrolysis mass spectrometry. Anal. Chem. 4., 587–590.

Nilsson, C.L. (1999). Fingerprinting of *Helicobacter pylori* strains by matrix-assisted laser desorption/ionization mass spectrometric analysis. Rapid Commun. Mass Spectrom. 13, 1067–1071.

Ornstein, L. (1964). Disc electrophoresis -1. Background and theory. Ann. N. Y. Acad. Sci. 121, 321–349.

Park, J.W., Song, J.Y., Lee, S.G., Jun, J.S., Park, J.U., Chung, M.J., Ju, J.S., Nizamutdinov, D., Chang, M.W., Youn, H.S., Kang, H.L., Baik, S.C., Lee, W.K., Cho, M.J., and Rhee, K.H. (2006). Quantitative analysis of representative proteome components and clustering of *Helicobacter pylori* clinical strains. Helicobacter 11, 533–543.

Pereira, D.R., Martins, D., Winck, F.V., Smolka, M.B., Nishimura, N.F., Rabelo-Gonçalves, E.M., Hara, N.H., Marangoni, S., Zeitune, J.M. and Novello, J.C. (2006). Comparative analysis of two-dimensional electrophoresis maps (2-DE) of *Helicobacter pylori* from Brazilian patients with chronic gastritis and duodenal ulcer: a preliminary report. Rev. Inst. Med. Trop. Sao Paulo 48, 175–177.

Reiner, E. (1965). Identification of bacterial strains by pyrolysis gas-liquid chromatography. Nature 200, 1272–1274.

Reynaud af Geijersstam, A., Culak, R., Molenaar, L., Chattaway, M., Røslie, E., Peciuliene, V., Haapasalo, M. and Shah, H.N. (2007). Comparative analysis of virulence determinants and mass spectral profiles of Finnish and Lithuanian endodontic *Enterococcus faecalis* isolates. Oral Microbiol. Immunol. 22, 87–94.

Schleifer, K.H. and Kandler, O. (1972). Peptidoglycan types of bacterial cell walls and their taxonomic implications. Bacteriol. Rev. 36, 407–477.

Schmid, O., Ball, G., Lancashire, L., Culak, R. and Shah, H.N. (2006). New approaches to identification of bacterial pathogens by surface enhanced laser desorption/ionisation time of flight mass spectrometry in concert with artificial neural networks, with special reference to *Neisseria gonorrhoeae*. J. Med. Microbiol. 54, 1–7.

Shah, H.N. and Collins, M.D. (1983). Genus *Bacteroides:* a chemotaxonomical perspective. J. Appl. Bacteriol. 55, 403–416.

Shah, H.N., Encheva, V., Schmid, O., Nasir, P., Culak, R.A., Ines, I., Chattaway, M.A., Keys, C.J., Jacinto, R.C., Molenaar, L., Ayenza, R.S., Hallas, G., Hookey, J.V. and Rajendram, D. (2005). Surface enhanced laser desorption/ionization time of flight mass spectrometry (SELDI-TOF-MS): a potentially powerful tool for rapid characterisation of microorganisms. In *Encyclopedia of Rapid Microbiological Methods*,Vol. 3, Ed. M.J. Miller, DHI Publishing, LLC, River Grove, IL, pp. 57–96.

Shah, H.N., Keys, C.J., Gharbia, S.E., Ralphson, K., Trundle, F., Brookhouse, I. and Claydon, M.A. (2000). The application of matrix-assisted laser desorption/ionisation time of flight mass spectrometry to profile the surface of intact bacterial cells. Microb. Ecol. Health Dis. 12, 241–246.

Shah, H.N., Keys, C.J., Schmid, O. and Gharbia, S.E. (2002). Matrix-assisted laser desorption/ionization time-of-flight mass spectrometry and proteomics, a new era in anaerobic microbiology. Clin. Infect. Dis. 35, S58–S64.

Shah, H.N., van Steenbergen, T.J.M., Hardie, J.M. and de Graaff, J. (1982). DNA base composition, DNA-DNA reassociation and isoelectric focusing of proteins of strains designated *Bacteroides oralis*. FEMS Microbiol. Lett. 13, 125–130.

Shrivas, K. and Wu, H.F. (2008). Modified silver nanoparticle as a hydrophobic affinity probe for analysis of peptides and proteins in biological samples by using liquid–liquid microextraction coupled to AP-MALDI-ion trap and MALDI-TOF mass spectrometry. Anal. Chem. 80, 2583–2589.

Spencer, M.T., Furutani, H., Oldenburg, S.J., Darlington, T.K. and Prather, K.A. (2008). Gold nanoparticles as a matrix for visible-wavelength single-particle matrix-assisted laser desorption/ionization mass spectrometry of small biomolecules. J. Phys. Chem. C 112, 4083.

Spigaglia, P. and Mastrantonio, P. (2004). Comparative analysis of *Clostridium difficile* clinical isolates belonging to different genetic lineages and time periods. J. Med. Microbiol. 53, 1129–1136.

Suzuki, T., Matsuo, K., Ito, H., Hirose, K., Wakai, K., Saito, T., Sato, S., Morishima, Y., Nakamura, S., Ueda, R. and Tajima, K. (2006). A past history of gastric ulcers *and Helicobacter pylori* infection increase the risk of gastric malignant lymphoma. Carcinogenesis 27, 1391–1397.

Szabo, D., Silveria, F., Hujer, A.M., Bonomo, R.A., Hujer, K.M., Marsh, J.W., Bethel, C.R., Doi, Y., Deeley, K. and Paterson, D.L. (2006). Outer membrane protein changes and efflux pump expression together may confer resistance to ertapenem in *Enterobacter cloacae*. Antimicrob. Agents Chemother. 50, 2833–2835.

Taira, S., Kitajima, K., Katayanagi, H., Ichiishi, E. and Ichiyanagi, Y. (2009). Manganese oxide nanoparticle-assisted laser desorption/ionization mass spectrometry for medical applications. Sci. Technol. Adv. Mater. 10, 034602.

Tanaka, K., Waki, H., Ido, Y., Akita, S., Yoshida, Y., Yoshida, T. and Matsuo, T. (1988). Protein and polymer analyses up to m/z 100,000 by laser ionization time-of-flight mass spectrometry. Rapid Commun. Mass Spectrom. 2, 151–153.

Utt, M., Nilsson, I., Ljungh, A. and Wadström, T. (2002). Identification of novel immunogenic proteins of *Helicobacter pylori* by proteome technology. J. Immunol. Methods. 259, 1–10.

Voth, D.E. and Ballard, J.D. (2005). *Clostridium difficile* toxins: mechanism of action and role in disease. Clin. Microbiol. Rev. 18, 247–263.

Wang, Z., Russon, L., Li, L., Roser, D.C. and Long, S.R. (1998). Investigation of spectral reproducibility in direct analysis of bacteria proteins by matrix-assisted laser desorption/ionization time-of-flight mass spectrometry. Rapid Commun. Mass Spectrom. 12, 456–464.

Welham, K.J., Domin, M.A., Scannell, D.E., Cohen, E. and Ashton, D.S. (1998). The characterization of micro-organisms by matrix-assisted laser desorption/ionization time-of-flight mass spectrometry. Rapid Commun. Mass Spectrom. 12, 176–180.

Woodford, N., Carattoli, A., Karisik, E., Underwood, A., Ellington, M.J. and Livermore, D.M. (2009). Complete nucleotide sequences of plasmids pEK204, pEK499, and pEK516, encoding CTX-M enzymes in three major *Escherichia coli* lineages from the United Kingdom, all belonging to the international O25:H4-ST131 clone. Antimicrob. Agents Chemother. 53, 4472–4482.

Work, E. and Dewey, E.L. (1953). The distribution of α,ε-diaminopimelic acid among various microorganisms. J. Gen. Microbiol. 9, 394–406.

Yakoob, J., Fan, X.G., Hu, G.L., Yang, H.X., Liu, L., Liu, S.H., Tan, D.M., Li, T.G. and Zhang, Z. (2001). Polycolonization of *Helicobacter pylori* among Chinese subjects. Clin. Microbiol. Infect. 7, 187–192.

Zanoni, P.G., Debruyne, L., Ross, M., Revez, J. and Vandamme, P. (2009). *Campylobacter cuniculorum* sp. nov. from rabbits. Int. J. Syst. Evol. Microbiol. 59, 1666–1671.

Zemany, P.D. (1952). Identification of complex organic materials by mass spectrometric analysis of their pyrolysis products. Anal. Chem. 24, 1709–1713.

2

Microbial Phylogeny and Evolution Based on Protein Sequences (The Change from Targeted Genes to Proteins)

Radhey S. Gupta

Department of Biochemistry, McMaster University, Hamilton, Canada

2.1 Introduction

In the current phylogeny of prokaryotes based on 16S rRNA, a number of important issues are not understood. For most bacterial groups, no distinctive molecular or biochemical characteristics are known that can be used to identify these groups in definitive terms. The branching order and interrelationships among various groups are also not clear. Analyses of protein sequences from microbial genomes are leading to the discovery of numerous molecular characteristics that are enabling characterization of different groups of microbes in clear terms and for understanding their evolution. This chapter describes some of these new approaches and their limited applications for understanding the microbial phylogeny of α-proteobacteria and *Bacteroidetes-Chlorobi* phyla. Studies of these new molecular markers should also lead to identification of novel biochemical and physiological properties that are unique to different groups of microbes.

2.2 Bacterial Phylogeny: Overview and Key Unresolved Issues

The *Bacteria* comprise the vast majority of known prokaryotes and hence an understanding of the evolutionary relationships among them constitutes the main objective of microbial

---

Mass Spectrometry for Microbial Proteomics   Edited by Haroun N. Shah and Saheer E. Gharbia
© 2010 John Wiley & Sons, Ltd

phylogeny. Our current understanding of the evolutionary relationships among *Bacteria* is largely based on 16S rRNA sequences (Woese, 1987; Olsen *et al.*, 1994; Maidak *et al.*, 2001; Ludwig and Klenk, 2005). Based on 16S rRNA trees, the second edition of *Bergey's Manual of Systematic Bacteriology* divides various cultured *Bacteria* into 23 main groups or phyla (Garrity *et al.*, 2005). Some of these groups (e.g. *Thermodesulfobacteria, Thermomicrobia, Chrysiogenetes, Dictyoglomi* and *Deferribacters*) consist of only a few species, whereas others (namely, *Proteobacteria, Cyanobacteria, Firmicutes, Actinobacteria* and *Bacteroidetes*) are made up of thousands of species accounting for more than 95% of cultured bacteria (Maidak *et al.*, 2001; Garrity *et al.*, 2005).

Most of these bacterial groups were first described when only a limited number of sequences were available and these groups could be clearly distinguished from each other by long naked branches in the 16S rRNA trees (Woese *et al.*, 1985; Woese, 1987; Olsen *et al.*, 1994). However, with the enormous increase in the number of sequences in recent years, most of the naked branches separating these groups have been filled, making it very difficult to clearly demarcate these groups in phylogenetic terms (Ludwig and Schleifer, 1999; Ludwig and Klenk, 2005). Except for their branching pattern in phylogenetic trees, for most of these bacterial groups, no molecular, biochemical or physiological properties are known that are unique to them. Hence, an important aspect of great importance to microbiology that remains to be understood is: 'In what aspects do different main groups of bacteria differ from each other and do species from these groups share any unique molecular, biochemical, structural or physiological properties that are distinctive of them?'

The division of *Bacteria* into the above 23 or so main phyla is at present largely arbitrary and it has no specific evolutionary or taxonomic significance (Ludwig and Klenk, 2005; Stackebrandt, 2006). This is due to the fact that currently there are no established criteria as to what constitutes a phylum or any other higher taxonomic groups (namely, Class, Order or Family) (Ludwig and Klenk, 2005; Stackebrandt, 2006). The arbitrariness of this classification is illustrated by the following example. The bacterial groups known as α-, β- and γ-proteobacteria are amongst the largest groups within prokaryotes accounting, respectively, for about 12, 8 and 26% of all cultured bacteria (Woese, 1987; Maidak *et al.*, 2001). Species from these groups are also clearly distinguished from all other bacteria, both in phylogenetic trees and based on many other discrete molecular characteristics (Gupta, 2000, 2005b, 2006; Kersters *et al.*, 2006; Gupta and Mok, 2007; Gao *et al.*, 2009). Yet, at present these major bacterial groups are recognized as subdivisions (taxonomic rank Class) of the Proteobacteria phylum (Murray *et al.*, 1990; Garrity *et al.*, 2005), whereas many other poorly characterized groups consisting of only a few species are recognized as the main phyla of *Bacteria* (Garrity *et al.*, 2005). Therefore, it is necessary to develop more objective criteria, based on sound evolutionary principles, for assignment of higher taxonomic ranks within microbes. In the phylogenetic trees based on 16S rRNA or other gene/protein sequences, the branching order or interrelationships among different bacterial groups are also not resolved (Olsen *et al.*, 1994; Ludwig and Schleifer, 1999; Ludwig and Klenk, 2005; Ciccarelli *et al.*, 2006). Hence, it is important to develop other means to clarify these critical aspects of microbial evolution (Gupta and Griffiths, 2002; Oren and Stackebrandt, 2002; Shah *et al.*, 2009).

The availability of complete genomes from a large number of microbes in recent years has provided an unprecedented opportunity for discovering novel molecular characteristics that are helpful in understanding these and other unresolved problems in microbial phylogeny (Nelson *et al.*, 2001; Gupta and Griffiths, 2002; Korbel *et al.*, 2005). The markers that are ideally suited for evolutionary/taxonomic studies should be homologous apomorphic characters that evolved only once (synapomorphy) during the course of evolution (Gupta, 1998; Stackebrandt, 2006). The presence or absence of these markers in orthologous sequences should also be readily discernible and it should be minimally affected by factors such as long-branch attraction effect, differences in evolutionary rates, lateral gene transfers, etc., which are known to confound inferences from phylogenetic trees (Felsenstein, 2004; Delsuc *et al.*, 2005). Comparative analyses of protein sequences from microbial genomes are indeed proving to be a rich resource for identifying novel molecular markers that satisfy these criteria and which are proving of great value in understanding phylogeny (Gupta, 1998, 2002, 2005a; Rokas and Holland, 2000; Gupta and Griffiths, 2002; Lerat *et al.*, 2005). A brief description of these novel protein-based markers is provided below.

2.3 New Protein-Based Molecular Markers for Systematic and Evolutionary Studies

Comparative analyses of genomic sequences have identified two main kinds of protein-based markers that are of importance for understanding microbial phylogeny. The first of these markers consists of conserved inserts and deletions (indels) in widely distributed proteins (Gupta, 1998; Gupta and Griffiths, 2002). Although homologs of these proteins are found in various organisms, the indels of interest are limited to particular groups of microbes. Often in a single protein multiple conserved indels are present in different locations that are specific for different bacterial groups (Gupta, 1998, 2000; Singh and Gupta, 2009). The indels that provide useful phylogenetic markers are of defined size and they are flanked on both sides by conserved regions to ensure that they constitute reliable characteristics, which are not resulting from sequence alignment artifacts (Gupta, 1998, 2000; Gupta and Griffiths, 2002). The genetic changes that give rise to a given conserved indel are highly specific and extremely rare in occurrence, so such changes are less likely to arise independently in different groups by convergent evolution (Gupta, 1998; Rokas and Holland, 2000). Hence, when a conserved signature indel (CSI) of defined size is uniquely found in a phylogenetically defined group(s) of species, the simplest explanation for this observation is that the genetic change responsible for this CSI occurred only once in a common ancestor of this group of species and then passed on to various descendents. Because the presence or absence of a given CSI in different species is not affected by factors such as differences in evolutionary rates, the CSIs that are restricted to particular clade(s) have provided good phylogenetic markers (synapomorphies) of common evolutionary descent. Figure 2.1 provides an example of a 3-aa conserved indel in the protein phosphoribosyl-aminoimidazole-succinocarboxamide synthase (PurC) that is specific for α-proteobacteria. This 3-aa indel is present in all available α-proteobacterial homologs (>100) but it is not found in any other bacteria (>700 sequences). The absence of this indel in all other bacterial groups indicates that it is an insert in the α-proteobacteria and the genetic change leading to this synapomorphy occurred in a common ancestor of this large group.

Because genetic changes that give rise to CSIs can occur at various stages in evolution, it is possible to identify CSIs in gene/protein sequences at different phylogenetic depths

```
                                                        173                                    205
       ┌ Agrobacterium tumefaciens      17935733   LVDFKIECGRLFEGD │MMR│ IILADEISPDSCRLW
       │  Sinorhizobium meliloti        15965525   -----V-----Y---  ---  -V-------------
       │  Bartonella henselae           49475721   -I---M-F---W-DE  A--  -V--------A---
       │  Brucella suis                 23501729   -----M-----W---  ---  -VV-------A---
       │  Bradyrhizobium japonicum      27380840   -----M------NE   ---  --V-----------
       │  Rhodopseudomonas palustris    39936882   -----M------SE   ---  --V-----------
       │  Rhodobacter sphaeroides       46192729   -A--R--V--IW---  Y--  L-V-----------
       │  Silicibacter sp. TM1040       52012003   -------I--IYD--  FQ-  LVV-----------
       │  Oceanicola batsensis          ZP_00999209 -I-----I--VYD--  FQ-  L-V-----------
       │  Roseobacter denitrificans     YP_682534   -------V--VYD--  FQ-  LVV-----------
       │  Paracoccus denitrificans      ZP_00631276 -I-----V--IWD--  F--  LVV-----------
  Alpha│  Caulobacter crescentus        16126730   -----V-F--VY---  FS-  V-------------
Proteo-│  Maricaulis maris              YP_756263   -----L-F--HY---  -V-  TV------------
bacteria  Parvularcula bermudensis      ZP_01017948 -I---L-F--QY---  ---  V-------------
       │  Novo. aromaticivorans         48851171   -----L-F--IWD--  YS-  V---------G----
       │  Erythrobacter litoralis       YP_457381   -----L-F--IYD--  FS-  V---------G----
       │  Sphingopyxis alaskensis       YP_615638   -----L-F---Y---  FS-  ----------G----
       │  Gluconobacter oxydans         YP_192689   -----L-F--IW--E  E--  -L--------N----
       │  Acidiphilium cryptum          YP_001233537 -------F---W-N-  D--  -V--------N----
       │  Mag. magnetotacticum          23013896   -------F---YN--  D-Q  -V------------
       │  Rhodospirillum centenum       YP_002299532 -----V-F---W-NE  E--  ----------N----
       │  Rickettsia prowazekii         15604091   --EC-L-F--V-N-E  ESI  -M-T------N----
       │  Rickettsia akari              52698606   --EC-L-F--VLN-E  ESI  -M-T------N----
       └  Ehrlichia canis               46308836   -----L-F-K-YNNK  ASD  LL--------T----
       ┌ Escherichia coli               16130401   -----L-F-LYKGE     VV-G--F---GS---
       │  Psedomona aeruginosa          15596210   -----L-F-LFHGQ     -V-G--F---G----
       │  Yersinia pestis               16123236   -----L-F-LFNGE     VV-G--F---GS---
       │  Coxiella burnetii             29654523   ---A-Y-F-VSNDE     -Y-G--------I-
       │  Helicobacter hepaticus        32267118   -----L-F-QDSA-N    ----------------
       │  Campylobacter jejuni          15791874   -I-----L-LTKDNE    LV-----------F-
       │  Nostoc sp. PCC 7120           17229760   -----L-F-LDSQQQ    -L--------T----
       │  Synechocystis PCC 6803        16330103   -----L-F-GDRQ-K    ----------T----
       │  Aquifex aeolicus              15607068   -----L-F---PS-E    LAIV-------M---
 Other │  Thermus thermophilus          46199453   -----L-F---GGE     VL--------TM---
Bacteria  Deinococcus radiodurans       NP_293950  -I---L-F-K-PS-E    -V--------T--F-
       │  Streptococcus pyogenes        50913417   -I---L-F-FDQ--T    ------L---N----
       │  Bifidobacterium longum        23465675   -------M--ATD-T    LL---T-------
       │  Rubrobacter xylanophilus      YP_643771   -------V--DAG-S    LV--------T----
       │  Listeria monocytogenes        16803812   -I---L-F--DAA-E    -L--------T----
       │  Lactococcus lactis            14195102   -----L-F---AD-S    ----------TS---
       │  Bacillus subtilis             16077713   -I---L-F-LDA--Q    VL--------T----
       │  Staphylococcus aureus         15924056   -------F-KTET-Q    -L--------T--I-
       │  Clostridium difficile         YP_001086687 -------F-KDS--N   ------V---T----
       └  Fusobacterium nucleatum       ZP_00144346 -------F-KNSK-E    -L-----T--T----
```

Figure 2.1   *Partial alignment of PurC protein sequences showing a 3-aa insert (boxed) that is uniquely found in various α-proteobacteria (Gupta, 2005b). Dashes in all sequence alignments indicate identity with the amino acid on the top line. Accession numbers for the sequences are given in the second column. Sequence information for only representative species is presented here*

corresponding to different taxonomic groupings (e.g. phylum, order, family or genus) (Figures 2.2 and 2.3). The CSIs, which are specific for different groups provide well-defined markers for identifying these groups of bacteria in definitive molecular terms. Additionally, these studies have also identified many CSIs that are commonly shared by species from a number of different phyla. These latter CSIs have provided valuable information regarding the branching order and interrelationships among different main groups of bacteria (Gupta, 1998, 2003; Gupta and Griffiths, 2002; Griffiths and Gupta, 2004b). To infer whether a given indel is an insert or a deletion in a given group, a reference point is necessary (Gupta, 1998; Gupta and Griffiths, 2002). For indels that are limited to a particular group or phylum of bacteria, its presence or absence in other bacterial groups serves as a reference point (Figure 2.1). If this indel is lacking in all other groups as in Figure 2.1, then the absence of this indel represents the ancestral character state and the

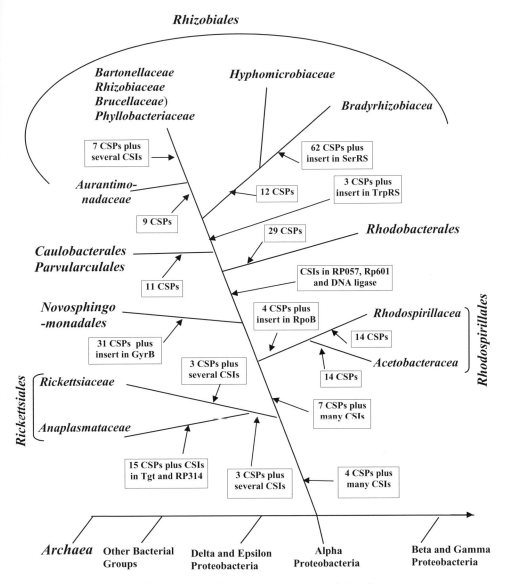

Figure 2.2 *A summary diagram showing the evolutionary relationships among
α-proteobacteria based on species distribution patterns of various CSIs and CSPs that are
specific for this group. The arrows mark the evolutionary stages where different identified
CSIs and CSPs were likely introduced (Gupta, 2005b; Gupta and Mok, 2007). The
placement of α-proteobacteria in between the δε-proteobacteria and βγ-proteobacteria
is based on earlier studies (Gupta, 2000; Gupta and Sneath, 2007)*

indel in question is an insert in the given group. On the other hand, when a sequence
region that is found in various other bacteria is lacking in a given group, then this indel
represents a deletion event in the indicated group. For protein sequences that are also found
in *Archaea*, presence or absence of the indel in the archaeal homologs can be used to

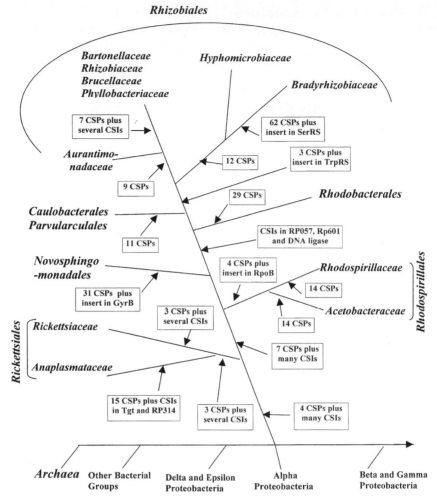

Figure 2.3 *A summary diagram showing the evolutionary relationships amongst Fibrobacter-Chlorobi and Bacteroidetes species based on species distribution patterns of various identified CSIs and CSPs that are specific for these groups (Gupta, 2004; Gupta and Lorenzini, 2007). The arrows mark the evolutionary stages where different identified CSIs and CSPs were likely introduced. The phylogenetic placement of these species is based on earlier work (Griffiths and Gupta, 2001, 2004b, 2007b; Gupta 2003, 2005a)*

determine whether the indel is an insert or a deletion in the given group(s) of bacteria (Woese *et al.*, 1990; Gupta, 1998; Gupta and Griffiths, 2002; Lake *et al.*, 2007).

The second kind of molecular markers that are proving very useful for systematic and phylogenetic studies are whole proteins that are uniquely found in particular groups or subgroups of bacteria (Gao *et al.*, 2006; Gupta and Lorenzini, 2007; Gupta and Mok, 2007). Analyses of genomic sequences have revealed that a large fraction of genes (open reading frames, ORFs) in different organisms encodes for proteins that are of unknown functions (Danchin, 1999; Siew and Fischer, 2003; Daubin and Ochman, 2004; Doerks *et al.*, 2004; Galperin and Koonin, 2004). These proteins are commonly referred to as 'ORFans' (i.e. ORFs that have no known homologs) or 'hypothetical proteins' (Daubin and Ochman, 2004; Lerat *et al.*, 2005; Kainth and Gupta, 2005). A large fraction of these ORFan proteins is unique to a given species or closely related strains (Siew and Fischer, 2003; Daubin and Ochman, 2004; Lerat *et al.*, 2005; Dutilh *et al.*, 2008). However, recent studies have revealed that many of these proteins are present in a highly conserved state at different phylogenetic depths (Gao *et al.*, 2006; Gao and Gupta, 2007; Gupta and Lorenzini, 2007; Gupta and Mok, 2007; Dutilh *et al.*, 2008). Such proteins will be referred to as conserved signature proteins (CSPs) in the present work. Recent studies show that many of these CSPs are uniquely found in all species from particular groups (Gao *et al.*, 2006; Gupta and Griffiths, 2006; Mulkidjanian *et al.*, 2006; Gao and Gupta, 2007; Gupta and Lorenzini, 2007; Gupta and Mok, 2007).

Table 2.1 presents the results of blast searches for a few proteins that are specific for the *Rickettsiales* group of α-proteobacteria (Kainth and Gupta, 2005; Gupta and Mok, 2007). The first of these proteins (WD0715) is specific for the entire *Rickettsiales* order, whereas the other two proteins are specific for the *Anaplasmataceae* and *Rickettsiaceae* families, respectively. For all three of these proteins, all significant blast hits are observed for the indicated taxonomic groups of species. These proteins are also present in all sequenced genomes from these groups, indicating that they represent distinctive characteristics of these groups. The genes for these proteins likely evolved in a common ancestor of these groups and then retained by all of the descendents. Because of their taxa specificity, CSPs such as these provide valuable molecular markers for identifying different groups of species in molecular terms. Further, similar to the CSIs, based upon species distribution patterns of these CSPs, it is again possible to draw robust phylogenetic inferences regarding interrelationships among various bacterial groups (see Figures 2.2 and 2.3).

Our recent work in this area has led to identification of large numbers of CSIs and CSPs that are distinctive characteristics of various main groups within *Bacteria* as well as *Archaea* (Table 2.2). Based on the identified CSIs and CSPs, all of the major groups within *Bacteria* and *Archaea* can now be described and circumscribed in molecular terms. These newly discovered molecular markers also provide powerful means for discovering novel biochemical and physiological characteristics that are unique properties of different groups of prokaryotes. To illustrate the usefulness of these new approaches for understanding microbial phylogeny and systematics, some work that has been done using them on α-proteobacteria and *Bacteroidetes-Chlorobi* phyla is reviewed below.

2.4 Molecular Markers Elucidating the Evolutionary Relationships among α-Proteobacteria

The α-proteobacteria form one of the largest groups within bacteria that includes numerous phototrophs, chemolithotrophs, chemoorganotrophs and aerobic photoheterotrophs (Kersters *et al.*, 2006). In the current taxonomic scheme based on 16S rRNA, α-proteobacteria are recognized as a class within the phylum Proteobacteria (Garrity *et al.*, 2005; Kersters *et al.*, 2006). They are subdivided into seven main subgroups or orders (namely, *Caulobacterales, Rhizobiales, Rhodobacterales, Rhodospirillales, Rickettsiales,*

Sphingomonadales and *Parvularculales*) (Garrity *et al.*, 2005). The α-proteobacteria and its various orders are presently distinguished from each other and from other bacteria primarily on the basis of their branching in phylogenetic trees (Woese *et al.*, 1984; Ludwig and Klenk, 2005; Kersters *et al.*, 2006). Presently, sequence information for >100 α-proteobacterial genomes covering all of its main subgroups are available. Comparative analyses of protein sequences from these genomes have identified a large number of CSIs and CSPs that are distinctive characteristics of α-proteobacteria and its various subgroups (Kainth and Gupta, 2005; Gupta, 2005b; Gupta and Mok, 2007). One example of a conserved CSI in the PurC protein that is specific for α-proteobacteria is shown in Figure 2.1. A summary of various other CSIs and CSPs that are specific for α-proteobacteria and/or its subgroups and the evolutionary relationships that emerge based on them is presented in Figure 2.2 (Kainth and Gupta, 2005; Gupta, 2005b; Gupta and Mok, 2007).

Table 2.1 *Bacterial phyla (groups) for which CSIs and CSPs have been described*

Bacterial group	CSPs/CSIs	Reference
Actinobacteria	Many CSIs and CSPs that are specific for the entire phyla and its various subgroups	Gao and Gupta, 2005;Gao *et al.*, 2006
Deinococcus-Thermus	Many CSIs and 65 CSPs that are specific for *Deinococcus-Thermus* as well as >200 CSPs, which are specific for *Deinococcci*	Griffiths and Gupta, 2004a, 2007a
Cyanobacteria	Many CSIs and CSPs thatare specific for all Cyanobacteria and its major clades	Gupta *et al.*, 2003; Mulkidjanian *et al.*, 2006; Gupta, 2009
Bacteroidetes, Chlorobi and Fibrobacter	Many CSIs and CSPs specific for the *Bacteroidetes* and *Chlorobi* phyla and their subgroups, or which are commonly shared by species from these groups (see text)	Griffiths and Gupta, 2001; Gupta, 2004; Gupta and Lorenzini, 2007
Chlamydiae	Many CSIs and CSPs that are specific for either all Chlamydiae or its main subgroups	Griffiths and Gupta, 2002; Griffiths *et al.*, 2005, 2006; Gupta and Griffiths, 2006
Aquificae	Many CSIs and CSPs that are specific for the entire phyla and provide information regarding its phylogenetic placement	Griffiths and Gupta 2004b, 2006
α-Proteobacteria	Many CSIs and CSPs, which are specific for all α-proteobacteria and its various main orders and families (see text and Figure 2.2)	Gupta, 2000, 2005b; Kainth and Gupta, 2005; Gupta and Mok, 2007
γ-Proteobacteria	Many CSIs and CSPs, which are specific for γ-proteobacteria and its main orders	Gupta, 2000; Gao *et al.*, 2009
ε-Proteobacteria	Many CSIs and CSPs, which are specific for ε-proteobacteria and its main orders	Gupta, 2006
Archaea	Many CSPs that are specific for either all Archaea or its various subgroups	Gao and Gupta, 2007

Table 2.2 Blast results for some CSPs that are specific for the Rickettsiales

Protein	Order or family	WD0715 [NP_966474]	WD0083 [NP_965909]	RP192 [NP_220581]
Length		94 aa	271 aa	128 aa
Possible function		Unknown	Unknown	Unknown
Wolbachia endo. (Drosophia)	Anaplasmataceae	1e-48 (94)	1e-137 (271)	—
Wolbachia endo. (Culex)		3e-44 (93)	1e-107 (267)	—
Ehrlichia ruminantium		2e-28 (90)	3e-38 (295)	—
Ehrlichia chaffeensis		2e-28 (90)	3e-34 (294)	—
Ehrlichia canis		8e-28 (89)	6e-38 (293)	—
Anaplasma marginale		3e-25 (94)	2e-29 (293)	—
Anaplasma phagocytophilum		5e-25 (93)	3e-27 (295)	—
Neorickettsia sennetsu		7e-22 (100)	6e-06 (262)	—
Rickettsia prowazekii	Rickettsiaceae	7e-14 (107)	—	7e-40 (128)
Rickettsia typhyi		1e-13 (107)	—	8e-39 (128)
Rickettsia massiliae		3e-13 (111)	—	2e-32 (136)
Rickettsia felis		3e-13 (105)	—	3e-31 (119)
Rickettsia akari		2e-12 (107)	—	4e-31 (119)
Rickettsia rickettsii		3e-13 (107)	—	2e-30 (119)
Rickettsia conorii		3e-13 (107)	—	7e-29 (119)
Rickettsia canadensis		2e-14 (109)	—	2e-28 (118)
Rickettsia bellii RML369-C		2e-17 (107)	—	2e-20 (122)
Rickettsia bellii OSU 85-389		2e-17 (107)	—	2e-20 (122)
Orientia tsutsugamushi		1e-14 (113)	—	—
First non-Rickettsiae hit		0.001 (727)	0.092 (1012)	No other hits observed
Species name		Trichoplax adhaerens	Debaryomyces hansenii	

Of the CSIs that have been identified, 10 of them in broadly distributed proteins (namely, cytochrome assembly protein Ctag, PurC, replicative DNA helicase, ATP synthase α, exonuclease VII, PLPG transferase, RP-400, puruvate phosphate dikinase, FtsK and Cyt b) are solely found in various α-proteobacteria (Figure 2.2) (Gupta, 2005b). Four CSPs (namely, CC1365, CC2102, CC3319 and CC3292) are also exclusively present in various α-proteobacteria and their homologs are not detected in other bacteria (Kainth and Gupta, 2005; Gupta and Mok, 2007). These CSIs and CSPs provide molecular markers that are distinctive characteristics of the α-proteobacteria and can be used to identify species from this large group in molecular terms. Several other CSIs and CSPs are exclusively found in various other α-proteobacteria but they are absent in all of the *Rickettsiales* (Figure 2.2). Because *Rickettsiales* form the deepest branching lineage within α-proteobacteria in phylogenetic trees based on 16S rRNA and various proteins (Gupta, 2005b; Kersters *et al.*, 2006; Williams *et al.*, 2007; Gupta and Mok, 2007), the genetic events leading to these CSIs and CSPs likely occurred in a common ancestor of other α-proteobacteria after the branching of *Rickettsiales*. Many other CSIs and CSPs shown in Figure 2.2 are specific for different orders (namely, *Rhizobiales, Rhodobacterales, Rhodospirillales, Rickettsiales, Sphingomonadales* and *Caulobacterales*) or families (namely, *Rickettsiaceae, Anaplasmataceae, Rhodospirillaceae, Acetobacteraceae, Bradyrhizobiaceae, Brucellaceae* and *Bartonellaceae*) of α-proteobacteria (Gupta and Mok, 2007).

Based upon these α-proteobacteria-specific CSIs and CSPs, it is now possible to define nearly all of the higher taxonomic groups (i.e. most orders and many families) within α-proteobacteria in definitive molecular terms based upon multiple characteristics (Figure 2.2). The species distribution profiles of these signatures also provide important information regarding branching order and interrelationships among various α-proteobacterial subgroups, and the relationships suggested based on both CSIs and CSPs are highly concordant with each other (Kainth and Gupta, 2005; Gupta and Mok, 2007). Importantly, the relationships that emerge based on these analyses are also in excellent agreement with the branching patterns of these species in different phylogenetic trees (Gupta, 2005b; Kersters *et al.*, 2006; Williams *et al.*, 2007; Gupta and Mok, 2007), giving a high degree of confidence in the derived inferences.

The CSIs in a number of other proteins provide information that Proteobacteria is a late branching phylum in comparison with other groups of *Bacteria* (Gupta, 2000). The CSIs that are informative in this regard include a 4-aa insert in alanyl-tRNA synthetase, a >100-aa insert in RNA polymerase β (RpoB) subunit, a >150-aa insert in DNA gyrase B, a 10-aa insert in CTP synthase, a 2-aa insert in inorganic pyrophosphatase, and a 2-aa insert in Hsp70 (DnaK) protein (Gupta and Griffiths, 2002; Gupta, 2003, 2005a; Griffiths and Gupta, 2004b, 2007b). The indicated CSIs in these proteins are found in all proteobacterial homologs, but they are absent from most other bacterial phyla (namely, *Firmicutes, Actinobacteria, Thermotogae, Deinococcus-Thermus, Cyanobacteria* and *Spirochetes*). For a number of these proteins, where their homologs are found in *Archaea* (namely, RpoB, Hsp70 and AlaRS), the archael homologs also lacked the indicated indels, indicating that the absence of these indels constitute the ancestral states and that these signatures (inserts) were introduced after branching of the groups lacking these indels (Gupta and Griffiths, 2002; Gupta, 2003; Griffiths and Gupta, 2004b). Several other CSIs, which include a 7-aa insert in SecA, 1-aa deletion in the Lon protease, 4-aa insert in DnaK, 36–34-aa insert in Gyrase A, 3-aa insert in Rho protein, 2-aa insert in biotin carboxylase,

37-aa insert in valyl-tRNA synthetase and 11-aa insert in ATP synthase α subunit, provide evidence that within the Proteobacteria phylum, α-proteobacteria have branched off after the δ,ε-subdivisions but prior to the evolution of β,γ-proteobacteria (Gupta, 1998, 2000; Gupta and Griffiths, 2002; Kersters *et al.*, 2006; Gupta and Sneath, 2007).

2.5   Molecular Markers for the *Bacteroidetes-Chlorobi* Phyla

The *Bacteroidetes* and *Chlorobi* are two other presently recognized main phyla of *Bacteria* (Woese, 1987; Ludwig and Klenk, 2005; Garrity *et al.*, 2005). The bacteria from the *Bacteroidetes* phylum [previously known as the Cytophaga-Flavobacteria-Bacteroides (CFB) group] exhibit a 'pot-pourri' of phenotypes including gliding behavior and their ability to digest and grow on a variety of complex substrates such as cellulose, chitin and agar (Reichenbach, 1992; Shah, 1992; Paster *et al.*, 1994; Shah *et al.*, 2005). They inhabit diverse habitats including the oral cavity of humans, the gastrointestinal tracts of mammals, saturated thalassic brines, soil and fresh water (Ohkuma *et al.*, 2002; Shah *et al.*, 2005, 2009). The *Bacteroides* species such as *B. thetaiotaomicron* and *B. fragilis* are among the dominant microbes in the large intestine of human and other animals and can cause soft tissue infections as well as diarrheal diseases (Salyers, 1984; Shah *et al.*, 1998). Other bacteroidetes species, such as *Porphyromonas gingivalis* and *Prevotella intermedia*, are major causative agents in the initiation and progression of periodontal disease in humans (Paster *et al.*, 2001; Duncan, 2003). In contrast to the *Bacteroidetes* species, bacteria from the phylum *Chlorobi* occupy a narrow environmental niche mainly consisting of anoxic aquatic settings in stratified lakes (chemocline regions), where sunlight is able to penetrate (Overmann, 2006). These bacteria, also commonly known as Green Sulfur bacteria, are all anoxygenic obligate photoautotrophs, which obtain electrons for anaerobic photosynthesis from hydrogen sulfide (Truper and Pfennig, 1992; Overmann, 2006).

In phylogenetic trees based on 16S rRNA as well as other gene/protein sequences, the *Bacteroidetes* and *Chlorobi* species exhibit a close relationship with each other (Gupta, 2004; Ludwig and Klenk, 2005; Garrity *et al.*, 2005). The species from the *Bacteroidetes* phyla are presently distinguished solely on the basis of their branching in phylogenetic trees (Olsen *et al.*, 1994). There is no molecular or biochemical characteristic known that is exclusively found in the *Bacteroidetes* species. The species from *Chlorobi* and *Chloroflexi* phyla contain unique membrane-attached, sac-like structures called 'chlorosomes', which contain the light harvesting pigments of these bacteria (Blankenship, 1992; Overmann, 2006). One of the proteins involved in the synthesis of chlorosomes, namely, the Fenna-Matthew-Olson protein, is uniquely found in *Chlorobi* species (Frigaard *et al.*, 2003). Sequence information for a large number of *Bacteroidetes* and *Chlorobi* species is now available. Comparative analyses of these and other genomes have identified a large number of CSIs and CSPs that are distinctive characteristics of the *Bacteroidetes* and *Chlorobi* phyla and provide important information regarding their evolution (Gupta, 2004; Gupta and Lorenzini, 2007). A summary of these results is presented in Figure 2.3 and briefly discussed below.

The *Bacteroidetes* species can be distinguished from all other bacteria based on two CSIs, a 4-aa insert in Gyrase B and a 45-aa insert in the SecA protein (Gupta, 2004), and 27 CSPs that are uniquely found in these bacteria (Gupta and Lorenzini, 2007). This phylum is comprised of three orders: *Bacteroidales, Flavobacteriales* and *Sphingobacte-riales*. Many CSPs that are specific for the *Bacteroidales* or *Flavobacteriales* species have been identified; additionally many other proteins are found to be exclusively shared by species from these two orders, but they are absent in *Sphingobacteriales* (Figure 2.3) (Gupta and Lorenzini, 2007). The unique shared presence of these CSPs by these two orders of *Bacteroidetes* provide evidence that they are more closely related to each other, which is also seen in phylogenetic trees (Gupta, 2004). Within the *Bacteroidales* order, large numbers of proteins (185) that are specific for the genus *Bacteroides* have also been identified. Similar analyses of *Chlorobi* genomes have identified 51 proteins and two CSIs, a 28-aa insert in the DnaE protein and a 12–14-aa insert in alanyl-tRNA synthetase, that are exclusively found in this group of bacteria (Figure 2.3) (Gupta and Lorenzini, 2007).

Importantly, these studies have also identified six CSPs as well as three CSIs in essential proteins (namely, FtsK, UvrB and ATP synthase α subunit) that are uniquely shared by all available *Bacteroidetes* and *Chlorobi* species (Gupta, 2004; Gupta and Mok, 2007). Figure 2.4 presents sequence information for a CSI consisting of an 8-aa insert in the FtsK protein that is commonly shared by these two groups of bacteria. The shared presence of these molecular characteristics by species from these two phyla provides evidence that species from these two groups shared a common ancestor exclusive of all other bacteria, which is also supported by phylogenetic analyses (Gupta, 2004). In phylogenetic trees based on 16S rRNA as well as a number of protein sequences, *Fiborbacter succinogenes* branches close to the *Bacteroidetes* and *Chlorobi* species and it generally forms an out-group of the latter two groups (Gupta, 2004). The comparative genomic studies have uncovered two large CSIs, a 5–7-aa insert in the RNA polymerase β' subunit and a 14–16-aa insert in serine hydroxymethyl transferase, as well as one CSP (PG0081; accession number NP_904430) that are exclusively shared by species from all these three groups of bacteria, but which are not found in any other bacteria (Figure 2.3) (Gupta, 2004; Gupta and Lorenzini, 2007). The unique shared presence of these molecular characteristics by various species from these three groups of bacteria provide evidence that they shared a common ancestor exclusive of all other bacteria and they should be placed in a single phylum (or superphylum) (Gupta, 2004; Gupta and Lorenzini, 2007).

2.6   Branching Order and Interrelationships among Bacterial Phyla

The comparative genomic studies have also been instrumental in the discovery of several CSIs that have provided valuable information regarding the branching order among the major bacteria phyla (Gupta, 1998, 2001, 2003; Griffiths and Gupta, 2001, 2004b; Gupta and Griffiths; 2002). One example of such a CSI that is found in the RNA polymerase ß subunit (RpoB) is provided in Figure 2.5. In this case, a prominent insert of about 100 aa (boxed) is present in all *Proteobacteria, Chlamydiae-Verrucomicrobia-Planctomycetes, Aquificae, Fibrobacter, Bacteroidetes* and *Chlorobi* species but it is not found in any other phyla of bacteria including various *Firmicutes, Actinobacteria, Thermotoga, Deinococcus-Thermus*, Cyanobacteria, *Chloroflexi, Spirochetes* as well as various *Archaea* (Gupta and Griffiths, 2002; Gupta, 2003, 2005a; Griffiths and Gupta 2004b). The absence of this indel in various *Archaea* provides evidence that lack of this indel is an ancestral charac-

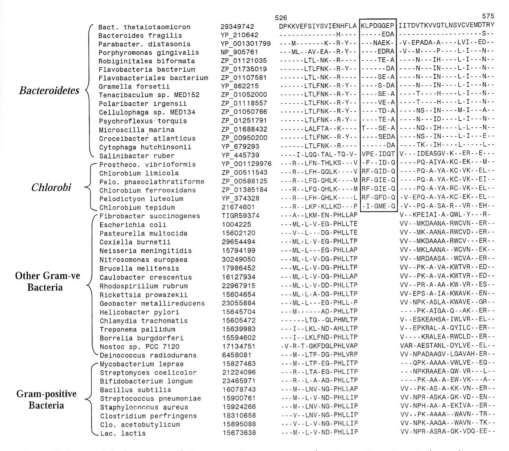

Figure 2.4 *Partial alignment of FtsK protein sequences showing a 8-aa insert (boxed), which is specific for Bacteroidetes and Chlorobi species (Gupta, 2004). Sequence information for only representative species is presented*

teristic and this indel is an insert in the *Proteobacteria, Chlamydiae-Verrucomicrobia-Planctomycetes, Aquificae, Fibrobacter, Bacteroidetes* and *Chlorobi* phyla, which branched off after the divergence of other bacterial phyla, as indicated in Figure 2.3. Based upon the species distribution patterns of many other CSIs of this kind in widely distributed proteins, it is now possible to logically deduce that various main groups (or phyla) within bacteria have evolved in the following order (from the earliest to the most recent): *Firmicutes* → *Actinobacteria* → *Fusobacteria-Thermotogae* → *Deinococcus-Thermus-Chloroflexi* → *Cyanobacteria* → *Spirochetes*→ *Fiborbacteres-Chlorobi-Bacteroidetes* → *Chlamydiae- Verrucomicrobia-Plnactomyctes* → *Aquificae* → *Proteobacteria* (δ and ε) → *Proteobacteria* (α) → *Proteobacteria* (β) and → *Proteobacteria* (γ) (Gupta, 1998, 2001, 2003, 2005a; Gupta and Griffiths, 2002; Griffiths and Gupta, 2004b, 2007b).

```
                                                    919                                              1058
              ⎧ E. coli              42818      RVPNGVSGTVIDVQVFTRD GVE(91aa)  KRRKIT QGDD LAPGVLKIVKVYLAVKR
              │ Pas. multocida       15603602   ----S-------------  ---(91aa)  Q-G--I ----  -------V---------
              │ Pse. aeruginosa      15599466   ---T-TK-----------  ---(103aa)-K--LQ ----  --------------I--
              │ V. cholerae          15640355   ------A-----------  ---(92aa)  ------ ----  ----------------
              │ Ral. solanaceum      17547753   ---S-M-----------E  --T(104aa)--K-L- ---E  P---I-M----------
Proteobacteria│ Nei. meningitidis    15676060   -M-T-M-----------E  -IQ(104aa)--KK-L- ---E  -Q---Q-M---FI-I--
              │ Ca. crescentus       16124757   -L-P--A--IV--R--N-H --D(103aa)-VD--Q R--E  -P---M-MV--FV----
              │ A. tumefaciens       15889250   -M-P-TF--IVE-R--N-H ---(104aa)-VE-VQ R--E  MP---M-M---FV----
              │ Geo. sulfurreducens  NP_953905  ---P--E----GAKI-S-K -AD(111aa)-VQ-LK R---  -P---I-M-----I-I--
              │ Hel. pylori          15645812   YC-PSLE------K---KK -Y-(108aa)EKLS-L EK--I-PN--I-K--L-I-T--
              ⎩ Camp. jejuni         15791842   YATASLE-V-V--KI--KK -Y-(108aa)EKLE-L EK--I-PS--I-L----I-T--
Chlamydiae,   ⎧ Aqu. aeolicus        15606949   -C-P--E-I---------K -TG(110aa)EKKETL LKRRD-P---ITL---FI-N--
Verrucomicrobia│ Hyd. marinus        AY188442   ---T--E-I-V-----A-K -I-(125aa)--IKEVE K-A- -K---NEL----I-Q--
Aquificae     ⎨ Cb. tepidum          AAM71403   H--A-MK-I--KTKL-S-K KKI(110aa)EKY--N V--E -P--IEELA---I-Q--
Bacteroidetes-│ Por. gingivalis      NP_904700  KATPSLR-V---TKL-SKA AKK(108aa)RKLDE- IG-E -PA-IVQMA---I-K--
Chlorobi      │ Chl. trachomatis     15605036   T--P-TE-V-M--K--S-K DRL(113aa)EAEH-K E--AD-DH--IRQ----V-S--
              │ Chlam. pneumoniae    16753076   T--P-TE-V-M--K--S-K DRL(113aa)EVEH-R E--AD-DH--IRQ----V-S--
              ⎩ Ver. spinosum        TIGR       --S-CT-I-M--R-SS-N -AD(110aa)GLDRLE --EE MES--I-Q---FI-S--
              ⎧ Bor. burgdorferi     15594734   K--H-TE-------RI-KE                  DVGN -S---EE-L---V-K--
Other Gram −ve│ Tre. pallidum        15639233   ---H--E--------RLR-S                 E----N---SEV---LI-T--
Bacteria      ⎨ Nostoc sp. PCC 7120  17229086   -----EK-R-V--RL---E                  ---E -P--ANMV-R--V-Q--
              │ Gloe. violaceus      NP_925229  -----EK-R-V--R----E                  ---E -P--ANMV-R----Q--
              │ D. radiodurans       15805937   --QS-QG-I-VKTVR-R-G                  DEGVD-K---REM-R--V-Q--
              ⎩ T. maritima          15643224   -L-H--E-R--R-D-YDQN                  DIAE -GA----L-R--V-SRK
              ⎧ Cor. glutamicum      19551731   K--H-ET-K--G-RH-S-E                  DD-- -----NEMIRI-V-Q--
              │ Myc. leprae          418765     K--H-E--K--GIR--SHE                  DD-E -PA--NEL-R--V-Q--
Gram −        │ Str. ceolicolor      21223036   K--H-EI-K--G-R--D-E                  E--E -P---NQL-R--V-Q--
positive      ⎨ Bac. subtilis        CAB11883   ---H-GG-IIH--K--N-E                  D--E -P---NQL-R--IVQ--
Bacteria      │ Clo. acetobutylicum  18311395   ---H-EA-IIV--K----E                  N--- -S---NEL-RC-I-Q--
              │ Sta. aureus          15923532   ---H-AG-I-L--K--N-E                  E---T-S---NQL-R--IVQ--
              ⎩ Strep. pneumoniae    15901784   ---H-AD-V-R--KI---V                  N--E -QS--NML-R--I-Q--
              ⎧ Halo. sp. NCR-1      AAG20693    TMRS-ED-VVDT-TLMEG-                  DGSK-AK-SVRDE-
Archaea       ⎨ Meth. barkeri        ZP_00079000 TMRSNET I DT-ILTE3I                 NGTRLA--KVRDE-
              ⎩ Pyr. aerophilum      AAL62934    A-RR-EK-I-DK-IITESP                 EGN-L---R-REL-
```

Figure 2.5 *Excerpts from RpoB sequence alignment showing a large (>100 aa) insert (boxed) that is a distinctive characteristic of all Proteobacteria and several Gram-negative phyla (Chlamydiae-Verrucomicrobia-Planctomycetes, Aquificales, Bacteroidetes-Chlorobi) but not found in other phyla of bacteria (Gupta, 2003, 2005a; Griffiths and Gupta, 2004b). Due to the large size of this insert, its entire sequence is not shown. The absence of this large insert in the archaeal homologs indicates that the bacterial groups containing this insert have diverged after the branching of those which do not contain this insert*

2.7   Importance of Protein Markers for Discovering Unique Properties for Different Groups of Bacteria

As indicated earlier, virtually all of the higher taxonomic groups within bacteria are currently identified only in phylogenetic terms. For most of them, no biochemical or physiological characteristics are known that are unique to them. In this context, identification of numerous CSIs and CSPs that are specific for different groups of bacteria is highly significant as they provide powerful means for discovering novel biochemical and physiological characteristics that are unique to these groups. Hence, functional studies on these lineage-specific CSIs and CSPs should prove of great value in filling this important gap in our knowledge regarding different groups of microbes. In this regard it should be noted that most of the discovered CSIs are present in widely distributed proteins that are involved in essential functions. The primary biochemical functions of these proteins are vital for cell survival and they are expected to remain the same in all organisms. Hence, the question arises: what is the functional significance of these evolutionarily preserved CSIs that are specific for different bacterial lineages? In our recent work, we have provided evidence that a number of conserved indels that are found in the Hsp60 (GroEL) and Hsp70 (DnaK) proteins are essential for the groups of species where they are found and deletions or most other changes in these CSIs resulted in failure to support cellular growth (Singh and Gupta, 2009). Because of the presence of these CSIs on the surfaces of various proteins, we have suggested that they confer new functional capabilities (i.e. ancillary functions) on these essential proteins (Singh and Gupta, 2009). These ancillary functions are expected to be important for the lineages where these CSIs are found and they could include ability of the protein(s) to interact with some other cellular proteins or ligands (with the CSI serving as a docking site) that either modulate the activity of these proteins or confer some new function(s) on them. Hence, further studies on understanding how these CSIs modify the cellular functions of various proteins are of much interest as they will likely reveal novel properties that are unique to different groups of bacteria.

Unlike the lineage-specific CSIs, which are found in widely distributed proteins, nearly all of the lineage-specific CSPs are of unknown functions. The discovery of these proteins points to our lack of knowledge regarding many fundamental aspects of cells, particularly functions that are specific for different bacterial groups. Hence, an important challenge for the future is to understand the cellular functions of these genes/proteins (Danchin, 1999; Galperin and Koonin, 2004; Roberts, 2004; Gupta and Griffiths, 2006), which should provide valuable insights into the biochemical and physiological characteristics that are unique to different groups of bacteria at various taxonomic levels.

2.8 Concluding Remarks

The availability of genome sequence data is enabling identification of numerous protein-based markers consisting of CSIs and CSPs that are specific for different groups of prokaryotes at various taxonomic levels. The discovery of these new protein markers is proving very useful in understanding many critical issues in prokaryotic phylogeny and systematics that are not resolved by 16S rRNA or other gene sequences. In particular, based on these molecular markers, it is now possible to identify and circumscribe most of the major phyla, as well as many of their subgroups, within *Bacteria* and *Archaea* in definitive molecular terms. Additionally, these molecular markers are also providing means to logically delineate the branching order of the main phyla within *Bacteria*. Because of their taxa specificity, functional studies on these newly discovered proteins, or protein characteristics, hold much promise for discovering novel biological properties that are distinctive characteristics of different groups of prokaryotes. Thus far, most of the work on identification of these taxa-specific CSIs and CSPs has been aimed at character-izing higher taxonomic groups within the prokaryotes. However, this approach should be equally applicable in characterizing clades at various phylogenetic depths including specific genera as well as individual species or subspecies. The application of these new approaches, which are based on the presence or absence of definitive molecular charac-teristics, also hold much promise for clarifying the species concept within prokaryotes, which has proven very difficult to resolve based on phylogenetic analysis of 16S rRNA or other genes (Ward and Fraser, 2005; Staley, 2006). The identified CSIs and CSPs, because of their unique presence in specific groups of bacteria, also provide powerful new means for identification/diagnostics of these bacteria by means of matrix assisted laser desorption/ionisation time-of-flight mass spectrometry (MALDI-TOF-MS).

Acknowledgements

The research work from the author's laboratory has been supported by research grants from the Canadian Institute of Health Research and Natural Sciences and Engineering Research Council of Canada.

References

Blankenship, R.E. (1992) Origin and early evolution of photosynthesis. Photosyn. Res., 33, 91–111.

Ciccarelli, F.D., Doerks, T., von Mering, C., Creevey, C.J., Snel, B., Bork, P. (2006) Toward automatic reconstruction of a highly resolved tree of life. Science, 311, 1283–1287.

Danchin, A. (1999) From protein sequence to function. Curr. Opin. Struct. Biol., 9, 363–367.

Daubin, V., Ochman, H. (2004) Bacterial genomes as new gene homes: the genealogy of ORFans in *E. coli*. Genome Res., 14, 1036–1042.

Delsuc, F., Brinkmann, H., Philippe, H. (2005) Phylogenomics and the reconstruction of the tree of life. Nat. Rev. Genet., 6, 361–375.

Doerks, T., von Mering, C., Bork, P. (2004) Functional clues for hypothetical proteins based on genomic context analysis in prokaryotes. Nucleic Acids Res., 32, 6321–6326.

Duncan, M.J. (2003) Genomics of oral bacteria. Crit. Rev. Oral Biol. Med., 14, 175–187.

Dutilh, B.E., Snel, B., Ettema, T.J., Huynen, M.A. (2008) Signature genes as a phylogenomic tool. Mol. Biol. Evol., 25, 1659–1667.

Felsenstein, J. (2004) Inferring Phylogenies. Sinauer Associates, Inc., Sunderland, MA.

Frigaard, N.U., Chew, A.G., Li, H., Maresca, J.A., Bryant, D.A. (2003) *Chlorobium tepidum*: insights into the structure, physiology, and metabolism of a green sulfur bacterium derived from the complete genome sequence. Photosynth. Res., 78, 93–117.

Galperin, M.Y., Koonin, E.V. (2004) 'Conserved hypothetical' proteins: prioritization of targets for experimental study. Nucleic Acids Res., 32, 5452–5463.

Gao, B., Gupta, R.S. (2005) Conserved indels in protein sequences that are characteristic of the phylum *Actinobacteria*. Int. J. Syst. Evol. Microbiol., 55, 2401–2412.

Gao, B., Gupta, R.S. (2007) Phylogenomic analysis of proteins that are distinctive of *Archaea* and its main subgroups and the origin of methanogenesis. BMC Genomics, 8, 86.

Gao, B., Mohan, R., Gupta, R.S. (2009) Phylogenomics and protein signatures elucidating the evolutionary relationships among the Gammaproteobacteria. Int. J. Syst. Evol. Microbiol., 59, 234–247.

Gao, B., Parmanathan, R., Gupta, R.S. (2006) Signature proteins that are distinctive characteristics of Actinobacteria and their subgroups. Antonie van Leeuwenhoek, 90, 69–91.

Garrity, G.M., Bell, J.A., Lilburn, T.G. (2005) The revised road map to the manual. In: Brenner, D.J., Krieg, N.R., Staley, J.T. (eds) Bergey's Manual of Systematic Bacteriology, Volume 2, Part A, Introductory Essays. Springer, New York, pp. 159–220.

Griffiths, E., Gupta, R.S. (2001) The use of signature sequences in different proteins to determine the relative branching order of bacterial divisions: evidence that *Fibrobacter* diverged at a similar time to *Chlamydia* and the *Cytophaga-Flavobacterium-Bacteroides* division. Microbiology, 147, 2611–2622.

Griffiths, E., Gupta, R.S. (2002) Protein signatures distinctive of chlamydial species: Horizontal transfer of cell wall biosynthesis genes *glmU* from Archaebacteria to Chlamydiae, and *murA* between Chlamydiae and *Streptomyces*. Microbiology, 148, 2541–2549.

Griffiths, E., Gupta, R.S. (2004a) Distinctive protein signatures provide molecular markers and evidence for the monophyletic nature of the *Deinococcus-Thermus* phylum. J. Bacteriol., 186, 3097–3107.

Griffiths, E., Gupta, R.S. (2004b) Signature sequences in diverse proteins provide evidence for the late divergence of the order *Aquificales*. Int. Microbiol., 7, 41–52.

Griffiths, E., Gupta, R.S. (2006) Molecular signatures in protein sequences that are characteristics of the Phylum Aquificales. Int. J. Syst. Evol. Microbiol., 56, 99–107.

Griffiths, E., Gupta, R.S. (2007a) Identification of signature proteins that are distinctive of the Deinococcus-Thermus phylum. Int. Microbiol., 10, 201–208.

Griffiths, E., Gupta, R.S. (2007b) Phylogeny and shared conserved inserts in proteins provide evidence that Verrucomicrobia are the closest known free-living relatives of chlamydiae, Microbiology, 153, 2648–2654.

Griffiths, E., Petrich, A., Gupta, R.S. (2005) Conserved indels in essential proteins that are distinctive characteristics of *Chlamydiales* and provide novel means for their identification. Microbiology, 151, 2647–2657.

Griffiths, E., Ventresca, M.S., Gupta, R.S. (2006) BLAST screening of chlamydial genomes to identify signature proteins that are unique for the *Chlamydiales, Chlamydiaceae, Chlamydophila* and *Chlamydia* groups of species. BMC Genomics, 7, 14.

Gupta, R.S. (1998) Protein Phylogenies and signature sequences: a reappraisal of evolutionary relationships among Archaebacteria, Eubacteria, and Eukaryotes. Microbiol. Mol. Biol. Rev., 62, 1435–1491.

Gupta, R.S. (2000) The phylogeny of Proteobacteria: relationships to other eubacterial phyla and eukaryotes. FEMS Microbiol. Rev., 24, 367–402.

Gupta, R.S. (2001) The branching order and phylogenetic placement of species from completed bacterial genomes, based on conserved indels found in various proteins. Int. Microbiol., 4, 187–202.

Gupta, R.S. (2002) Phylogeny of Bacteria: Are we now close to understanding it? ASM News, 68, 284–291.

Gupta, R.S. (2003) Evolutionary relationships among photosynthetic bacteria. Photosynth. Res., 76, 173–183.

Gupta, R.S. (2004) The phylogeny and signature sequences characteristics of *Fibrobacters, Chlorobi* and *Bacteroidetes*. Crit. Rev. Microbiol., 30, 123–143.

Gupta, R.S. (2005a) Molecular sequences and the early history of life. In: Sapp, J. (ed.) Microbial Phylogeny and Evolution: Concepts and Controversies. Oxford University Press, New York, pp. 160–183.

Gupta, R.S. (2005b) Protein signatures distinctive of Alpha proteobacteria and its subgroups and a model for Alpha proteobacterial evolution. Crit Rev. Microbiol., 31, 135.

Gupta, R.S. (2006) Molecular signatures (unique proteins and conserved Indels) that are specific for the epsilon proteobacteria (Campylobacterales). BMC Genomics, 7, 167.

Gupta, R.S. (2009) Protein signatures (molecular synapomorphies) that are distinctive characteristics of the major cyanobacterial clades. Int. J. Syst. Evol. Microbiol., 59, 2510–2526.

Gupta, R.S., Griffiths, E. (2002) Critical issues in bacterial phylogenies. Theor. Popul. Biol., 61, 423–434.

Gupta, R.S., Griffiths, E. (2006) Chlamydiae-specific proteins and indels: novel tools for studies. Trends Microbiol., 14, 527–535.

Gupta, R.S., Lorenzini, E. (2007) Phylogeny and molecular signatures (conserved proteins and indels) that are specific for the Bacteroidetes and Chlorobi species. BMC Evol. Biol., 7, 71.

Gupta, R.S., Mok, A. (2007) Phylogenomics and signature proteins for the alpha Proteobacteria and its main groups. BMC Microbiol., 7, 106.

Gupta, R.S., Sneath, P.H.A. (2007) Application of the character compatibility approach to generalized molecular sequence data: Branching order of the proteobacterial subdivisions. J. Mol. Evol., 64, 90–100.

Gupta, R.S., Pereira, M., Chandrasekera, C., Johari, V. (2003) Molecular signatures in protein sequences that are characteristic of Cyanobacteria and plastid homologues. Int. J. Syst. Evol. Microbiol., 53, 1833–1842.

Kainth, P., Gupta, R.S. (2005) Signature proteins that are distinctive of alpha proteobacteria. BMC Genomics, 6, 94.

Kersters, K., Devos, P., Gillis, M., Swings, J., Vandamme, P., Stackebrandt, E. (2006) Introduction to the proteobacteria. In: Dworkin, M., Falkow, S., Rosenberg, E., Schleifer, K.H., Stackebrandt, E. (eds) The Prokaryotes: A Handbook on the Biology of Bacteria. Springer, New York, pp. 3–37.

Korbel, J.O., Doerks, T., Jensen, L.J., Perez-Iratxeta, C., Kaczanowski, S., Hooper, S.D., Andrade, M.A., Bork, P. (2005) Systematic association of genes to phenotypes by genome and literature mining. PLoS Biol., 3, e134.

Lake, J.A., Herbold, C.W., Rivera, M.C., Servin, J.A., Skophammer, R.G. (2007) Rooting the tree of life using nonubiquitous genes. Mol. Biol. Evol., 24, 130–136.

Lerat, E., Daubin, V., Ochman, H., Moran, N.A. (2005) Evolutionary origins of genomic repertoires in bacteria. PLoS. Biol., 3, e130.

Ludwig, W., Klenk, H.-P. (2005) Overview: A phylogenetic backbone and taxonomic framework for prokaryotic systamatics. In: Brenner, D.J., Krieg, N.R., Staley, J.T., Garrity, G.M. (eds) Bergey's Manual of Systematic Bacteriology. Springer-Verlag, Berlin, pp. 49–65.

Ludwig, W., Schleifer, K.H. (1999) Phylogeny of *Bacteria* beyond the 16S rRNA standard. ASM News, 65, 752–757.

Maidak, B.L., Cole, J.R., Lilburn, T.G., Parker, C.T., Jr, Saxman, P.R., Farris, R.J., Garrity, G.M., Olsen, G.J., Schmidt, T.M., Tiedje, J.M. (2001) The RDP-II (Ribosomal Database Project). Nucleic Acids Res., 29, 173–174.

Mulkidjanian, A.Y., Koonin, E.V., Makarova, K.S., Mekhedov, S.L., Sorokin, A., Wolf, Y.I., Dufresne, A., Partensky, F., Burd, H., Kaznadzey, D., Haselkorn, R., Galperin, M.Y. (2006) The cyanobacterial genome core and the origin of photosynthesis. Proc. Natl Acad. Sci. USA, 103, 13 126–13 131.

Murray, R.G.E., Brenner, D.J., Colwell, R.R., De Vos, P., Goodfellow, M., Grimont, P.A.D., Pfennig, N., Stackebrandt, E., Zavarzin, G.A. (1990) Report of the Ad Hoc Committee on Approaches to Taxonomy within the Proteobacteria. Int. J. Syst. Bacteriol., 40, 213–215.

Nelson, K.E., Paulsen, I.T., Fraser, C.M. (2001) Microbial genome sequencing: A window into evolution and physiology. ASM News, 67, 310–317.

Ohkuma, M., Noda, S., Hongoh, Y., Kudo, T. (2002) Diverse bacteria related to the bacteroides subgroup of the CFB phylum within the gut symbiotic communities of various termites. Biosci. Biotechnol. Biochem., 66, 78–84.

Olsen, G.J., Woese, C.R., Overbeek R. (1994) The winds of (evolutionary) change: breathing new life into microbiology. J. Bacteriol., 176, 1–6.

Oren, A., Stackebrandt, E. (2002) Prokaryote taxonomy online: challenges ahead. Nature, 419, 15.

Overmann, J. (2006) The family Chlorobiaceae. In: Dworkin, M., Falkow, S., Rosenberg, E., Schleifer, K.H., Stackebrandt, E. (eds) The Prokaryotes, Vol. 7, Proteobacteria: Delta and Epsilon Subclass. Springer, New York, pp. 359–378.

Paster, B.J., Boches, S.K., Galvin, J.L., Ericson, R.E., Lau, C.N., Levanos, V.A., Sahasrabudhe, A., Dewhirst, F.E. (2001) Bacterial diversity in human subgingival plaque. J. Bacteriol., 183, 3770–3783.

Paster, B.J., Dewhirst, F.E., Olsen, I., Fraser, G.J. (1994) Phylogeny of *bacteroides, prevotella*, and *porphyromonas* spp. and related bacteria. J. Bacteriol., 176, 725–732.

Reichenbach, H. (1992) The Order *Cytophagales*. In: Balows, A., Truper, H.G., Dworkin, M., Harder, W., Schleifer, K.H. (eds) The Prokaryotes. Springer-Verlag, New York, pp. 3631–3675.

Roberts, R.J. (2004) Identifying protein function–a call for community action. PLoS. Biol., 2, E42.

Rokas, A., Holland, P.W. (2000) Rare genomic changes as a tool for phylogenetics. Trends Ecol. Evol., 15, 454–459.

Salyers, A.A. (1984) Bacteroides of the human lower intestinal tract. Annu. Rev. Microbiol., 38, 293–313.

Shah, H.N. (1992) The genus *Bacteroides* and related taxa. In: Balows, A., Truper, H.G., Dworkin, M., Harder, W., Schleifer, K.H. (eds) The Prokaryotes. Springer-Verlag, New York, pp. 3593–3607.

Shah, H.N., Gharbia, S.E., Duerden, B.I. (1998) Bacteroides, Prevotella and Porphyromonas. In: Balows, A., Duerden, B.I. (eds) Topley & Wilson's Microbiology and Microbial Infections, vol. 2. Systematic Bacteriology. Arnold, London, pp. 1305–1330.

Shah, H.N., Gharbia, S.E., Olsen, I. (2005) Bacteroides, Prevotella, and Porphyromonas. In: Borrelio, S.P., Murray, P.R., Funke, G. (eds) Topley & Wilson's Microbiology and Microbial Infections. Hodder Arnold, London, pp. 1913–1944.

Shah, H.N., Olsen, I., Bernard, K., Finegold, S.M., Gharbia, S.E., Gupta, R.S. (2009) Approaches to the study of the systematics of anaerobic, Gram-negative, non-spore-forming rods: current status and perspectives. Anaerobe, 15, 179–194.

Siew, N., Fischer, D. (2003) Analysis of singleton ORFans in fully sequenced microbial genomes. Proteins, 53, 241–251.

Singh, B., Gupta, R.S. (2009) Conserved inserts in the Hsp60 (GroEL) and Hsp70 (DnaK) proteins are essential for cellular growth. Mol. Genet. Genom., 281, 361–373.

Stackebrandt, E. (2006) Defining taxonomic ranks. In: Dworkin, M., Falkow, S., Rosenberg, E., Schleifer, K.-H., Stackebrandt, E. (eds) The Prokaryotes. Springer, New York, pp. 29–57.

Staley, J.T. (2006) The bacterial species dilemma and the genomic-phylogenetic species concept; Philos. Trans. R. Soc. London, Ser. B, 361, 1899–1909.

Truper, H.G., Pfennig, N. (1992) The family *Chlorobiaceae*. In: Balows, A., Truper, H.G., Dworkin, M., Harder, W., Schleifer, K.H. (eds) The Prokaryotes. Springer-Verlag, New York, pp. 3583–3592.

Ward, N., Fraser, C.M. (2005) How genomics has affected the concept of microbiology. Curr. Opin. Microbiol., 8, 564–571.

Williams, K.P., Sobral, B.W., Dickerman, A.W. (2007) A robust species tree for the Alphaproteobacteria. J. Bacteriol., 189, 4578–4586.

Woese, C.R. (1987) Bacterial evolution. Microbiol. Rev., 51, 221–271.

Woese, C.R., Kandler, O., Wheelis, M.L. (1990) Towards a natural system of organisms: proposal for the domains Archaea, Bacteria, and Eucarya. Proc. Natl Acad. Sci. USA, 87, 4576–4579.

Woese, C.R., Stackebrandt, E., Macke, R.J., Fox, G.E. (1985) A phylogenetic definition of the major eubacterial taxa. System. Appl. Microbiol., 6, 143–151.

Woese, C.R., Stackebrandt, E., Weisburg, W.G., Paster, B.J., Madigan, M.T., Fowler, C.M.R., Hahn, C.M., Blanz, P., Gupta, R., Nealson, K.H., Fox G.E. (1984) The phylogeny of purple bacteria: the alpha subdivision. System. Appl. Microbiol., 5, 315–326.

Part II

Proteomics Tools and Biomarker Discovery

3

Overview of Proteomic Tools and Their Links to Genomics

Raju Misra

*Department for Bioanalysis and Horizon Technologies, Health Protection
Agency Centre for Infections, London, UK*

3.1   Introduction

Protein identification is the correct assignment of a mass spectrometry spectrum to a peptide and is essential for proteomic data processing. It is possible to perform proteomic studies without identification, for example clustering and classification of samples based upon pattern matching using gel based protein separations. However, to determine the biological process leading to any observed differences between samples, identification of the proteins of interest is crucial.

There are a variety of different methods to explore complex protein samples, in which mass spectrometry (MS) has proven invaluable for protein identification. MS is complex as are the data it generates, which are in the form of a mass spectrum (Figure 3.1). The mass spectrum is a graph of ion intensity as a function of the mass-to-charge ratio and it is used to determine the molecular weight and structure of the compounds being analysed and is commonly represented as a histogram or a line graph.

Sample preparation, instrument choice, instrument configuration (tuning and calibration), data acquisition and peak finding all contribute to the quality of mass spectra used for protein identification (McHugh and Arthur, 2008). To highlight the importance of good preparation before performing a MS experiment, that is the proteomic processes used to generate mass spectra, up to 90% of tandem mass spectra in a 'typical' liquid chromatography mass spectrometry (LC-MS) analysis cannot be identified using database search algorithms due to poor quality spectra (Resing and Ahn, 2005; McHugh and Arthur, 2008).

Mass Spectrometry for Microbial Proteomics   Edited by Haroun N. Shah and Saheer E. Gharbia
© 2010 John Wiley & Sons, Ltd

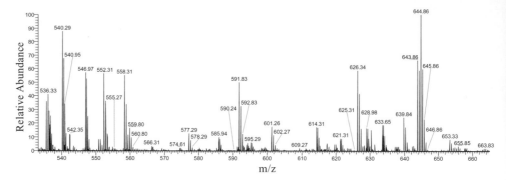

Figure 3.1   *An example of a mass spectrum describing the relationship between the relative abundance (y-axis) and the mass-to-charge ratio (m/z; x-axis). Each vertical line, labelled with an m/z value, represents a detected ion*

Therefore it is essential for the downstream bioinformatic analyses, that both sample and instrument are optimised to the highest standards.

3.2   Protein Identification

The rapid and accurate identification of proteins/peptides from mass spectra can only be achieved through bioinformatic analyses, for which there are a wide range of software tools and algorithms (Table 3.1). The basic identification algorithm is shown in Figure 3.2, whereby the mass spectrum for an unidentified protein is compared with theoretical data from known proteins in a database, and a score is assigned based on 'closeness of fit' between the two datasets. A short list of candidate peptide matches is generated and through user defined parameters and thresholds, for example mass tolerance, proteolytic enzyme constraints and post translational modifications, the list of candidate peptides can be refined. The resulting peptide list is ranked in order of similarity between the experimental and theoretical spectra. Only the top ranking peptides are further analysed using more rigorous statistical analyses to determine the best match ('hit') and if there are no suitable matches that satisfy the user defined parameters and thresholds, then the protein remains unidentified. The threshold is a defined value, which more recently through algorithms such as peptide prophet has become data dependent, and is used to resolve correct from incorrect identifications. Unfortunately, as shown in Figure 3.3, there is no clear cut-off, therefore when applying a threshold it is important to minimise the number of false positives (incorrectly identified 'correct matches') and false negatives (correct matches wrongly rejected). The following sections describe commonly used bioinformatic, proteomic identification strategies.

3.2.1   Peptide Mass Fingerprint (PMF)

A variety of MS methods and instrument configurations have been applied for the analysis of microorganisms. Commonly used methods include matrix assisted laser desorption/

Table 3.1   *A list of commonly used MS tools*

	Reference
Database searching	
MASCOT	Perkins *et al.*, 1999
SEQUEST	Eng *et al.*, 1994
Phenyx	Colinge *et al.*, 2003
TANDEM	Craig and Beavis, 2004
OMSSA	Geer *et al.*, 2004
SpectrumMill	
ProFound	Zhang and Chait, 2000
Aldente	Tuloup *et al.*, 2002
ProteinProspector	Clauser *et al.*, 1999
Spectral matching	
SpectrST	Lam *et al.*, 2007
X! P3	Craig *et al.*, 2005
De novo sequencing and sequence tag	
PEAKS	Ma *et al.*, 2003
pepNOVO	Frank and Pevzner, 2005
Lutefisk	Johnson and Taylor, 2002
GutenTag	Tabb *et al.*, 2003
DirecTag	Tabb *et al.*, 2008
Inspect	Tanner *et al.*, 2005
Popitam	Hernandez *et al.*, 2003
Statistical validation of MS data	
Scaffold	
Peptide prophet	Keller *et al.*, 2002
Protein prophet	Nesvizhskii *et al.*, 2003
Protein quantification	
MSQuant	
EXPRES	
ASAPRatio	Li *et al.*, 2003
Libra	
Proteomics data storage	
Proteus	
PRIDE	Martens *et al.*, 2005
PeptideAtlas	Desiere *et al.*, 2005
SBEAMS	
Global Proteome Machine (GPM)	Craig *et al.*, 2004

ionization (MALDI) and electro/nano-spray ionization mass spectrometry (ESI/NSI-MS). The instrument configuration and method applied to a sample will influence the choice of algorithm used to analyse the resultant data. A variety of scoring algorithms have been developed; the earliest methods were termed peptide mass fingerprinting and are still widely used. As a generic PMF method, the reference protein database is subjected to *in silico* enzymatic digestion. A theoretical spectrum is generated comprising a list of masses expected by the enzymatic digestion. The experimental spectrum and masses are compared with this theoretical dataset and the closest match is identified. Popular PMF packages include Mascot (Perkins *et al.*, 1999), Aldente (Tuloup *et al.*, 2002), MS-fit (Clauser *et al.*, 1999) and ProFound (Zhang and Chait, 2000). Each of these packages attempts to

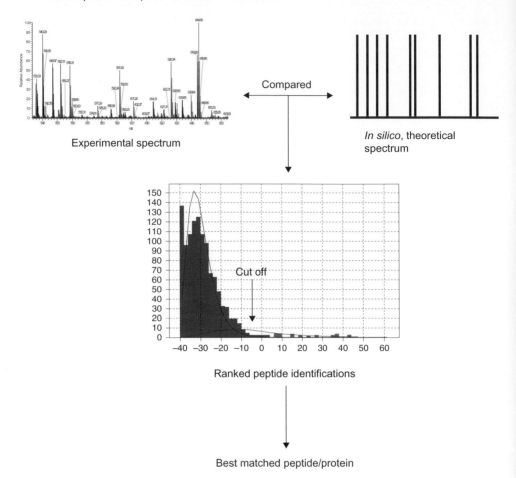

Experimental spectrum

Compared

In silico, theoretical spectrum

Ranked peptide identifications

Best matched peptide/protein

Figure 3.2   *Simplified peptide identification pathway by peptide fragment fingerprint (PFF) database searching. An experimentally generated mass spectrum is compared with a theoretical spectrum generated in silico from the reference database. A preliminary score and ranking is performed to short list potential matching peptides, from which a more rigorous set of multivariate statistical analyses is performed to determine the best match or 'hit'*

calculate the probability that the observed match between the experimental and the *in silico* datasets occurred by chance. Therefore, when calculating the score, a multivariate statistical approach is used, combining the probability a match was a chance event with the similarity measures. There are two caveats to this approach; the first is that no sequence information is presented for the experimental data. Secondly, it is now required by some publishers that at least one peptide identified by PMF should be confirmed by MS/MS.

3.2.2   Peptide Fragment Fingerprint

An alternative to the PMF method is the more data rich peptide fragment fingerprinting technique. PFF (MS/MS) pipelines are now common place when performing high-through-

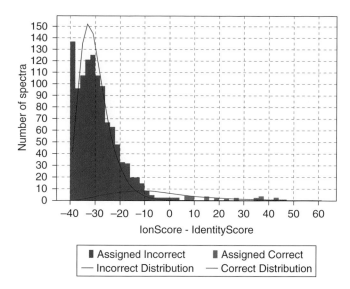

Figure 3.3 *Using Scaffold implementation of the peptide prophet algorithm, a distribution profile of spectra versus ion score is calculated from which an experiment specific cut-off between correct and incorrect distributions can be determined*

put proteomic experiments. In common with PMF, proteins are digested with an enzyme to produce peptide fragments; however, unlike PMF, the resultant peptides are subjected to further fragmentation to yield PFF or MS/MS spectra. These spectra contain not only PMF mass data but also amino acid sequence information. Although PMF data, if of a high enough accuracy, theoretically should be sufficient to make an accurate protein identification, the lack of corroboratory information increases the likelihood of false positives (Kall *et al.*, 2008). In addition, the secondary fragmentation reduces the impact of potential errors borne through protein modifications, therefore the additional peptide sequence information, allows for a far greater statistical confidence when performing protein identifications. A number of PFF packages are available and are listed in Table 3.1. Arguably the most popular of these are MASCOT (Perkins *et al.*, 1999) and SEQUEST (Eng *et al.*, 1994), both of which are commercially available.

A variety of scoring schemes have been applied, such as spectral correlation functions, as used by SEQUEST (Eng *et al.*, 1994). Related to spectral counting are concepts such as shared fragment counts and dot product, which are used by MASCOT (Perkins *et al.*, 1999), TANDEM (Craig and Beavis, 2004) and OMSSA (Geer *et al.*, 2004). Alternatively, statistically derived fragmentation frequencies have been used by other MS scoring tools, for example, PHENYX (Colinge *et al.*, 2003). Each of these statistical algorithms measure how well an experimental spectrum matched or 'hit' a peptide in the reference sequence database. In addition to the score, a statistical confidence is often applied to determine the likelihood of the match being correct (Choi and Nesvizhskii, 2008). To achieve this, a number of different algorithms have been applied to determine the statistical confidence, the most common methods being p and E values. These values are generated by comparing the protein/peptide search score against a distribution of all peptide scores in the search database. The further away the observed score is located from the core of the null distribution, the more significant, that is the more likely the 'hit' will be correct. The *p* value is essentially a measure of the tail probability in the distribution of random matches, whereas the *E* value is a measure of the likelihood the peptides of interest matched the experimental spectrum by random chance.

3.2.3 Peptide Sequencing

Thus far PMF and PFF based protein identifications have been discussed, both of which identify the protein/peptide sequence through comparisons with a reference dataset. Although this approach yields a wealth of information, it is still based on comparative analyses and is hugely dependent on the reference database and scoring algorithms (Pitzer *et al.*, 2007). If the database does not contain the correct sequence information, no identification can be made, hence to resolve this peptide *de novo* sequencing approaches may be applied.

Peptide *de novo* sequencing algorithms attempt to extract the peptide sequence directly from the peptide (MS/MS) fragmentation spectra. Due to the complexity of MS/MS spectra, only short sequences are generated and are often referred to as 'tags' (Nesvizhskii *et al.*, 2007). A list of commonly used tag based programs is given in Table 3.1. To perform a protein/peptide identification using the short length tag sequences, they are submitted to an alignment program such as variants of the BLAST and FASTA algorithms, which through sequence homology make a match to an annotated sequence (Graves and Haystead, 2002). Due to the sequence tags being short in length, they may share high homology and align against many protein sequences, however, it is expected that the correct protein sequence will have many different tag sequences aligned to it (Nesvizhskii *et al.*, 2007). In terms of performance, tag based methods have been demonstrated to perform much faster than many PFF search tools, due to more efficient filtering steps which exclude proteins unlikely to match to the tag sequences (Tabb *et al.*, 2008). However, the total number of identifications made is often far lower than PFF algorithms. Ideally both PFF and peptide sequencing should be performed, although this may not be feasible for large projects as it can be very time consuming to perform both and cross correlate the datasets.

3.2.4 False Discovery Rate (FDR)

Recently, there has been a call by many scientific journals for greater stringency when reporting database search results. Therefore for publications, as well as good practice the FDR should be calculated for experiments. The aim of an FDR calculation is to measure the significance of a hit and is defined by Choi and Nesvizhskii (2008) as the 'expected' proportion of incorrect assignments at the global level. One of the most popular approaches to determine the FDR is by performing a search against two databases, the first the reference database and the second a decoy database. A decoy database is usually the reference database in which the sequences have either been reversed or randomized (Choi and Nesvizhskii, 2008; Kall *et al.*, 2008). There has been much discussion on which is better, a concatenated reversed or randomized database but when using MASCOT a randomized database is the preferred method. In MASCOT, by simply choosing the 'Decoy' checkbox, the decoy database and FDR will automatically be calculated. This can also be done for SEQUEST searches via the Thermo Scientific software package Proteome Discoverer.

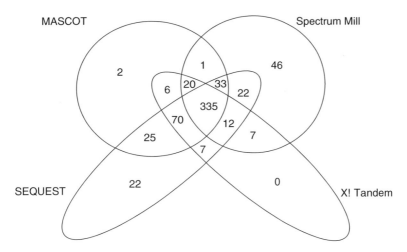

Figure 3.4   *Four-way Venn diagram showing the overlap between four different MS/MS search algorithms, MASCOT, SEQUEST, Spectrum Mill and X!Tandem. The number of correctly identified peptides by one or more algorithms is indicated, for example, 335 peptide hits are correctly identified based on consensus of all four algorithms (intersection), whilst 608 peptide hits are correctly identified by all of the algorithms (union). (Reproduced with permission, from Kapp et al., 2005)*

3.2.5   Validating Protein Identifications

Comparisons between various PFF packages have been performed in an attempt to determine the selectivity and sensitivity of each of the packages tested (Kapp *et al.*, 2005;
Resing and Ahn, 2005). These results clearly demonstrated that no one package was able
to identify all proteins in a known mixture, moreover the proteins identified by each
package varied as demonstrated in Figure 3.4. Statistically more confidence can be given
to those proteins identified by a variety of search algorithms, for example protein 'A' being
identified in both MASCOT and SEQUEST, as opposed to just being identified by
SEQUEST. This does not necessarily mean that the proteins identified by just one of the
packages are incorrect, but statistically they should be given a lower confidence. To take
advantage of this observation, statistical algorithms have been developed to make use of
the data from two or more PFF packages. One such algorithm, MSPlus, has been developed
to improve reproducibility and sensitivity in identifying proteins from shotgun proteomic
data. MSPlus uses a heuristic set of rules applied to MASCOT and SEQUEST results,
followed by a least-squares fitting step to produce a sum-score for each candidate protein
(McHugh and Arthur, 2008).

  An alternative to MSPlus is the commercial implementation of the peptide prophet
algorithm (Keller *et al.*, 2002), Proteome Software Scaffold. When performing a database
search using the raw mass spectral data, users typically search their data using one of the
many protein identification programs, determine the FDR and apply a threshold. Anything
above that threshold is accepted as a 'good' identification. However, it was demonstrated
that the threshold was data dependent, that is the threshold is affected by the sample,

database searched and the instrument type (Qian *et al.*, 2005). Therefore, despite an identification being made, the score may not be a true reflection of the quality of the identification. The peptide prophet algorithm attempts to resolve these variables by automatically adjusting the threshold to the characteristics of the data (Keller *et al.*, 2002). Peptide prophet as implemented in Scaffold, rescores the protein/peptide identifications into a single discriminate score, a *D* value, this is done for all the spectra in a sample. It creates a histogram of these scores and using Bayesian statistics calculates the probability that a match is correct. Scaffold is able to perform these calculations for different search engines, such as SEQUEST, MASCOT as well as X-Tandem and displays them in a single graphical user interface (GUI). However, the limiting factor to analysing large volumes of data is memory. The application is java based, which inherently makes it memory intensive, therefore on 32-bit systems it struggles to handle datasets in excess of 500 000 spectra. Interfacing via the command line, using Scaffold Batch, helps to resolve this, but still large datasets will require more memory (>2 GB) and much analysis time.

3.2.6   Reference Database

The sequence database used as the reference set is critical to MS based protein identifications. If the correct protein sequence is not entered in the database then the algorithms will be unable to determine the correct identification. It must be noted that many of the publicly available reference databases are prone to errors therefore annotations must be corroborated with other reference datasets to ensure their accuracy. In addition to database curation, longer sequences are subjected to an array of modifications, making it difficult to determine accurately the theoretical mass spectra. In recent years, MS has become more accessible to researchers and a wide range of prokaryotes have been submitted for MS analysis, but often no suitable reference sequence was available to search against. With the advent of high throughput genome sequencers, so-called next generation sequencers, more genome information is available for MS, however the sequences are often unfinished as a collection of contigs or scaffolds. Despite this, using gene finding algorithms such as Glimmer[*] (Salzberg *et al.*, 1998), the resulting coding sequences can be translated into protein sequences and used as a reference database for MS protein identification.

Probability based protein identification algorithms, such as Mascot (Perkins *et al.*, 1999), are database dependent, that is, the final score will vary according to database size. If a small database is used, for example a genus specific nonredundant protein database, it would generally require lower cut-off confidence scores than for searches performed using much larger databases, such as the NCBInr database (Pruitt *et al.*, 2007; McHugh and Arthur, 2008). There is no specific cut-off for what is considered a small or large database, but when interpreting protein identification data, the probability scores alone should not be used to delineate good identifications from bad, the FDR, inspection of the spectra and further validation should also be performed.

Many of the publicly available databases such as the NCBInr, Swissprot or Uniprot resources contain redundant sequences; that is, there may be two or more identical

---

[*]Gene finding is complex and a variety of gene finding algorithms are available to correctly identify a coding sequence. In addition to gene finding, annotation pipelines are publicly available to annotate coding sequences.

sequences in a database, which leads to multiple hits between an experimental spectrum and the reference dataset. Although this is not catastrophic as the correct identification would still be made, it would slow search times and further complicate the search results output. To resolve this, it is best practice to create a nonredundant database, that is remove all duplicate entries, which can be done either through simple in-house programs or publicly available programs such as CD-Hit (Li and Godzik, 2006).

3.2.7   Data Storage

A wide variety of data analysis tools have been described, although powerful in their analyses, interpretation can be difficult particularly when attempting to integrate the outputs from the different data analysis tools. A typical proteomics pipeline to identify proteins in a sample could produce many different datasets, which ideally should be integrated as part of the interpretation process. Example data outputs include 2D gel images, gel comparisons, spot cutting information, MS spectrum files, PFF identifications, sample comparisons using Scaffold and *de novo* sequencing data. Since it is difficult to integrate these data sources, in house databases and scripts can be implemented, but these are local solutions, which many laboratories cannot easily replicate. Fortunately there are software solutions which can integrate many of the different data types listed above and implemented by different laboratories. One such solution is Genologics Proteus, which is a GUI driven data storage and analysis system. It is capable of storing data from many of the popular proteomics analysis tools. Its purpose is not to replicate the algorithms but to integrate data from these different tools; its GUI driven reporting tool allows for direct comparison of statistical as well as image data, thus allowing for a far more efficient method of interpreting data. Many of the tools listed above produce transient outputs, that is, the results are static, representing the data submitted at that time. However, when new data become available, they may require the complete reanalysis of all data old and new combined. To overcome this, data stored using Proteus can be archived and as new data become available they can be integrated with them, which may appear insignificant and trivial, but for long term projects, it is hugely beneficial.

3.3   Applications

3.3.1   Biomarker Discovery

Thus far, a background of commonly used bioinformatic tools and methods has been described; the following section demonstrates how these tools may be applied. An important area of research, which has gained much momentum in recent years, is biomarker discovery of pathogenic microorganisms. Proteins/peptides that form significant clusters, representing distinct taxonomic divisions or phenotypes can be used to identify an organism or a specific property; these are referred to as biomarkers.

The advantage of using proteomics for biomarker discovery, in particular MS, over many other techniques is the ability to successfully characterize individual microorganisms based on protein and nonprotein features (Demirev and Fenselau, 2008). It is a difficult process, but the *in silico* processes can be split into clearly defined stages: discovery, validation, identification, validation (again), and assay implementation.

The discovery stage necessitates the generation of vast amounts of mass spectra, data that reflect as many of the expressed proteins as possible for the organism of interest. Ideally, as stated above, the additional information acquired from PFF analyses would greatly improve the quality of downstream analyses. On acquisition of the MS spectra, protein/peptide identifications are performed using the database search engine(s) listed in Table 3.1. Typically, from a gel-C experiment, one would identify a small percentage of the potential proteome, therefore to ensure as many of the potential proteins are identified from the spectra as possible, two or more search engines should be used. The resulting data from the search engines are refined using the appropriate cut-offs for the search engine (ensuring that publication standards are met). The datasets are then merged to produce a single list of distinct protein identifications representing the characterized proteome of a sample. It is important to ensure every effort is made to identify as many proteins as possible, whether by different experimental conditions or different *in silico* tools, as there is a huge amount of data loss during downstream analyses, when trying to identify a distinct set of protein features for use as biomarkers. Much data is generated from these approaches and will contain some redundancy, for example MASCOT and SEQUEST will undoubtedly identify the same peptides and proteins. To simplify this, data redundancy needs to be reduced as much as possible. Following this, the first attempt at validating the data can be performed, using MS data comparison tools, such as Scaffold, Proteus and Nonlinear Progenesis in which replicate datasets can be analysed to ensure that they are reproducible and controls were successful.

The proteins/peptides detected represent the pool of data from which candidate biomarkers can be identified. The simplest approach is to take the sequences or spectra from the proteins/peptides identified and through a series of statistical filters compare them with a reference set from which any datasets which cluster uniquely to the organism of interest can be short listed as a candidate biomarker. Sequence based comparisons can be performed using sequence homology programs, for example BlastP (Altschul *et al.*, 1990). It should be noted that when using BlastP, parameters such as the word length, expect value, matrix and gap penalties need to be altered for short length peptides sequences. Large public protein sequence databases, such as the NCBInr (Pruitt *et al.*, 2007) or EBI (Emmert *et al.*, 1994) database can be used as the reference database, against which the identified proteins can be compared. The resultant data are parsed and filtered to retrieve only those sequences that are unique to the organism of interest. Alternatively, using the program Progenesis LC-MS, the spectrum files can be analysed and clustered through the use of multivariate statistics to reveal features that can be used as candidate biomarkers.

To ensure that the most significant biomarkers are taken forward, further validation is required and this entails extensive experimental work. At present there is much debate as to what is considered a good validation pipeline and no specific pipeline has been standardized or widely adopted. Consequently, the number of biomarkers implemented for clinical use is low (Rifai *et al.*, 2006). However, this is not to imply that protein biomarkers are insignificant, simply a more coherent, standardized pipeline is required. The experimental details for biomarker validation are briefly described in Chapter 4, but in short it may require assaying the candidate biomarker(s) against potentially many hundreds of samples, performed in different laboratories to measure the robustness of the biomarker (Rifai *et al.*, 2006). Quantitative work would be advisable in which sensitivity assays that entail selected reaction monitoring (SRM)/multiple reaction monitoring (MRM) methodologies can be performed. Software tools have been developed to make use of quantitative datasets and some of the more popular free and commercial tools are listed in Table 3.1. It is only after extensive and successful validation can the time and economic commitment be made in developing the 'biomarker' into a diagnostic assay.

3.3.2 Integrating Genomics with Proteomics

With the accumulation of vast amounts of DNA sequences in databases, researchers are realizing that merely having complete sequences of genomes is not sufficient to elucidate biological function. A cell is normally dependent upon a multitude of, metabolic and regulatory pathways for its survival, therefore, the study of a cell's protein content (or proteome) is complementary to genomic and transcriptomic investigations (Pandey and Mann, 2000; Dongre *et al.*, 2001).

In recent years, the advent of new high throughput sequencing technologies has led to the transition of genome sequencing from being a very costly and challenging exercise to being a relatively routine endeavour. This can be highlighted by the number of sequenced (partial or complete) submissions made to the NCBI genome database since the first complete free-living microbial genome for *Haemophilus influenzae* was published in 1995 (Fleischmann *et al.*, 1995). By 1997 the number of completed genome sequences had risen to six organisms: *Haemophilus influenzae* (Fleischmann *et al.*, 1995), *Mycoplasma genitalium* (Fraser *et al.*, 1995), *Mycoplasma pneumoniae* (Himmelreich *et al.*, 1996), *Synechocystis sp.* (Kaneko *et al.*, 1996), *Methanococcus jannaschii* (Bult *et al.*, 1996) and *Saccharomyces cerevisiae* (Goffeau *et al.*, 1996). Since then over 9500 completed genomes have been submitted and many more hundreds have been sequenced by the so-called 'next generation', high throughput sequencers. Interestingly, approximately 5100 different species have been investigated, highlighting a bias towards certain genera, borne out of industrial or medical interest. Having the genome sequence, is only the first step albeit an important one, however to better understand the phenotype of an organism, it has to be annotated. Annotation is a complex process involving the identification of genes and then assigning a function for them. A range of automated bioinformatic pipelines have been established to annotate a newly sequenced genome, but they are all primarily dependent on comparative genomics to elucidate the function of a gene. Where there is no information regarding the function of a gene, except an *in silico* prediction, it is referred to as a hypothetical coding sequence (CDS) and can represent between 20% and 50% of the identified genome. There are a number of algorithms that are used to identify CDSs, for example Glimmer (Salzberg *et al.*, 1998) and GeneMark (Besemer and Borodovsky, 2005), however when the data from these different algorithms are combined, it is not unusual to find contradictory results. One algorithm may identify a CDS that the others do not, although corroboration between them suggests a prediction is correct, differing results do not necessarily mean that the predicted CDS is incorrect. It is difficult to annotate these genes through *in silico* approaches alone therefore to address this problem high throughput proteomics have been applied. Proteomic approaches combined with MS provide a robust method to determine whether a protein coding region is expressed as a protein product. As described earlier, MS based search tools rely on a well curated database to infer protein/peptide identifications from a MS spectrum. Therefore, as first demonstrated by Yates and colleagues (Yates III *et al.*, 1995), the newly sequenced genome can be translated into the corresponding protein sequence and used as the reference database against which searches are performed. This approach has greatly aided in confirming that hypothetical CDSs are genuine and can be annotated as a functionally unknown (FUN) gene rather than as hypothetical. In addition to annotating hypothetical genes, MS based annotations can be used to confirm CDS annotations, specifically ensuring the predicted start and end sites as well as reading frames of a CDS. MS based proteomics is a very powerful and robust method to identify protein/peptides from a sample and when combined with *de novo*/tag based methods, it can be used to identify CDSs that may have been missed during *in silico* annotation.

An example of this can be demonstrated from our own work. A common approach to perform functional annotations is to determine the subcellular location of a gene product and having this information can aid in deducing the function of a gene. Fortunately, this can be predicted using a variety of subcellular prediction algorithms. Some are 'general' prediction algorithms able to assign a location to any one of the main prokaryote features, for example PsortB (Gardy *et al.*, 2005) and CELLO (Yu *et al.*, 2006) can assign a gene product to either the cytoplasm, outer/inner membrane, periplasm or extracellular subcellular locations. Other algorithms are designed to assign a protein to a single feature, for example TMHMM (Emanuelsson *et al.*, 2007) determines the presence and number of transmembrane helices. lipoP (Juncker *et al.*, 2003) and BOMP (Berven *et al.*, 2004) can assign proteins to the outer membrane and SignalP (Emanuelsson *et al.*, 2007) identifies signal peptides which are indicative of proteins ejected from the cytoplasm. However, well cited subcellular prediction tools are only predictions and require experimental verification. Only high throughput proteomics can achieve this but it is dependent on the method used to fractionate a sample into distinct subcelluar fractions (see Chapters 8–10). Once separated to enrich for a fraction, the samples can be analysed and the proteins identified using MS. It must be noted that biochemical fractionation procedures do not provide 100% separation, thus to accurately annotate a CDS a combination of genomic, bioinformatic and proteomic approaches is needed.

There is a need to integrate data, as each 'omics' technology individually yields much information. However, to better understand an organism from DNA to proteins, an integrative systems biology approach is required and bioinformatic approaches are essential in achieving this.

References

Altschul SF, Gish W, Miller W, Myers EW and Lipman DJ (1990) Basic Local Alignment Search Tool. *J Mol Biol* **215**: pp 403–410.

Berven FS, Flikka K, Jensen HB and Eidhammer I (2004) BOMP: a Program to Predict Integral Beta-Barrel Outer Membrane Proteins Encoded Within Genomes of Gram-Negative Bacteria. *Nucleic Acids Res* **32**: pp W394–W399.

Besemer J and Borodovsky M (2005) GeneMark: Web Software for Gene Finding in Prokaryotes, Eukaryotes and Viruses. *Nucleic Acids Res* **33**: pp W451–W454.

Bult CJ, White O, Olsen GJ, Zhou L, Fleischmann RD, Sutton GG, Blake JA, FitzGerald LM, Clayton RA, Gocayne JD, Kerlavage AR, Dougherty BA, Tomb JF, Adams MD, Reich CI, Overbeek R, Kirkness EF, Weinstock KG, Merrick JM, Glodek A, Scott JL, Geoghagen NS and Venter JC (1996) Complete Genome Sequence of the Methanogenic Archaeon, *Methanococcus jannaschii*. *Science* **273**: pp 1058–1073.

Choi H and Nesvizhskii AI (2008) False Discovery Rates and Related Statistical Concepts in Mass Spectrometry-Based Proteomics. *J Proteome Res* **7**: pp 47–50.

Clauser KR, Baker P and Burlingame AL (1999) Role of Accurate Mass Measurement (+/– 10 ppm) in Protein Identification Strategies Employing MS or MS/MS and Database Searching. *Anal Chem* **71**: pp 2871–2882.

Colinge J, Masselot A, Giron M, Dessingy T and Magnin J (2003) OLAV: Towards High-Throughput Tandem Mass Spectrometry Data Identification. *Proteomics* **3**: pp 1454–1463.

Craig R and Beavis RC (2004) TANDEM: Matching Proteins with Tandem Mass Spectra. *Bioinformatics* **20**: pp 1466–1467.

Craig R, Cortens JP and Beavis RC (2004) Open Source System for Analyzing, Validating, and Storing Protein Identification Data. *J Proteome Res* **3**: pp 1234–1242.

Craig R, Cortens JP and Beavis RC (2005) The Use of Proteotypic Peptide Libraries for Protein Identification. *Rapid Commun Mass Spectrom* **19**: pp 1844–1850.

Demirev PA and Fenselau C (2008) Mass Spectrometry in Biodefense. *J Mass Spectrom* **43**: pp 1441–1457.

Desiere F, Deutsch EW, Nesvizhskii AI, Mallick P, King NL, Eng JK, Aderem A, Boyle R, Brunner E, Donohoe S, Fausto N, Hafen E, Hood L, Katze MG, Kennedy KA, Kregenow F, Lee H, Lin B, Martin D, Ranish JA, Rawlings DJ, Samelson LE, Shiio Y, Watts JD, Wollscheid B, Wright ME, Yan W, Yang L, Yi EC, Zhang H and Aebersold R (2005) Integration with the Human Genome of Peptide Sequences Obtained by High-Throughput Mass Spectrometry. *Genome Biol* **6**: p R9.

Dongre AR, Opiteck G, Cosand WL and Hefta SA (2001) Proteomics in the Post-Genome Age. *Biopolymers* **60**: pp 206–211.

Emanuelsson O, Brunak S, Von Heijne G and Nielsen H (2007) Locating Proteins in the Cell Using TargetP, SignalP and Related Tools. *Nat Protoc* **2**: pp 953–971.

Emmert DB, Stoehr PJ, Stoesser G and Cameron GN (1994) The European Bioinformatics Institute (EBI) Databases. *Nucleic Acids Res* **22**: pp 3445–3449.

Eng JK, McCormack AL and Yates JR (1994) An Approach to Correlate Tandem Mass Spectral Data of Peptides with Amino Acid Sequences in a Protein Database. *J Am Soc Mass Spectrom* **5**: pp 976–989.

Fleischmann RD, Adams MD, White O, Clayton RA, Kirkness EF, Kerlavage AR, Bult CJ, Tomb JF, Dougherty BA and Merrick JM (1995) Whole-Genome Random Sequencing and Assembly of *Haemophilus influenzae* Rd. *Science* **269**: pp 496–512.

Frank A and Pevzner P (2005) PepNovo: De Novo Peptide Sequencing Via Probabilistic Network Modeling. *Anal Chem* **77**: pp 964–973.

Fraser CM, Gocayne JD, White O, Adams MD, Clayton RA, Fleischmann RD, Bult CJ, Kerlavage AR, Sutton G, Kelley JM, Fritchman RD, Weidman JF, Small KV, Sandusky M, Fuhrmann J, Nguyen D, Utterback TR, Saudek DM, Phillips CA, Merrick JM, Tomb JF, Dougherty BA, Bott KF, Hu PC, Lucier TS, Peterson SN, Smith HO, Hutchison CA, III and Venter JC (1995) The Minimal Gene Complement of *Mycoplasma genitalium*. *Science* **270**: pp 397–403.

Gardy JL, Laird MR, Chen F, Rey S, Walsh CJ, Ester M and Brinkman FS (2005) PSORTb V.2.0: Expanded Prediction of Bacterial Protein Subcellular Localization and Insights Gained from Comparative Proteome Analysis. *Bioinformatics* **21**: pp 617–623.

Geer LY, Markey SP, Kowalak JA, Wagner L, Xu M, Maynard DM, Yang X, Shi W and Bryant SH (2004) Open Mass Spectrometry Search Algorithm. *J Proteome Res* **3**: pp 958–964.

Goffeau A, Barrell BG, Bussey H, Davis RW, Dujon B, Feldmann H, Galibert F, Hoheisel JD, Jacq C, Johnston M, Louis EJ, Mewes HW, Murakami Y, Philippsen P, Tettelin H and Oliver SG (1996) Life with 6000 Genes. *Science* **274**: pp 546, 563–567.

Graves PR and Haystead TA (2002) Molecular Biologist's Guide to Proteomics. *Microbiol Mol Biol Rev* **66**: pp 39–63.

Hernandez P, Gras R, Frey J and Appel RD (2003) Popitam: Towards New Heuristic Strategies to Improve Protein Identification from Tandem Mass Spectrometry Data. *Proteomics* **3**: pp 870–878.

Himmelreich R, Hilbert H, Plagens H, Pirkl E, Li BC and Herrmann R (1996) Complete Sequence Analysis of the Genome of the Bacterium *Mycoplasma pneumoniae*. *Nucleic Acids Res* **24**: pp 4420–4449.

Johnson RS and Taylor JA (2002) Searching Sequence Databases via De Novo Peptide Sequencing by Tandem Mass Spectrometry. *Mol Biotechnol* **22**: pp 301–315.

Juncker AS, Willenbrock H, Von Heijne G, Brunak S, Nielsen H and Krogh A (2003) Prediction of Lipoprotein Signal Peptides in Gram-Negative Bacteria. *Protein Sci* **12**: pp 1652–1662.

Kall L, Storey JD, MacCoss MJ and Noble WS (2008) Assigning Significance to Peptides Identified by Tandem Mass Spectrometry Using Decoy Databases. *J Proteome Res* **7**: pp 29–34.

Kaneko T, Sato S, Kotani H, Tanaka A, Asamizu E, Nakamura Y, Miyajima N, Hirosawa M, Sugiura M, Sasamoto S, Kimura T, Hosouchi T, Matsuno A, Muraki A, Nakazaki N, Naruo K, Okumura S, Shimpo S, Takeuchi C, Wada T, Watanabe A, Yamada M, Yasuda M and Tabata S (1996) Sequence Analysis of the Genome of the Unicellular *Cyanobacterium synechocystis* sp. Strain PCC6803. II. Sequence Determination of the Entire Genome and Assignment of Potential Protein-Coding Regions. *DNA Res* **3**: pp 109–136.

Kapp EA, Schutz F, Connolly LM, Chakel JA, Meza JE, Miller CA, Fenyo D, Eng JK, Adkins JN, Omenn GS and Simpson RJ (2005) An Evaluation, Comparison, and Accurate Benchmarking of Several Publicly Available MS/MS Search Algorithms: Sensitivity and Specificity Analysis. *Proteomics* **5**: pp 3475–3490.

Keller A, Nesvizhskii AI, Kolker E and Aebersold R (2002) Empirical Statistical Model to Estimate the Accuracy of Peptide Identifications Made by MS/MS and Database Search. *Anal Chem* **74**: pp 5383–5392.

Lam H, Deutsch EW, Eddes JS, Eng JK, King N, Stein SE and Aebersold R (2007) Development and Validation of a Spectral Library Searching Method for Peptide Identification From MS/MS. *Proteomics* **7**: pp 655–667.

Li W and Godzik A (2006) Cd-Hit: a Fast Program for Clustering and Comparing Large Sets of Protein or Nucleotide Sequences. *Bioinformatics* **22**: pp 1658–1659.

Li XJ, Zhang H, Ranish JA and Aebersold R (2003) Automated Statistical Analysis of Protein Abundance Ratios from Data Generated by Stable-Isotope Dilution and Tandem Mass Spectrometry. *Anal Chem* **75**: pp 6648–6657.

Ma B, Zhang K, Hendrie C, Liang C, Li M, Doherty-Kirby A and Lajoie G (2003) PEAKS: Powerful Software for Peptide De Novo Sequencing by Tandem Mass Spectrometry. *Rapid Commun Mass Spectrom* **17**: pp 2337–2342.

Martens L, Hermjakob H, Jones P, Adamski M, Taylor C, States D, Gevaert K, Vandekerckhove J and Apweiler R (2005) PRIDE: the Proteomics Identifications Database. *Proteomics* **5**: pp 3537–3545.

McHugh L and Arthur JW (2008) Computational Methods for Protein Identification from Mass Spectrometry Data. *PLoS Comput Biol* **4**: p e12.

Nesvizhskii AI, Keller A, Kolker E and Aebersold R (2003) A Statistical Model for Identifying Proteins by Tandem Mass Spectrometry. *Anal Chem* **75**: pp 4646–4658.

Nesvizhskii AI, Vitek O and Aebersold R (2007) Analysis and Validation of Proteomic Data Generated by Tandem Mass Spectrometry. *Nat Methods* **4**: pp 787–797.

Pandey A and Mann M (2000) Proteomics to Study Genes and Genomes. *Nature* **405**: pp 837–846.

Perkins DN, Pappin DJ, Creasy DM and Cottrell JS (1999) Probability-Based Protein Identification by Searching Sequence Databases Using Mass Spectrometry Data. *Electrophoresis* **20**: pp 3551–3567.

Pitzer E, Masselot A and Colinge J (2007) Assessing Peptide De Novo Sequencing Algorithms Performance on Large and Diverse Data Sets. *Proteomics* **7**: pp 3051–3054.

Pruitt KD, Tatusova T and Maglott DR (2007) NCBI Reference Sequences (RefSeq): a Curated Non-Redundant Sequence Database of Genomes, Transcripts and Proteins. *Nucleic Acids Res* **35**: pp D61-D65.

Qian WJ, Liu T, Monroe ME, Strittmatter EF, Jacobs JM, Kangas LJ, Petritis K, Camp DG and Smith RD (2005) Probability-Based Evaluation of Peptide and Protein Identifications from Tandem Mass Spectrometry and SEQUEST Analysis: the Human Proteome. *J Proteome Res* **4**: pp 53–62.

Resing KA and Ahn NG (2005) Proteomics Strategies for Protein Identification. *FEBS Lett* **579**: pp 885–889.

Rifai N, Gillette MA and Carr SA (2006) Protein Biomarker Discovery and Validation: the Long and Uncertain Path to Clinical Utility. *Nat Biotechnol* **24**: pp 971–983.

Salzberg SL, Delcher AL, Kasif S and White O (1998) Microbial Gene Identification Using Interpolated Markov Models. *Nucleic Acids Res* **26**: pp 544–548.

Tabb DL, Ma ZQ, Martin DB, Ham AJ and Chambers MC (2008) DirecTag: Accurate Sequence Tags From Peptide MS/MS Through Statistical Scoring. *J Proteome Res* **7**: pp 3838–3846.

Tabb DL, Saraf A and Yates JR, III (2003) GutenTag: High-Throughput Sequence Tagging Via an Empirically Derived Fragmentation Model. *Anal Chem* **75**: pp 6415–6421.

Tanner S, Shu H, Frank A, Wang LC, Zandi E, Mumby M, Pevzner PA and Bafna V (2005) InsPecT: Identification of Posttranslationally Modified Peptides from Tandem Mass Spectra. *Anal Chem* **77**: pp 4626–4639.

Tuloup M, Hoogland C, Binz P-A and Appel RD (2002) A New Peptide Mass Fingerprinting Tool on ExPASy: ALDentE. *Swiss Proteomics Soc Congr Conf Proc*.

Yates JR, III, Eng JK and McCormack AL (1995) Mining Genomes: Correlating Tandem Mass Spectra of Modified and Unmodified Peptides to Sequences in Nucleotide Databases. *Anal Chem* **67**: pp 3202–3210.

Yu CS, Chen YC, Lu CH and Hwang JK (2006) Prediction of Protein Subcellular Localization. *Proteins* **64**: pp 643–651.

Zhang W and Chait BT (2000) ProFound: an Expert System for Protein Identification Using Mass Spectrometric Peptide Mapping Information. *Anal Chem* **72**: pp 2482–2489.

4

Tandem Mass Spectrometry-Based Proteomics, Protein Characterisation and Biomarker Discovery in Microorganisms

Min Fang

Department for Bioanalysis and Horizon Technologies, Health Protection Agency Centre for Infections, London, UK

4.1 Introduction

Proteomics, in particular, mass spectrometry (MS)-based proteomics has established itself as an indispensable technology in rapid detection and characterisation of microorganisms (1, 2). Two predominant approaches, utilising MS spectra and tandem MS (MS/MS) spectra generated by MS and tandem MS, respectively, have been successfully applied for characterisation of microorganisms at the species and, in some instances, at the subspecies levels. The former, utilising MS spectra generated mainly by matrix assisted laser desorption/ionisation time-of-flight (MALDI-TOF)-based instruments, has been the subject of several in-depth reviews (3–6). Highlighted in this chapter is the latter, utilising MS/MS spectra generated by electrospray ionisation (ESI)-based tandem MS instruments with ever increasing performance.

As the most abundant class of molecules in microorganisms (7), it is expected that a wealth of information can be derived from a comprehensive study of proteins. Indeed, protein biomarkers, including intact proteins, their proteolytic peptides and non-ribosomal peptides, have all been applied to successful detection, identification and characterisation of microorganisms during the last few decades. The rationale for using protein biomarkers for the rapid detection and characterisation of microorganisms relies on the high specificity

Mass Spectrometry for Microbial Proteomics   Edited by Haroun N. Shah and Saheer E. Gharbia
© 2010 John Wiley & Sons, Ltd

and sensitivity of naturally inherited biomarkers. By detecting individual, organism-specific protein biomarker(s), the source microorganism is subsequently identified.

The use of MS to identify molecular markers can be traced back several decades. For example, Anhalt and Fenselau (8) demonstrated that biomolecules from unprocessed intact bacteria, when introduced into a mass spectrometer, could be vaporised and ionised directly by electron impact to generate signature mass spectra. However, the resurgence and success of MS-based proteomics in the characterisation of microorganisms has been driven by several factors. These include expanding genome and proteome databases derived from whole genome sequencing, technical and conceptual advances in MS instrumentation, including the development of the 'soft' ionisation techniques MALDI (9, 10) and ESI (11), increasing accuracy, resolution and dynamic range of mass analysers, the rise of 'hybrid' instruments (12–16), and the application of computational techniques to analyse, organise and interpret data in a biologically meaningful manner, i.e. bioinformatics (17).

Tandem MS generates sequence-specific fragments from intact proteins or proteolytic peptides in a data-dependent manner. Peptide identity may be assigned to the MS/MS spectra through database searching using various computer algorithms (18, 19) from which individual biomarker proteins are identified, then the source microorganism. At present, the vast majority of proteomic data are generated by tandem MS, in particular, ESI-based tandem MS, by taking advantage of its high sensitivity, specificity and greater efficiency in data acquisition, and greater capacity for high throughput analysis of complex protein samples (20–23). A further dimension is added when coupled with different liquid chromatographic separation systems.

4.2   Mass Spectrometry

MS measures the most fundamental property of molecules, i.e. the mass or, more precisely, the mass-to-charge ratio (m/z). A typical mass spectrometer consists of three principal components; the ion source, the mass analyser and the detector. In the ion source, molecules are transferred into the gas phase and subsequently charged. The ionised ions are then separated according to their m/z ratios in the mass analyser, and the relative abundance of each ion is recorded by the detector to generate a mass spectrum.

4.2.1   MALDI Versus ESI

The development of two 'soft' ionisation methods, MALDI and ESI, allows the routine analysis of thermally labile and nonvolatile biomolecules, such as peptides and proteins.

MALDI has gained its popularity in proteomics because of its simplicity, robustness and sensitivity. The preparation of a sample for MALDI analysis involves mixture of the analyte with a saturated solution of an energy absorbing matrix in an organic solvent. The analyte to matrix molar ratio varies between laboratories and the type of samples to be analysed but the general principle is the same. The sample is placed on a stainless steel target plate with an admixture of the matrix solution. Rapid evaporation of the organic solvent enables the matrix to crystallise and incorporate the analyte into a relatively homogeneous, microcrystalline matrix layer. Once dried, the target plate is loaded into the instrument under vacuum and the laser fired. MALDI then generates pulses of ions out of this matrix layer at a rate of 1–100 pulses per second and uses a TOF analyser (several variants exist) to measure the *m/z* ratios of very large masses. These ions normally carry only a few charges, with the singly charged ions being predominant (24, 25).

By contrast, ESI ionises the analyte out of a solution, generating ions by applying a potential to a flowing liquid delivered by liquid chromatography (LC), e.g. nano-LC at a flow rate of $0.3 \mu l \min^{-1}$ (26). Upon application of a high potential difference (with respect to the counter electrode), the liquid is charged and subsequently sprayed into a fine aerosol. These fine droplets are then repelled from the needle towards the vacuum sample cone of the mass spectrometer, wherein the evaporation of solvent and repeated droplet explosion under the influence of high temperature, leads to the formation of multiple-charged ions, enabling the analysis of large molecules (27).

Thus, unlike MALDI, there are two distinctive features associated with ESI. First, it is naturally compatible with many types of liquid-based separation techniques; secondly, there is no limitation when coupling of ESI with mass analysers. Indeed, various mass analysers such as a quadrupole, TOF, ion trap, and Fourier transform ion cyclotron resonance (FTICR) instruments have all been successfully coupled with the ESI source.

4.2.2   Tandem Mass Spectrometry and Hybrid Mass Spectrometers

Tandem MS is a multi-step mass selection and ionisation process. Initially ions of interest generated in an ion source are mass-selected by the first mass analyser, and subjected to further activation and dissociation to generate characteristic secondary fragment ions. The resulting ions, rich in structural information, are further analysed by the second mass analyser. Two key developments that govern the type and quality of MS/MS data obtained, the rise of hybrid mass spectrometers and the invention of different types of ion dissociation method, have contributed to the success of tandem MS.

The new generation of complex multistage instruments, the 'hybrids' are mass spectrometers that use different types of mass analyser for the first and second stages of mass analysis in tandem MS (28). The strength of a 'hybrid' lies in its ability to integrate systems with different operating principles into a single machine, and therefore multiply the strengths of different mass analysers. High mass accuracy, high resolution, high sensitivity and large dynamic range can be expected from the 'hybrids'. As the central component of hybrid mass spectrometers, mass analysers that are currently available can be classified into four categories, namely, TOF, quadrupole, ion trap, and FTICR. The differences reside in their underlying physical principles, hence their different analytical performances and applications. Each has strengths and weaknesses, and their development is continuously being driven by the increasing needs of protein research.

There are numerous collections of 'hybrids' for tandem MS commercially available. They can be classified into different categories by using different parameters such as the kinetic energy of the ions, resolving power and mass accuracy, and whether the analysis and excitation events are separated spatially or temporally. For example, for the beam technique, every stage of mass analysis is done in different mass analysers that are physically separated in space. All sector and TOF instruments, such as triple-quadrupole (QqQ), quadrupole-TOF (qTOF) and TOF-TOF, fall into this category. For the second category of technique, in contrast, all stages of the analysis and excitation are performed in the same analyser, with different stages being separated in time rather than in space.

Table 4.1 *Comparison of performance characteristics of commonly used mass spectrometers for proteomics. Reprinted from Mass Spectrometry for Proteomics, 12, Current opinion in Chemical Biology, 483–49 Copyright 2008. With permission from Elsevier*

Instrument	Mass resolution	Mass accuracy (ppm)	Sensitivity	m/z range
QIT	1000[a]	100–1000	Picomole	50–2000; 200–4000
LTQ	2000[a]	100–500	Femtomole	50–2000; 200–4000
Q-q-Q	1000	100–1000	Attomole to femtomole	10–4000
Q-q-LIT	2000[a]	100–500	Femtomole	5–2800
TOF	10 000–20 000	10–20[b]; <5[c]	Femtomole	No upper limit
TOF-TOF	10 000–20 000	10–20[b]; <5[c]	Femtomole	No upper limit
Q-q-TOF	10 000–20 000	10–20[b]; <5[c]	Femtomole	No upper limit
FTICR	50 000–750 000	<2	Femtomole	50–2000; 200–4000
LTQ-Orbitrap	30 000–100 000	<5	Femtomole	50–2000; 200–4000

Reprinted with permission from Han *et al.* (29).
[a] Mass resolution achieved at normal scan rate; higher resolution achievable at slower scan rate.
[b] With external calibration.
[c] With internal calibration.
[d] $n > 2$, up to 13.
[e] Fragmentation achievable by post-source-decay.

All trapping instruments, such as the linear trap quadrupole (LTQ) and more recently the LTQ-Orbitrap and the LTQ-FTICR, fall into this category. The analytical characteristics and capabilities of commonly used 'hybrids' have been summarised by Han *et al.* (29) in Table 4.1

4.2.3   Fragmentation in Tandem Mass Spectrometry

As stated above, a crucial aspect of tandem MS is to induce protein/peptide dissociation in the gas phase to generate comprehensive structural fragmentation information which in turn provides information on peptide sequences, and thus protein identity. Consequently,

Scan rate	Dynamic range	MS/MS capability	Ion source	Main applications
Moderate	1E3	MS^{nd}	ESI	Protein identification of low complex samples; PTM identification
Fast	1E4	MS^{nd}	ESI	High throughput large scale protein identification from complex peptide mixtures by on-line $LC–MS^n$; PTM identification
Moderate	6E6	MS/MS	ESI	Quantification in selective reaction monitoring (SRM) mode; PTM detection in precursor ion and neutral loss scanning modes
Fast	4E6	MS^{nd}	ESI	Quantification in SRM mode; PTM detection in precursor ion and neutral loss scanning modes
Fast	1E4	n/a[e]	MALDI	Protein identification from in-gel digestion of gel separated protein band by peptide mass fingerprinting
Fast	1E4	MS/MS	MALDI	Protein identification from in-gel digestion of gel separated protein band by peptide mass fingerprinting or sequence tagging via CID MS/MS
Moderate to fast	1E4	MS/MS	MALDI; ESI	Protein identification from complex peptide mixtures; intact protein analysis; PTM identification
Slow	1E3	MS^{nd}	ESI; MALDI	Top-down proteomics; high mass accuracy PTM characterization
Moderate to fast	4E3	MS^{nd}	ESI; MALDI	Top-down proteomics; high mass accuracy PTM characterization; protein identification from complex peptide mixtures; quantification

the development of efficient fragmentation models to fulfil different requirements has been an area of active research. There are a variety of approaches for peptide/protein activation and dissociation available at present. For example, black-body infrared radiative dissociation (30), infrared multiphoton dissociation (31), surface-induced dissociation (32), electron-capture dissociation (ECD) (33, 34), electron transfer dissociation (ETD) (35, 36), collision-induced dissociation (CID) (37, 38), and more recent pulsed Q collision-induced dissociation (PQD) developed by Thermo Scientific, have all be applied for peptide/protein sequencing and identification. Of these, CID, ECD and ETD have received the most attention.

4.2.3.1   Collision-Induced Dissociation

CID, also referred to by some as collision-activated dissociation (CAD), is the most widely used gas-phase activation and dissociation technique. Although the mechanisms of CID activation and dissociation are still poorly understood, as an energetic dissociation process, CID relies on the activation arising from the collision of ions with neutral gas, e.g. helium or argon. CID can be conducted in the front stages of the mass spectrometer, known as 'in-source CID', or in a specifically designed collision cell. In the latter approach, ions of interest can be selected by *m/z* for dissociation, thereby, enabling data-dependent fragmentation; the process is often referred to as tandem MS. In a typical MS/MS CID experiment, a preselected ion is accelerated into an inert collision gas by applying acceleration voltage. Upon collision with the inert gas atoms or molecules, the kinetic energy is then converted into internal energy. In fact, thousands of such collisions occur over tens of microseconds in the limited space of a collision cell. As a result, internal energy is accumulated gradually that eventually leads the preselected ion to undergo fragmentation (39, 40).

Upon collision activation, gas-phase peptides/proteins are vibrationally excited and lead to sequence-specific peptide backbone breakage at the carbonyl-carbon-amide nitrogen bonds (CO-N bonds). The breakage of the CO-N bonds produces the *b-type* and *y-type* fragments. Strikingly, the vast majority of fragments observed in a typical peptide MS/MS spectrum are *b-type* and *y-type*.

In CID, the quality and the information content of a MS/MS spectrum depend on the collision energy. The 'ergodic' energetic feature of CID results in sequence-specific peptides on the one hand but seldom provides complete MS/MS sequence information for large peptides and intact proteins on the other hand, thus the sequence information obtained is nowhere near completion and *de novo* sequencing of a whole protein is virtually impossible. In addition, substituents that are added in post-translational modifications (PTMs), in most cases, dissociate with even lower activation energies than those of backbone cleavages, and get lost during CID excitation. Alternative techniques for peptide and protein ion fragmentation are required, in order to obtain more extensive sequence coverage and preserve PTMs.

4.2.3.2   Electron Capture Dissociation

ECD was introduced by McLafferty and co-workers in 1998 (33). As a nonergodic process that does not involve intramolecular vibrational energy redistribution, ECD involves the direct introduction of low energy electrons into the gas phase where they are captured by multiple protonated polypeptide cations. Attaching an electron to a protonated polypeptide cation generates an excited radical cation. This then undergoes very rapid rearrangement; the process liberates the electric potential energy carried by the excited radical cation, resulting in bond-specific fragmentation. In ECD, backbone fragmentation of larger polypeptides is abundant, homogeneous and dominated by *c-type* and *z-type* ions, products of breakage at the amide nitrogen-alpha carbon bonds (N-Cα bonds), in contrast to the energetic CID that mostly involves backbone cleavage at the amide bonds. The nonergodic feature of ECD causes dissociation of larger molecules and provides more extensive sequence coverage in polypeptides, as well as preserving labile PTMs and noncovalent bonds. As a result, ECD fragmentation chemistries are rich, specific, and complementary to those of protonated peptides in CID. These unique features make ECD a method of choice in top-down sequence characterisation (41), *de novo* sequencing, disulfide bond analysis (42) and the most important PTM studies (43, 44). However, ECD has its limitations in that it is costly, technically challenging and restricted to a FTICR mass spectrometer.

4.2.3.3 Electron Transfer Dissociation

ETD is analogous to ECD and was developed in 2004 by Hunt and co-workers (35). ETD, a method of fragmenting multiple protonated peptides, has achieved an ECD-like dissociation but can be coupled with low cost, widely accessible mass spectrometers such as the LTQ and more recently the LTQ-Orbitrap.

In ETD, instead of directly introducing electrons, they are delivered using anions, such as fluorethene radical ions, as vehicles. Electron transfer via ion/ion reactions of multiple protonated peptide cations with singly charged anions induces backbone fragmentation via similar nonergodic processes as in ECD, producing a complete or near complete series of *c-type* and *z-type* ions and thus extensive peptide sequence information (35, 36).

The underlying chemistry of the complex gas-phase dissociation of peptides in ECD/ ETD (shortened to ExD) is not well understood, however, in most cases, ExD-derived information is complementary to that obtained using traditional CID fragmentation (45). It is worth mentioning that PQD was developed by Thermo Scientific complementary to CID. PQD generates spectra similar to CID. Its ability to activate precursor ions at high Q values (high energies) and collect fragments at low Q values not only generates more fragments in the high mass range from higher energy fragmentation pathways, but also extends the mass range to the lower *m/z* range (<100) that is usually excluded from CID spectra. Therefore, their combination can provide more detailed structural information about proteins/peptides, both globally and locally.

4.3 Proteomic Strategies for Protein Identification

Complex experimental strategies referred to 'bottom-up', 'top-down' and, in some cases, 'middle-down' approaches have been developed that integrate various separation techniques with a wide variety of MS instrumentation to ionise, fragment and analyse peptides efficiently and effectively (46–49). Matching the peptide MS/MS spectra generated by these approaches to theoretical predicted peptide sequences, known as *in silico* peptide sequences, in protein sequence databases, is central to MS-based proteomics, particularly in large-scale proteomics studies.

4.3.1 Bottom-Up Proteomics

The 'bottom-up' approach has been used extensively to identify proteins from MS/MS spectra of their proteolytic peptides. A typical bottom-up proteomics experiment consists of several steps. This may be: (i) up-front protein fractionation by one-dimensional sodium dodecyl sulfate-polyacrylamide gel electrophoresis (1D-SDS-PAGE) or two-dimensional, two-dimensional, i.e. isoelectric-focusing SDS-PAGE (2D-SDS-PAGE) (see Chapters 6 and 20); or more recently one-dimensional (LC) or multidimensional (LC/LC) liquid chromatography; (ii) in-gel or in-solution proteolytic digestion of pre-fractionated proteins to obtain a peptide mixture; (iii) separation of the peptide mixture further using a LC or LC/LC system; (iv) introduction of the peptide eluates directly into a mass spectrometer via electrospay; (v) acquisition of intact tryptic peptide masses to generate peptide mass fingerprinting (PMF) data; or in most scenarios now, fragmentation of data-dependent MS/MS masses in order to produce sequence-specific CID type of MS/MS spectra (22).

Intact peptide masses allow their identification and that of the parent proteins using PMF. However, at the centre of large-scale MS-based protein identification is the assignment of MS/MS spectra to peptide sequences via comprehensive protein sequence database searching using various search algorithms which depend upon the unique pieces of sequence information carried by the MS/MS spectra.

A large number of computational algorithms and software programs have been developed to automatically assign *in silico* generated peptides to experimental MS/MS spectra (see Chapter 3). Among them, MASCOT, SEQUENCE and X! TANDEM are simple to use and have gained wide acceptance. They may be used individually but it should be noted that by combining the results from MASCOT, SEQUEST and X! TANDEM, better and cross-validated protein identifications have been achieved (50).

However, it should be borne in mind that *in silico* peptide sequences are generated from computer digestion and fragmentation of proteins in databases that are derived from translating genomic sequencing, using fragmentation models that are currently understood. Given the fact that the mechanisms of CID are not fully understood and our understanding of peptide fragmentation is still incomplete, assignments of peptide sequences are prone to errors (51). It is argued that the fragmentation models based on the most basic *b-type* and *y-type* of ions and scoring systems currently used are oversimplified. More sophisticated algorithms that take into account different fragmentation pathways, ion patterns and abundances, etc., should improve the reliability of MS/MS spectra assignment. In addition, in most cases, the proportion of identifiable peptides is low and many MS/MS spectra remain unassigned, i.e. there are far too many missing pieces to assemble a protein from MS/MS data. In general, incorrect identification using CID is higher than using ECD and ETD.

Sequence tag (52) and *de novo* sequencing (53) are two alternatives to find the correct amino acid sequence from MS/MS spectra, in addition to the development of methods assessing the statistical confidence in MS/MS-derived peptide and protein identification. Each of these approaches, is currently an active area of research but is beyond the scope of this chapter.

4.3.2   Top-Down Proteomics

Top-down proteomics, in contrast to a bottom-up approach, involves direct analysis of intact proteins without previous proteolytic digestion (54–56). When performing a 'top-down' analysis, intact proteins are introduced into the gas phase and ionised in the mass spectrometer in two steps. The first step involves high accuracy mass measurement of the ions of the intact protein and the second step involves direct fragmentation of these ions. The MS/MS level of information generated in the second stage is the 'top-down' equivalent of the tryptic digestion typically used in a 'bottom-up' approach.

The power of the 'top-down' approach lies in its ability to generate a fairly complete picture of the proteins presented by obtaining information on their masses, prior to dissecting out each for complete characterisation of its primary structure sequence in a single experiment. Han *et al.* (57) has extended 'top-down' MS to proteins with masses from 50 to >200 kDa. As a result, the 'top-down' approach generates fragments that cover a much larger range of mass values but more importantly all these fragments originate from a single protein that makes their mass values more specific than masses of peptides from protein digestion. The 'top-down' protein identification approach may be orders of magnitude greater in terms of confidence. The approach is analogous to solving a jigsaw puzzle in which the expected outcome is known. Here, the MS/MS spectra of intact protein(s) are equivalent to the relatively smaller numbers of big jigsaw pieces. When assembled together, one can expect a high level of confidence in the assignments of the intact protein(s) identities.

There are two key factors for success in 'top-down' analysis. Effective dissociation methods are those able to fragment intact proteins in the mass spectrometers and mass analysers with extremely high resolution and can perform extremely high mass accuracy measurement. The development of ECD and ETD techniques has greatly improved the dissociation efficiency for intact protein molecules. Indeed, historically, 'top-down' MS has mostly been performed on FTICR instruments equipped with ECD. The development of ETD has facilitated tandem MS experiments for intact proteins of high mass. There are increasing reports of the success of the 'top-down' approach on instruments other than for FTICR, such as the LTQ-Orbitrap hybrid, ion trap and TOF (58).

So far, the top-down proteomic approach has been used as a powerful tool in the characterisation of sequence and PTMs, and has demonstrated its potential to provide a panoramic view of proteins. New technological innovations in instrumentation as well as robust data deciphering algorithms are being developed. However, at present ProSight PTM is the only search engine available for identifying proteins from MS/MS spectra of intact proteins.

'Bottom-up' and 'top-up' analytical approaches are complementary to each other. A more recent method referred to as 'middle-down' combines some of the benefits of both approaches. This was recently demonstrated by Xu and Peng (59) who analysed peptide fragments with molecular weight exceeding 3 kDa with considerable success; an approach that is likely to increase in the future.

4.4   Multidimensional Protein Identification

Proteins span a concentration range several orders of magnitude that exceeds the dynamic range of any single analytical method or instrument currently available. Furthermore, most informative proteins such as those used as biomarkers are often present in low concentrations. The challenge of proteomic analysis is therefore to separate, identify and analyse biologically relevant, normally low abundant proteins, in highly complex protein mixtures.

2D-PAGE still remains a powerful technique for protein separation, more than three decades after it was first introduced by O'Farrell (60). However, there are a number of limitations to this technology, such as poor reproducibility, insufficient dynamic range, low sensitivity and its time-consuming nature which restricts its application for high throughput protein analysis. Consequently, alternative approaches have been pursued. One such technique, first described by Link *et al.* in 1999 (61) and referred to as multidimensional protein identification technology (MudPIT), utilises two-dimensional liquid chromatography (LC/LC) to separate a peptide mixture prior to mass spectral analysis. Using MudPIT, Yates and colleagues reported that a total of 1484 proteins were detected and identified in a large-scale yeast proteome study (62). Gygi's laboratory extended this by collecting >162 000 MS/MS spectra of which 26 815 matched to yeast peptides with 7537 being unique peptides. As a result, 1504 yeast proteins were unambiguously identified in a single experiment using completely automated analysis (63).

The strength of MudPIT separation lies in the orthogonality of the two chromatographic phases, namely, strong cation exchange (SCX) and reversed-phase (RP). Utilising their unique physical properties, i.e. the charge and the hydrophobicity, respectively, separation of peptides is achieved. In a standard LC/LC-MS/MS approach, peptides are first separated on a SCX column, then on a RP column (64). To address the challenge created by the extraordinary complexity and dynamic range of proteins in biological matrices, a few approaches have been developed to successfully integrate orthogonal separation steps, including capillary isoelectric focusing (CIEF), capillary electrophoresis (CE) and multiple, orthogonal dimensions of chromatography at the intact protein level with standard LC/LC-MS/MS. For example, intact proteins are first separated on a strong anion exchange (SAX) column and fractions collected for further separation on a RP column to achieve a second-dimensional separation. Following an in solution trypsin digestion, tryptic peptides are then analysed by RP-LC-MS/MS. Tryptic peptides may also be subjected to a SCX separation before RP-LC-MS/MS analysis to add a further dimension to peptide separation.

Multidimensional LC has been shown to overcome the sensitivity and dynamic limitations of commonly used 2D-PAGE to separate complex protein/peptide mixtures before mass spectrometric analysis. Proteins of relatively low abundance, extreme hydrophobicity or pI values, and large molecular weight proteins have been successfully identified. It is expected that a combination of extensive multidimensional separation with tandem MS will contribute to the number of accessible peptides as well as proteins.

4.5   Mass Spectrometry-Based Targeted Protein Quantification and Biomarker Discovery

The output from various search algorithms is a list of MS/MS spectra matched to peptide sequences and ranked according to the search scores. The list of peptide and protein identifications from the sample is compiled and compared using computer software tools, such as PEARLS, Scaffold and its new version Proteus-Analytics, to find the candidate peptide/protein biomarkers that are specific to the microorganism of interest. Another approach, based on microbial genome sequence information also identifies a list of candidate protein biomarkers. Integration of the two approaches dramatically increases the likelihood of finding unique biomarkers.

Following biomarker discovery, approaches now logically shift to developing target-focused, target-driven and quantitative molecular-based identification systems that can precisely characterise endogenous protein forms at the molecular level. Figure 4.1 summarises the process of proteomics-based protein biomarkers discovery, from biomarker identification to targeted quantitative mass spectrometric analysis of identified biomarkers (65).

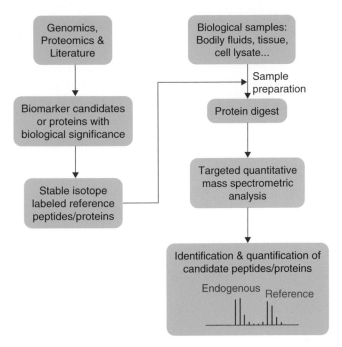

Figure 4.1   *General strategy of mass spectrometry-based targeted quantitative proteomics. Reprinted with permission from Journal of Proteome Research, Copyright 2009 American Chemical Society*

MS is not inherently quantitative; therefore various methods are being developed to attain this. One means of achieving this involves the use of synthetic stable isotope-labelled internal standards. This approach was first described by Desiderio *et al.* (66) in the early 1980s, and is based on introducing a stable isotope-labelled internal standard at a known concentration to a test sample. As the stable isotope-labelled molecule mimics its native counterpart precisely, the two molecules share identical physical and chemical properties, it is expected that they co-elute in chromatography, have the same ionisation efficiency, and are fragmented in an identical manner in a mass spectrometer. The only difference between a stable isotope-labelled internal standard and its corresponding native counterpart is the mass that can be unambiguously measured by MS, resulting in the differentiation of the two. Absolute quantification is subsequently achieved by comparison of their respective signal intensities.

For several years its application has been restricted to quantification of small molecules until the development of the *a*bsolute *qua*ntification of proteins (AQUA) in 2003 by Gerber *et al.* (67). This is an MS-based strategy that relies on the use of a synthetic stable isotope-labelled peptide as an internal standard which is used to measure the levels of protein expression and PTMs. In a typical experiment, synthetic internal standard peptides are customised with an incorporated stable isotope labelling of ^{13}C and ^{15}N on one selected amino acid in the peptide sequence. For example, a mass difference of 10.008 is expected if arginine (R) is fully labelled, while a mass different of 7.017 is expected if leucine (L) is fully labelled.

However, AQUA is not a strategy that can achieve quantification at the protein level. Instead quantification of proteins made by the AQUA strategy relies on measurements made at the peptide level. Indeed, Janecki *et al.* (68) reported an approach to obtain absolute quantification of the human liver alcohol dehydrogenase ADH1C1 isoenzyme using a stable isotope-labelled intact protein as an internal standard.

4.5.1   Selected Reaction Monitoring

Stable isotope-labelled peptides as internal standards coupled with selected reaction monitoring (SRM) has established itself as a powerful approach for the targeted and quantitative phases of protein biomarker development (69–71). SRM, also refereed to by some as multiple reaction monitoring (MRM), is normally performed on a triple quandrupole mass spectrometer which provides the sensitivity, selectivity and high throughput required at this stage of biomarker development. In addition, its quantitative nature also adds another dimension of attraction for its development.

In the SRM mode, an ion of interest (parent ion) is preselected by the first quadrupole (Q1), and fragmented via CID in the second quadrupole (Q2) to produce daughter ions. The daughter ions are then analysed by the third quadrupole (Q3). In contrast to traditional MS or the tandem MS mode that measures parent ions or daughter ions, respectively, the unique feature for typical SRM lies in its ability to measure the transition between a parent ion and one of its daughter ions, and more than one transition can be selected to achieve the measurement of the same parent; hence its superior selectivity and sensitivity (72). The utility of stable isotope-labelled peptides as internal standards to confirm initial findings from the biomarker discovery phase, and to quantify the findings is highlighted in studies undertaken in our laboratory in which the search for potential biomarkers of *Bacillus anthracis* (*B. anthracis*) is in progress (unpublished work). Here, protein mixtures from cell extracts of *B. anthracis* were initially separated on a 1D-PAGE, followed by protein band excision and in-gel trypsin digestion. The tryptic peptides were then analysed using RP-LC-MS/MS, which generated thousands of peptide spectra. Dozens of these peptides were identified as unique to *B. anthracis* both at the proteome and genome levels, and deemed as potential biomarkers. From these, five of the most abundant peptides were verified and validated on an LTQ Orbitrap using isotope-labelled peptides as internal standards.

Stable isotope-labelled peptides enable the determination of the retention times of the peptides, their fragmentation patterns and selection of the most appropriate ion(s) for the next stage of MS analysis. Using these parameters, the five selected peptides were confirmed and their levels determined. One of the peptides, with a molecular weight of 1396.7434, was derived from trypsin digestion of a protein with a molecular weight of 128.59 kDa. Using 1D-PAGE, bands between 62 kDa and 188 kDa (as indicated by protein standards) were excised and in-gel trypsin digestions were performed in the presence of internal standard labelled on L with ^{13}C and ^{15}N resulting in a peptide with a molecular weight of 1403.7614 (at a concentration of 100 fmol μl^{-1}). Figure 4.2 shows that the endogenous native peptide was detected by monitoring the doubly charged ions of the internal standard ($m/z = 702.8807$) and its expected native counterpart ($m/z = 699.3734$) using

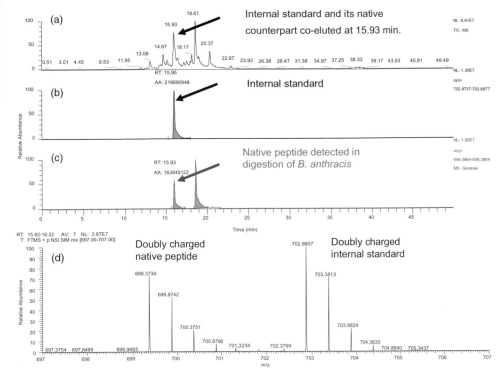

Figure 4.2 *Validation and quantitation of a newly identified biomarker of B. anthracis (unpublished). The targeted endogenous peptide was unambiguously identified and subsequently quantified using the corresponding synthetic stable isotope-labelled peptide as an internal standard. (a) Total ion current spectrum; (b) selected ion with m/z = 702.8802; (c) selected ion with m/z = 699.3734; (d) mass spectrum of doubly charged stable isotope-labelled internal standard and its endogenous native peptide of B. anthracis*

RP-LC separation coupled with selective ion monitoring (SIM) on an LTQ Orbitrap. The concentration of the endogenous native peptide was measured as 75.6 fmol μl^{-1}. It is anticipated that the second peak which eluted at 18.61 min with the same mass as the native endogenous peptide could be eliminated if an appropriate SRM approach is employed.

Sensitivity is as important as selectivity in the detection of biomarkers in a complex biological matrix. Thus, following confirmation of the initial selected peptides, assessment of the sensitivity of the method was pursued using a spiked recovery experimental strategy. This consisted of measuring a series of dilutions of the stable isotope labelled peptides in a peptide mixture obtained from trypsin digestion of a mixture of five proteins using the RP-LC-SIM method described above. Results indicated that the five peptides selected at 0.1 fmol all gave a signal-to-noise ratio (S/N) of >3, i.e. the limit of detection of the method for the peptides selected was estimated to be 0.1 fmol in the biomatrix chosen (in this case, the tryptic peptide mixture).

4.6   Conclusions

Tandem MS-based proteomics combined with a variety of sample preparation and separation techniques is now used widely and has become an essential platform for large scale protein characterisation and identification and is used in the search for novel biomarkers in microorganisms. Success in the field is due to the development of the two 'soft' ionisation techniques, namely, MALDI and ESI, the innovation of hybrids, new fragmentation techniques, and advances in chromatography and more appropriate computer algorithms.

Currently, the search for protein biomarkers using MS-based detection is mainly from the *in vitro* culture of microbes in which biomasses are plentiful. In contrast, specificity and sensitivity become extremely challenging *in vivo* when it is necessary to detect trace levels of microbial biomarkers in complex biological matrices such as infected body fluids or tissues. There is growing evidence to suggest that development of MS-based targeted biomarker quantification, utilising stable isotope-labelled peptides as internal standards and SRM approaches, will provide the high specificity and sensitivity required. Thus, while PCR-based detection may be feasible, MS-based methods will provide evidence of expression, and enable large numbers of microorganisms to be screened for diversity and levels of expression.

References

1. R.D. Holland, J.G. Wiles, F. Raffi, J.B. Sutherland and C.C. Persons, Rapid identification of intact whole bacteria based on spectral patterns using matrix-assisted laser desorption/ionisation with time-of-flight mass spectrometry, *Rapid Commun. Mass Spectrom.*, **10**, 1227–1232 (1996).
2. K.J. Welham, M.A. Domin, D.E. Scannell, E. Cohen and D.S. Ashton, The characterisation of microorganisms by matrix-assisted laser desorption/ionisation with time-of-flight mass spectrometry, *Rapid Commun. Mass Spectrom.*, **12**, 176–180 (1998).
3. C. Fenselau and P. Demirev, Characterisation of intact microorganisms by MALDI mass spectrometry, *Mass Spectrom. Rev.*, **20**, 157–171 (2001).
4. J. Lay, MALDI-TOF mass spectrometry of bacteria, *Mass Spectrom. Rev.*, **20**, 172–194 (2001).
5. P.A. Demirev and C. Fenselau, Mass spectrometry in biodefense, *J. Mass Spectrom.*, **43**, 1441–1457 (2008).
6. P. Demirev and C. Fenselau, Mass spectrometry for rapid characterisation of microorganisms, *Annu. Rev. Anal. Chem.*, **1**, 71–93 (2008).
7. J. Frantz and R. McCallum, Changes in macromolecular-composition and morphology of *Bacteroides fragilis*, cultured in a complex medium, *Appl. Environ. Microb.*, **39**, 445–448 (1980).
8. J.P. Anhalt and C. Fenselau, Identification of bacteria using mass spectrometry, *Anal. Chem.*, **47**, 219–225 (1975).
9. K. Tanaka, H. Waki, Y. Ido, S. Akita and Y. Yoshida, Protein and polymer analysis up to m/z 100,000 by laser ionisation time-of-flight mass spectrometry, *Rapid Commun. Mass Spectrom.*, **2**, 151–153 (1988).
10. M. Karas and F. Hillenkamp, Laser desorption ionization of proteins with molecular masses exceeding 10000 daltons, *Anal. Chem.*, **60**, 2299–2301 (1988).
11. J.B. Fenn, M. Mann, C.K. Meng, S.F. Wong and C.M. Whitehouse, Electrospray ionization for mass spectrometry of large biomolecules, *Science*, **246**, 64–71 (1989).
12. J.W. Hager, A new linear ion trap mass spectrometer, *Rapid Commun. Mass. Spectrom.*, **16**, 512–526 (2002).
13. J.C. Schwartz, M.W. Senko and J.E. Syka, A two-dimensional quadrupole ion trap mass spectrometer, *J. Am. Soc. Mass Spectrom. Rev.*, **13**, 659–669 (2002).

14. A.G. Marshall, C.L. Hendrickson and G.S. Jackson, Fourier transform ion cyclotron resonance mass spectrometry: a primer, *Mass Spectrom. Rev.*, **17**, 1–35 (1998).
15. M. Hardman and A. Makarov, Interfacing the orbitrap mass analyser to an electrospray ion source, *Anal. Chem.*, **75**, 1699–1705 (2003).
16. Q.Z. Hu, R.J. Noll, H.Y. Li, A. Marakov, M. Hardman and R.G. Cooks, The orbitrap: a new mass spectrometer, *J. Mass Spectrom.*, **40**, 430–443 (2005).
17. T. Reichhardt, It's sink or swim as a tidal wave of data approaches, *Nature*, **399**, 517–520 (1999).
18. J.K. Eng, A.L. McCormack and J.R. Yates, An approach to correlate MS/MS data to amino acid sequence in a protein database, *J. Am. Soc. Mass Spectrom. Rev.*, **5**, 976–989 (1994).
19. D.N. Perkins, D.J. Pappin, D.M. Creasy and J.S. Cottrell, Probability-based protein identification by searching databases using mass spectrometry data, *Electrophoresis*, **20**, 3551–3567 (1999).
20. D.F. Hunt, Characterisation of peptides bound to the class I MHC molecule HLA-A2.1 by mass spectrometry, *Science*, **255**, 1261–1263 (1992).
21. B.F. Cravatt, G.M. Simon and J.R. Yates III, The biological impact of mass-spectrometry-based proteomics, *Nature*, **450**, 991–1000 (2007).
22. R. Aebersold and M. Mann, Mass spectrometry-based proteomics, *Nature*, **422**, 198–207 (2003).
23. G.L. Glish and D.J. Burinsky, Hybrid mass spectrometry for tandem mass spectrometry, *J. Am. Soc. Mass Spectrom.*, **19**, 161–172 (2008).
24. F. Hillenkamp and M. Karas, Mass spectrometry of peptides and proteins by matrix-assisted ultraviolet laser desorption/ionisation, *Methods Enzymol.*, **193**, 280–295 (1990).
25. P.B. O'Connor and F. Hillenkamp, MALDI mass spectrometry instrumentation, in *MALDI MS: A Practical Guide to Instrumentation, Methods and Applications*, F. Hillenkamp and J. Peter-Katalinic (Eds), Wiley-VCH, Weinheim, 2007.
26. M. Wilm and M. Mann, Analytical properties of the nanoelectrospray ion source, *Anal. Chem.* **68**, 1–8 (1996).
27. S. Nguyen and J.B. Fenn, Gas-phase ions of solute species from charged droplets of solutions, *Proc Natl Acad Sci USA*, **104**, 1111–1117 (2007).
28. G.L. Glish and D.J. Burinsky, Hybrid mass spectrometers for tandem mass spectrometry, *J. Am. Soc. Mass Spectrom.*, **19**, 161–172 (2008).
29. X.M. Han, A. Aslanian and J.R. Yates III, Mass spectrometry for proteomics, *Curr. Opin. Chem. Biol.*, **12**, 483–490 (2008).
30. W.D. Price, P.D. Schnier and E.R. Williams, Blackbody infrared radiative dissociation of bradykinin and its analogues: energetics, dynamics, and evidence for salt-bridge structures in the gas phase, *Anal. Chem.*, **68**, 859–866 (1996).
31. D.P. Little, J.P. Speir, M.W. Senko, P.B. O'Connor and F.W. McLafferty, Infrared multiphoton dissociation of large multiply charged ions for biomolecule sequencing, *Anal. Chem.*, **66**, 2809–2815 (1994).
32. R.A. Chorush, D.P. Little, S.C. Beu, T.D. Wood and F.W. McLafferty, 'Surface-induced' dissociation of multiply-protonated proteins, *Anal. Chem.*, **67**, 1042–1046 (1995).
33. R.A. Zubarev, N.L. Kelleher and F.W. McLafferty, ECD of multiply charged protein cations. A non-ergodic process, *J. Am. Chem. Soc.*, **120**, 3265–3266 (1998).
34. R.A. Zubarev, Electron-capture dissociation tandem mass spectrometry, *Curr. Opin. Biotechnol.*, **15**, 12–16 (2004).
35. J.E. Syka, J.J. Coon, M.J. Schroeder, J. Shabanowitz and D.F. Hunt, Peptide and protein sequence analysis by electron transfer dissociation mass spectrometry, *Proc. Natl. Acad. Sci. USA*, **101**, 9528–9533 (2004).
36. L.M. Mikesh, B. Ueberheide, A. Chi, J.J. Coon, J.E.P. Syka, J. Shabanowitz and D.F. Hunt, The utility of ETD mass spectrometry in proteomic analysis, *Biochim. Biophys. Acta*, **1764**, 1811–1822 (2006).
37. K.P. Jennings, The changing impact of the collision-induced decomposition of ions on mass spectrometry, *Int. J. Mass Spectrom.*, **200**, 479–493 (2000).
38. A.K. Shukla and J.H. Futrell, Tandem mass spectrometry: dissociation of ions by collisional activation, *J. Mass Spectrom.*, **35**, 1069–1090 (2000).
39. L. Slendo and D.A. Volmer, Ion activation methods for tandem mass spectrometry, *J. Mass Spectrom.*, **39**, 1091–1112 (2004).

40. J.L.P. Benesch, Collisional activation of protein complexes: picking up the pieces, *J. Am. Soc. Mass Spectrom.*, **20**, 341–348 (2009).
41. Y. Ge, B.G. Lawhorn, M. EiNaggar, E. Strauss, J.H. Park, T. Begley and F.W. McLafferty, Top down characterisation of larger proteins (45 kDa) by electron capture dissociation mass spectrometry, *J. Am. Soc. Mass Spectrom.*, **124**, 672–678 (2002).
42. R.A. Zubarev, N.A. Kruger, E.K. Fridriksson, M.A. Lewis, D.M. Horn, B.K. Carpenter and F.W. McLafferty, Electron capture dissociation of gaseous multiply-charged proteins is favoured at disulfide bonds and other sites of high hydrogen atom affinity, *J. Am. Chem. Soc.*, **121**, 2857–2862 (1999).
43. E. Mirgorodskaya, P. Roepstorff and R.A. Zubarev, Localization of O-glycosylation sites in peptides by electron capture dissociation in a Fourier transform mass spectrometer, *Anal. Chem.*, **71**, 4431–4436 (1999).
44. J. Reinders and A. Sickmann, State-of-the-art in phosphoproteomics, *Proteomics*, **5**, 4052–4061 (2005).
45. J.M. Hogan, S.J. Pitteri, P.A. Chrisman and S.A. McLuckey, Complementary structural information from a trypic N-linked glycopeptide via electron transfer ion/ion reactions and collision-induced dissociation, *J. Proteome Res.*, **4**, 628–632 (2005).
46. E.M. Marcotte, How do shotgun proteomics algorithms identify proteins? *Nat. Biotechnol.*, **25**, 755–757 (2007).
47. D.F. Hunt, J.R. Yates III, J. Shabanowitz, S. Winston and C.R. Hauer, Protein sequencing by tandem mass spectrometry, *Proc. Natl. Acad. Sci. USA*, **83**, 6233–6237 (1986).
48. J.R. Yates, Mass spectral analysis in proteomics, *Annu. Rev. Biophys. Biomol. Struct.*, **33**, 297–316 (2004).
49. B. Domon and R. Aebersold, Mass spectrometry and protein analysis, *Science*, **312**, 212–217 (2006).
50. G. Alves, W.W. Wu, G. Wang, R.F. Shen and Y.K. Yu, Enhancing peptide identification confidence by combining search methods, *J. Proteome Res.*, **7**, 3102–3113 (2008).
51. M. Mann and M.S. Wilm, Error tolerant identification of peptides in sequence databases by peptide sequence tags, *Anal. Chem.*, **66**, 4390–4399 (1994).
52. N. Bandeira, D. Tsur, A. Frank and P.A. Pevzner, Protein identification by spectral networks analysis, *Proc. Natl Acad. Sci. USA*, **104**, 6140–6145 (2007).
53. A. Frank, M. Savitski, M. Nielsen, R.A. Zubarev and P.A. Pevzner, De novo peptide sequencing and identification with precision mass spectrometry, *J. Proteome Res.*, **6**, 114–123 (2007).
54. B.T. Chait, Mass spectrometry: bottom-up or top-down? *Science*, **314**, 65–66 (2006).
55. P.A. Demirev, A.B. Feldman, P. Kowalski and J.S. Lin, Top-down proteomics for rapid identification of intact microorganisms, *Anal. Chem.*, **77**, 7455–7461 (2005).
56. F.W. Mclafferty, K. Breuker, M. Jin, X. Han, G. Infusini, H. Jiang, X. Kong and T.P. Begley, Top-down MS, a powerful complement to the high capabilities of proteolysis proteomics, *BEBS J.*, **274**, 6256–6268 (2007).
57. X. Han, M. Jin, K. Breuker and F.W. McLafferty, Extending top-down mass spectrometry to protein with masses greater than 200 kilodaltons, *Science*, **314**, 109–112 (2006).
58. B. Macek, L.F. Waanders, J.V. Olsen and M. Mann, Top-down protein sequencing and MS^3 on a hybrid linear quadrupole ion trap-orbitrap mass spectrometer, *Mol. Cell. Proteomics*, **5**, 949–958 (2006).
59. P. Xu and J. Peng, Characterisation of polyubiquitin chain structure by middle-down mass spectrometry, *Anal. Chem.*, **80**, 3438–3444 (2008).
60. P.H. O'Farrell, High resolution two-dimensional electrophoresis of proteins, *J. Biol. Chem.*, **250**, 4007–4021 (1975).
61. A.J. Link, J. Eng, D.M. Schieltz, E. Carmack, G.J. Mize, D.R. Morris, B.M. Garvik and J.R. Yates III, Direct analysis of protein complex using mass spectrometry, *Nat. Biotechnol.*, **17**, 676–682 (1999).
62. M.P. Washburn, D. Wolters and J.R. Yates III, Large-scale analysis of the yeast proteome by multidimensional protein identification technology, *Nat. Biotechnol.*, **19**, 242–247 (2001).
63. J. Peng, J.E. Elias, C.C. Thoreen, L.J. Licklider and S.P. Gygi, Evaluation of multidimensional chromatography coupled with tandem mass spectrometry (LC/LC-MS/MS) for large-scale protein analysis: the yeast proteome, *J. Proteome Res.*, **2**, 43–50 (2003).

64. M.L. Fournier, J.M. Gilmore, S.A. Martin-Brown and M.P. Washburn, Multidimensional separations-based shotgun proteomics, *Chem. Rev.*, **107**, 3654–3686 (2007).

65. S. Pan, R. Aebersold, R. Chen, J. Rush, D.R. Goodlett, M.W. McIntosh, J. Zhang and T.A. Brentnall, Mass spectrometry based targeted protein quantification: methods and applications, *J. Proteome Res.*, **8**, 787–797 (2009).

66. D.M. Desiderio, M. Kai, F.S. Tanzer, J. Trimble and C. Wakelyn, Measurement of enkephalin peptides in canine brain regions, teeth, and cerebrospinal fluid with high-performance liquid chromatography and mass spectrometry, *J. Chromatogr.* **297**, 245–260 (1984).

67. S.A. Gerber, J. Rush, O. Stemman, M.W. Kirchner and S.P. Gygi, Absolute quantification of proteins and phosphoproteins from cell lysates by tandem MS, *Proc. Nat. Acad. Sci. USA*, **100**, 6940–6945 (2003).

68. D.J. Janecki, K.G. Bemis, T.J. Tegeler, P.C. Sanghani, L. Zhai, T.D. Hurley, W.F. Bosron and M. Wang, A multiple reaction monitoring method for absolute quantification of the human liver alcohol dehydrogenase ADH1C1 isoenzyme, *Anal. Biochem.*, **369**, 18–26 (2007).

69. L. Anderson and C.L. Hunter, Quantitative mass spectrometric multiple reaction monitoring assays for major plasma proteins, *Mol. Cell. Proteomics*, **5**, 573–588 (2006).

70. H. Keshishian, T. Addona, M. Burgess, E. Kuhn and S.A. Carr, Quantitative, multiplexed assays for low abundance proteins in plasma by targeted mass spectrometry and stable isotope dilution, *Mol. Cell. Proteomics*, **6**, 2212–2229 (2007).

71. D.S. Kirkpatrick, S.A. Gerber and S.P. Gygi, The absolute quantification strategy: a general procedure for the quantification of proteins and post-translational modifications, *Methods*, **35**, 265–273 (2005).

72. B. Han and R.E. Higgs, Proteomics: from hypothesis to quantitative assay on a single platform. Guidelines for developing MRM assays using ion trap mass spectrometers, *Brief Funct. Genomic Proteomics*, **7**, 340–354 (2008).

5

MALDI Mass Spectrometry Imaging, a New Frontier in Biostructural Techniques: Applications in Biomedicine

Simona Francese and Malcolm R. Clench
Biomedical Research Centre, Sheffield Hallam University, Sheffield, UK

5.1 Introduction

Matrix assisted laser desorption/ionisation mass spectrometric imaging (MALDI-MSI) is a relatively new imaging technology pioneered by Richard Caprioli and colleagues in the late 1990s (Caprioli *et al.*, 1997). In its most common implementation, the sample to be imaged is coated with an energy absorbing matrix material and then moved under a stationary laser. A two-dimensional array of sample positions is defined and the laser is fired at each of them in order to create a set of sample mass spectra. If positional information is stored along with the mass-to-charge and intensity values then two-dimensional images can be constructed by plotting the *x*- and *y*-coordinates versus the abundance of a selected ion or ions, represented as a grey or colour scale. This is shown schematically in Figure 5.1(a).

An alternative to this 'mass microprobe' approach, the 'mass microscope' has been pioneered by Ron Heeren and co-workers at FOM in Amsterdam (Luxemborg *et al.*, 2004). In this approach, rather than the laser beam being highly focused, a mass spectrometer that accepts a 150–300 μm diameter ion beam is used to map a magnified image of the spatial distribution of a selected m/z value onto a two-dimensional detector. This approach is shown in Figure 5.1(b). To image larger areas than the diameter of the laser beam

Mass Spectrometry for Microbial Proteomics   Edited by Haroun N. Shah and Saheer E. Gharbia
© 2010 John Wiley & Sons, Ltd

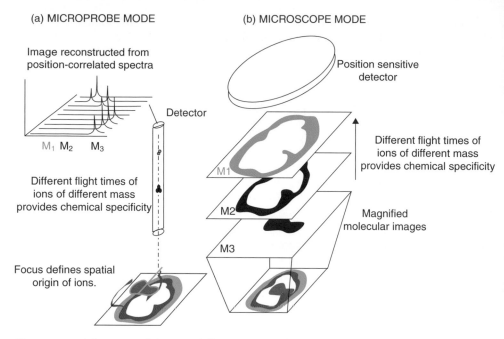

Figure 5.1 *Schematic of the two different approaches to imaging mass spectrometry. Microprobe mode imaging (a) collects mass spectra from an array of designated positions to reconstruct a molecular image after completion of the experiment. In microscope imaging (b), magnified images of the ion distributions are directly acquired using a two-dimensional detector. Reprinted from Anal. Chem., 76 (18), 5339–5344, 2004. Stefan L.Luxembourg, Todd H. Mize, Liam A. McDonnell, and Ron M. A. Heeren. High-Spatial Resolution Mass Spectrometric Imaging of Peptide and Protein Distributions on a Surface. Copyright 2004 American Chemical Society*

adjacent images are 'stitched' together in software and, by using ion gates, images of differing m/z values can be obtained. Using such instrumentation a spatial resolution of $4\,\mu m$ has been demonstrated.

A closely related technique to MALDI-MSI is MALDI-MS profiling (MALDI-MSP). In MALDI-MSP mass spectra are recorded directly from different regions of the surface of the sample of interest, e.g. a tissue section. These spectra are then available for visual comparison and interpretation or for a multi-variant statistical analysis (Caldwell and Caprioli, 2006).

The early work of the Caprioli group described the application of MALDI-MSI for the imaging of protein distribution in a range of different biological tissue types. These initial studies developed into a detailed body of work in oncology which is discussed in detail in Section 5.3.2. However, it is the breadth of applications of MALDI-MSI that has developed since 1997 that leads to the great excitement surrounding the technique. The ability to image the distribution of hundreds of proteins, xenobiotics and their biotransformation products and endogenous small molecules such as lipids and cellular metabolites without the requirement for labelling make this technique unique amongst imaging technologies.

In this chapter we intend to give a brief introduction to the methodology of MALDI-MSI and then to review its applications with a focus on the potential for its use in microbial investigations.

5.2 Practical Aspects of MALDI-MSI

5.2.1 Instrumentation for MALDI-MSI

MALDI-MSI experiments have now been performed on all types of modern mass spectrometers and commercial systems are available from all major vendors. The factors that influence the choice of instrumentation for a particular imaging experiment are a combination of those that are normally considered in mass spectrometry (MS), i.e. mass range, mass resolution, sensitivity and whether or not tandem MS is required. For imaging experiments however, there is one extra factor to consider, namely, the spatial resolution required.

In MALDI-MSI the spatial resolution can be defined by several parameters, the laser spot size, the smallest sample plate movement possible and the size of the matrix crystals. The lasers used for MALDI-MSI are typically focused down to somewhere between $50\,\mu m$ and $200\,\mu m$ diameter. A recent paper (Holle *et al.*, 2006) discussed some of the important features of lasers for MALDI-MSI. In the article the limitations of the 337 nm nitrogen lasers widely used for conventional MALDI-MS for imaging purposes are described, i.e. the short lifetime in terms of number of laser shots of such laser (approximately 2×10^7 shots) and low laser repetition rates (20 Hz). This latter issue can mean that images take several hours to record. Solid state lasers (commonly frequency tripled Nd:YAG lasers with operating wavelengths of 355 nm) are also widely used in MALDI-MS. These can have repetition rates up to 20 kHz and typically have lifetimes of 1×10^9 shots. Holle *et al.* suggested that the excellent Gaussian beam profiles produced by such lasers were in fact a disadvantage for MALDI since it meant that only a very small part of the total area irradiated by the laser produced ions. They addressed these issues by developing a solid state laser with a structured beam profile and demonstrated the improved ion generation arising from its use. This device has been commercialised by Bruker Daltonics as the 'Smartbeam' laser and is promoted as being capable of generating good quality images at spatial resolutions as high as $10\,\mu m$.

Conventional matrix assisted laser desorption/ionisation time-of-flight mass spectrometry (MALDI-TOF-MS) instrumentation remains the platform of choice for the original MALDI-MSI application of intact protein profiling. TOF instruments offer high sensitivity and potentially unlimited mass range. The limitations of such systems become apparent however when MALDI-MSI is employed for small molecule imaging. Of particular concern is the absence of the means to introduce specificity into the experiment. The use of tandem MS techniques to introduce specificity in MALDI-MSI is now generally accepted and instruments based on TOF-TOF, QTOF and QTRAP technologies are commercially available and their use has been widely reported (Hsiesh *et al.*, 2006). A standard small molecule MALDI-MSI experiment would target the compound(s) of interest by setting up one or more multiple reaction monitoring experiments and creating images from the intensities of characteristic product ions of the species of interest. The use of ultra-high resolution to introduce specificity is a very recent innovation (Cornett *et al.*, 2008).

Atmospheric pressure MALDI (AP-MALDI) has been employed as a method for coupling MALDI and MALDI imaging to a variety of mass spectrometer types (Laiko *et al.,* 2000). Here sample ions are generated by normal MALDI processes at atmospheric pressure. They are subsequently transferred into the mass spectrometer; either through a capillary inlet, that forms part of an existing electrospray or other atmospheric pressure ionisation source, or through a nozzle skimmer arrangement. The early AP-MALDI sources were reported to suffer from sensitivity issues but the introduction of pulsed dynamic focussing (Tan *et al.,* 2004) to entrap and focus ions into the mass spectrometer has addressed this. For imaging MALDI-MSI applications AP-MALDI would clearly have the advantage that the tissue could be kept at standard laboratory conditions or frozen (if required) rather than under vacuum. AP-MALDI-MSI has been combined with an infrared laser (Li *et al.,* 2007, 2008; Shrestha *et al.*, 2008) for the study of endogenous metabolites and is discussed in Section 5.3.3. The use of AP-MALDI-MSP, for the identification of bacterial spores of the genus *Bacillus* has also been reported and is discussed in Section 5.4.2 (Madonna *et al.,* 2003; Pribil *et al.,* 2005).

5.3   Applications

MALDI-MSI has been continuously evolving since 1997 and has been applied to a variety of lifescience fields. It is, in fact, a very versatile technology and this is due to a number of features, namely, the wide mass range it is applicable to (theoretically from 0 to 500 000 Da), the ability to analyse complex mixtures with little purity, easy data interpretation, the sensitivity (low femtomole to attomole levels for proteins and peptides) and the qualitative and structural aspects of the information provided. Its proven analytical capabilities make it a valid complement to existing techniques such as histopathology or whole body autoradiography (WBA) and in some circumstances, discussed below, even a 'stand alone' technique. Here is an overview of the applications in four main lifescience areas, namely, Biotechnology, Medicine, Pharmaceutical Science and Microbiology.

5.3.1   Pharmaceuticals

Reyzer and co-workers (Reyzer *et al.,* 2003) first reported the study of drugs in biological tissue by MALDI-MSI. In these experiments, matrix was applied to intact tissue either by spotting small volumes of the matrix in selected areas or by coating the entire surface by pneumatic spraying. A QTOF type instrument was used and MALDI images were created by selected reaction monitoring (SRM) to specifically monitor the drug under study. Such an approach minimises the potential for ions arising from either endogenous compounds or the MALDI matrix to interfere with the analyte signals and this has been the approach most commonly adopted for the study of drug and metabolite distribution.

In order to understand this approach to targeted small molecule imaging it is perhaps easiest to discuss one application example in detail. Hsiesh and co-wokers (Hsiesh *et al.,* 2006) have described the detection and imaging of the antipsychotic drug clozapine in rat brain. The test animal was dosed with $5\,mg\,kg^{-1}$ of the drug and sacrificed 45 min post-administration. After careful removal, the brain was snap frozen. Subsequently $12\,\mu m$ sections were taken using a cryostat for MALDI-MSI and in this instance carefully coated with matrix material (α-cyano-hydroxycinnamic acid $25\,mg\,ml^{-1}$ in $80:20$ acetonitrile : water) using a TLC reagent sprayer. [It should be noted that a number of approaches to matrix deposition are employed for MALDI-MSI and these have been discussed in detail in a recent review by Francese and colleagues (Francese *et al.*, 2009).] An issue with small molecule MALDI-MSI is the possibility of interferences in the signals of interest arising from either endogenous compounds or the MALDI matrix. In the SRM approach this is overcome by recording in the imaging experiment the intensity of a product ion specific to the compound of interest. In this example it was found that under collision induced dissociation clozapine produces an intense product ion at *m/z* 270. The tuning/set up of the hybrid quadrupole time-of-flight mass spectrometer used for the study was optimised specifically for the detection of this product ion and its intensity was measured as the laser was rastered at $150\,\mu$m increments across the surface. Figure 5.2 shows the results from the experiment. As can be seen, clozapine can be detected in the brain and the data obtained on its distribution are in good agreement with a comparative study carried out by autoradiography of radiolabeled compound.

5.3.2   MALDI-MSI and Medicine

When Richard Caprioli reported the invention of MALDI-MSI to obtain molecular maps of protein and peptides from biological tissues (Caprioli *et al.*, 1997), it was immediately very clear that clinical proteomics, preclinical studies, medical diagnosis and prognosis would be important applications areas. The literature shows that this has, in fact, occurred with the oncological field being the most pursued target, as the numerous publications in this field demonstrate. The goal of such studies is to locate both biomolecules more highly expressed in tumours and those that exhibit a down regulation, compared with normal tissues. This is the key principle of differential proteomics as well as of biomarker discovery, where biomarkers are defined as 'indicators of normal biological processes, pathogenic processes, or pharmacologic responses to a therapeutic intervention' (Atkinson *et al.*, 2001). The individuation of these species opens up the possibility to implement non-invasive investigation protocols (blood or fluid collection and direct analysis by specific kits) for early diagnosis.

The first applications in the oncological field were reported by Stoeckli and co-workers in 2001 (Stoeckli *et al.*, 2001). The authors were interested in human glioblastoma as it had been reported that, at the time, brain tumors accounted for about 11 000 deaths annually in the USA and gliomas accounted for 50% of all primary brain tumours, with glioblastomas accounting for half of those (Nelson *et al.*, 1993). In their work a xenograft glioblastoma was investigated through MALDI-MSI. More than 150 ions were detected and localised with many of them being very specific to the tumour area, such as Thymosine β4 (Figure 5.3).

These data showed for the first time the potential of MALDI-MSI to be used in intraoperative assessment of surgical margins of tumours and hence that it might provide the basis for a treatment decision (surgical or pharmacological) before the patient leaves the operating room. Provided that the adequate image resolution is possible, clinicians might be then able to evaluate, at a molecular level, tumour biopsies with the potential to identify sub-populations that are not evident, based on the cellular phenotype determined microscopically (Chaurand *et al.*, 2002). This opportunity would greatly support the current procedures for rapid decision making involving frozen sections and light microscopy. In

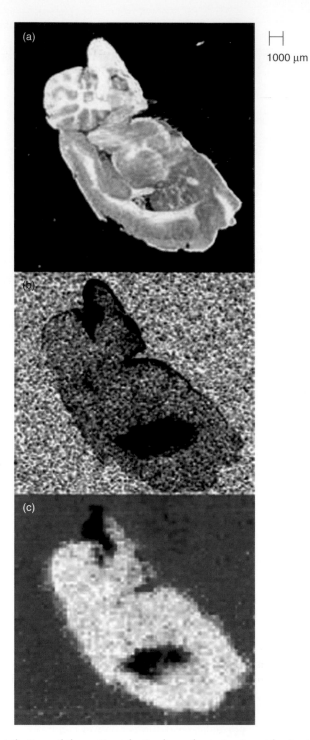

Figure 5.2 *Distribution of the antipsychotic drug clozapine in rat brain studied by MALDI-MSI. (a) Optical image; (b) radioautographic image; and (c) MALDI-MS/MS image from the study rat brain tissue section. Reproduced from Hsiesh, Y., Casale, R., Fukuda, E., Chen, J., Knemyer, I., Wingate, J., Morrison, R., Korfmacher, W., Matrix-assisted laser desorption/ionization imaging mass spectrometry for direct measurement of clozapine in rat brain tissue, Rapid Commun. Mass Spectrom., 20, 965–972 (2006). Copyright 2005 John Wiley & Sons, Ltd*

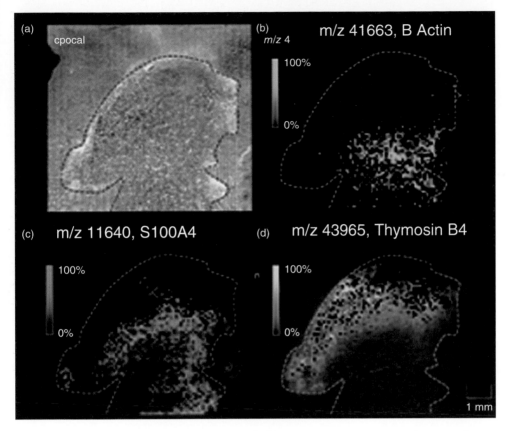

Figure 5.3   *Selected protein images from a glioblastoma section. (a) Optical image of human glioblastoma slice coated with matrix. (b–d) Mass spectrometric images of proteins showing high concentration in the proliferating area of the tumour (d) and other proteins present specifically in the ischemic and necrotic areas (b and c). Reprinted by permission from Macmillan Publishers Ltd, Nature Methods from Stoeckli et al., 7, 493–496, 2001*

some borderline cases these rely on the experience of the pathologist and therefore are subjective and potentially inaccurate.

Immunohistochemistry is another way to formulate a diagnosis. It is a very specific technique as specific antibodies target proteins that are recognised to be tumoural markers. This principle constitutes also the limitation of the technique; it requires an *a priori* knowledge of the target as well as the availability of the corresponding antibody to use. On the contrary, performing a MALDI-MSI analysis on these tissues does not require a previous knowledge of the target or antibodies as it relies simply on the detection of the *m/z* of the ions present in the tissue. The combination of histopathological and MALDI-MSI analyses has proved to have a great potential in disease diagnosis, prognosis and even patient specific therapeutic treatment. Chaurand and collaborators demonstrated this in 2004 by examining via MALDI-MSI snap-frozen normal brain and brain tumour specimens of different grades (Chaurand *et al.*, 2004). Peptide and protein expression were

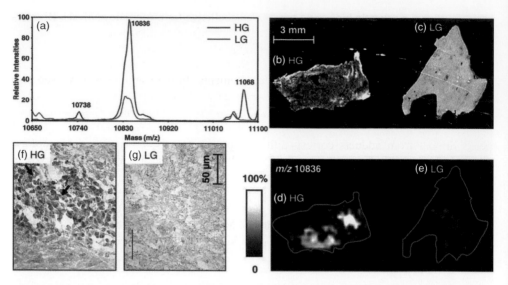

Figure 5.4   *Localization of the S100B protein in human low-grade and high-grade glioma biopsies by MALDI-MSI and immunohistochemistry. Partial survey MALDI-MS protein profiles (a) obtained from low-grade (LG, orange trace) and high-grade (HG, blue trace) glioma sections. The signal at m/z 10836 was identified as the S100B protein. Panels (b) and (c) show photomicrographs of high-grade and low-grade gliomas, respectively. Also shown are S100B ion intensity maps obtained from high-grade (d) and low-grade (e) glioma sections and high magnification photomicrographs obtained from high-grade (f) and low-grade (g) glioma sections after immunostaining for the S100B protein. Reprinted from Am. J. Pathol. 2004, 165, 1057–1068 with permission from the American Society for Investigative Pathology*

compared and the patterns assessed through hierarchical cluster analysis. The authors could reliably distinguish between gliomas and nontumour brain tissue as well as subclassify grade IV gliomas from grades II and III. As an example of their results, Figure 5.4 shows the simultaneous analysis by MALDI-MSI of two sections obtained from grade II (low-grade) and grade IV (high-grade) resected human glioma biopsies [Figure 5.4(a)].

In an initial MALDI TOF profiling analysis, amongst the up-regulated ions in the high grade tumour, they found the protein S100β [Figure 5.4(a)]. The corresponding density maps clearly distinguished between high- and low-grade tumours [Figure 5.4(b) and (c)] and this biomarker was validated by the immunoistochemistry analysis showing a number of astrocytes in the high-grade tumour with pronounced S100β immunoreactivity in the cytosol [Figure 5.4(d), arrowhead] and weak immunopositivity in the oligodendrocytes in the low-grade tumor [Figure 5.4(e)].

A recent application of MALDI-MSI has been reported in the analysis of formalin fixed paraffin embedded (FFPE) tissues. The group of Isabelle Fournier was the first to demonstrate in this context the feasibility of the technology (Lemaire *et al.*, 2007), which they later defined as 'the new frontier of histopathology proteomics' (Fournier *et al.*, 2008). This development provides access to massive amounts of archived clinical pathology samples which previously have presented a real challenge for proteomic analysis. Biomarker discovery requires a systematic and reproducible approach to tissue collection and analysis, as well as a statistically significant number of clinical cases and tissue collection; the possibility of using MALDI-MSP and -MSI to access these huge archives, even many years old, instantly addresses those two requirements. In their approach they used two methods according to the FFPE tissue age. For tissues less than 1 year old, they demonstrated the effective use of a reactive matrix called 2,4-dinitrophenylhydrazine (2,4-DNPH). This matrix acts by neutralizing unreacted formalin molecules and also suppresses peaks arising from adducts corresponding to protein-N=CH$_2$ ions. This methodology allows MALDI-MSI of proteins in up to 1 year old FFPE tissues. The second methodology is applicable for all FFPE tissues regardless of conservation time and is based on protein detection and identification through *in situ* micro-enzymatic digestion after paraffin removal. Microdigestion combined with either *in situ* extraction prior to classical nanoLC/MS-MS analysis or automated microspotting of MALDI matrix enables protein identifications with both nanoLC-nanoESI and MALDI-MSI.

5.3.2.1 *Proteomic Insights*

MALDI-MSI is being successfully used in conjunction with classical proteomic approaches for biomarker discovery. The literature shows numerous applications in this context, such as in drug toxicity on non-target organs (Meistermann *et al.*, 2006), clinical conditions such as Fabry's disease (Touboul *et al.*, 2007) and human gliomas of different grades (Chaurand *et al.*, 2004; Schwartz *et al.*, 2005). Proteomic investigation of tissues imaged by MALDI-MSI was first reported to perform a parallel homogenization of the tissue, protein extraction and proteolysis and MALDI-MS/MS and/or HPLC-ESI-MS/MS analysis (Stoeckli *et al.*, 2001). The group of Mitsutoshi Setou took a leap forward by mounting the tissue on a PVDF membrane and blotting it on another PVDF membrane (Shimma *et al.*, 2006). The blot was then spotted with trypsin by using a piezoelectric-based automatic dispenser and, following incubation and matrix application, the MALDI-MS and MS/MS analysis was performed directly on the blotted membrane without any extraction steps. They also reported the first ever example of 'on tissue' digestion and shotgun proteomics. Trypsin and then matrix were deposited in the same fashion as on the membranes directly onto rat brain tissues. Shotgun proteomics was carried out by performing MALDI-MS and MS/MS analysis on the same tissues and yielded numerous peptide ion signals and sequence information, which were searched against databases for protein identification which could be correlated with previously detected intact protein ion signals. This demonstrated that MALDI-MSI can be regarded as a unique and comprehensive tool for analysing the spatial distribution of peptides and proteins throughout tissue sections, providing a huge amount of data with minimal sample preparation and that it is thus a valid alternative to classical proteomic approaches.

A successful workflow nowadays may consist of: (a) MALDI-MSP and -MSI analysis; (b) selection of possible biomarker candidates through rigorous statistical software and methods; (c) the bottom-up *in situ* proteomic approach; (d) MALDI-MS and MS/MS of the digested material; (e) database search of the MS and MS/MS spectra through engines like MASCOT or Sequest for protein identification; and (f) MALDI-MSI of the peptides of interest related to the candidate protein (Figure 5.5). Biomarker validation is then typically performed by immunohistochemistry using antibodies directed against it.

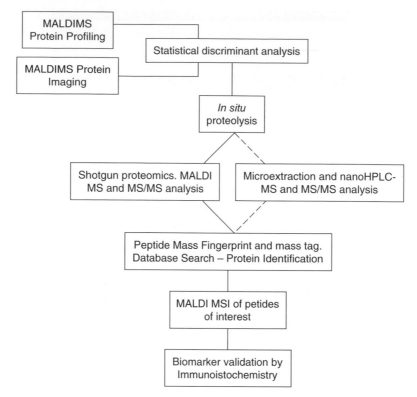

Figure 5.5   *Workflow of a thorough proteomic investigation by MALDI-MS analysis of intact tissues. Dashed lines indicate parallel non in situ analyses that can be performed*

The direct analysis of a tissue by MALDI-MSI implies addressing the presence of a wide range of proteins. Typically biomarkers are at the bottom of the wide concentration range exhibited by proteins. Moreover, on tissue direct analysis gives rise to ion competition and suppression events more than in conventional MALDI-TOF-MS analysis. This may lead to the idea that MALDI-MSI suffers a very important loss of information; this can certainly happen but the extent of this loss may not be higher than that from more classical proteomic approaches. In fact, although *on tissue* MALDI-MSI analysis has generally a lower sensitivity compared with conventional MALDI analysis, gel extraction and separation protocols, preceding sample preparation for conventional MALDI-MS analysis, imply an inevitable analyte loss. In some cases, it has been observed that direct homogenate analyses yield by far less intense signals and lower ion abundance if compared with a direct *on tissue* analysis (Dani *et al.*, 2008). Also, classical proteomic studies involving tissue or organ homogenates necessarily lead to loss of correlating information between biological function and localization, the importance of which is demonstrated by the fact that several diseases are associated with altered molecular distributions (Roher *et al.*, 2002). Despite the possible limitations of MALDI-MSI in terms of amount of qualitative information, protocols are being developed to improve sensitivity. Besides tissue washing procedures which greatly reduce interferences from endogenous lipids, salts and debris

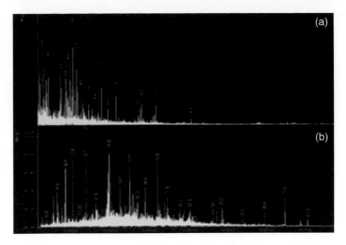

Figure 5.6 *Evaluation of method improvement for in situ digestion. (a) Observed peptide profiles after an in situ digestion performed with trypsin in water at room temperature. (b) In situ digested protein profile when using a trypsin solution containing 0.1% OcGlc and incubating the section at 37 °C for 2 h. Reproduced and adapted by permission of John Wiley & Sons, Ltd from Djidja et al., Proteomics, 9, 1–15 (2009)*

(Seeley *et al.*, 2008), the use of non-ionic detergents has proven to be very effective. Djidja and collaborators (Djidja *et al.*, 2009) reported the use of octyl α/β glucoside (OcGlc) for tryptic digestion of proteins in breast tumour tissue section samples. The incorporation of this detergent increased the yield of tryptic peptides for both fresh frozen and FFPE tumour tissue sections. Figure 5.6 shows an example of such improvement.

To further increase sensitivity and confidence in the identification, *in situ* proteolysis was followed by a novel approach combining MALDI-MSI and ion mobility separation MSI. The addition of the ion mobility separation, described elsewhere (Verbeck *et al.*, 2002), proved to be very effective for the detection and identification of low abundant or high mass proteins which are difficult to detect solely by MALDI-MSI. Numerous peptide signals were detected and some proteins including histone H3, H4 and Grp75 that were abundant in the tumour region were identified (Figure 5.7).

5.3.3 Biotechnology

One important aspect of MALDI-MSI that has not been widely explored is its potential to provide a metabolite profile at the single cell level, thus allowing not only the metabolite profile to be determined but also more importantly the known heterogeneity of tissues to be examined.

Subcellular MALDI imaging is not possible at present. Animal cells are of the order of 10 μm in diameter, plant cells of the order of 50 μm in diameter and hence at best MALDI-MSI, as generally used, would only yield one pixel per cell. Several groups have however begun to examine the low mass region of MALDI spectra obtained from biological tissue for the presence of primary and secondary metabolites. The challenges of small molecule MALDI-MS, namely, matrix interference/suppression effects and overlapping signals, are

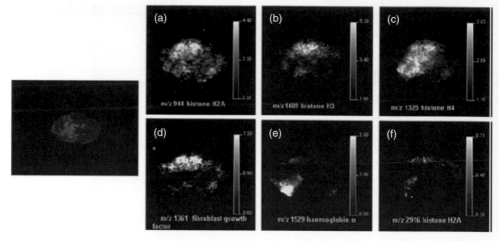

Figure 5.7   *MALDI-MS images of peptide distribution within an MCF7 xenograft tissue section. On the left is shown an optical image of an MCF7 xenograft tissue section. (a–f) MALDI-MS images of the distribution of some of the identified peptides within an MCF7 tissue section. Reproduced and adapted by permission of John Wiley & Sons, Ltd from Djidja et al., Proteomics, 9, 1–15 (2009)*

exacerbated when on-tissue shotgun metabonomics is the aim of the experiment. The total number of metabolites present in plants is thought to exceed 200 000 and the detection and identification of this number of compounds in a single experiment is the ultimate goal of such analyses. There are two possible MS approaches to this problem: one is the use of high resolution and accurate mass measurement to target the accurate mass of ions arising from the analytes of interest; the other is to perform multiple MS/MS experiments. There are obvious limitations to each of these approaches; if accurate mass measurement is used, then isomeric compounds will not be distinguished and the technology to perform several thousand MS/MS experiments in the few seconds of signal that a single MALDI sample spot generates simply does not exist at present.

Several groups however have felt it useful at this stage of the development of the technique to begin to examine the low molecular mass region of MALDI mass spectra obtained from biological surfaces for the presence of peaks arising from primary and secondary metabolites and to start to address the problems that occur in their use for imaging, i.e. low abundance and matrix interference. Vaidyanathan and co-workers (Vaidyanathan *et al.*, 2006) used conventional MALDI to study a cocktail of metabolites spiked into a microbial extract of *Escherichia coli* and demonstrated the feasibility of such an approach. Burrell and co-workers (Burrell *et al.*, 2007) imaged the distribution of a number of metabolites in wheat seeds using accurate mass measurement.

Figure 5.8(a) shows the distribution of m/z 381.0799 +/– 0.05 in a number of wheat seeds developed at different temperatures. The accurate m/z chosen corresponds to the $[M+K]^+$ ion for a hex_2 oligosaccharide (probably in this case sucrose). The mass accuracy was achieved in this experiment by recalibrating the data post-acquisition using known matrix peaks. Figure 5.8(b) shows the full MALDI mass spectrum acquired from the

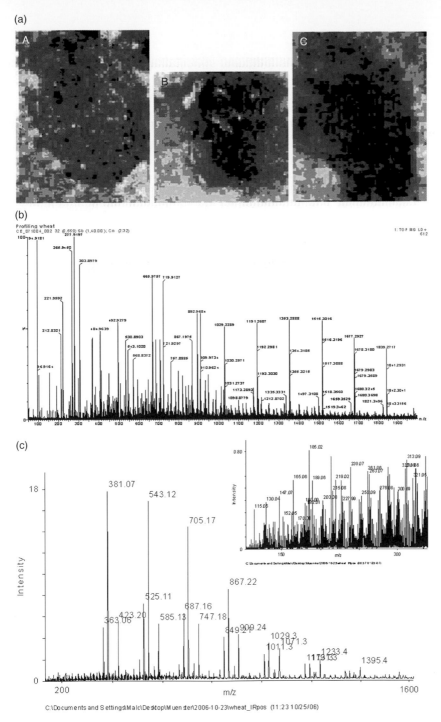

Figure 5.8 *Direct MALDI imaging and profiling of metabolites in wheat seeds. (a) Images of the distribution of sucrose (as its [M+K]⁺ adduct) in developing seeds grown at different temperatures: (A) 25 °C; (B,C) 20 °C. (Note, the darker the colour, the greater the signal.) (b) MALDI-MSP of a wheat seed recorded using a-CHCA as matrix by conventional UV MALDI. (c) MALDI-MSP of a wheat seed recorded using infrared LDI with a 2.94 μm laser. (a) Reproduced from Burrell, M.M., Earnshaw, C.J. and Clench, M.R., Imaging matrix assisted laser desorption ionisation mass spectrometry: a technique to map plant metabolites within tissues at high spatial resolution, J. Exp. Bot., 58, 757–763 (2007). Copyright (c) 2007 by the Society for Experimental Biology. (c) Courtesy of Dr Klaus Dreisewerd, University of Munster*

surface of a wheat seed and as can be clearly seen many images could be generated using the accurate mass measurement approach.

In order to overcome interferences arising from the MALDI matrix, infrared laser desorption ionisation (IR-LDI) has been proposed for this application. For plant metabolomic studies it has been shown (Dreisewerd *et al.,* 2007; Li *et al.,* 2008) that a laser of wavelength 2.94 *μ*m can be used to heat residual water in the tissue and that, as a consequence, desorption ionisation of plant metabolites occurs with them being observed in a similar manner to the wheat example shown as alkali metal adduct ions [Figure 5.8(c)].

There is little doubt that metabolomic profiling/imaging is likely to be fruitful areas for future research using any or a combination of the techniques described above.

5.4 Microbial Molecular Investigation by MALDI-TOF-MS

The feasibility of microorganism studies at a molecular level via MS was first demonstrated in 1975 (Anhalt and Fenselau, 1975). In this approach, electron ionisation (EI) was used to detect species which were characteristic of individual microorganisms. Through the years, a few mass spectrometric techniques, such as FAB and plasma desorption, were employed in the classification of microorganisms with considerable success especially when combined with tandem MS (Fenselau and Demirev, 2001). Microorganism typing via MALDI-TOF-MS, first proposed in 1996 (Claydon *et al.,* 1996; Holland *et al.,* 1996), has showed since then many advantages in characterising viruses, bacterial and fungal vegetative cells, and spores, because of its soft mode of ionisation and straightforward data interpretation. MALDI-TOF-MS profiling overcomes some of the 'similarities' problems linked to phenotypic identification as characterisation solely relies on the detection of the *m/z* value of the ions. The opportunity to explore the molecular content of microorganisms via MALDI-TOF-MS can be used for their classification with rapid (about 5 min) and minimal sample preparation (lysis by exposure of the microorganism to a strong organic acid) thus avoiding tedious and time-consuming protocols based on extraction, separation, or amplification. The associated high-throughput with automation further reduces the screening time. The wider dynamic range of the technology when compared with other MS methods, has allowed the classification based on a variety of biomolecule 'sizes' ranging from lipids to proteins and nucleic acids, although proteins remain the favourite biomarker target. Protein biomarkers from different microorganisms have been shown to be readily accessible with MALDI-TOF-MSP and spectral fingerprints have been obtained from the Gram-negative vegetative cells of *Helicobacter pylori* (Demirev *et al.,* 2001), intact spores of *Bacillus cereus T* (Ryzhov *et al.,* 2000), *Aspergillus flavus* cells and spores (Li *et al.,* 2000) and single cell parasites such as *Cryptosporidium parvum* (Magnuson *et al.,* 2000).

5.4.1 Microbial MALDI-TOF-MSI

A microscopic insight into cell organisation is one of the very ambitious aims of microbial imaging studies and MALDI-TOF imaging technology has been evolving through the years towards this goal. Its ongoing improvement will allow a deeper investigation of a wide range of problems in contemporary microbiology such as strain selection, understanding the correlation between microbial function and structure in relation to pathophysiology, as well as in the development of antimicrobial agents and vaccines.

5.4.1.1 Instrumental Requirements

An imaging analysis enabling localisation of molecules in specific cellular compartments requires instrumentation capable of achieving high resolution. The image resolution is limited by two key factors, crystal size and laser beam diameter. Once the first one is optimised, the laser beam diameter remains the critical parameter (Francese *et al.*, 2009). Many commercially available MALDI-TOF instruments, equipped with N_2 (337 nm) or tripled Nd:YAG (355 nm) lasers, have a laser beam size of ~100 μm. The typical diameter of a mammalian cell is around 20 μm, a bacterial cell's diameter is about 1–4 μm, a virus is even smaller with a diameter ranging from 20 to 40 nm and fungi range in diameter between 1 μm and 100 μm according to species; therefore if a bacterial cell which is 1 μm in diameter has to be imaged using a laser with a spot size of 100 μm, we would be effectively imaging about 100 cells and therefore no compartmentalised molecular localisation can be achieved but only spectral profiles of the molecular content. A few laboratories in the USA have been trying to customise their lasers and in one case a new MALDI instrument was developed able to image at a micrometric resolution by using a highly focused laser. Unfortunately, in that case and in general, the improvement in resolution has a cost in terms of sensitivity; in fact, high laser fluence was required to ionise molecules from such a small spot giving rise to extensive analyte fragmentation. Nonetheless, Caprioli and collaborators have achieved a remarkable spatial resolution by using a custom built scanning MALDI-TOF-imaging mass spectrometer (Chaurand *et al.*, 2007), optimised for the imaging of peptide and protein ions from thin mammalian tissue sections. This instrument is capable of achieving irradiation areas (spots) as small as ~7 μm in diameter, and therefore having imaging capabilities at a cellular level. As reported in Section 5.1, the group of Heeren and collaborators (Luxembourg *et al.*, 2004) also developed a MALDI instrument working in 'microscope mode' capable of achieving a spatial resolution of 4 μm in a field of view of 200 μm. In terms of spatial resolution, with a long history of imaging of surface elements and organic compounds, secondary ion mass spectrometry (SIMS) has no equal as it can achieve submicrometric resolution. A beautiful example using the TOF-SIMS technique comes from the work of Debois and collaborators (Debois *et al.*, 2008) showing the localisation of surfactins in a *Bacillus subtilis* swarming community with a spatial resolution of 2 μm. Surfactins were mainly located in the central mother colony (the site of initial inoculation) in a 'ring' surrounding the pattern and along the edges of the dendrite (Figure 5.9).

Nonetheless MALDI-TOF is far more used for imaging purposes both for its superior sensitivity and its wider exploitable mass range which enables the analysis of macromolecules such as proteins; mass spectral interpretation in SIMS is less straightforward compared with MALDI, where mainly intact molecular ions are produced, and many SIMS instruments do not operate in MS/MS mode thus preventing molecular structural characterisation. This is the reason why mass spectrometer manufacturers are investing their efforts in improving imaging capabilities in terms of resolution in MALDI mass spectrometers. It can be concluded then that, at present, MALDI-MSI at subcellular level is not possible although the technologies described and their constant evolution is certainly very promising.

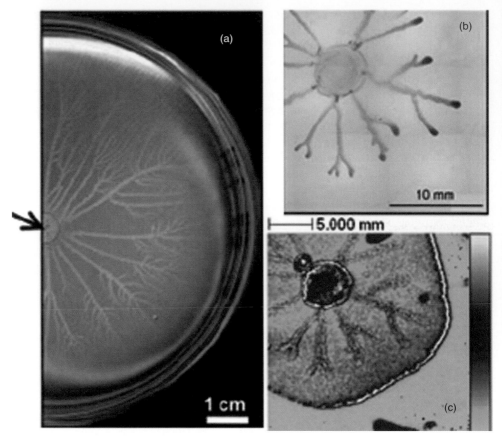

Figure 5.9 *TOF-SIMS imaging of surfactins from a B. subtilis swarm community. (a) The swarming pattern after 30h of incubation at 30 °C with the black arrow indicating the site of the initial inoculation. (b) The low-definition microscope scan of the surfactin distribution. (c) The TOF-SIMS image of the sum of surfactin ions. Reproduced and adapted by permission of John Wiley & Sons, Ltd from Debois et al., Proteomics, 8, 3682–3691 (2008)*

5.4.1.2   Applications

Although MALDI-MS is not at present able to offer imaging at below 10 μm spatial resolution, some interesting microbial applications are reported in the literature which try to stretch the current limitations. Esquenazi and co-workers for example, focused on marine cyanobacteria, as they are prolific sources of natural products with therapeutic applications (Esquenazi *et al.*, 2008). In their paper they underline the importance of this technology as a means to allow the isolation and characterisation of genomes via single cell genomic sequencing, to aid the discovery of new therapeutic agents as well as to unravel host–microbe interactions. In their work they used MALDI-MSI to characterise the spatial distribution of natural products from single strands of cyanobacteria. In the case of *Lyngbya majuscule JHB*, working at a spatial resolution of 100 μm, they estimated to be

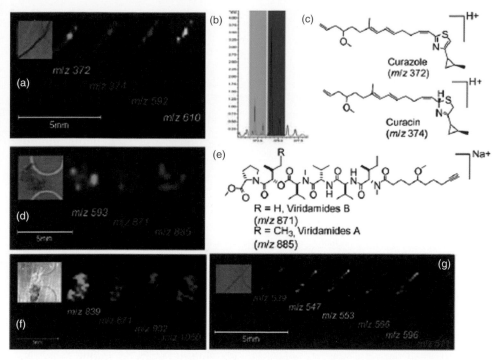

Figure 5.10   *The spatial distribution of selected ions observed to co-localise with* Lyngbya majuscula *3L (a),* Oscillatoria nigro-viridis *(d), a* Phormidium *species (f), and* Lyngbya bouillonii *(g). The average mass spectral trace showing curacin and curazole and their respective colours is shown in (b). The structures of curacin, curazole (c) and viridamides (e) are also shown. From Esquenazi et al., Molecular Biosystems, 4, 562–570 (2008). Reproduced by permission of the Royal Society of Chemistry*

analysing from 25–40 individual cyanobacterial cells at one time from a single filament as the cells are 20–50 μm in width and 2–4 μm in length. In this way they managed to map the natural products Jamaicamide B and Yanucamide B within the single bacterial filament. To demonstrate feasibility of the methodology, other cyanobacteria such as *Lyngbya majuscula 3L, Oscillatoria nigro-viridis, Lyngbya bouillonii,* and a *Phormidium* species were imaged and many detected ions were in agreement with previously reported natural products such as curacin A and curazole (Figure 5.10).

A proof of principle experiment was also conducted to demonstrate that spatial resolution is not lost when imaging heterogeneous colonies such as a mixture of *Lyngbya majuscula 3L* and *JHB, Oscillatoria nigro-viridis* and *Lyngbya bouillonii*. Results showed that the technology could readily distinguish ion masses known to be associated with each cyanobacterium (Figure 5.11).

5.4.2   Microbial Proteomic Characterisation and Classification via MALDI-TOF-MS and MS/MS

Microbial proteomic characterisation may have a remarkable impact in many biomedicine and biotechnological applications. For example, it may aid in phyloproteomics which aims

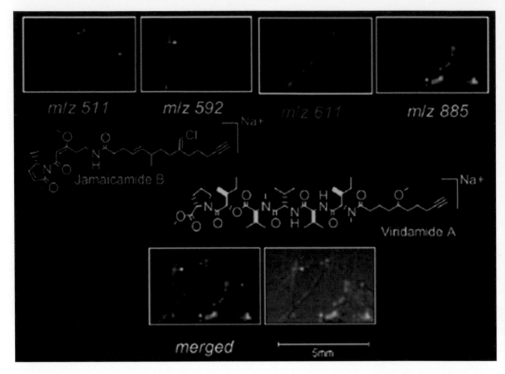

Figure 5.11   *MALDI-MSI of a complex mixture of cyanobacteria: a single MALDI-MSI run on a mixture of* Lyngbya majuscula *JHB (orange), and 3L (green),* Lynbya bouillonii *(red) and* Oscillatoria nigro-viridis *(blue). The top panels represent detection of two known masses – Jamaicamide B (orange structure shown) and viridamide A (blue structure shown) as well as two unknown masses (red and green), each specifically and differentially locates to a particular organism. The bottom panel emphasises the scale and spatial resolution. From Esquenazi* et al., *Molecular Biosystems, 4, 562–570 (2008). Reproduced by permission of the Royal Society of Chemistry*

at classifying the relationships between microorganisms by the mass of their respective proteins or in the rapid diagnosis of dermatophytoses, which are fungal infections. Bacteria have been the primary target for intact microbial investigation at a molecular level through MALDI-TOF technology. Some time has been given to viruses whilst fungi remain largely unexplored in this context. For the past 35 years, a number of MS-based methods have been developed to quickly differentiate bacteria. Most of these methods have relied on spectral fingerprints to distinguish between them at a species and even at a strain level. Nonetheless these signatures have remained unknown. The identification of these biomarkers instead would represent a leap forward towards the thorough understanding of the microbial biomolecular patrimony which we can use to refine therapies and diagnosis as well as in bioremediation and bioterrorism prevention. A comprehensive strategy would include intact protein profiling to generate microbial fingerprints, to be used for differentiation, and a bottom-up proteomic approach upon enzymatic digestion, enabling species determination from the identification of peptides (and therefore proteins) representative of a species by MS and MS/MS analysis. Whilst there are many examples in the literature of microbial classification following extraction and purification or two-dimensional gel separation (Holland *et al.*, 1999; Hollemeyer *et al.*, 2005; Vödisch *et al.*, 2009), in particular of metabolites and lipids, only a few examples of proteomic identification from intact microorganisms (shotgun proteomics) have been shown so far, especially by using MALDI-MS and MS/MS technology. The group of Fenselau and Demirev has always been amongst the most active with numerous and advanced publications in many microbial investigative fields in biomedicine and biodefence such as the detection of *P. falciparum* (malaria parasite) in pregnant women, rapid virus identification and the detection of biological weapons such as *B. anthracis Sterne* and *B. thuringiensis* (Yao *et al.*, 2002; Demirev, 2004; Demirev *et al.*, 2005). Madonna and co-workers as well as Pribil and collaborators have demonstrated the use of AP-MALDI in the proteomic investigation of bacteria relevant to biodefence (Pribil *et al.*, 2005). Madonna and co-workers reported the use of a novel proteomic approach involving the use of AP-MALDI to detect intact *Bacillus* species (Madonna *et al.*, 2003). In particular, by coupling an AP-MALDI source with a quadrupole ion trap mass spectrometer, the authors obtained the amino acid sequence for the lipopeptide biomarkers they were investigating. They also showed that airborne *Bacillus* spores could be captured from the atmosphere onto double-sided tape and then successfully interrogated for fengycin biomarkers. Since the AP-MALDI source does not have to be maintained under vacuum there is perhaps an opportunity for the development of an integrated aerosol collection/detection system to screen infectious microorganisms in atmospheric samples.

Recently, in the context of the fight against bioterrorism, Dugas and collaborators (Dugas *et al.*, 2008) developed a methodology that could be implemented in portable mass spectrometers for the rapid identification of bioaerosols which arc highly dangerous even in small quantities. They conducted a proteomic analysis on collected microbial aerosols by developing a new on-target protein digestion system. First, proteolysis occurred on 'impacted' bioaerosols; bioaerosols were generated using a pneumatic nebulizer and infused into a chamber for sampling. A single-stage impactor was then used to collect the bioaerosols on a MALDI target. The proteolysis was carried out inside removable mini-wells, acting as miniature reactors, placed directly over an impacted bioaerosol spot on the MALDI target (Figure 5.12). Trypsin could then be added to the well, avoiding enzyme immobilisation and humid chambers.

Once the reaction solvent evaporated, the mini-well was removed and the target was analysed. By coupling information from the MALDI- TOF-MS analysis of the intact protein species [Figure 5.13(a)] and those derived from peptide mass fingerprinting on spectra of the digested *E. coli* aerosol [Figure 5.13(b)], 19 unique proteins were identified.

In addition to the bottom-up approach, the top-down approach has also been reported where the intact protein species are submitted to MALDI-MS and MS/MS experiments and readily identified by deducing the amino acid sequence from its fragments and thus avoiding the proteolytic digestion step, as shown in Figure 5.14 (Demirev *et al.*, 2008).

Performing top-down proteomics by MALDI-MS/MS has a clear advantage over other MS techniques such as ESI, which requires microbial lysis, protein extraction, purification and separation on a one- or multidimensional fractionation platform.

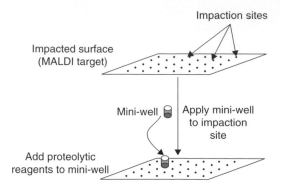

Figure 5.12 Schematic of mini-wells used on a MALDI target for 'in situ' tryptic digestion. Reprinted from Anal. Chim. Acta, 627, A. J. Dugas and K. K. Murray, On-target digestion of collected bacteria for MALDI mass spectrometry, 154–161. Copyright Elsevier (2008)

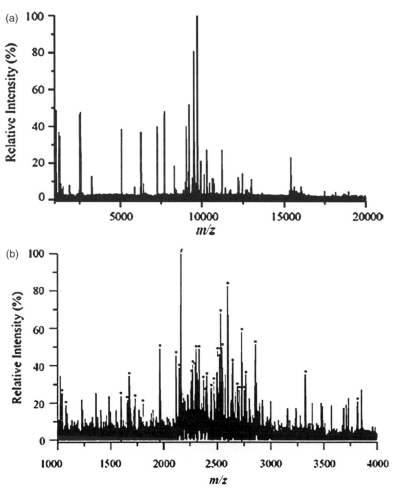

Figure 5.13 MALDI-TOF mass spectra of impacted E. coli showing intact protein (a) and peptide profiles of the digested E. coli aerosol (b). Reprinted from Anal. Chim. Acta, 627, A. J. Dugas and K. K. Murray, On-target digestion of collected bacteria for MALDI mass spectrometry, 154–161. Copyright Elsevier (2008)

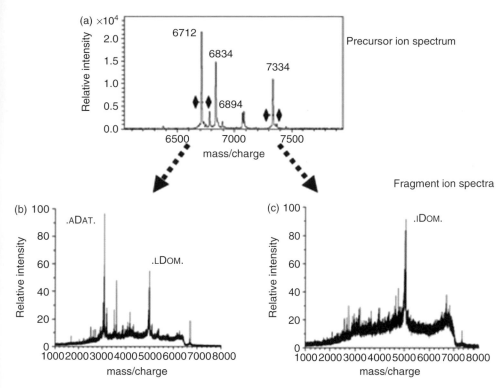

Figure 5.14 *Top-down proteomics for spore identification: (a) precursor ion mass spectrum of a mixture of B. cereus and B. globigii intact spores (MALDI-TOF); (b, c) fragment spectra of the isolated precursor ions at (b) m/z 6712 and (c) m/z 7334 (MALDI-TOF/TOF). Two SASP proteins, one originating from B. cereus and the other from B. globigii, are identified as the most plausible precursor candidates, with probabilities for a random match of $1.8 * 10^{-30}$ and $8.1 * 10^{-16}$, respectively. The partial sequences at major cleavage sites are denoted. Reproduced by permission of John Wiley & Sons, Ltd from Demirev and Fenselau, J. Mass Spectrom., 2008, 43, 1441–1457*

5.4.2.1 Data Analysis: Use of Fingerprint Libraries and Statistical Methods

The major problems to overcome in the classification of microorganisms and biomarker discovery are due to factors such as cell growth and even the different instruments and operators that recorded the data identification. These factors may have a significant influence on the robustness and reproducibility of the methodology. Efforts have been invested to design a protocol and an analysis system which can be used routinely and reliably independently from the operator and the MALDI instrument. In contrast to the initial approaches, it is currently a general view that such classification must be based on both reproducibility of the specific *m/z* value and signal intensity (Fenselau and Demirev, 2001). Different signal intensity may in fact, in some cases, indicate different stages of development of the microorganism. Researchers all over the world and some MS instrument manufacturers have outlined search algorithms/statistical methods which match mass spectra obtained from bacterial cells with tabulated fingerprint libraries for microbial classification. The criteria for comparing and estimating similarity between the experimental and the reference spectra vary amongst these algorithms. Hayek and co-workers (Hayek *et al.*, 1999) have described an algorithm to identify potential biological threat agents. The measured spectrum was compared with a small library of stored mass spectra of threat agents and in both cases, the representative parameters of the spectrum were represented by both *m/z* and intensity values. A multivariate linear least-squares regression algorithm was used for finding the best match as well as establishing a criterion for the probability of false matches. Another interesting approach was undertaken by Demirev and collaborators (Demirev *et al.*, 2001) which classifies microorganisms based on matching protein molecular masses in the spectrum with protein molecular masses predicted from sequenced genomes; one limitation of this approach is the restriction to those microorganisms whose genomes are sequenced. Bright and colleagues observed that the matching approach based only on the *m/z* values and corresponding intensities excludes a lot of other information that may be very helpful for a more accurate identification and discrimination especially at a strain level. Another important observation they made was that a successful biomarker approach would require all the genus-specific, species-specific and strain-specific markers to be present in all strains and at that point, common biomarkers had been determined from a very limited number of strains (Bright *et al.*, 2002). Bright and co-workers made a leap forward by developing a hybrid neural network software package (Manchester Metropolitan University Search Engine MUSE™) that stores all the information in MALDI-TOF spectra of intact cells, but maps this to a single point vector in an 'n'-dimensional space. This allowed them to rapidly build and search databases. In particular, a database of 35 strains, representing 20 species and 12 genera, was built with MUSE™ in 26 s, loaded in 10 s and was ready to search and identify 212 isolates, taking less than 1 s per isolate. They report that correct matches were made in 79, 84 and 89% of the 212 samples at strain, species and genus levels, respectively. At least 50% of the replicates of 42 of the 45 isolates matched the correct strain, and the most commonly identified species for 43 of the 45 isolates was the correct one. MS manufacturers such as Waters and Bruker also offer packages for bacterial classification. As an example, in the Bruker system, BioTyper™, the library spectra have been generated by several measurements of known bacterial species and strains in slightly different conditions with the reference spectra made of ribosomal proteins as they are highly expressed and stable. The system is reported to be robust and reproducible as it is unaffected by parameters such as bacterial state, medium growth and the type of the MALDI-TOF instrument operated. When identifying microorganisms, the matching score obtained is based on identified masses and their intensity correlation is generated and used for ranking the result (Maler *et al.*, 2006). For de-replication, clustering and generation of family trees it is possible to build dendrograms and to use an unsupervised multivariate analysis based on principle component analysis.

5.5   Conclusions

With over 380 publications in 11 years, MALDI-MSI has demonstrated its capability to significantly impact numerous lifescience fields. Pharmaceutical science is benefiting from its specificity and sensitivity to get quicker insights into pharmacodynamics; metabolomics is also being tackled with success particularly in plant biotechnology investigation; and medicine has now another powerful investigative tool in the biomarker discovery for disease prevention. Microbial investigation has been aided by several MALDI-MSP studies especially for classification purposes in fields such as pathology of infectious diseases, bioremediation and bioterrorism. MALDI-MSI has also been attempted on intact microorganisms, with the best example so far being the MALDI-MSI of a single filament of cyanobacteria estimated to contain 25–40 individual cyanobacterial cells. Current working resolution of the technology, in fact, cannot, even at its best, compete with techniques such as SIMS for mapping molecules in the different cellular compartments. Continuous technological advances which have led to the recent development of optics and lasers having a spot size of as little as $10\,\mu m$ are promising for future employment of MALDI-MSI in single cell analysis thus opening up new investigative opportunities in the field of microbial Imaging and proteomics.

References

Anhalt, J.P.; Fenselau, C., Identification of bacteria using mass spectrometry; Anal. Chem., 47, 219–225 (1975).

Atkinson, A.J.; Colburn, W.A.; DeGruttola, V.G.; DeMets, D.L.; Downing, G.J.; Hoth, D.F.; Oates, J.A.; Peck, C.C.; Schooley, R.T.; Spilker, B.A.; Woodcock, J.; Zeger, S.L., Biomarkers and surrogate endpoints: Preferred definitions and conceptual framework; Clin. Pharmacol. Ther., 69, 89 (2001).

Bright, J.J.; Claydon, M.A.; Soufian, M.; Gordon, D.B., Rapid typing of bacteria using matrix-assisted laser desorption ionisation time-of-flight mass spectrometry and pattern recognition software; J. Microbiol. Methods, 48, 127–138 (2002).

Burrell, M.M.; Earnshaw, C.J.; Clench, M.R., Imaging Matrix Assisted Laser Desorption Ionisation Mass Spectrometry: a technique to map plant metabolites within tissues at high spatial resolution; J. Exp. Bot., 58, 757–763 (2007).

Caldwell, R.L.; Caprioli, R.M., Tissue profiling by mass spectrometry – a review of methodology and applications; Mol. Cell Proteom., 4, 394–401 (2006).

Caprioli, R.M.; Farmer, T.B.; Gile, J., Molecular imaging of biological samples: localization of peptides and proteins using MALDI-TOF MS; Anal. Chem., 69, 4751–4760 (1997).

Chaurand, P.; Sanders, M.E.; Jensen, R.A.; Caprioli, R.M., Proteomics in diagnostic pathology: profiling and imaging proteins directly in tissue sections; Am. J. Pathol., 165, 1057–1068 (2004).

Chaurand, P.; Schriver, K.E.; Caprioli, R.M., Instrument design and characterization for high resolution MALDI-MS imaging of tissue sections; J. Mass Spectrom., 42, 476–489 (2007).

Chaurand, P.; Schwartz, S.A.; Caprioli, R.M., Imaging mass spectrometry: a new tool to investigate the spatial organization of peptides and proteins in mammalian tissue sections; Curr. Opin. Chem. Biol., 6, 676–681 (2002).

Claydon, M.A.; Davey, S.N.; Edwards-Jones, V.; Gordon, D.B., The rapid identification of intact microorganisms using mass spectrometry; Nat. Biotechnol., 14, 1584–1586 (1996).

Cornett, D.S.; Frappier, S.L.; Caprioli, R.M., MALDI-FTICR imaging mass spectrometry of drugs and metabolites in tissue; Anal. Chem., 80, 5648–5653 (2008).

Dani, F.R.; Francese, S.; Mastrobuoni, G.; Felicioli, A.; Caputo, B.; Simard, F.; Pieraccini, G.; Moneti, G.; Coluzzi, M.; della Torre, A.; Turillazzi, S., Exploring proteins in *Anopheles gambiae* male and female antennae through MALDI mass spectrometry profiling; PLoS ONE, 3, e2822 (2008).

Debois, D.; Hamze, K.; Guérineau, V.; Le Caër, J.P.; Holland, I.B.; Lopes, P.; Ouazzani, J.; Séror, S.J.; Brunelle, A.; Laprévote, O., In situ localisation and quantification of surfactins in a *Bacillus subtilis* swarming community by imaging mass spectrometry; Proteomics, 8, 3682–3691 (2008).

Demirev, P.A., Mass spectrometry for malaria diagnosis; Expert Rev. Mol. Diagn., 4, 821–829 (2004).

Demirev, P.A.; Fenselau, C.J. Mass Spectrom., Mass Spectrometry in Biodefense, 43, 1441–1457 (2008).

Demirev, P.A.; Feldman, A.B.; Kowalski, P.; Lin, J.S., Top-down proteomics for rapid identification of intact microorganisms; Anal Chem., 77, 7455–7461 (2005).

Demirev, P.A.; Lin, J.S.; Pineda, F.J.; Fenselau, C., Bioinformatics and mass spectrometry for microorganism identification: proteome-wide post-translational modifications and database search algorithms for characterization of intact *H. pylori*; Anal. Chem., 73, 4566–4573 (2001).

Demirev, P.A.; Ramirez, J.; Fenselau, C., Tandem mass spectrometry of intact proteins for characterization of biomarkers from *Bacillus cereus* T spores; Anal. Chem., 73, 5725–5731 (2001).

Djidja, M.-C.; Francese, S.; Loadman, P.M.; Sutton, C.W.; Scriven, P.; Claude, E.; Snel, M.F.; Franck, J.; Salzet, M.; Clench, M.R., Detergent addition to tryptic digests and ion mobility separation prior to MS/MS improves peptide yield and protein identification for in situ proteomic investigation of frozen and formalin-fixed paraffin-embedded adenocarcinoma tissue sections; Proteomics, 9, 1–15 (2009).

Dreisewerd, K.; Draude, F.; Kruppe, S.; Rohfling, A.; Berkenkamp, S.; Pohlentz, G., Molecular analysis of native tissue and whole oils by infrared laser mass spectrometry; Anal. Chem., 79, 4514–4520 (2007).

Dugas, A.J.; Murray, K.K., On-target digestion of collected bacteria for MALDI mass spectrometry; Anal. Chim. Acta, 627, 154–161 (2008).

Esquenazi, E.; Coates, C.; Simmons, L.; Gonzalez, D.; Gerwick, W.H.; Dorrestein, P.C., Visualizing the spatial distribution of secondary metabolites produced by marine cyanobacteria and sponges via MALDI-TOF imaging; Mol. BioSyst., 4, 562–570 (2008).

Fenselau, C.; Demirev, P.A., Characterization of intact microorganisms by MALDI mass spectrometry; Mass Spectrom. Rev., 20, 157–171 (2001).

Fournier, I.; Wisztorski, M.; Salzet, M., Tissue imaging using MALDI-MS: a new frontier of histopathology proteomics; Expert Rev. Proteomics, 5, 413–424 (2008).

Francese, S.; Dani, F.R.; Traldi, P.; Mastrobuoni, G.; Pieraccini, G.; Moneti, G., MALDI Imaging Mass Spectrometry, from its origins up to today: the state of the art; J. Combinat. Chem. High Throughput Screening, 12, 156–174 (2009).

Hayek, C.S.; Pineda, F.J.; Doss, O.W.; Lin, J.S., Computer assisted interpretation of mass spectra; APL Technical Digest; 20, 363–371 (1999).

Holland, R.D.; Duffy, C.R.; Rafii, F.; Sutherland, J.B.; Heinze, T.M.; Holder, C.L.; Voorhees, K.J.; Lay, J.O., Identification of bacterial proteins observed in MALDI TOF mass spectra from whole cells; Anal. Chem., 71, 3226–3230 (1999).

Holland, R.D.; Wilkes, J.G.; Rafii, F.; Sutherland, J.B.; Persons, C.C.; Voorhees, K.J.; Lay, J.O., Rapid identification of intact whole bacteria based on spectral patterns using matrix-assisted laser desorption/ionization with time-of-fight mass spectrometry; Rapid Commun. Mass Spectrom., 10, 1227–1232 (1996).

Holle, A.; Haase, A.; Kayser, M.; Höhndorf, J., Optimizing UV laser focus profiles for improved MALDI performance; J. Mass Spectrom., 41, 705–716 (2006).

Hollemeyer, K.; Jager, S.; Altmeyer, W.; Heinzle, E., Proteolytic peptide patterns as indicators for fungal infections and non fungal affections of human nails measured by matrix-assisted laser desorption/ionization time-of-flight mass spectrometry; Anal. Biochem., 338, 26–31 (2005).

Hsiesh, Y.; Casale, R.; Fukuda, E.; Chen, J.; Knemyer, I.; Wingate, J.; Morrison, R.; Korfmacher, W., Matrix-assisted laser desorption/ionization imaging mass spectrometry for direct measurement of clozapine in rat brain tissue; Rapid Commun. Mass Spectrom., 20, 965–972 (2006).

Laiko, V.V.; Baldwin, M.A.; Burlingame, A.L., Atmospheric pressure matrix-assisted laser desorption/ionization mass spectrometry; Anal. Chem., 72, 652–657 (2000).

Lemaire, R.; Desmons, A.; Tabet, J.C.; Day, R.; Salzet, M.; Fournier, I., Direct analysis and MALDI imaging of formalin-fixed, paraffin-embedded tissue sections; J. Prot. Res., 6, 1295–1305 (2007).

Li, T.Y.; Liu, B.H.; Chen, Y.C., Characterization of Aspergillus spores by matrix-assisted laser desorption/ionization time-of-flight mass spectrometry; Rapid Commun. Mass Spectrom., 14, 2393–2400 (2000).

Li, Y.; Shrestha, B.; Vertes, A., Atmospheric pressure molecular imaging by infrared MALDI mass spectrometry; Anal. Chem., 79, 523–532 (2007).

Li, Y.; Shrestha, B.; Vertes, A., Atmospheric pressure infrared MALDI imaging mass spectrometry for plant metabolomics; Anal. Chem., 80, 407–420 (2008).

Luxemborg, S.L.; Mize, T.H.; McDonnell, L.A.; Heeren, R.A., High-spatial resolution mass spectrometric imaging of peptide and protein distributions on a surface; Anal. Chem., 76, 5339–5344 (2004).

Madonna, A.J.; Voorhees, K.J.; Taranenko, N.I.; Laiko, V.V.; Doroshenko, V.M., Detection of cyclic lipopeptide biomarkers from Bacillus species using atmospheric pressure matrix-assisted laser desorption/ionization mass spectrometry; Anal. Chem., 75, 1628–1637 (2003).

Magnuson, M.L; Owens, J.H.; Kelty, C.A., Characterization of *Cryptosporidium parvum* by matrix-assisted laser desorption/ionization time of flight mass spectrometry; Appl. Environ. Microbiol., 66, 4720–4724 (2000).

Maler, T.; Klepel, S.; Renner, U.; Kostrzewa, M., Fast and reliable MALDI TOF-MS based microorganism identification; Nat. Methods, 3, i-ii (2006).

Meistermann, H.; Norris, J.L.; Aerni, H.R.; Cornett, D.S.; Friedlein, A.; Erskine, A.R.; Augustin, A.; De Vera Mudry, M.C.; Ruepp, S.; Suter, L.; Langen, H.; Caprioli, R.M.; Ducret, A., Biomarker discovery by imaging mass spectrometry: transthyretin is a biomarker for gentamicin-induced nephrotoxicity in rat; Mol. Cell Proteomics, 5, 1876 (2006).

Nelson, D.F.; McDonald, J.V.; Lapham, L.W.; Quazi, R.; Rubin, P., Central nervous system tumors, in Clinical Oncology: A Multidisciplinary Approach for Physicians and Students; ed. Rubin, P.; W.B. Saunders, Philadelphia, 1993.

Pribil, P.A.; Patton, E.; Black, G.; Doroshenko, V.; Fenselau, C., Rapid characterization of Bacillus spores targeting species-unique peptides produced with an atmospheric pressure matrix-assisted laser desorption/ionization source; J. Mass Spectrom., 40, 464–474 (2005).

Reyzer, M.L.; Hsieh, Y.; Ng, K.; Korfmacher, W.A.; Caprioli, R.M., Direct analysis of drug candidates in tissue by matrix-assisted laser desorption/ionization mass spectrometry; J. Mass Spectrom., 38, 1081–1092 (2003).

Roher, A.E.; Weiss, N.; Kokjohn, T.A.; Kuo, Y.M.; Kalback, W.; Anthony, J.; Watson, D.; Luehrs, D.C.; Sue, L.; Walker, D.; Emmerling, M.; Goux, W.; Beach, T., Increased A beta peptides and reduced cholesterol and myelin proteins characterize white matter degeneration in Alzheimer's disease; Biochemistry, 41, 11 080–11 090 (2002).

Ryzhov, V.; Hathout, Y.; Fenselau, C., Rapid characterization of spores of *Bacillus cereus* group bacteria by matrix assisted laser desorption/ionization time of flight mass spectrometry; Appl. Environ. Microbiol., 66, 3828–3834 (2000).

Schwartz, S.A.; Weil, R.J.; Thompson, R.C.; Shyr, Y.; Moore, J.H.; Toms, S.A.; Johnson, M.D.; Caprioli, R.M., Proteomic-based prognosis of brain tumor patients using direct-tissue matrix-assisted laser desorption ionization mass spectrometry; Cancer Res., 65, 7674–7681 (2005).

Seeley, E.H.; Oppenheimer, S.R.; Chaurand, P.; Caprioli, R.M., Enhancement of protein sensitivity for MALDI imaging mass spectrometry after chemical treatment of tissue sections; Am. Soc. Mass Spectrom., 19, 1069–1077 (2008).

Shimma, S.; Furuta, M.; Ichimura, K.; Yoshida, Y.; Setou, M., A novel approach to in situ proteome analysis using chemical inkjet printing technology and MALDI-QIT-TOF tandem mass spectrometer; J. Mass Spectrom. Soc. Jap., 54, 133–140 (2006).

Shrestha, B.; Li, Y.; Vertes, A., Rapid analysis of pharmaceuticals and excreted xenobiotic and endogenous metabolites with atmospheric pressure infrared MALDI mass spectrometry; Metabolomics, 4, 297–311 (2008).

Stoeckli, M.; Chaurand, P.; Hallahan, D.E.; Caprioli, R.M., Imaging mass spectrometry: a new technology for the analysis of protein expression in mammalian tissues; Nat. Med., 7, 493–496 (2001).

Tan, P.V.; Laiko, V.V.; Doroshenko, V.M., Atmospheric pressure MALDI with pulsed dynamic focusing for high-efficiency transmission of ions into a mass spectrometer; Anal. Chem., 76, 2462–2469 (2004).

Touboul, D.; Roy, S.; Germain, D.P.; Chaminade, P.; Brunelle, A.; Laprévote, O., MALDI-TOF and cluster-TOF-SIMS imaging of Fabry disease biomarkers; Int. J. Mass Spectrom., 260, 158–165 (2007).

Vaidyanathan, S.; Gaskell, S.; Goodacre, R., Matrix-suppressed laser desorption/ionization mass spectrometry and its suitability for metabolome analysis; Rapid Commun. Mass Spectrom., 20, 1192–1198 (2006).

Verbeck, G.F.; Ruotolo, B.T.; Sawyer, H.A.; Gillig, K.J.; Russell, D.H., A fundamental introduction to ion mobility mass spectrometry applied to the analysis of biomolecules; J. Biomol. Tech., 13, 56–61 (2002).

Vödisch, M.; Albrecht, D.; Lessing, F.; Schmidt, A.D.; Winkler, R.; Guthke, R.; Brakhage, A.A.; Kniemeyer, O., Two-dimensional proteome reference maps for the human pathogenic filamentous fungus *Aspergillus fumigatus*; Proteomics, 5, 1407–1415 (2009).

Yao, Z.P.; Fenselau, C., Mass spectrometry-based proteolytic mapping for rapid virus identification; Anal. Chem., 74, 2529–2534 (2002).

Part III

Protein Samples: Preparation Techniques

6

Conventional Approaches for Sample Preparation for Liquid Chromatography and Two-Dimensional Gel Electrophoresis

Vesela Encheva[1] and Robert Parker[2]

Department for Bioanalysis and Horizon Technologies, Health Protection Agency Centre for Infections, London, UK

[1] Present address: Research and Technology Department, Laboratory of the Government Chemist (LGC), Teddington, UK

[2] Present address: University of British Columbia, Center for High-Throughput Biology, Vancouver, Canada

6.1 Introduction

The term 'proteomics' was first used in 1995 and was defined as the large-scale characterisation of the entire protein complement of a cell, tissue or organism. Today the growth of proteomics is a direct result of advances made in large-scale nucleotide sequencing and in the development of highly sensitive methods for protein identification such as mass spectrometry (MS). However, the greatest challenge for proteomic technologies is the inherently complex nature of the cellular proteomes. To uncover this complexity two main technologies for protein separation are used in conjunction with MS: two-dimensional gel electrophoresis (2D GE) and liquid chromatography (LC).

The application of 2D GE for protein profiling of complex mixtures is what marked the beginning of the proteomics era and the first 2D GE profiles of bacteria were created even before the first fully sequenced genome became available. The first protein study that could be referred to as proteomics of microorganisms was reported in 1966 by O'Farrell

Mass Spectrometry for Microbial Proteomics Edited by Haroun N. Shah and Saheer E. Gharbia
© 2010 John Wiley & Sons, Ltd

who attempted to characterise the proteins expressed by *E. coli*. However, prior to the availability of MS, 2D GE was mainly a descriptive technique as the identities of the separated proteins could not be revealed. Early application of 2D GE used the method simply as a sensitive technique for the differentiation of closely related microbial isolates on the basis of protein charge and molecular weight. 2D GE has been widely used to discriminate between related isolates of several bacterial species, e.g. *Haemophilus* spp (1), *Lysteria monocytogenes* (2) *Mycoplasma arthitidis* (3) and, more recently, *Neisseria meningitidis* (4).

The advances made in the field of MS for identification of electrophoretically separated proteins allowed the technique of 2D GE to develop further. Nowadays, 2D GE is rarely used as a tool for epidemiological studies and instead is making its mark on elucidating the mechanisms of bacterial pathogenicity, antibiotics resistance, physiology and vaccine development. 2D GE is a state-of-the-art technology which is not only the oldest but also the most widely used approach for protein separation. However, the technique has several limitations especially for the characterisation of hydrophobic, basic or high-molecular weight proteins. Therefore, non-gel based approaches such as LC are becoming the focus of interest for the characterisation of protein species which are underrepresented on 2D gels. However, non-gel based techniques fail to correlate changes observed on the peptide level to individual protein isoforms which is accomplished by 2D GE. Therefore, with the increased popularity of non-gel based separations, the traditional 2D gel approach will be more and more supplemented but not replaced by new technologies.

Recent advances in LC-MS have enabled the development of highly robust platforms for the rapid analysis of complex mixtures of proteins and peptides. Developments in the speed of the chromatographic separation and duty cycle of modern tandem mass spectrometers now allows high throughput processing of hundreds of samples on a weekly basis. Even with the rapid advancement in these technologies, accurate and reproducible results are still highly reliant on good sample preparation procedures being adopted prior to injection. With LC-MS platforms now widely accessible to microbiologists in both basic research and clinical environments it is essential that procedures are available to meet the needs of this dynamic research community.

Regardless of the separation method used, the sample preparation is a fundamental first step in any proteomics workflow. Both 2D GE and LC are strongly dependent on the quality of the protein extract in order to produce satisfactory results. However, it appears that the process of sample preparation has been somehow overlooked and the majority of the effort from academia and industry has gone into the development of the instrumentation for MS and protein separation.

Furthermore, there are several publications reviewing the subject of sample preparation for protein studies but the emphasis is largely on eukaryotic cells and tissues, while sample preparation from microorganisms is barely discussed (5, 6). Nevertheless, there are numerous proteomic investigations on pathogenic microorganisms reported in the literature and various sample preparation methods have been developed independently by different research groups. Similarly in our laboratory we have optimised our own procedures for cell disruption, protein solubilisation and trypsin digestion for the analysis of different bacterial species, some of which we have published over the years.

However, it is difficult to find a comprehensive compilation of methods and this chapter attempts to partially fill this gap. Here various strategies and protocols for proteome characterisation of Gram-positive and Gram-negative microorganisms are reviewed and several procedures favoured by us due to their robustness and relative ease of use are discussed.

6.2   Cell Lysis Methods

6.2.1   Mechanical Lysis

6.2.1.1   Cell Homogenisation

Mechanical homogenisation can be used to prepare any biological material as a sample for proteomic analysis. There are many different types of homogenisers used in the literature including the mickle machine (Mickle laboratories), the ribolyser (Hybaid) and more recently the FastPrep system (MP Biomedicals). Cell lysis is usually achieved in the presence of beads of various materials and sizes. It appears that each research group uses its own protocol where different cell numbers, lysis solutions or duration of homogenisation steps are applied. This makes it impossible to compare equipment or the methodology in order to recommend the best solution. We favour the use of the FastPrep system with a combination of glass beads (<105 μm in size) and solubilisation cocktail containing detergents/chaotropes. This has proven to be a robust universal method which gives satisfactory result for both Gram-positive and Gram-negative microorganisms.

Tip: The standard lysis solution which is well suited for the majority of samples contains 7M urea, 2M thiourea, 4% 3-[(3-cholamidopropyl) dimethylammonio]-1-propanesulfonate (CHAPS), 50 mM DTT and 20 mM Tris. However, when performing homogenisation in the presence of urea overheating of the samples should be avoided. Therefore, we usually homogenise the lysates for 1 min and cool them on ice for 1 min. For cells difficult to lyse this can be repeated up to three times. An additional step of enzymatic lysis (described below) can also be advantageous for Gram-positive microorganisms.

6.2.1.2   Freeze–Thaw

In this extraction method the formation of ice crystals during the freezing process is utilised to disintegrate the cell envelope and release the cellular contents. The method is rapid, does not require any specialised equipment and, also importantly, does not introduce any impurities into the sample. In our experience, several cycles of freezing and thawing are required and the method is usually combined with other mechanical or chemical lysis procedures to achieve satisfactory results. However, in the case of *Streptococcus pneumoniae* due to the presence of a thick cell wall composed of several peptidoglycan layers the method could not produce enough material for 2D GE analysis. Despite that, three repeated cycles of freezing and thawing were performed in the presence of CHAPS, urea and thiourea; the method failed to lyse the majority of the cells (7). It has also been reported that some microbial cells, preconditioned in starvation medium, are resistant to the freeze–thaw cycles, as in the case of *Vibrio parahaemolyticus* (8).

6.2.1.3   Sonication (Ultrasonication)

Sonication involves applying sound energy (usually ultrasound) in order to agitate particles in a sample. In the laboratory, it is usually applied using an *ultrasonic bath* or an *ultrasonic probe*. Sonication has various applications including degassing, homogenising and emulsifying liquids. For biological applications ultrasonic homogenisers, also known as sonificators or sonicators, are mainly used to disrupt the cell membranes and release the cellular contents. Cell disruption via sonication has been the method of choice for many proteomic investigations when working with cells that are difficult to lyse. Sonication has been successfully used to characterise the soluble and membrane fractions of *Mycobacterium leprae* (9, 10), *Bacillus subtilis* (11) and *Bacillus anthracis* resulting in the detection of several hundred bacterial proteins. In the case of the latter a full reference map was created containing a total of 534 identified proteins (12). In a different study, sonication was used to selectively extract the membrane fraction of *B. anthracis* which allowed identification of more than 80 protein spots including a large number of S-layer proteins (13). Sonication has also been applied for the preparation of exosporium samples of the fully virulent *B. anthracis* Ames strain. The sonication was performed for 6–7 min causing partial fragmentation of the exosporium without disturbing the spores (14). The prepared extracts were further characterised by sodium dodecyl sulphate polyacrylamide gel electrophoresis (SDS-PAGE) and the protein bands identified using N-terminal sequencing or MS which revealed several spore coat proteins. Sonication has undoubtedly proven very useful for the preparation of protein extracts as it can achieve a high percentage of cell lysis and very good protein yields even from Gram-positive microorganisms.

Tip: In one of our recent investigations we tested the performance of several lysis strategies on S. pneumoniae. Despite the fact that the highest protein yield was obtained when sonicating the samples, the reproducibility of the method was questionable. This is most likely due to the overheating of the extracts during sonication which is difficult to prevent or keep constant from sample to sample. As a result, a high standard deviation in the number of spots detected was observed amongst the sonicated samples and another more reproducible method was chosen for the analysis (7).

6.2.1.4   French Pressure Cell

The pressure homogenisation system, also known as the French pressure cell, is one of the most effective systems for cell disruption of bacterial cells. It has been successfully used to extract proteins from Gram-positive bacteria such as *Staphylococcus aureus* and *Bacillus anthracis*. The French pressure cell is very often applied for the preparation of cell membrane fractions, e.g. for the study of *E. coli* K^+/H^+ transmembrane transport.

A high-percentage of cell lysis can be achieved using the French pressure cell but the approach has several drawbacks. The equipment is relatively expensive and, from our experience, is not particularly easy to use. The technique is also not amenable to automation or high throughput analysis as only one sample at a time can be passed through the pressure cell. Furthermore, a significant amount of sample loss can occur and there is a potential for cross contamination. Health and safety considerations are also an issue, especially when working with pathogenic microorganisms.

> *Tip: For higher protein yield and more efficient cell lysis multiple passages through the French pressure cell are recommended. Performing multiple (2–6) passages can significantly increase the percentage of broken cells and the total protein yield for cells that are difficult to lyse such as S. aureus.*

6.2.2   Chemical and Osmotic Lysis

Osmotic and chemical lysis are both gentle lysis methods which are well suited for mammalian cells which lack a cell wall and capsule but have limited application in microbiology. There are a few proteomic studies utilising strong solubilising reagents to lyse Gram-negative bacteria and a satisfactory level of cell lysis appeared to be achieved (10).

> *Tip: In our experience an additional step of mechanical cell disruption or enzymatic lysis is recommended even when working with Gram-negative bacteria such as S. enterica.*

6.2.3   Enzymatic Lysis

The difficulties of performing lysis of bacterial cells arise from the presence of the peptidoglycan cell wall. This is particularly true for Gram-positive microorganisms which possess a very thick cell wall built of multiple layers of peptidoglycan. Therefore, the use of enzyme to digest and remove the cell wall greatly facilitates the subsequent lysis steps. The most commonly used enzyme is lysozyme, also known as muramidase or N-acetylmuramide glycanhydrolase. The lysoszyme functions by hydrolysing the glycosidic bond that connects *N*-acetylmuramic acid with the fourth carbon atom of *N*-acetylglucosamine, the building units of the bacterial peptidoglycan. As a result the cells are converted into protoplast surrounded only by a plasma membrane. We have also described the successful use of another muralitic enzyme, mutanolysin (which has the same substrate specificity as lysozyme) for *S. pneumoniae* and the combined use of mutanolysin and lysozyme for the lysis of *S. mutans* has also been reported (7, 15). However, it appears that the use of lysoszyme is not suitable for *S. aureus* as it requires close to 12 h to exert its action (16). For *S. aureus,* due to the complex nature of its peptidoglycam, the use of lysostaphin is a better alternative.

> *Tip: The removal of the above mentioned enzymes from the sample can be a difficult and time consuming process. In an attempt to remove lysozyme from B. cereus lysates 30 washes were performed resulting in the loss of large proportion of the harvested protoplasts. However, a faint lysozyme band was still present on the resulting Nu PAGE profiles. Using less of the enzyme can potentially minimise the level of contamination.*

6.3   Sample Preparation for 2D GE

6.3.1   Removal of Interfering Substances

Upon cell disruption, all cellular contents are released into the extraction medium resulting in a crude lysate containing a vast mixture of components. Unless present in concentrations that are low enough to be negligible, interfering substances can have a detrimental effect on the subsequent separation steps such as isoelectric focusing (IEF) or gel electrophoresis. While insoluble material and particulates can easily be removed by a single centrifugation step, many other nonproteinaceous substances remain in solution. Therefore, compounds such as salts, proteases, nucleic acids, polysaccharides and lipids must be removed prior to further analysis.

6.3.1.1   *Protection from Proteases*

The rupture of cell structures provokes the liberation of various proteases which begin to exert their action on the extracted proteins. This is less of an issue with bacterial cells, in comparison with eukaryotic cells, due to the absence of cellular organelles such as the lysosomes. However, bacteria do contain proteases and the addition of protease inhibitors is recommended in order to keep the integrity of the extracted proteins. If no preventive measures are taken common effects of proteolysis are the presence of artifactual spots and loss of high molecular weight proteins (17). Some proteases are resistant to denaturation, although solubilising the proteins in a strong denaturing agent may prevent their action. It has been demonstrated that while active in a high concentration of urea the proteases are inhibited by the addition of thiourea to the extraction buffer (18). They are also less active at lower temperatures so keeping the samples cool throughout the extraction process is advisable. In addition, proteolysis can often be inhibited by preparing the sample in Tris base, basic carrier ampholytes or by precipitation in trichloroacetic acid (TCA) or TCA/ acetone (6).

Protease inhibitors are usually added when preparing protein extracts from bacteria. Commonly used inhibitors include phenylmethylsulfonyl fluoride (PMSF), aminoethyl benzylsulfonyl fluoride (AEBSF), EDTA, pepstatin, aprotinin, benzamidine and leupeptin. Microbial proteases are predominantly extracellular and can be classified into four groups based on the catalytic residue of their active site. The four groups are: serine proteases, cysteine proteases, aspartate proteases and metalloproteases. However, aspartate proteases are rare in bacteria and to date none have been reported in pathogenic microorganisms. Metalloproteases, on the other hand, seem to be a common feature in most bacterial pathogens (19). However, a quick review of the literature reveals that the preferred protease inhibitor in bacterial proteomics is PMSF, which inhibits serine and cysteine proteases rather than EDTA, which inhibits metalloproteases by chelating their metal ions. Only a few studies use EDTA in the extraction buffer (20). Finally, protease inhibitors must be used with caution as they can modify proteins and introduce charge artifacts thus hampering data interpretation and further peptide analysis (21).

6.3.1.2   *Removal of Salts*

In our experience bacterial cell lysates do not contain excessive amounts of salts unless working with halophilic microorganisms. High amounts of salt will increase conductivity of the IEF gel thus prolonging the focusing time. In extreme cases the IEF process may stop due to salt fronts. This is particularly a problem when samples are applied via the in-gel rehydration method, whereas higher salt concentrations are better tolerated by the cup-loading method. In such cases low voltages (~150 V) must be applied for several hours and the filter electrode pads need to be replaced three to four times during the duration of the IEF (17). Salt removal can also be achieved by spin dialysis or precipitation but both methods are accompanied by protein losses.

6.3.1.3   Removal of Nucleic Acids

Nucleic acids are charged molecules that bind to proteins and carrier ampholytes via electrostatic interactions thus preventing the process of IEF. As big polymers they increase the viscosity of the sample and may clog the pores of the acrylamide gels. Their presence also interferes with the recovery of DNA or RNA binding proteins (17). The addition of protease-free nucleases (DNAases and RNAases) is the most successful approach for nucleic acid removal. Precipitation in TCA or ultracentrifugation are other alternative methods, but high-speed centrifugation can remove high molecular weight proteins as well (22).

Tip: DNAase and RNAase are most commonly added to a final concentration of 1 and 0.25 mg ml^{-1}, respectively. In order to exert its action DNase requires Mg^{2+} ions so the addition of up to 20 mM $MgCl_2$ is recommended.

6.3.1.4   Removal of Lipids

Lipids are widely present in bacteria as the main component of the outer and inner membranes. Bacterial membranes consist of a double layer of phospholipids which have negatively charged polar heads and account for about 20–30% of the dry weight of the cell. The outer membrane of Gram-negative bacteria also contains large amounts of a unique lipid, lipopolysaccharide (LPS). LPS almost completely replaces the phospholipids in the surface layer of the membrane and is chemically quite different to a phospholipid. LPS is a very complex molecule containing six or seven saturated fatty acid residues and has a molecular weight of over 10000 Da. The presence of such high molecular weight charged lipids can reduce the protein solubility and alter their pI and molecular weight. Their presence can also deplete the sample from detergents and therefore their removal is essential (17). Precipitation with acetone or TCA/acetone removes lipids sufficiently. Other organic solvents used for delipidation include ethanol and, more recently, a combination of acetonitrile and 1% trifluoroacetic acid (TFA) (23). However, with this approach severe protein losses may be experienced either because some proteins are soluble in organic solvent or the precipitated proteins are difficult to resolubilise (5). The use of centrifugal filter devices in the presence of CHAPS allows for successful lipid and salt removal and appears to be advantageous to the more classical approach of boiling with SDS and dialysis (24).

6.3.1.5   Removal of Polysaccharides

The presence of contaminating polysaccharides in the protein extracts of bacteria is a common problem due to the release of cell wall and capsule material during cell lysis.

The major component of the bacterial cell wall is peptidoglycan. Peptidoglycan is a heteropolymer made up of substituted sugars and amino acids and synthesised uniquely by prokaryotes. The cell wall of Gram-positive microorganisms contains multiple layers of peptidoglycan which not only makes their lysis difficult to achieve (see above) but also results in high amounts of polysaccharide in their cell extracts. The cell walls of Gram-negative bacteria are thinner but also contain peptidoglycan.

In addition, some prokaryotes synthesise an outer polysaccharide layer referred to as a 'capsule'. The structure of the polysaccharide capsule varies widely amongst different species, serotypes and strains which makes the design of a universal enzyme digestion protocol not feasible.

The removal of polysaccharides is essential as they are huge molecules that can clog the pores of PAGE gels resulting in streaky and distorted profiles. Similarly to nucleic acids, the bacterial polysaccharides are negatively charged allowing them to bind proteins via electrostatic interactions. However, their lesser charge density makes them more difficult to precipitate and their removal less simple compared with nucleic acids (25). As a simple rule, ultracentrifugation or precipitation with TCA will remove high molecular weight polysaccharides (5).

Tip: *In our experience the interaction of polysaccharides with proteins can be prevented by solubilisation in SDS or high pH (26).*

6.3.2   Solubilisation Strategies

Another obstacle for obtaining good quality and reproducible 2D GE profiles is in maintaining the proteins in a soluble state during IEF. Solubilisation can be defined as the process of disrupting the noncovalent or covalent interactions of the proteins with other substances which may or may not be proteinaceous. Such noncovalent interactions are hydrogen bonding, ionic interactions, dipole moments, hydrophobic interactions and van der Waals forces (27). For this reason, special attention has to be paid to the cell lysis conditions, the choice of adequate detergents, chaotropes and the amount of reducing agent used.

The role of the chaotropes is to disrupt hydrogen bonding, leading to protein unfolding and denaturation. Urea has become the universally accepted component included in the rehydration solution to aid the solubility of the proteins during the process of IEF. As a chaotrope, urea decreases the strength of the hydrogen bonds and simultaneously increases the water solubility of the hydrophobic compounds (28). Therefore the presence of urea considerably decreases the precipitation close to the pI. Another chaotrope added to the reswelling solution in order to aid the solubility is thiourea. Thiourea, when used in combination with a high concentration of urea, dramatically increased the recoveries of less soluble proteins, such as integral membrane proteins (28, 29). However, the use of urea and thiourea also induces denaturation and thus exposes the hydrophobic groups of the proteins to the solvent. This effect can initiate hydrophobic interactions between the proteins so urea by itself is not sufficient to solubilise the proteins completely. Therefore detergents are almost always added to the urea based reswelling solutions (27).

Detergents are included in the reswelling solution for IEF to act synergistically with chaotropes. They help by preventing the hydrophobic interactions that occur due to the exposure of hydrophobic domains generated by the action of the chaotropes (29). The detergents used for IEF should not have an electric charge and therefore only nonionic and zwitterionic detergents are compatible. Therefore, the use of SDS, an ionic detergent, is not recommended. However, SDS is a very strong solubilising agent and sometimes can be used for the initial solubilisation prior to IEF (27, 29). Low amounts of SDS can be tolerated during IEF provided that a high concentration of urea, or nonionic or zwitterionic detergents are present to ensure complete removal of SDS from the proteins during IEF (27). The detergents commonly used are sulfobetaines. Standard sulfobetaines with linear alkyl tails have received limited application because of their incompatibility with high concentrations of urea. Therefore the most widely used type of sulfobetaine detergent is CHAPS which consists of a bulky, polycyclic tail with a sulfobetaine head and has been shown to be much more efficient than triton-type detergents (30). More recently, new types of sulfobetaines have been developed such as ASB-14 and SB 3–11, which have been used for the solubilisation of highly hydrophobic membrane proteins (31, 32).

One of the ways to improve solubility is by adding reducing agents to break the existing disulfide bonds. This is often a critical step in the solubilisation of proteins with numerous disulfide bonds that stabilise their folding. Cyclic reducing agents such as dithiothreitol (DTT) or dithioerythritol (DTE) are the most commonly used agents for protein solubilisation during IEF. The mode of action of these reagents is an equilibrium reaction, so the loss of the reducing agent due to migration in the pH gradient (DTT/DTE are weakly acidic) can lead to reoxidation of the disulfides and contribute to horizontal streaking (29). This problem can be solved by increasing the concentration of DTT (27) or by the use of an alternative type of reducing agent, such as phosphines. Phosphines offer an alternative to thiol reducing agents because they operate in a stoichiometric reaction, allowing the use of a low concentration of the reagent (2 mM). The use of tributyl phosphine (TBP) has been demonstrated to enhance the resolution of keratin proteins (33). Unfortunately, short chain length phosphines, including TBP, are volatile, toxic and highly flammable in concentrated stocks.

Although ampholytes are, in theory, required only for the IEF process, their presence during sample preparation procedures is beneficial as they help with the solubilisation process and inhibit the interaction between sample proteins and the immobilines present in the immobilised pH gradient (IPG) strips. Their concentration is calculated considering the pI range used and the chosen procedure (6).

6.3.3   Sample Preparation for Difference in Gel Electrophoresis (DIGE)

A bottleneck for performing high-throughput 2D GE experiments is the time consuming nature of the technique especially when image analysis is concerned (see Chapter 18). To shorten this laborious procedure and to improve the accuracy of the analysis Ünlu *et al.* (34) have developed a method called Difference in Gel Electrophoresis (DIGE) (see Chapter 1). DIGE is a significant development in 2D GE, which utilises fluorescent tagging of three protein samples with three different fluorescent dyes. The labelled proteins are then simultaneously run on the same gel and post-run fluorescence imaging of the gel is used to create three superimposed images. This technique circumvents the need to compare several 2D gels and a trademark software program DeCyder has been developed to perform the data analysis. The key benefit of this approach is that it enables the incorporation of an internal standard with every gel which compensates for gel-to-gel variability and allows for the detection of even subtle expression changes. The DeCyder software utilises the internal standard to produce high quality quantitative data with statistical confidence reflecting true biological changes rather then experimental variation. The potential of the technique and the software was first demonstrated in a study using *E.coli*, where 179 differentially expressed proteins were identified following treatment with benzoic acid (35).

Initially applied to characterise the adaptive mechanisms of bacteria, DIGE has since found numerous applications in all areas of science. The technology is promising for the identification of expression changes and regulatory networks which is an area of great interest in microbiology. Therefore, it is quite surprising that DIGE has been utilised in only a handful of microbiological investigations. To date DIGE has proven successful for determining expression differences between morphological variants of *Spirulina platensis* (36, 37), plasmid bearing and cured strains of *B. anthracis* (37) and biofilm against planctonic cultures. It has also been applied to detect protein expression changes associated with different phase growth or growth on different substrates (38, 39). Furthermore, DIGE can be potentially useful in quality control studies. This has been exploited by Vipond and associates in a study to quantify the porin proteins in the vaccine of *Neisseria meningitidis* (40). The wide linear dynamic range of the fluorescent dyes and the multiplexing nature of the technology provided the means to assess the uniformity of vaccine batches manufactured by different suppliers and revealed quantitative variation undetected previously.

However, DIGE cannot simply be integrated in every proteomics workflow as not every sample is suitable for labelling with the cyanine (Cy) dyes. The sample preparation is tricky and often a time consuming process, which partially explains the limited use of the technology. In order to successfully label all proteins in the sample several requirements need to be met. The pH during labelling has to be between 8.0 and 9.0. Lower pH will result in little or no labelling while higher pH will result in more than two lysine residues per protein being labelled. Therefore a suitable lysis buffer is required capable of maintaining the pH at 4 °C, as the labelling is carried out at this temperature. Usually Tris, HEPES or bicarbonate based buffer is used in a concentration no higher than 30 mM, so it does not interfere with the IEF process. The sample lysis buffer should be free of any primary amines such as ampholytes as they will deplete the sample from the dyes resulting in fewer labelled proteins. Thiol reagents such as DTT are also detrimental for the labelling and need to be excluded from the lysis buffer or removed prior to labelling. The 2D clean up kit (GE Healthcare) will sufficiently remove DTT at concentrations up to 60 mM. However, amounts higher than this cannot be fully removed and therefore should be avoided.

To avoid background when scanning the images, all solutions used for DIGE need to be prepared with high quality reagents (HPLC grade if possible). Using low quality reagents results in interfering signal from contaminants which usually emit in the Cy3 channel.

The protein concentration of the sample needs to be between 5 mg ml^{-1} and 10 mg ml^{-1} and there are a number of methods to achieve this when more diluted extracts are encountered. Precipitation of the proteins is a common solution and in our laboratory we have favoured the 2D clean up kit provided by GE Healthcare. It uses a proprietary protocol to precipitate and solubilise the extracted proteins in DIGE compatible reagents and has proven to be quick, easy and highly reproducible. Other possibilities are TCA or TCA/

acetone precipitation but difficulties with protein resolubilisation in DIGE solubilisation buffer have been encountered (40). The problem can also be overcome by using centrifugal filter devices such as Centricon YM-3. This approach has been applied for the preparation of outer membrane vesicles of *N. meningitidis* and has resulted in good quality DIGE profiles (40). However, this method is time consuming as it requires 1 h for the centrifugation step at 4500 rpm which also takes place at room temperature, thus increasing the possibility of proteolysis.

Another requirement when performing labelling is to keep the Cy dyes as the limiting component in the reaction. For this reason, a ratio of 50 µg of protein to 400 pmol of Cy dye is recommended. This is to ensure that only 1–2% of lysine residues are labelled (minimal labelling). In this way, high sensitivity as well as a linear dynamic range of five orders of magnitude is achieved.

Recently Cy dyes with different chemistry have been developed which label a higher percentage of lysine residues, also known as saturation labelling. This method is aimed at applications with few samples which require high sensitivity but has not been used for microbiological applications.

Tip: Most often than not background signal is caused by low quality SDS or Tris base. We have also observed a strong interfering signal in the Cy5 channel when using low quality HCl for pH adjustment. In addition, methanol can also interfere with detection so should not be used for cleaning the low-fluorescent glass plates.

6.3.4   Preparation of Environmental Samples

Preparation of protein extracts from environmental samples has proven challenging and impeded the progress of this field. Two methods for extracting proteins from water, soil and sediments have been developed (41). In the first method, microbial proteins were extracted from 1 g of soil or sediment by boiling the samples while in the second method the same quantities of environmental sample are incubated for 1 h at 0 °C followed by four 10-min freeze–thaw cycles. It appeared that the first method was suitable when working with wastewater while the second method performed better with soil and sediments (42, 43). A multi-step protocol for the extraction and purification of the entire proteome from a laboratory-scale activated sludge system has been developed. The protocol involves a range of washing, solubilisation and precipitation steps including lysis with the French pressure cell. The extracted proteins were separated using 2D GE (44).

A limiting factor when preparing soil samples is their high humic acid content. A method combining 0.1 M NaOH treatment and phenol extraction has been published which successfully separates the proteins from the humic organic matter. The treatment with NaOH appears to release the humic acids and soil minerals and simultaneously disrupts the microorganisms. The resulting samples were successfully separated using SDS-PAGE and 2D GE and several proteins identified. The identification of enzymes such as chlorocatechol dioxygenases is consistent with the metabolic pathways expected to be expressed in those conditions (45).

6.4   Fractionation Strategies

6.4.1   Surface Associated Proteins

Proteins from bacterial membranes are notoriously difficult to characterise using the traditional solubilisation and extraction protocols and therefore are underrepresented in the majority of published proteome maps. However, surface proteins mediate bacterial communication with the environment and facilitate the uptake of nutrients as well as the disposal of toxic by-products of the metabolic activities of the cell. They are also involved in crucial pathogenic processes such as adhesion, colonisation, motility, injection of toxin and proteases into the host cells as well as the removal of antibiotics thus conferring resistance to their action. As such they are regarded as being of central importance as potential drug targets and as the basis for the development of novel vaccines. Several methods compatible with 2D GE have been developed for the selective enrichment of bacterial membrane proteins using differential solubility, chemical enrichment or surface labelling via biotinylation (46). For a more detailed description of the various approaches for bacterial surface proteomics the reader is referred to an excellent review by Cordwell (46). Additionally, Chapter 8 in this book describes a novel proprietary technology developed by Nanoxis (Sweden) specifically designed for the extraction and characterisation of membrane proteins.

6.4.2   Secreted Proteins

The analysis of secreted proteins presents a different challenge to current proteomic techniques. Unlike membrane proteins secreted proteins are readily soluble and therefore relatively easy to separate by IEF. However, as they are usually secreted in very low concentrations in the culture media their recovery is difficult to achieve. In addition the preparations are usually heavily contaminated with cytosolic proteins liberated in the culture media upon lysis of dead cells or during the centrifugation steps. Furthermore, the culture media is usually rich in salts and many other compounds that interfere with most proteomic techniques, therefore selective precipitation of proteins is mandatory for the success of their further analysis.

The most common method used for the purification of secreted proteins is precipitation with TCA. The culture supernatant is mixed with TCA to a final concentration of 10–13% and incubated at $4\,^{\circ}C$ overnight with constant stirring (47). Increasing concentrations of TCA, up to 40%, have been tested but do not deliver higher protein yield (47). The optimal concentration of TCA is therefore 10% as it fits the U-shaped precipitation curve as described by Sivaraman *et al.* (48). Good results are also obtained when combining TCA with acetone or with sodium deoxycholate (DOC). The optimal yield is achieved when using 2.5 volumes of acetone and 10% w/v TCA but higher concentrations of the precipitating agents do not seem to result in higher protein yield. The most efficient precipitation method appears to be a mixture of 0.03% DOC and 10% TCA (49). DOC supports the precipitation effect of TCA by hydrophobic interactions with the proteins and is appropriate especially at low protein concentrations, e.g. $1\,\mu g\,ml^{-1}$ (49). Polyethylene glycol (PEG) precipitation has also been described for the proteome of *C. acetobutylicum* (47). The efficiency of PEG increases with the increasing size of the polymer and PEG 6000 has been shown to perform well but the protein yield was lower in comparison with TCA/ DOC or acetone based precipitation.

Tip: When precipitating the secretome from the culture media large starting volumes (0.5–2 l) are required. Therefore, despite the fact that acetone precipitation is efficient it is not always practical as at least 2.5 volumes of acetone need to be added. Therefore, TCA is our preferred precipitation agent when working with large volumes of culture. However, a few washing steps with methanol, ethanol or acetonitrile are recommended to remove the acid followed by a brief wash with distilled water.

6.5 Sample Preparation for Liquid Chromatography coupled with Mass Spectrometry (LC-MS)

This section describes some of the basic principles for successfully preparing protein samples from microorganisms for LC-MS focusing on some of the key concepts of sample preparation for nano-LC scale separation and electrospray source ionisation (ESI)-MS (LC-MS) of peptides.

6.5.1 Brief Background to Protein Identification by LC-MS

Protein identification by LC-MS is most commonly carried out by a process referred to as bottom-up proteomics (50) (see Chapter 4). In this process isolated proteins are initially fragmented by enzymatic digestion with a highly specific protease or chemical process. The resultant peptides are then characterised by ESI coupled with a tandem mass spectrometer. The development of ESI-MS revolutionised the analysis of proteins by enabling the production of multiply charged ions and its relatively easy coupling with high pressure liquid chromatography (HPLC) apparatus (51). LC provides a front end solution to protein separation, allowing the online separation and characterisation of complex samples. Direct coupling of peptide separation to ESI-MS is limited to a few liquid based chromatographic techniques, the most common are reversed and normal phase chromatography and capillary electrophoresis (CE) (52). The most commonly performed separation is reversed phase separation using a stationary phase comprised of a nonpolar sorbant which mediates hydrophobic interactions with amino acid side chains. The mobile phase is a polar organic solvent which is applied as a gradient to disrupt these electrostatic interactions. Commonly used solvents are acetonitrile, methanol and isopropanol. During the gradient peptides elute in discreet populations (peaks) determined by hydrophobicity (most polar first to least polar species last). Subsequently, peptides undergo phase transition to form gaseous ions by the process of ESI, enabling them to be manipulated by a mass spectrometer (52). Characterisation by tandem MS is an information rich method resulting in two types of mass spectra. Initially survey scanning results in a broad mass/charge (*m/z*) spectrum in which the charged peptides can be observed. The mass/charge ratio and intensity information acquired in the survey scan are then used to make data dependent decisions about which ions are to be further characterised (53, 54). Subsequently ions are isolated and fragmented to produce a characteristic series of daughter ions which are then scanned over a predetermined mass range. This type of acquisition is known as MS^n as it can be repeated for several generations on *de novo* daughter ions. Generally only MS^1 is required to gain sufficient structural information for identification of peptides (55). For other studies regarding post translation modification further fragmentation may be required to accurately characterise these peptides (56). Protein identification is achieved by an error tolerant matching process, where the raw or preprocessed precursor and fragment mass information is aligned with theoretical spectra for known proteins and amino acid combinations. The confidence of this match at a peptide and protein level is assigned using various cross correlation and homology base algorithms and probability scores (57) (see Chapters 3 and 17).

6.5.2 Pitfalls of Poor Sample Preparation in LC-MS of Peptides

Confident protein identification and quantification requires high quality tandem mass spectra. Accurate and intense spectra of both the precursor peptide and subsequent daughter fragments are essential prerequisites to successful identification. If signal or ion current for the peptide is poor then weak or no identification is possible due to low intensity fragment spectra and low signal to noise ratio of the precursor peptide. Poor signal to noise ratio for eluting species also prevents good data dependent decisions being made by the acquisition software, and can drastically reduce the number of peptides identified. Ion suppression (IS) is a major problem limiting the sensitivity of LC-MS and effecting accurate quantitation (58–60). IS occurs due to the limited energy available at the electrospray source and the finite number of charges that can be stored in the mass analyser, thus if one species dominates it will suppress the ionisation and detection of lower abundant species. Ion suppression can also result during the HPLC phase and is mainly due to overloading of analytes or nonpolar contaminants binding to ligands on the stationary phase. The result is an overall reduction in column capacity and a subsequent reduction in dynamic range for the analytes of interest. Thus problems associated with ion suppression should be taken seriously when reviewing data, especially if quantitation is required. Poor sample preparation ultimately leads to reduced sensitivity and poor signal to noise ratio for both MS and MS^n spectra due to the introduction of unwanted contaminants into the system. In the worst of cases and unfortunately all too common, lack of adequate sample preparation procedures can result in lengthy instrument down times due to excessive contamination of the mass spectrometer or LC apparatus. In such cases extensive cleaning of the flow path in the LC or ion transfer lenses of the mass spectrometer is required. Contamination can be an expensive process, not only due to instrument downtime but the skills to perform the necessary restorative tasks may be beyond the scope of many microbiologists, thus requiring a specialist engineer. Awareness of the need for adequate sample clean up and applying simple procedures prior to analysis can save time, money and greatly enhance results of any experiment.

> *Tip: Blank runs are a good source of information; run blanks to ensure a clean system. Run blanks between samples to check for carryover and sources of potential contaminants. Avoid plastics and acids! The mobile phases of LC systems usually contain acid. Acids pipetted/stored with plastics are the main cause of systemic polymer contamination.*

6.5.3 General Sample Lysis Consideration

Several methods are commonly used for the preparation of microorganisms for proteomic analysis by LC-MS. The lysis method has to be carefully developed empirically and the selection of the method will have a great influence on the subsequent purification steps required. The types of lysis methods available have been described in the previous section so no great detail will be given here. The choice of the method is largely dependent on the microorganism under investigation and each can be used with a plethora of different solubilising/denaturing reagent cocktails. However, a different set of rules and considerations apply when preparing samples for LC-MS analysis in comparison with 2D GE. Many of the reagents for 2D GE are not necessary and in some cases can interfere with the LC-MS analysis.

If protease inhibitors are to be included they must be removed or inactivated as they may reduce the effectiveness of exogenous proteases (e.g. trypsin) essential in subsequent steps (discussed below). Generally, methods employ a surfactant based solubilisation, using either ionic detergents such as SDS, sodium deoxycholate, or zwitterionic detergents such as CHAPS, or nonionic detergents such as octyl-β-glucopyranoside (OG) and octyl-phenylpolyethylene glycol (NP-40, Igepal CA-630). In some cases where protein denaturation is required or beneficial to extraction, chaotropes such as urea, thiourea and guanidine hydrochloride can be utilised. It also aids protein extraction to include salt, either sodium chloride or potassium chloride, to break weak ionic interactions some proteins may form. At this point a reducing agent such as dithiothreitol should be included to break existing disulfides. Buffing of protein extraction is key and a weak base such as tris(hydroxymethyl) methylamine of pK_a 8.06 will buffer effectively within a pH range of 7.5–9.0 titrated with a conjugate acid such as hydrochloric acid. Other buffers are also effective depending on which pH is found optimum for extraction. In all cases it is important that the extraction method chosen is effective for the organism being studied and omission of particular substances is only necessary if its removal prior to LC-MS is particularly challenging. The selection of reagents which are either compatible or easily removed prior to analysis is one method to greatly streamline and enhance the throughput of experiments, though it should be noted that omitting detergents can significantly reduce the recovery of many proteins.

Tip: Check which detergents effectively solubilise different classes of proteins. If the targets are potentially insoluble then choose an effective detergent! Make sure the detergent can be effectively removed prior to injection.

6.5.4 Crude Protein Purification

Following the cell lysis it is essential to perform a purification step in order to remove non-protein cellular components and significantly reduce the presence of chemicals used in lysis that may interfere with downstream processing/analysis of samples. In order to choose an effective method a good knowledge of the makeup of the sample and the chemical properties of the reagents is vital. Whilst detergents are essential for cell lysis and protein solubilisation they present a problem in LC-MS as their hydrophobic portions will effectively bind to reversed phase material used in most applications. At high concentrations this will saturate the column preventing peptides from binding and/or suppressing detection by MS. Also, a high lipid, DNA and carbohydrate content can be deleterious;

phospholipids and carbohydrates are readily ionisable and can be observed during reversed phase HPLC. A quick and relatively simple method for purification is protein precipitation and there are a number of different options available (61). TCA can be used to precipitate proteins with high efficiency and can effectively remove polysaccharide from the sample; it is also highly effective at preventing endogenous protease activity (62). Organic solvents such as acetone, methanol, and ethanol are also very good choices especially when working with stable samples at relatively high protein concentrations (63, 64). Organic solvents will also reduce the amount of nonionic detergents such as CHAPS and NP-40 that are carried over and will solubilise the majority of lipids. Thus, with this knowledge one can now make use of these detergents in upstream processes such as cell lysis. All precipitation methods provide significant enhancement in protein purity but can incur problems due to inefficient precipitation and poor resolubilisation. Acetone and ethanol base precipitation are effective and achieve around 90–95% recovery of proteins. TCA is the most effective but precipitates suffer from poor resolubilisation due to high acidity. This can be overcome by washing a TCA precipitate with ice cold acetone or ethanol and using resolubilisation solutions of high buffering capacity or titrating the sample using sodium hydroxide to ensure a pH between 7 and 8.

Tip: Keep preciptations cold! Mix immediately and well to prevent conglomeration and ensure effective precipitation. Denaturation can be avoided prior to precipitation as denatured proteins may be poorly precipitated.

6.5.5   Protein Resolubilsation and In-Solution Digestion

Protein resolubilisation prepares proteins for enzymatic digestion. For effective cleavage it is essential that polypeptides exhibit only primary structure. Chaotropes such as urea and thiourea, ionic detergents such as SDS and sodium deoxycholate, nonionic detergents such as n-octylglucoside (OG) or patented surfactants such as the acid labile sodium 3-[(2-methyl-2-undecyl-1,3-dioxolan-4-yl)methoxyl]-1-propanesulfonate (RapiGest, Waters Micromass), Invitrosolve (Invitrogen) and Silent Surfactant (PPS, Protein Discovery) are effective at denaturing proteins and have been successfully used during digestion (65, 66). Also, of recent note is the use of some organic compounds such as acetonitrile, methanol and trifluroethanol either alone or in combination with more classical reagents (67–71). For effective resolubilisation and denaturation, samples can be sonicated using a water-bath or heated. However, if urea is used heating must not be carried out as this may result in extensive protein modification. In general the use of these chemicals is permitted because they are somewhat compatible with digestion and can be effectively removed prior to injection (72). Table 6.1 describes effective concentrations commonly used for digestion and how these reagents can be removed prior to LC-MS. A second aspect of protein resolubilisation is the irreversible blocking of sulfahydryl groups by reduction and alkylation. This step effectively breaks and prevents the reformation of disulfide bridges in cysteine containing proteins. This two-step reaction is most commonly performed using dithiolthreitol and iodoacetamide though phosphine based reducing agents tributyl phosphine and

Table 6.1 *Maximum concentrations of commonly used reagents for protein solubilisation during in-solution digestion with trypsin and their effective removal prior to LC-MS*

Reagent	Compatible concentration	Method of removal
Urea, thiourea	1 M	SPE with C18 sorbant
Deoxycholate	1%	Acidification and centrifugation
SDS	0.5%	SPE with strong cation exchange sorbant, ethyl acetate extraction
Rapigest	1%	Acidification
Acetonitrile, methanol	40%	Evaporation

tris(2-carboxyethyl)phosphine and the alkylating reagents iodoacetic acid and iodoacetamide are also commonly used. Generally this reaction is highly effective at low millimolar concentrations, though it is important to ensure the correct ratio of each reagent is used as they can cross react. Reduced and alkylated cysteines will have a different mass so this must be accounted for when performing a database search for protein identification.

Tip: A good way of performing this step is to include the reducing agent in the resolubilsation cocktail and subsequently add the alkylating reagent. Also when performing protein separation by gel electrophoresis samples can be alkylated prior to electrophoresis greatly speeding up the in-gel digestion!

6.5.6 Protein Digestion

Protein digestion is a critical step in most proteomic experiments. Briefly, protein identification is based on the creation of peptides by the enzymatic digestion of intact proteins followed by mass spectrometric sequencing and subsequent rematching to the source protein. The sole reason for digestion is purely to create molecules of a suitable size for analysis by MS. The analytical challenges that digestion causes in terms of increased sample complexity and subsequent informatics are huge and will not be discussed here. Digestion is generally performed with the serine protease trypsin. Trypsin cleaves with high specificity at the C-terminal peptide bond of unmodified arginine and lysine residues (73, 74). It is essential to ensure that the trypsin used is at least of sequencing grade and that it has been modified by reductive methylation of lysines to prevent autolysis. Trypsin should also be chromatographically purified and treated with L-(tosylamido-2-phenyl) ethyl chloromethyl ketone (TPCK) to inhibit contaminating chymotryptic activity (75). The purity of the trypsin used for digestion is vital as unspecific cleavages can greatly complicate the analysis and identification of proteins. The 'Kiel' rule indicates that normal tryptic cleavage is inhibited if the amino acid proline is found C-terminal to either arginine or lysine residues (76). For the majority of proteomic experiments this rule is used. However, recent studies suggest that the Kiel rule should not be as strictly applied as first thought (77). However, any deviation from the rules towards unspecific cleavages should be considered with a high amount of scepticism unless high mass accuracy has been achieved when measuring these fragments.

Tip: To check the specificity of digestion perform database searches with semi-tryptic, chemotryptic or no specificity criteria. (Note, some of these searches will take a long time and increase the rate of false positive identifications.)

There are a host of different enzymes and methods for the digestion of proteins into usable sized peptides. This section will focus on the use of trypsin, as it is most commonly used in proteomics today and still provides the most reliable method for effective and accurate digestion. Cleavage with trypsin is highly effective as long as the protein is fully denatured and contains sufficient numbers of unmodified lysine and arginine residues. In solution, proteins may exist is differing states of denaturation and in order to promote denaturation, chatropes and surfactants are generally required for efficient digestion (discussed earlier). Unfortunately trypsin activity is restricted in the presence of these substances and their concentrations should be limited if digestion is to be effective (Table 6.1). In order to overcome this, the endoproteinase Lys-C from *Lysobacter enzymogenes*, which cleaves at the carboxyl side of lysine residues can be used (78). Lys-C can be utilised with only a small drop in activity at concentrations of up to 8 M urea and 0.5% (w/v) SDS (79). It is most commonly employed in a double digestion strategy where proteins are initially digested with Lys-C in the presence of urea (6–8 M for 3 h) or 0. 5% (w/v) SDS before the sample is diluted fivefold and trypsin is added to complete the digestion overnight. Although Lys-C alone can produce fragments suitable for MS its use with trypsin is optimal for most applications. Lys-C will initially cleave proteins creating smaller fragments less likely to form secondary or tertiary structure when the sample is diluted for trypsinisation. Due to the presence of urea, digestion should not be carried out at elevated temperatures for extended periods of time to prevent protein modification (80). Typically both trypsin and Lys-C are used at a ratio of 1 µg of enzyme to 50–100 µg of protein and optimal pH for digestion is approximately 8.0. This is usually achieved with either a freshly prepared 50 mM ammonium bicarbonate or appropriate buffer. Although most protein digests are carried out overnight to ensure completeness, shorter digests have also found to be sufficient (68, 81). Recently, several publications have emerged where digestion has been completed in under 1 h using methods such as micro and sonic wave irradiation sources to enhance the kinetics of digestion (82, 83). Once digestion is complete the remaining enzyme activity can be inhibited by lowering the pH, temperature or performing solid phase extraction (SPE).

Tip: Store your enzymes in the appropriate buffer in single use aliquots. If you omit urea and use other detergents you can boil the sample and perform digestion at 30 °C. Stop your digest by adding reagents to mimic the solution used for equilibration of the stationary phase used for SPE. Your sample is now ready for online or offline SPE (see Section 6.5.7).

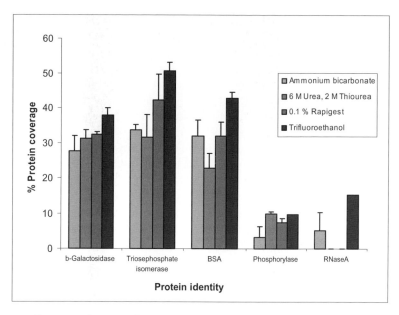

Figure 6.1   *Efficiency of tryptic digestion with different denaturants. An equal molar mixture of five proteins was denatured using the four different chemical mixtures shown. Then 50 fmol of each protein digest was analysed by RP-LC-MS, spectra were interpreted using Bioworks Browser 3.10 (ThermoFisher Scientific, Bremen, Germany) and the percentage coverage calculated. Results are kindly provided by N. V. J. Harpham and L. Bowler, University of Sussex Proteomics Centre, Trafford Centre for Medical Research, University of Sussex, Brighton, UK*

The efficiency of tryptic digestion with different denaturants has been investigated by LC-MS (Figure 6.1). The results presented indicate the importance of effective denaturation, with higher sequence coverage observed for some proteins in the presence of certain denaturing reagents. Denaturation with trifluoroethanol was found to be particularly effective for all proteins tested. These results only give a snapshot of how important effective denaturation is, and thus these observations are likely to be magnified in real samples. It is therefore important that a suitable method is developed for specific sample needs.

6.5.7   Microscale Clean Up Prior to LC-MS

It is necessary to significantly deplete detergents and salts from the sample prior to LC-MS (72, 84). High levels of salts and detergents will significantly reduce the detection sensitivity for peptides by reducing column capacity resulting in ion suppression. Detergents such as CHAPS and SDS will usually result in a major detergent peak which masks the elution of peptides and may take extensive cleaning to remove from the system. Salts are extremely problematic in LC-MS; initially high concentrations of salt can result in clogged emitter needles and reduce flow to the detector and increase the back pressure above the limits of the LC pumping system. At lower concentrations the problems of salt are due to their ability to replace protons as the main source of charge during peptide ionisation. This may not be evident without detailed examination of the spectra, where adducts of sodium and potassium can be observed for the same peptide. Salt adducts will reduce the overall intensity of peptides and increase the complexity of the sample at the spectrum level. Salt adducts will be distinct in mass and therefore analysed separately but ultimately result in the same identification. This will negatively impact on the sampling efficiency of the mass spectrometer reducing the number of identified proteins.

In order to overcome these issues nano-microscale SPE can be performed either online or offline depending on the instrumentation available and the choice of SPE. For high throughput ultrasensitive studies it is preferred to use a HPLC system fitted with a separate SPE trap or precolumn with a stationary phase similar to that of the analytical column (85). After injection sample can be bound to the trap, concentrated and desalted with the liquid flow path out of line with the analytical column, thus flushing polar contaminants to waste. After this process, the trap is switched in line with the analytical column and the mobile phase solvent gradient is passed through the entire flow path. Peptides elute from the trap and are separated by the analytical column. This method is effective for the removal of salts and allows the direct analysis of in-gel and in-solution digests where sensitivity is vital, providing there are no major nonpolar contaminants present. SPE can also be carried out offline using specialised SPE devices, in which the choice of stationary phase is more flexible (86–88). It can either be of similar material and is analogous to what is achieved online or a specially selected stationary phase can be used to remove a specific nonpolar contaminant. For example, strong cation exchange chromatography (SCX) will effectively retain acidified peptides leaving SDS in the mobile phase. In such a study, peptide elution is achieved in relatively high salt concentrations and thus a second SPE step is required 'to desalt' prior to injection. There are many devices available on the market either in the form of loose bed spin cartridges or various tip based devices based on fixed porous, monolithic or membrane immobilised stationary phases. Membranes are particularly advantageous as they provide a filtration step effectively removing particulates from samples. If this is not done it is highly recommended to remove particulates by centrifugation or filtration as particulates can easily block various parts of nano-LC systems.

6.6   Conclusion

Sample preparation is one of the critical elements in proteomics and while seemingly somewhat mundane, has been the Achilles' heel that has impeded the success of many exciting and otherwise promising studies. However, in studies where efficient and reproducible methods for sample preparation have been established, good quality data are obtained. Similarly, overlooking this first and most important step can lead to poor chromatograms or 2D gel profiles or can interfere with protein identification using MS. Therefore understanding the chemistry behind protein solubilisation and the principles behind separation methods such as IEF or LC is extremely important when designing the sample preparation protocols. Furthermore, when it comes to bacteria, good knowledge of the structure of the bacterial cell envelope as well as the technology available to aid with cell lysis can be helpful. By providing theoretical discussion as well as basic protocols, we hope that this chapter will enable microbiologists not only to perform successful 2D GE and LC-MS experiments but understand and develop their own methodology.

References

(1) Cash P, Argo E, Bruce KD. Characterisation of *Haemophilus influenzae* proteins by two-dimensional gel electrophoresis. Electrophoresis 1995; 16(1): 135–148.

(2) Gormon T, Phan-Thanh L. Identification and classification of Listeria by two-dimensional protein mapping. Res. Microbiol. 1995; 146(2): 143–154.

(3) Andersen H, Birkelund S, Christiansen G, Freundt EA. Electrophoretic analysis of proteins from *Mycoplasma hominis* strains detected by SDS-PAGE, two-dimensional gel electrophoresis and immunoblotting. J. Gen. Microbiol. 1987; 133(1): 181–191.

(4) Bernardini G, Braconi D, Santucci A. The analysis of *Neisseria meningitidis* proteomes: Reference maps and their applications. Proteomics 2007; 7(16): 2933–2946.

(5) Bodzon-Kulakowska A, Bierczynska-Krzysik A, Dylag T, Drabik A, Suder P, Noga M, Jarzebinska J, Silberring J. Methods for samples preparation in proteomic research. J. Chromatogr., 2007; 849(1–2): 1–31.

(6) Canas B, Pineiro C, Calvo E, Lopez-Ferrer D, Gallardo JM. Trends in sample preparation for classical and second generation proteomics. J. Chromatogr. 2007; 1153(1–2): 235–258.

(7) Encheva V, Gharbia SE, Wait R, Begum S, Shah HN. Comparison of extraction procedures for proteome analysis of *Streptococcus pneumoniae* and a basic reference map. Proteomics 2006; 6(11): 3306–3317.

(8) Wong HC, Chang CN. Hydrophobicity, cell adherence, cytotoxicity, and enterotoxigenicity of starved *Vibrio parahaemolyticus*. J. Food Prot. 2005; 68(1): 154–156.

(9) Marques MA, Espinosa BJ, Xavier da Silveira EK, Pessolani MC, Chapeaurouge A, Perales J, Dobos KM, Belisle JT, Spencer JS, Brennan PJ. Continued proteomic analysis of *Mycobacterium leprae* subcellular fractions. Proteomics 2004; 4(10): 2942–2953.

(10) Coldham NG, Woodward MJ. Characterization of the *Salmonella typhimurium* proteome by semi-automated two dimensional HPLC-mass spectrometry: detection of proteins implicated in multiple antibiotic resistance. J. Proteome Res. 2004; 3(3): 595–603.

(11) Antelmann H, Scharf C, Hecker M. Phosphate starvation-inducible proteins of *Bacillus subtilis*: proteomics and transcriptional analysis. J. Bacteriol. 2000; 182(16): 4478–4490.

(12) Wang J, Ying T, Wang H, Shi Z, Li M, He K, Feng E, Wang J, Yuan J, Li T, Wei K, Su G, Zhu H, Zhang X, Huang P, Huang L. 2-D reference map of *Bacillus anthracis* vaccine strain A16R proteins. Proteomics 2005; 5(17): 4488–4495.

(13) Chitlaru T, Ariel N, Zvi A, Lion M, Velan B, Shafferman A, Elhanany E. Identification of chromosomally encoded membranal polypeptides of *Bacillus anthracis* by a proteomic analysis: prevalence of proteins containing S-layer homology domains. Proteomics 2004; 4(3): 677–691.

(14) Redmond C, Baillie LW, Hibbs S, Moir AJ, Moir A. Identification of proteins in the exosporium of *Bacillus anthracis*. Microbiology 2004; 150(Pt 2): 355–363.

(15) Len AC, Harty DW, Jacques NA. Proteome analysis of *Streptococcus mutans* metabolic phenotype during acid tolerance. Microbiology 2004; 150(Pt 5): 1353–1366.

(16) Ginsburg I. The role of bacteriolysis in the pathophysiology of inflammation, infection and post-infectious sequelae. APMIS 2002; 110(11): 753–770.

(17) Gorg A, Weiss W, Dunn MJ. Current two-dimensional electrophoresis technology for proteomics. Proteomics 2004; 4(12): 3665–3685.

(18) Castellanos-Serra L, Paz-Lago D. Inhibition of unwanted proteolysis during sample preparation: evaluation of its efficiency in challenge experiments. Electrophoresis 2002; 23(11): 1745–1753.

(19) Firdaus RM, Ahmad HA, Sharum MY, Azizi N, Mohamed R. ProLysED: an integrated database and meta-server of bacterial protease systems. Appl. Bioinformatics 2005; 4(2): 147–150.

(20) Antelmann H, Williams RC, Miethke M, Wipat A, Albrecht D, Harwood CR, Hecker M. The extracellular and cytoplasmic proteomes of the non-virulent *Bacillus anthracis* strain UM23C1–2. Proteomics 2005; 5(14): 3684–3695.

(21) Dunn MJ. Gel Electrophoresis of Proteins. Bios Scientific Publisher Alden Press, Oxford, 1993.

(22) O'Farrell PH. High resolution two-dimensional electrophoresis of proteins. J. Biol. Chem. 1975; 250(10): 4007–4021.

(23) Lawless MK, Hopkins S, Anwer MK. Quantitation of a 36-amino-acid peptide inhibitor of HIV-1 membrane fusion in animal and human plasma using high-performance liquid chromatography and fluorescence detection. J. Chromatogr., 1998; 707(1–2): 213–217.

(24) Joo WA, Lee DY, Kim CW. Development of an effective sample preparation method for the proteome analysis of body fluids using 2-D gel electrophoresis. Biosci. Biotechnol. Biochem. 2003; 67(7): 1574–1577.

(25) Rabilloud T. Solubilisation of proteins for electrophoretic analysis. Electrophoresis 1996; 17(5): 813–829.

(26) Encheva V, Wait R, Gharbia SE, Begum S, Shah HN. Proteome analysis of serovars Typhimurium and Pullorum of *Salmonella enterica* subspecies I. BMC Microbiol. 2005; 5: 42.

(27) Rabilloud T. Detergents and chaotropes for protein solubilization before two-dimensional electrophoresis. Methods Mol. Biol. 2009; 528: 259–267.

(28) Rabilloud T, Luche S, Santoni V, Chevallet M. Detergents and chaotropes for protein solubilization before two-dimensional electrophoresis. Methods Mol. Biol. 2007; 355: 111–119.

(29) Molloy MP, Herbert BR, Slade MB, Rabilloud T, Nouwens AS, Williams KL, Gooley AA. Proteomic analysis of the *Escherichia coli* outer membrane. Eur. J. Biochem. 2000; 267(10): 2871–2881.

(30) Holloway PJ, Arundel PH. High-resolution two-dimensional electrophoresis of plant proteins. Anal. Biochem. 1988; 172(1): 8–15.

(31) Tastet C, Charmont S, Chevallet M, Luche S, Rabilloud T. Structure-efficiency relationships of zwitterionic detergents as protein solubilizers in two-dimensional electrophoresis. Proteomics 2003; 3(2): 111–121.

(32) Luche S, Santoni V, Rabilloud T. Evaluation of nonionic and zwitterionic detergents as membrane protein solubilizers in two-dimensional electrophoresis. Proteomics 2003; 3(3): 249–253.

(33) Wilkins MR, Gasteiger E, Gooley AA, Herbert BR, Molloy MP, Binz PA, Ou K, Sanchez JC, Bairoch A, Williams KL, Hochstrasser DF. High-throughput mass spectrometric discovery of protein post-translational modifications. J. Mol. Biol. 1999; 289(3): 645–657.

(34) Ünlu M, Morgan ME, Minden JS. Difference gel electrophoresis: a single gel method for detecting changes in protein extracts. Electrophoresis 1997; 18(11): 2071–2077.

(35) Yan JX, Devenish AT, Wait R, Stone T, Lewis S, Fowler S. Fluorescence two-dimensional difference gel electrophoresis and mass spectrometry based proteomic analysis of *Escherichia coli*. Proteomics 2002; 2(12): 1682–1698.

(36) Hongsthong A, Sirijuntarut M, Prommeenate P, Thammathorn S, Bunnag B, Cheevadhanarak S, Tanticharoen M. Revealing differentially expressed proteins in two morphological forms of *Spirulina platensis* by proteomic analysis. Mol. Biotechnol. 2007; 36(2): 123–130.

(37) Park SH, Oh HB, Seong WK, Kim CW, Cho SY, Yoo CK. Differential analysis of *Bacillus anthracis* after pX01 plasmid curing and comprehensive data on *Bacillus anthracis* infection in macrophages and glial cells. Proteomics 2007; 7(20): 3743–3758.

(38) Durrschmid K, Reischer H, Schmidt-Heck W, Hrebicek T, Guthke R, Rizzi A, Bayer K. Monitoring of transcriptome and proteome profiles to investigate the cellular response of *E. coli* towards recombinant protein expression under defined chemostat conditions. J. Biotechnol. 2008; 135(1): 34–44.

(39) Lopez-Campistrous A, Semchuk P, Burke L, Palmer-Stone T, Brokx SJ, Broderick G, Bottorff D, Bolch S, Weiner JH, Ellison MJ. Localization, annotation, and comparison of the *Escherichia coli* K-12 proteome under two states of growth. Mol. Cell Proteomics 2005; 4(8): 1205–1209.

(40) Vipond C, Suker J, Jones C, Tang C, Feavers IM, Wheeler JX. Proteomic analysis of a meningococcal outer membrane vesicle vaccine prepared from the group B strain NZ98/254. Proteomics 2006; 6(11): 3400–3413.

(41) Keller M, Hettich R. Environmental proteomics: a paradigm shift in characterizing microbial activities at the molecular level. Microbiol. Mol. Biol. Rev 2009; 73(1): 62–70.

(42) Ogunseitan OA. Protein Profile variation in cultivated and native freshwater microorganisms exposed to chemical environmental pollutants. Microb. Ecol. 1996; 31(3): 291–304.

(43) Ogunseitan OA. Protein method for investigating mercuric reductase gene expression in aquatic environments. Appl. Environ. Microbiol. 1998; 64(2): 695–702.

(44) Wilmes P, Bond PL. The application of two-dimensional polyacrylamide gel electrophoresis and downstream analyses to a mixed community of prokaryotic microorganisms. Environ. Microbiol. 2004; 6(9): 911–920.

(45) Benndorf D, Balcke GU, Harms H, von Bergen M. Functional metaproteome analysis of protein extracts from contaminated soil and groundwater. ISME J. 2007; 1(3): 224–234.

(46) Cordwell SJ. Technologies for bacterial surface proteomics. Curr. Opin. Microbiol. 2006; 9(3): 320–329.

(47) Schwarz K, Fiedler T, Fischer RJ, Bahl H. A Standard Operating Procedure (SOP) for the preparation of intra- and extracellular proteins of *Clostridium acetobutylicum* for proteome analysis. J. Microbiol. Methods 2007; 68(2): 396–402.

(48) Sivaraman T, Kumar TK, Jayaraman G, Yu C. The mechanism of 2,2,2-trichloroacetic acid-induced protein precipitation. J. Protein Chem. 1997; 16(4):291–297.

(49) Peterson GL. Determination of total protein. Methods Enzymol 1983; 91:95–119.

(50) Chen G, Pramanik BN. LC-MS for protein characterization: current capabilities and future trends. Expert Rev. Proteomics 2008; 5(3): 435–444.

(51) de Hoog CL, Mann M. Proteomics. Annu Rev Genomics Hum Genet 2004; 5: 267–293.

(52) Fenn JB, Mann M, Meng CK, Wong SF, Whitehouse CM. Electrospray ionization for mass spectrometry of large biomolecules. Science 1989; 246(4926): 64–71.

(53) Link AJ, Eng J, Schieltz DM, Carmack E, Mize GJ, Morris DR, Garvik BM, Yates JR, 3rd. Direct analysis of protein complexes using mass spectrometry. Nat. Biotechnol. 1999; 17(7): 676–682.

(54) Washburn MP, Wolters D, Yates JR, 3rd. Large-scale analysis of the yeast proteome by multidimensional protein identification technology. Nat. Biotechnol. 2001; 19(3): 242–247.

(55) Hunt DF, Yates JR, 3rd, Shabanowitz J, Winston S, Hauer CR. Protein sequencing by tandem mass spectrometry. Proc. Natl Acad. Sci USA 1986; 83(17): 6233–6237.

(56) Larsen MR, Trelle MB, Thingholm TE, Jensen ON. Analysis of posttranslational modifications of proteins by tandem mass spectrometry. Biotechniques 2006; 40(6): 790–798.

(57) Kapp E, Schutz F. 2007. Overview of tandem mass spectrometry (MS/MS) database search algorithms. Curr. Protoc. Protein Sci. 49:25.2.1–25.2.19.

(58) Annesley TM. Ion suppression in mass spectrometry. Clin. Chem. 2003; 49(7): 1041–1044.

(59) Sandra K, Moshir M, D'Hondt F, Verleysen K, Kas K, Sandra P: Highly efficient peptide separations in proteomics. Part 1. Unidimensional high performance liquid chromatography. J. Chromatogr., 2008; 866(1–2): 48–63.

(60) Matuszewski BK, Constanzer ML, Chavez-Eng CM. Matrix effect in quantitative LC/MS/MS analyses of biological fluids: a method for determination of finasteride in human plasma at picogram per milliliter concentrations. Anal. Chem. 1998; 70(5): 882–889.

(61) England S, Seifter S. Precipitation techniques. Methods Enzymol. 1990; 182: 285–300.

(62) Sivaraman T, Kumar TK, Jayaraman G, Yu C. The mechanism of 2,2,2-trichloroacetic acid-induced protein precipitation. J. Protein Chem. 1997; 16(4): 291–297.

(63) Jiang L, He L, Fountoulakis M. Comparison of protein precipitation methods for sample preparation prior to proteomic analysis. J. Chromatogr., 2004; 1023(2): 317–320.

(64) Zellner M, Winkler W, Hayden H, Diestinger M, Eliasen M, Gesslbauer B, Miller I, Chang M, Kungl A, Roth E, Oehler R: Quantitative validation of different protein precipitation methods in proteome analysis of blood platelets. Electrophoresis 2005; 26(12):2481–2489.

(65) Nomura E, Katsuta K, Ueda T, Toriyama M, Mori T, Inagaki N. Acid-labile surfactant improves in-sodium dodecyl sulfate polyacrylamide gel protein digestion for matrix-assisted laser desorption/ionization mass spectrometric peptide mapping. J. Mass Spectrom. 2004; 39(2): 202–207.

(66) Yu YQ, Gilar M, Lee PJ, Bouvier ES, Gebler JC. Enzyme-friendly, mass spectrometry-compatible surfactant for in-solution enzymatic digestion of proteins. Anal. Chem. 2003; 75(21): 6023–6028.

(67) Wang H, Qian WJ, Mottaz HM, Clauss TR, Anderson DJ, Moore RJ, Camp DG, 2nd, Khan AH, Sforza DM, Pallavicini M, Smith DJ, Smith RD, Development and evaluation of a micro- and nanoscale proteomic sample preparation method. J Proteome Res 2005; 4(6):2397–2403.

(68) Strader MB, Tabb DL, Hervey WJ, Pan C, Hurst GB. Efficient and specific trypsin digestion of microgram to nanogram quantities of proteins in organic-aqueous solvent systems. Anal. Chem. 2006; 78(1): 125–134.

(69) Hervey WJ, Strader MB, Hurst GB. Comparison of digestion protocols for microgram quantities of enriched protein samples. J. Proteome Res. 2007; 6(8): 3054–3061.

(70) Chen EI, Cociorva D, Norris JL, Yates JR, 3rd. Optimization of mass spectrometry-compatible surfactants for shotgun proteomics. J. Proteome Res. 2007; 6(7): 2529–2538.

(71) Zhang N, Chen R, Young N, Wishart D, Winter P, Weiner JH, Li L. Comparison of SDS- and methanol-assisted protein solubilization and digestion methods for *Escherichia coli* membrane proteome analysis by 2-D LC-MS/MS. Proteomics 2007; 7(4): 484–493.

(72) Yeung YG, Nieves E, Angeletti RH, Stanley ER. Removal of detergents from protein digests for mass spectrometry analysis. Anal. Biochem. 2008; 382(2): 135–137.

(73) Brown WE, Wold F. Alkyl isocyanates as active-site-specific reagents for serine proteases. Identification of the active-site serine as the site of reaction. Biochemistry 1973; 12(5): 835–840.

(74) Brown WE, Wold F. Alkyl isocyanates as active-site-specific reagents for serine proteases. Reaction properties. Biochemistry 1973; 12(5): 828–834.

(75) Kostka V, Carpenter FH. Inhibition of chymotrypsin activity in crystalline trypsin preparations. J. Biol. Chem. 1964; 239: 1799–1803.

(76) Keil BR. Specificity of Proteolysis. Springer-Verlag, Berlin, 1992.

(77) Rodriguez J, Gupta N, Smith RD, Pevzner PA. Does trypsin cut before proline? J. Proteome Res. 2008; 7(1): 300–305.

(78) de Souza GA, Godoy LM, Mann M: Identification of 491 proteins in the tear fluid proteome reveals a large number of proteases and protease inhibitors. Genome Biol 2006; 7(8):R72.

(79) Fischer S, Brunner U, Galaboff E, Geuß U, Schäfer M, Kreße GB, Proteinases sequencing grade endoproteinases LysC, GluC, trypsin and AspN and carboxypeptidases P and Y. Fresenius' Journal of Analytical Chemistry 1988; 330(4):358–359.

(80) McCarthy J, Hopwood F, Oxley D, Laver M, Castagna A, Righetti PG, Williams K, Herbert B: Carbamylation of proteins in 2-D electrophoresis–myth or reality? J Proteome Res 2003; 2(3):239–242.

(81) Terry DE, Umstot E, Desiderio DM: Optimized sample-processing time and peptide recovery for the mass spectrometric analysis of protein digests. J Am Soc Mass Spectrom 2004; 15(6):784–794.

(82) Sun W, Gao S, Wang L, Chen Y, Wu S, Wang X, Zheng D, Gao Y. Microwave-assisted protein preparation and enzymatic digestion in proteomics. Mol. Cell Proteomics 2006; 5(4): 769–776.

(83) Lopez-Ferrer D, Capelo JL, Vazquez J. Ultra fast trypsin digestion of proteins by high intensity focused ultrasound. J. Proteome Res. 2005; 4(5): 1569–1574.

(84) Zhang N, Li L. Effects of common surfactants on protein digestion and matrix-assisted laser desorption/ionization mass spectrometric analysis of the digested peptides using two-layer sample preparation. Rapid Commun. Mass Spectrom. 2004; 18(8): 889–896.

(85) Shen Y, Zhao R, Berger SJ, Anderson GA, Rodriguez N, Smith RD. High-efficiency nanoscale liquid chromatography coupled on-line with mass spectrometry using nanoelectrospray ionization for proteomics. Anal. Chem. 2002; 74(16): 4235–4249.

(86) Rappsilber J, Ishihama Y, Mann M. Stop and go extraction tips for matrix-assisted laser desorption/ionization, nanoelectrospray, and LC/MS sample pretreatment in proteomics. Anal. Chem. 2003; 75(3): 663–670.

(87) Ishihama Y, Rappsilber J, Mann M. Modular stop and go extraction tips with stacked disks for parallel and multidimensional peptide fractionation in proteomics. J. Proteome Res. 2006; 5(4): 988–994.

(88) Rappsilber J, Mann M, Ishihama Y. Protocol for micro-purification, enrichment, prefractionation and storage of peptides for proteomics using StageTips. Nat. Protoc. 2007; 2(8): 1896–1906.

7

Isolation and Preparation of Spore Proteins and Subsequent Characterisation by Electrophoresis and Mass Spectrometry

Nicola C. Thorne, Haroun N. Shah and Saheer E. Gharbia
Department of Bioanalysis and Horizon Technologies, Health Protection Agency Centre for Infections, London, UK

7.1 Introduction

Bacterial spores provide a vehicle for widespread contamination in environments such as care homes and hospital wards where they can persist in the environment for months. They are notoriously difficult to eradicate as they can colonise virtually any habitat and remain viable even after a long period of dormancy (1).

Endospores are produced by certain bacteria in response to severe external stress. They are unusually dehydrated, impervious, highly refractile cells that have much resistance to heat, desiccation, radiation and many toxic chemicals. In addition, they are not only resistant to certain surface disinfectants, but have in fact been shown to germinate on contact (2). They are not reproductive and have no metabolic activity. Their main ecological role is to survive and prolong the life of the cell in hostile environments and upon return to favourable conditions dormant spores can readily convert into actively growing vegetative cells via a process termed germination when nutrient returns to the environment (3), e.g. on entry into a host. Only two groups of bacteria, namely, *Bacillus* and *Clostridium,* form spores and most of the research related to spores has focused on pathogenic *Bacillus* species to date.

Mass Spectrometry for Microbial Proteomics   Edited by Haroun N. Shah and Saheer E. Gharbia
© 2010 John Wiley & Sons, Ltd

7.1.1   The Model Organism: *Bacillus subtilis*

Bacilli (i.e. of the genus *Bacillus*), like clostridia, form endospores in response to hostile environments as a means to survive. One species in particular, *B. subtilis*, has proven highly amenable to genetic manipulation and has thus become widely adopted as the Gram-positive model organism in laboratory studies, including studies of sporulation. As such it is the most widely studied Gram-positive endospore-forming bacterium and its mechanism of spore formation and the endospore structure of this organism have been well examined (4–9). Conversely, sporulation and germination in clostridia are not as well studied. This is mainly due to the fact that until recently there was a lack of effective methods for site-directed mutagenesis for inactivation of specific genes in the clostridia. However, the development of a mutagenesis system based on the mobile group II intron from the *ltrB* gene of *Lactococcus lactis* that has been adapted to function in clostridial hosts, has been shown to work in *C. difficile* (10). This paves the way for researchers to create 'sporulation mutants' of *C. difficile* by knocking out specific genes in order to characterise those that have a functional role in the sporulation process.

7.1.2   Sporulation

Spores are formed via the process termed 'sporulation' that involves asymmetric cell division during which a copy of the genome is partitioned into each of the two sister cells: the outer 'mother cell' and an inner maturing spore, known as the 'forespore'. The initiation of sporulation in the model organism *B. subtilis* uses a phosphorelay system of five sensory histidine kinases (KinA–KinE). These orphan kinases activate (by phosphorylation) the Spo0A protein which is the master regulator of sporulation in *Bacillus* (11) (Figure 7.1). Spo0A is thought to act as an activator of stationary phase genes or as a repressor of early expressed genes (4). In clostridia the two main processes that lead to spore production markedly differ from the known *Bacillus* system and thus the wealth of the data available on *Bacillus* spores is of limited relevance to understanding *C. difficile* spores.

Although morphological changes during sporulation are very similar in the clostridia and bacilli, a genomic comparison of clostridia found large differences in the early stages of the sporulation pathway, mainly that hardly any early sporulation genes of *B. subtilis* are present in the genomes of the clostridia (4).

7.1.3   The Spore Structure

The spore coat is what affords the spore its hardy characteristics due to its multilayered proteinaceous form. In *B. subtilis*, this was found to consist of >70 different cross-linked proteins, a few of which also have significant roles in coat assembly (12). The spore of *B. subtilis* is formed inside the mother cell, and hence termed an 'endospore', and the same is true of clostridia. Coat proteins comprise about 70% of total spore protein (13). They are synthesised in the cell cytoplasm and layered onto the cell surface and the spore is released by lysis of the mother cell (3). The mechanism whereby the coat protects the spores against reactive chemicals is not known but coat proteins may possibly detoxify such compounds before they gain access to more sensitive structures of the spore (12). The coat layer is semi-permeable, allowing passage of small molecules (<5 kDa) to the inner layers where receptors are located in the spore's inner membrane. These receptors

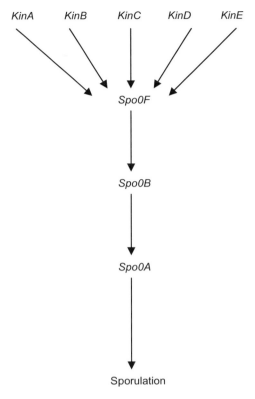

Figure 7.1 *Sporulation cascade of* B. subtilis. *Adapted from reference (11)*

can sense the presence of nutrient molecules from the external cell environment and trigger spore germination (12, 14). Importantly for spore survival, larger molecules such as potential lytic enzymes, are prevented from accessing the spore's peptidoglycan cortex beneath the coat layer and the cell wall beneath this (12). Surrounding the spore coat is the 'exosporium', the outermost spore layer, which comprises only about 3% of the total coat protein (15). Underneath the coat lies a thick 'cortex' containing peptidoglycan which surrounds the 'core', the spore interior, and helps to keep the core relatively dry. The core houses the spore chromosome and also contains a complex of calcium and dipicolinic acid (DPA) and the small acid soluble proteins (SASPs). It is the interaction between these proteins and the DNA which is thought to be responsible for the spores' ability to resist stresses such as high temperatures and UV radiation (16).

7.1.4 *C. difficile* and Disease

C. difficile has in the recent past become a major cause of hospital acquired infections and is able to persist and withstand hospital disinfectants by virtue of its spores. The organism is an anaerobic, Gram-positive spore-forming bacterium that was first described in 1935 when it was isolated from the faecal flora of healthy neonates (17). It is normally considered a harmless environmental organism but has been emerging as an important nosocomial agent since broad-spectrum antibiotics (especially cephalosporins, ampicillin/amoxacillin and clindamycin) were introduced into clinical practice in the 1960s–1970s (18). *C. difficile* infection (CDI) causes symptoms ranging from mild to severe diarrhoea and abdominal pain in susceptible individuals, particularly after antibiotic therapy which disrupts the protective normal flora of the gut to allow colonisation by *C. difficile*, if exposed to the organism or its spores from endogenous or exogenous origins. Production of the two bacterial toxins *tcdA* and *tcdB* can lead to further complications including the inflammation and ulceration of the intestinal wall, termed pseudomembranous colitis (PMC).

CDI is a major problem in healthcare environments that can culminate in patient isolation, ward closures and, in extreme circumstances, hospital closure. In 2006 reported hospital cases in the elderly alone in the UK exceeded 55 000, and 6500 deaths were attributed to CDI (19).

7.1.5   Bacterial Spores of Clostridia

The main obstacles to the study of the *C. difficile* spore proteome is the process of spore purification from cellular debris resulting from lysis of vegetative cells; and solubilisation of spore proteins for analysis due to the reported hydrophobicity of proteins within the coat layers.

Despite spores having such a vital role in contamination and disease transmission, little is known about the structure and biology of the spore coat or the molecular mechanisms that lead to sporulation in *C. difficile*. To understand how such structures are built and how they function, it is crucial to identify and characterise their component polypeptides. Proteomics provides a powerful platform for the analysis of macromolecular components of subcellular structures through separation of complex protein mixtures and identification by mass spectrometry (MS).

7.2   Experimental

7.2.1   Sporulation Media

Previous methods to induce high spore yields for *Bacillus* species and *C. difficile* seem to favour nutrient exhaustion by culturing for long periods (usually 5–10 days) in different broth or solid media. A sporulation medium (SM) consisting of 9% trypicase peptone, 0.5% proteose peptone no. 3, 0.15% Tris and 0.1% ammonium sulfate (pH 7.4), was found to yield greater than 10^7 *C. difficile* spores per millilitre after incubation for 48 h in initial studies (20). In a more recent study similar levels of *C. difficile* sporulation ($\sim 10^7$ spores ml^{-1}) were observed between SM and Duncan-Strong (DS) medium (21): a sporulation medium developed and routinely used for *Clostridium perfringens* that consists of 0.4% yeast extract, 1.5% proteose peptone, 0.4% soluble starch, 0.1% sodium thioglycolate and 1.0% Na$_2$HPO$_4$.7H$_2$O (22). Brain heart infusion supplemented with yeast extract (5 mg ml^{-1}) and L-cysteine (0.1%) (BHIS) media has also been used to induce sporulation of *C. difficile* (23, 24). In addition, a recently published study found that culturing in SM broth produced 10- to 100-fold more spores than culturing in brain heart infusion (BHI) broth (25), suggesting that either SM or DS media are the preferred choices for producing high titres of *C. difficile* spores.

7.2.2   Spore Purification

In some earlier studies on *Bacillus*, multiple cold water washes were reportedly sufficient to produce spore preparations that were >98% free of vegetative cells and cellular debris (26). This was also reported for *C. perfringens* spores (27). Alternatively a lysozyme incubation step followed by salt washes, sodium dodecyl sulfate (SDS) and water was used (7–9).

Density centrifugation has often been used instead of, or in addition to, the above to enhance spore purification from any remaining vegetative or germinating cells and cellular debris. Previous studies have used gradients of Renografin (alternatively named 'Urographin', 'Nycodenz', 'Renocal-76' or 'Histodenz') (6, 23, 28–31) for this purpose. However, for proteomic studies of the spore structure and components involved in sporulation of *C. difficile*, from our own experience and that of others (25), it is apparent that a more involved and prolonged purification process is required to avoid contamination with free cytosolic/cellular proteins in the media. Lawley and colleagues found multiple cold water washes and a subsequent density gradient centrifugation step (an established method for purifying *B. subtilis* spores) left significant levels of dead vegetative cells and cellular matter behind, suggesting purification of *C. difficile* spores requires a more thorough procedure. The same study developed a robust protocol involving subsequent treatments with detergent, lysozyme, sonication, and proteinase K followed by centrifugation through 50% sucrose to purify spores from cellular debris (25).

Early experiments in our laboratory found that a spore purification method comprising lysozyme treatment and salt washes used previously for *B. subtilis* spores (7, 9) (Figure 7.2) was not sufficient at purifying *C. difficile* spores of associated cellular debris. One-dimensional sodium dodecyl sulfate-polyacrylamide gel electrophoresis (1D SDS-PAGE) analysis showed that 'purified' spore preparations were still highly contaminated with non-spore related protein, when compared with a vegetative cell protein extract (unpublished data) (Figure 7.3). This purification step is critical to reduce the complexity of the protein extracts for MS by removing non-spore related proteins that may be free in solution or, more critically, adhered to the spore outer surface due to their 'stickiness' or trapped between the spores as they have the tendency to form clumps in aqueous solution. However, it should be noted that with multiple washings, readily soluble coat polypeptides may be lost, and indeed many of the steps that are useful for removing cells and debris may lead to activation and subsequent germination of spores, including sucrose and Renografin gradient centrifugation (15).

7.2.3   Spore Protein Extraction and Solubilisation

Due to their structure and composition, spores are particularly difficult to lyse. Various chemical, enzymatic and physical methods have been reported in the literature to solubilise the spore coat proteins from pure spore preparations. Challenges arise from the thick layers of extensively cross-linked coat proteins and also from the complexity of the protein sample that can be solubilised from wild-type coats. Approximately a third of total coat protein in wild-type spores is resistant to extraction procedures normally employed to solubilise most other proteins (29). Differences have been also been found in the solubility of different coat fractions (15), which could complicate spore protein extraction further.

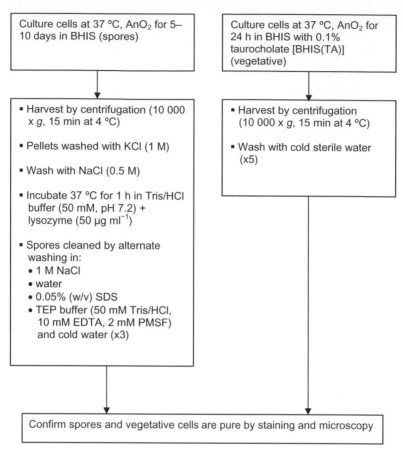

Figure 7.2   *A flow diagram of methods used for harvesting and purification of* C. difficile *spores and vegetative cells in our laboratory. AnO₂: anaerobic conditions. Spore methods adapted from reference (9)*

B. subtilis spore coat proteins, for example, have a relatively high content of cysteine and hydrophobic amino acids, and are highly resistant to conditions such as heat, pressure, chemicals, dyes and lytic enzymes (32). However recent advances in protein solubilisation techniques coupled with significant increases in MS sensitivity plus the availability of genomic sequence data have made spore protein identification more realistic (33).

Apart from the spore coat, other spore proteins of interest include the SASPs. These are present in large amounts in the core region of the spore, comprising up to 20% of total spore core protein (34). They are of great interest to researchers as they are reported to have a role in the resistance of spores to chemicals, UV radiation and heat (16, 35, 36). They are of additional interest in *Bacillus* species because they can be used as reliable biomarkers for differentiating the closely related species *B. anthracis* and *B. cereus* using MS (37).

In order to attempt to solubilise the whole spore proteome, one approach is to use multiple solubilisation methods on the same sample, as was done recently for *C. difficile* (25).

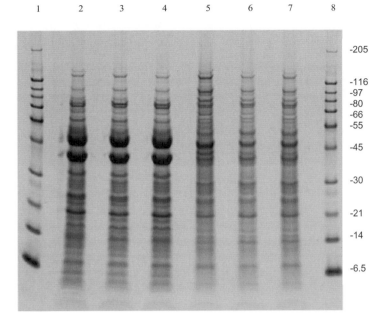

1   2   3   4   5   6   7   8

-205

-116
-97
-80
-66
-55

-45

-30

-21

-14

-6.5

Figure 7.3 *A comparison of protein extracts from C.* difficile *vegetative cells and spores 'cleaned' using a method previously used for B.* subtilis *spores. Lanes 1 and 8: protein standard; lanes 2–4: biological replicates of vegetative cell solubilised proteins; lanes 5–7: biological replicates of spore solubilised proteins. Protein standard sizes (shown on the right) are in kDa (unpublished data). Gel profiles show much overlap in proteins between spores and vegetative cells suggesting spore extracts remained significantly contaminated with cellular debris post-purification*

7.2.3.1   Chemical Methods

Isolation of spore coat proteins of *B. subtilis* was done by boiling in the presence of SDS/ dithiothreitol (DTT) (pH 6.7). Also, an alkali extraction at 0°C in sodium hydroxide (NaOH) was performed. Both of these methods are known to solubilise coat proteins. However, the alkali extraction solubilised less protein than the SDS/DTT method (7, 29) yet its utility was demonstrated by the more efficient extraction of polypeptides of interest in one case (29) (Figure 7.4). The spores are reported to remain intact and refractile during the extraction and thus the polypeptides were concluded to have been derived solely from the spore coat (7). A NaOH incubation at 0°C was also used in an earlier study to solubilise spore coat polypeptides, and again reportedly extracted less proteins than a simultaneous detergent/reducing agent extraction (9).

Boiling spores in SDS-PAGE loading buffer was successfully used to extract coat proteins from *B. subtilis* and *B. anthracis* (30, 33, 38). Similarly, boiling in lithium dodecyl sulfate (LDS) sample buffer solubilised distinct spore-specific proteins in a recent study of the *C. difficile* spore proteome (25) (Figure 7.5).

The spore coat can also be removed for analysis by extraction using detergents at high pH. A method from the *B. subtilis* spore literature used a 2-h incubation in 0.1 M DTT,

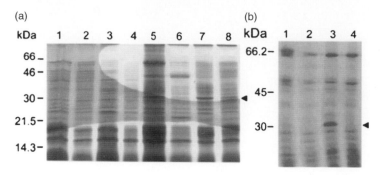

Figure 7.4   *Spore coat proteins extracted from wild-type and deletion mutants of* B. subtilis *using a buffer containing SDS-DTT (a), or with alkali (b) and resolved by SDS-PAGE using 12.5% PAGE gels. Figure taken from reference (29). Reproduced with permission from American Society for Microbiology*

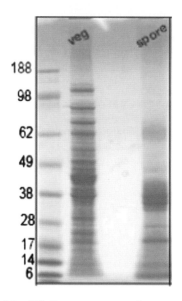

Figure 7.5   *SDS-PAGE gel of* C. difficile *vegetative cells (veg) and pure spores (spore) after solubilisation with LDS. Figure taken from reference (25). Reproduced with permission from American Society for Microbiology*

0.1 M NaCl, 0.1 M NaOH, 1% SDS (pH 10) at 70°C to remove much of the *B. subtilis* spore coat protein as well as the spore's outer membrane, effectively decoating the spore and allowing for subsequent rupture of the spores with a physical method such as sonication (12, 39). Ghosh *et al.* (12) found this method useful for comparing the coat protein content of wild type spores and spores with coat protein gene deletions (*cotE* and *gerE*) (Figure 7.6).

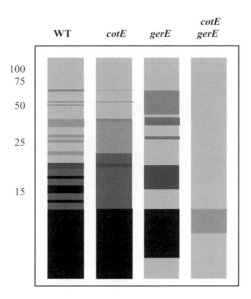

Figure 7.6 *Diagrammatic representation of the SDS-PAGE of extracted spore coats of*
B. subtilis *using various purified spore preparations using detergents at high pH. WT, wild
type; cotE, deletion mutant for coat gene 'cotE'; gerE, deletion mutant for coat gene 'gerE';
cotE gerE, deletion mutant for both coat genes. Adapted from reference (12)*

As mentioned previously, a recent study on *C. difficile* spores used subsequent extraction methods on the same sample to solubilise maximal protein in order to identify as
much of the spore proteome as possible. After an LDS buffer extraction, the insoluble
pelleted material was further treated with urea and sodium carbonate and the soluble fraction extracted for MS analysis. The remaining insoluble material was treated with a strong
solution of formic acid prior to trypsin digestion for MS analysis along with the other two
fractions. From the LDS fraction, 233 proteins were identified by liquid chromatography-
tandem mass spectrometry (LC-MS/MS); the subsequent urea extraction identified another
71 proteins, and the final formic acid treatment led to the identification of an additional
32 proteins. (25).

Various other methods from the literature include boiling purified spores in Laemmli
buffer (Tris, SDS, 2-mercaptoethanol, DTT, bromophenol blue, glycerol at pH 6.8) (29)
and treatment with other combinations of detergents (e.g. CHAPS, SDS), reducing agents
(e.g. DTT, 2-mercaptoethanol) and other protein denaturants (e.g. urea, thiourea, Tween)
as well as protease inhibitors (e.g. phenylmethylsulfonyl fluoride (PMSF)) to try to solubilise as much of the protein component of the sample as possible for analysis.

7.2.3.2   Mechanical Methods

In a study of spore protein components of the germination response in *B. subtilis* (the *Ger*
proteins), spore protein fractions were produced by spore breakage using a FastPrep cell
disintegrator (MP Biomedicals), where glass beads were added to spores suspended in a

breakage buffer containing Tris, ethylenediaminetetraacetic acid (EDTA) and a protease inhibitor and subjected to short bursts of agitation (26).

Breakage with glass beads in a Bronwill cell disintegrator has been used previously for the disruption *B. subtilis* spores. A previous study reported 7 min of shaking was sufficient for total spore disruption (40). An earlier study reported a 3 min treatment achieved greater than 99% breakage (8). A dental amalgamator combined with glass beads has also been used to dry-rupture lyophilised *B. subtilis* spores (39).

Sonication is a commonly used method that uses pulsed, high frequency sound waves to shear bacterial cells and spores. Sound waves are delivered to the sample using a vibrating probe that is immersed in the cell suspension. The mechanical energy from the probe causes microscopic bubbles to form momentarily and then implode causing shock waves to radiate through the sample, which lyses the cells. As well as for cell lysis, sonication has also been used to physically remove remnants of digested or lysed cellular material from the spore surface (41).

An alternative method to bead beating, sonication and the other variations mentioned above is Pressure Cycling Technology (PCT) (Pressure BioSciences Inc.). PCT uses rapid cycling between high and low pressure to rupture cells in a temperature controlled chamber.

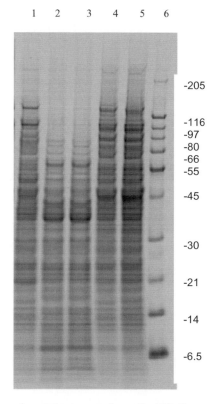

Figure 7.7   *A comparison of protein extracts from* C. difficile *spores using: lane 1, FastPrep with an in-house lysis buffer; lanes 2 and 3, ProteoSolv-SB (Pressure BioSciences Inc.) and PCT; and lanes 4 and 5, PCT with an in-house lysis buffer. Lane 6, protein standard. Protein standard sizes (shown on the right) are in kDa (unpublished data)*

Although at the time of publication no published bacterial spore protein data were available, from studies of the non-sporing, Gram-negative bacterium *Escherichia coli*, PCT was found to extract 14.2% more protein compared with using a bead mill as determined by two-dimensional gel electrophoresis, and revealed several low-abundant species that were not detected in the bead mill lysate (42). This method is deemed particularly useful for low abundant, high molecular weight, hydrophobic and basic proteins over other conventional extraction methods (32). In our laboratory, PCT was found to extract more of the higher molecular weight protein species from spores compared with the FastPrep extraction method using the same extraction buffer and the comparative abundance of some of the weaker bands increased (unpublished data) (Figue 7.7). The ProteoSolv-SB kit (Pressure BioSciences Inc.), a detergent-free extraction kit for PCT, gave very low protein concentrations, as determined by the Bradford method (43), however loss of protein may have occurred during the precipitation step of the process. A higher abundance of some of the lower molecular weight proteins and potentially some novel protein bands between 14 kDa and 21 kDa were observed (Figure 7.8).

For many years the French Press has been regarded as the most efficient mechanical method for disrupting cells. It works by applying hydraulic pressure to a cell suspension using a piston under high pressure to force a liquid sample through a tiny hole in the press. The sudden change in pressure from 8000 to 20000 psi (typically) to atmospheric pressure causes shearing of cells.

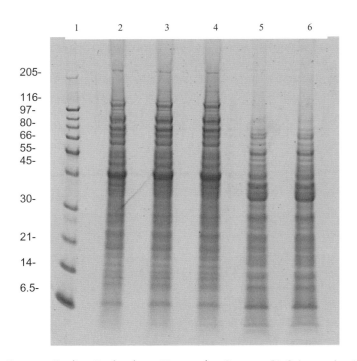

Figure 7.8 *Pressure Cycling Technology (Barocycler, Pressure BioSciences Inc.) protein extraction. Lane 1: protein standard; lanes 2–4: biological replicates of C.* difficile *spore PCT protein extracts in cell lysis buffer; lanes 5 and 6: spore replicates of PCT using the ProteoSolve-SB kit. Protein standard sizes (shown on the left) are in kDa (unpublished data)*

A major drawback of using mechanical methods over chemical or enzymatic is the former are more likely to damage the released subcellular contents. Ultimately the energy absorbed by the cell suspension will cause localised heating within the sample, leading to protein denaturation and degradation, but can be minimised by pre-chilling equipment or buffers, processing the sample on ice or interspersing repeated processing with frequent, short incubations on ice. Alternatively, for the French Press, the entire stainless steel pressure cell may be immersed in liquid nitrogen so that when the cell suspension is added to the sample chamber it is frozen immediately. Even after it is forced through to the collecting chamber by the piston, the cell extract is again immediately frozen and so minimises any degradation due to heat.

7.3 Conclusion

Spores are vital for the long-term survival of *Bacillus* and *Clostridium*. Some of the most dangerous pathogens known to man (e.g. *B. anthracis*, *C. tenani*, *C. botulinum*, etc.) belong to these genera. Hence their persistence and mechanisms of spore germination *in vivo* are of considerable importance in human infections. *C. difficile* acting as a pathogen is a relatively recent problem, which is exacerbated by the ability of its spores to thrive in hospital environments even after decontamination procedures. The structure, composition and mechanisms of bacterial spores and their potential to initiate disease in man have long been studied. Difficulties in lysis, the nature of the macromolecules and the problems of being able to clearly differentiate between vegetative cells and spore proteins remain major obstacles to fully understanding the biology of these structures. More recently, the availability of several full genomes of multiple strains, new genetic modification systems for expression and analysis, together with the high resolution and sensitivity of mass spectral analysis, has led to a resurgence of work in this field and more detailed mechanisms are unfolding in different species. Once these are available it may be possible to design therapeutic strategies to specifically disrupt key processes in their germination and reduce their infectious disease risk. Already, unique molecules within their structure are being coupled to nanoparticles and analysed by MS to detect their presence directly in biological fluids (Chapter 1).

References

1. Henriques AO, Moran Jr CP. Structure, assembly, and function of the spore surface layers. Annu. Rev. Microbiol. 2007; 61: 555–588.
2. Wilcox MH, Fawley WN. Hospital disinfectants and spore formation by *Clostridium difficile*. The Lancet 2000; 356: 1324.
3. Driks A. Maximum shields: the assembly and function of the bacterial spore coat. Trends Microbiol. 2002; 10(6): 251–254.
4. Paredes CJ, Alsaker KV, Papoutsakis ET. A comparative genomic view of Clostridial sporulation and physiology. Nature Rev. Microbiol. 2005; 3: 969–978.
5. Eichenberger P, Fujita M, Jensen ST, Conlon EM, Rudner DZ, Wang ST, Ferguson C, Haga K, Sato T, Liu JS, Losick R. The program of gene transcription for a single differentiating cell type during sporulation in *Bacillus subtilis*. PLoS Biol. 2004; 2: 1664–1683.
6. DelVecchio VG, Connolly JP, Alefantis TG, Walz A, Quan MA, Patra G, Ashton JM, Whittington JT, Chafin RD, Liang X, Grewel P, Khan AS, Mujer CV. Proteomic profiling and

identification of immunodominant spore antigens of *Bacillus anthracis*, *Bacillus cereus*, and *Bacillus thuringiensis*. Appl. Environ. Microbiol. 2006; 72(9): 6355–6363.

7. Donovan W, Zheng L, Sandman K, Losick R. Genes encoding spore coat polypeptides from *Bacillus subtilis*. J. Mol. Biol. 1987; 196: 1–10.

8. Goldman RC, Tipper DJ. *Bacillus subtilis* spore coats: complexity and purification of a unique polypeptide component. J. Bacteriol. 1978; 135(3): 1091–1106.

9. Jenkinson HF, Sawyer WD, Mandelstam J. Synthesis and order of assembly of spore coat proteins in *Bacillus subtilis*. J. Gen. Microbiol. 1981; 123: 1–16.

10. Heap JT, Pennington OJ, Cartman ST, Carter GP, Minton NP. The ClosTron: A universal gene knock-out system for the genus *Clostridium*. J. Microbiol. Methods 2007; 70: 452–464.

11. Piggot PJ, Hilbert DW. Sporulation of *Bacillus subtilis*. Curr. Opin. Microbiol. 2004; 7: 579–586.

12. Ghosh S, Setlow B, Wahome PG, Cowan AE, Plomp M, Malkin AJ, Setlow P. Characterisation of spores of *Bacillus subtilis* that lack most coat layers. J. Bacteriol. 2008; 190(20): 6741–6748.

13. Murrell WG. Chemical composition of spores and spore structures. In: Gould GW, Hurst A, editors. The Bacterial Spore. New York: Academic Press Inc.; 1969. p. 215.

14. Moir A, Corfe BM, Behravan J. Spore germination. Cell. Mol. Life Sci. 2002; 59(3): 403–409.

15. Aronson AI, Fitz-James P. Structure and morphogenesis of the bacterial spore coat. Bacteriol. Rev. 1976; 40(2): 360–402.

16. Setlow P. Spores of *Bacillus subtilis*. Their resistance to and killing by radiation, heat and chemicals. J. Appl. Microlbiol. 2006; 101(3): 514–525.

17. Hall IC, O'Toole E. Intestinal flora in newborn infants with a description of a new pathogenic anaerobe, *Bacillus difficilis*. Am. J. Dis. Child 1935; 49: 390–402.

18. Farrell RJ, LaMont JT. Pathogenesis and clinical manifestations of *Clostridium difficile* diarrhea and colitis. In: Aktories K, Wilkins TC, editors. Current Topics in Microbiology and Immunology 50: *Clostridium difficile*. Berlin: Springer-Verlag; 2000. p. 109.

19. Dawson LF, Valiente E, Wren BW. *Clostridium difficile* – a continually evolving and problematic pathogen. Infect. Genet. Evol. 2009; 9(6): 1410–1417.

20. Wilson KH, Kennedy MJ, Fekety FR. Use of sodium taurocholate to enhance spore recovery on a medium selective for *Clostridium difficile*. J. Clin. Microbiol. 1982; 15(3): 443–446.

21. Paredes-Sabja D, Bond C, Carman RJ, Setlow P, Sarker MR. Germination of spores of *Clostridium difficile* strains, including isolates from a hospital outbreak of *Clostridium difficile*-associated disease (CDAD). Microbiol. 2008; 154: 2241–2250.

22. Duncan CL, Strong DH. Improved medium for sporulation of *Clostridium perfringens*. Appl. Microbiol. 1968; 16(1): 82–89.

23. Sorg JA, Sonenshein AL. Bile salts and glycine as cogerminants for *Clostridium difficile* spores. J. Bacteriol. 2008; 190(7): 2505–2512.

24. Haraldsen JD, Sonenshein AL. Efficient sporulation in *Clostridium difficile* requires disruption of the sigma k gene. Mol. Microbiol. 2003; 48(3): 811–821.

25. Lawley TD, Croucher NJ, Yu L, Clare S, Sebaihia M, Goulding D, Pickard DJ, Parkhill J, Choudhary J, Dougan G. Proteomic and genomic characterisation of highly infectious *Clostridium difficile* 630 spores. J. Bacteriol. 2009; 191(17): 5377–5386.

26. Hudson KD, Corfe BM, Kemp EH, Feavers IM, Coote PJ, Moir A. Localisation of GerAA and GerAC germination proteins in the *Bacillus subtilis* spore. J. Bacteriol. 2001; 183(14): 4317–4322.

27. Paredes-Sabja D, Setlow B, Setlow P, Sarker MR. Characterisation of *Clostridium perfringens* spores that lack SpoVA proteins and dipicolinic acid. J. Bacteriol. 2008; 190(13): 4648–4659.

28. Buchanan CE, Neyman SL. Correlation of penicillin-binding protein composition with different functions of two membranes in *Bacillus subtilis* forespores. J. Bacteriol. 1986; 165(2): 498–503.

29. Serrano M, Zilhao R, Ricca E, Ozin AJ, Moran Jr CP, Henriques AO. A *Bacillus subtilis* secreted protein with a role in endospore coat assembly and function. J. Bacteriol. 1999; 181(12): 3632–3643.

30. Bauer T, Little S, Stover AG, Driks A. Functional regions of the *Bacillus subtilis* spore coat morphogenetic protein CotE. J. Bacteriol. 1999; 181(22): 7043–7051.

31. Coleman WH, Chen D, Li Y, Cowan AE, Setlow P. How moist heat kills spores of *Bacillus subtilis*. J. Bacteriol. 2007; 189(23): 8458–8466.

32. Tao F, Li C, Smejkal G, Lazarev A, Lawrence N, Schumacher RT. Pressure Cycling Technology (PCT) applications in extraction of biomolecules from challenging biological samples. In: Proceedings of the 4th International Conference on High Pressure Bioscience and Biotechnology 2007. Tsukba, Japan. Vol. 1. pp. 166–173. Available from http://www.pressurebiosciences.com/downloads/publications/HPBB-proceeding-paper.pdf.

33. Lai E-M, Phadke ND, Kachman MT, Giorno R, Vazquez S, Vazquez JA, Maddock JR, Driks A. Proteomic analysis of the spore coats of *Bacillus subtilis* and *Bacillus anthracis*. J. Bacteriol. 2003; 185(4): 1443–1454.

34. Moeller R, Setlow P, Reitz G, Nicholson WL. Roles of small, acid-soluble spore proteins and core water content in survival of *Bacillus subtilis* spores exposed to environmental solar UV radiation. Appl. Environ. Microbiol. 2009; 75(16): 5202–5208.

35. Paredes-Sabja D, Raju D, Torres JA, Sarker MR. Role of small, acid-soluble proteins in the resistance of *Clostridium perfringens* spores to chemicals. Int. J. Food Microbiol. 2008; 122(3): 333–335.

36. Raju D, Waters M, Setlow P, Sarker MR. Investigating the role of small, acid-soluble proteins (SASPs) in the resistance of *Clostridium perfringens* spores to heat. BMC. Microbiol. 2006; 6: 50.

37. Castanha ER, Fox A, Fox KJ. Rapid discrimination of *Bacillus anthracis* from other members of the *B. cereus* group by mass and sequence of 'intact' small acid soluble proteins (SASPs) using mass spectrometry. J. Microbiol. Methods 2006; 67: 230–240.

38. Little S, Driks A. Functional analysis of the *Bacillus subtilis* morphogenetic spore coat protein CotE. Mol. Microbiol. 2001; 42(4): 1107–1120.

39. Bagyan I, Noback M, Bron S, Paidhungat M, Setlow P. Characterisation of *yhcN*, a new forespore-specific gene of *Bacillus subtilis*. Gene 1998; 212: 179–188.

40. Pandey NK, Aronson AI. Properties of the *Bacillus subtilis* spore coat. J. Bacteriol. 1979; 137(3): 1208–1218.

41. Plomp M, McCaffery JM, Cheong I, Huang X, Bettegowda C, Kinzler KW, Zhou S, Vogelstein B, Malkin AJ. Spore coat architecture of *Clostridium novyi* NT spores. J. Bacteriol. 2007; 189(17): 6457–6468.

42. Smejkal G, Robinson MH, Lawrence NP, Tao F, Saravis CA, Schumacher RT. Increased protein yields from *Escherichia coli* using Pressure-Cycling Technology. J. Biomol. Tech. 2006; 17: 173–175.

43. Kruger NJ. The Bradford method for protein quantitation. Methods Mol. Biol. 1994; 32: 9–15.

8

Characterization of Bacterial Membrane Proteins Using a Novel Combination of a Lipid Based Protein Immobilization Technique with Mass Spectrometry

Roger Karlsson[1], Darren Chooneea[2], Elisabet Carlsohn[3], Vesela Encheva[2,4] and Haroun N. Shah[2]

[1] *Department of Chemistry, University of Gothenburg, Gothenburg, Sweden*
[2] *Department for Bioanalysis and Horizon Technologies, Health Protection Agency Centre for Infections, London, UK*
[3] *Proteomics Core Facility, Sahlgrenska Academy, University of Gothenburg, Gothenburg, Sweden*
[4] *Present address: Research and Technology Department, Laboratory of the Government Chemist (LGC), Teddington, UK*

8.1 Introduction

The Gram stain of 1884 (Gram, 1884) led to bacterial species being placed into two broad groups that reflect the structure of their cell wall (Bergey *et al.*, 1994; Beveridge, 2001). The Gram-positive bacterium has a thick layer of peptidoglycan in the cell envelope [Figure 8.1(a)]. Beneath this is the periplasmic space, followed by the plasma membrane (the cytoplasmic membrane). Gram-negative bacteria possess an outer membrane on the surface of the cell wall, which contains the lipopolysaccharides and outer membrane proteins (e.g. porins), that allow acquisition of nutrients; lipoproteins that are anchored to the outer membrane by a lipid tail; and integral outer membrane proteins with several transmembrane-spanning regions. Beneath the outer membrane is a periplasmic space in

Mass Spectrometry for Microbial Proteomics   Edited by Haroun N. Shah and Saheer E. Gharbia
© 2010 John Wiley & Sons, Ltd

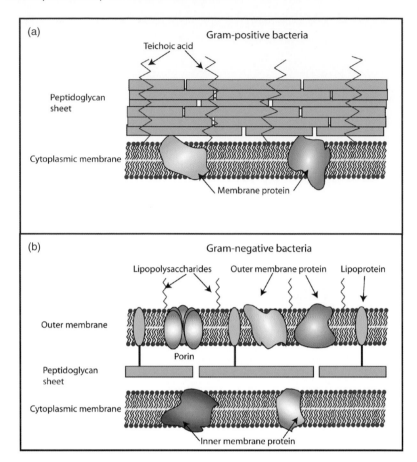

Figure 8.1 *Schematic overview of the structure of the cell envelope of Gram-positive and Gram-negative bacteria. (a) Gram-positive bacteria have a thick peptidoglycan sheet covering the outer membrane of the bacterial cell. Membrane proteins with various functions are embedded into this cytoplasmic membrane. (b) Gram-negative bacteria have two membranes that separate the cytosol of the cell from the surrounding environment. The outer membrane contains membrane proteins (e.g. porins), integral membrane proteins and lipoproteins. Beneath the outer membrane is a peptidoglycan sheet connected to the membrane by the lipoproteins. The periplasmic space surrounds the peptidoglycan sheet. The inner membrane (the cytoplasmic membrane) also contains membrane proteins*

which a much thinner peptidoglycan layer is embedded, and finally the plasma membrane which surrounds the cytosol of the bacterium [Figure 8.1(b)].

8.2   The Surface Proteome

Surface proteins including outer membrane proteins (OMPs) are of utmost importance for pathogenic bacteria. For example, bacteria use surface proteins for sensing their surrounding environment, for acquiring essential nutrition, and for communication purposes

(Koebnik *et al.*, 2000). Motility, adherence and colonization of host cells are highly important pathogenic processes governed by surface proteins (Koebnik *et al.*, 2000). Some outer membrane proteins form channels which helps the bacterium to remove antibiotics and toxic by-products from their cells as well as facilitate a means to inject toxins and extracellular proteases into their host. Most of the OMPs of Gram-negative bacteria have a β-barrel structure, whereas the integral membrane proteins of the inner (cytoplasmic) membrane contains transmembrane α-helices to a much higher degree (Koebnik *et al.*, 2000; Delcour, 2002; Schulz, 2002) (Table 8.1). Some key proteins are conserved throughout many different kinds of bacteria, for example the ompA outer membrane protein, which is important for membrane structural integrity by linking the outer membrane to the peptidoglycan layer (Table 8.2). In order to gain insight into host–pathogen interactions and therefore detailed mechanisms of disease, it is imperative to elucidate the surface proteome (surfacesome) including the OMPs. Furthermore, most OMPs are surface-exposed which makes them potentially useful as candidates for establishing subunit vaccines or for use as drug targets.

8.3   Proteomics of Pathogenic Bacteria

The genomes of a large number of pathogenic bacteria have been completed, however the function of many of the predicted proteins are still unknown. Importantly, a substantial part of the unknown proteins are predicted to be membrane-associated, possibly exposed on the surface of the bacteria. During the last few years, proteomic tools have become available to investigate the structure and function of the predicted proteins and especially observe which proteins are actually expressed by the organism. For example, the genomes of *Salmonella* Typhi and *Salmonella* Typhimurium are very similar, having an approximate genetic overlap of 98% (McClelland *et al.*, 2001; Parkhill *et al.*, 2001; Edwards *et al.*, 2002; Baker and Dougan, 2007). However, infection by *S.* Typhimurium normally leads to mild forms of diarrhoea, whereas infection by *S.* Typhi may induce fatal complications. However, it can be hypothesized that variation in expression levels of the proteins encoded in both genomes accounts for the diversity in their pathogenicity. In conjunction with investigation of expression levels, the spatial and temporal distributions of proteins need to be fully understood in order to elucidate the mechanisms of survival and virulence.

Traditionally, a standard proteomic analysis of an organism begins with the isolation of soluble proteins, followed by the separation and visualization of the protein mixture by two-dimensional gel electrophoresis (2D GE) (Washburn and Yates, 2000) (see Chapter 6). Although powerful, there are severe limitations associated with 2D GE. Besides being time-consuming and laborious, 2D gels rarely identify low-abundance and hydrophobic proteins, the latter mainly due to poor solubility in water-based buffers (see Chapter 9). Membrane proteins associated with lipid bilayers are generally hydrophobic and often contain multiple trans-membrane domains (Washburn and Yates, 2000). For pathogenic Gram-negative bacteria, it has been shown that many of the OMPs span membranes through the formation of β-barrels with the hydrophobic residues oriented outwards (Cordwell, 2008). If several β-barrels are present, as is the case with integral membrane proteins with trans-membrane-spanning regions, solubilisation is almost impossible even with the most aggressive ionic detergents (Cordwell, 2008).

Table 8.1 *Structural and functional features of prototype OMPs from E. coli. Examples of OMPs sharing the β-barrel structure. Some proteins function as membrane anchors, some as enzymes or receptors, and others as general or specific porins*

Protein family	Small β-barrel membrane achors	Small β-barrel membrane achors	Membrane-integral enzymes	General (nonspecific) porins	Substrate-specific Porins	TonB-dependent receptors
Prototype protein	ompA	ompX	pldA (OMPLA)	ompF	lamb	fhuA
Protein function	Physical linkage between OM and peptidoglycan	Neutralizing host defence mechanisms	Hydrolysis of phospho-lipids	Diffusion pore for ions and other small molecules	Maltose and malto-dextrin uptake	Uptake of iron-siderophore complexes; signal transduction
Size of membrane domain	171 residues	148 residues	269 residues	340 residues	421 residues	714 residues
Number of trans-membrane β-strands	8	8	12	16	18	22

Adapted from Koebnik *et al.* (2000).

Table 8.2 Examples of overlapping protein families found in some pathogenic bacteria. The table illustrates how the OMP's ompA, fadL, tsx and ompW are distributed among some human Gram-negative pathogens. The ompA and fadL are conserved among the species mentioned here; however the tsx and ompW are not present in all species

Protein family	ompA	fadL	tsx	ompW
Function	Physical linkage between OM and peptidoglycan	Transport of long-chain fatty acids; receptor for colicin K and Bacteriophage T6	Substrate-specific channel for nucleosides; receptor for colicin K and Bacteriophage T6	This protein may form the receptor for S4 colicins in *E. coli*
Species				
Escherichia coli K12-MG1655	X	X	X	X
Pseudomonas aeruginosa PAO1	X	X	X	X
Salmonella typhi CT18	X	X	X	X
Salmonella Typhimurium LT2	X	X	X	X
Haemophilus influenzae KW20	X	X		
Helicobacter pylori 26695	X	X		
Neisseria meningitidis MC58	X	X		

New methods and techniques for analysis of bacterial surface proteins have recently been developed (Macher and Yen, 2007: Cordwell, 2008), including improved sample preparation prior to 2D GE, combination of sodium dodecyl sulfate polyacrylamide gel electrophoresis (SDS-PAGE) [one-dimensional (1D) gel] together with liquid chromatography-tandem mass spectrometry (LC-MS/MS), shot-gun proteomics and 2D LC methods. For example, the traditional 2D GE has been used in conjunction with optimized separation and solubilization protocols in order to analyse membrane proteins of several kinds of bacteria such as *Escherichia coli* and *Pseudomonas aeruginosa* (Molloy *et al.*, 2000; Nandakumar *et al.,* 2005; Schindler *et al.*, 2006; Zahedi *et al.*, 2005; Zuobi-Hasona and Brady, 2008; Braun *et al.*, 2007; Mujahid *et al.*, 2007; Cordwell, 2008; Molloy, 2008; Zuobi-Hasona *et al.*, 2005). These sample preparation protocols rely on sequentially removing hydrophilic proteins mainly from the cytosol by using solubilising reagents, such as strong detergents and sodium carbonate, and centrifugation steps. The final insoluble pellet assumes enrichment of membrane proteins, however many of the identified proteins are the highly abundant ones including the relatively hydrophilic membrane proteins, such as porins. For Gram-negative bacteria, various protocols for the selective extraction of membrane proteins have been developed. Examples include the use of the detergent lauroyl sarcosine which disrupts the inner membrane whilst leaving the outer membrane more or less intact. In this way, a membrane preparation containing both outer and inner membrane can be processed to enrich for the outer membrane fraction. This method has been used successfully for elucidating the outer membrane proteome of several bacteria (Baik *et al.*, 2004; Rhomberg *et al.*, 2004; Peng *et al.*, 2005; Carlsohn *et al.*, 2006a; Liu *et al.*, 2008; Rivas *et al.*, 2008). A drawback of this method is that much of the sample may be lost in the purification step. Another strategy for enrichment of surface-exposed membrane proteins involves use of the biotinylation agent (Ge and Rikihisa, 2007; Harding *et al.*, 2007; Smither *et al.*, 2007). The surface proteins are labelled with biotin and then purified by affinity chromatography. This technique has been used for *H. pylori* (Sabarth *et al.*, 2002b) and *Streptococcus pyogenes* (Cole *et al.,* 2005). However, it has been shown that unwanted permeation of the biotinylation agent through the bacterial membrane may occur, leading to the labelling of cytosolic proteins (Cordwell, 2008). All the above-mentioned methods normally use 2D GE for analysis, whereby highly hydrophobic and low abundant proteins were still missing from the analysis.

Instead of 2D GE, 1D SDS-PAGE has been used in conjunction with LC-MS/MS analysis for the elucidation of bacterial membrane proteomes (Carlsohn *et al.*, 2006a; Liu *et al.*, 2008). Examples include the membrane proteome of *Mycobacterium tuberculosis* (Xiong *et al.*, 2005) and *Listeria monocytogenes* (Wehmhoner *et al.*, 2005). Enriched membrane fractions are prepared by different solubilisation protocols and centrifugations steps followed by solubilisation with SDS and separation of the extracted proteins using SDS-PAGE. Subsequently, gel pieces are excised and the proteins are eluted and enzymatically digested by trypsin and the resulting peptides analysed using LC-MS/MS. A drawback of this approach is that it is not suitable for quantitative/comparative type of experiments.

8.4   Lipid-Based Protein Immobilization Technology

The Lipid-based Protein Immobilization (LPI) technique was developed by Nanoxis AB (Gothenburg, Sweden) with the aim of studying membrane proteins (integral or associated) in their natural lipid bilayer environment. The technique is based on specifically designed surfaces created to attract cellular membrane material in the form of membrane vesicles containing membrane associated proteins. The work-flow starts with purification of membrane preparations and production of small vesicles containing the proteins of interest. The purified membrane vesicles are injected into the LPI™ FlowCell, allowing attachment to the surfaces. The flow cell format enables fast exchange of fluid around the stationary surface-attached proteoliposomes. The carrier flow makes various treatments possible, e.g. enzymatic digestion of proteins, in one or multiple steps using various enzymes sequentially (Figure 8.2). By using proteases such as trypsin, the surface exposed parts of the membrane associated proteins are digested into smaller peptide fragments which can be

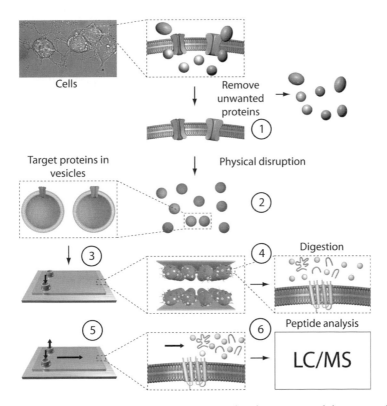

Figure 8.2   *Schematic overview showing the principle of operation of the LPI™ FlowCell. (1) Cells are homogenized and the membranes are washed in order to remove unwanted proteins (for example, highly abundant cytosolic proteins). (2) The washed membranes are physically disrupted into small vesicles by ultrasonication protocols. (3) The LPI™ FlowCells are loaded with the solution containing the vesicles which are allowed to attach to the membrane-attracting surfaces. (4) The stationary membrane proteins can subsequently be processed by various procedures, for example enzymatic digestion using different proteases. (5) As illustrated, trypsin digests external loops exposed on the surface of the membranes into smaller peptide fragments. (6) These peptides are eluted from the FlowCell and analysed by LC-MS/MS*

eluted from the flow cell and analysed by LC-MS/MS. In order to retrieve parts of the membrane proteins embedded in the lipid bilayer, other proteases have also been used in sequential digestion steps following trypsin digestion. Using this multi-step, multi-protease method very high sequence coverage has been achieved on transmembrane proteins (Figure 8.3). When analysing the outer membrane proteome of *S.* Typhimurium, a multi-step digestion procedure resulted in 70% sequence coverage of the membrane-embedded β-barrel structure of the outer membrane protein OmpA.

Originally developed for eukaryotic cell lines, LPI™ technology has been used in a wide variety of applications, involving, for example, stem cells and breast cancer cells. LPI™ technology can also be used for proteins purified and reconstituted into lipid bilayer vesicles. The concept was shown by using human aquaporin 1 (Nyblom *et al.*, 2007). This transmembrane protein was overexpressed in yeast, solubilised, purified and reconstituted in lipid bilayer vesicles. Membrane protein analysis on the different membrane preparations (crude, purified and reconstituted) was performed by subjecting the proteoliposomes with aquaporins to enzymatic digestion in the LPI™ FlowCell. Apart from anaimal cells, plant cell membrane proteins (unpublished data) have been analysed using the LPI™ FlowCell. Additional examples include e.g. plant cell membrane proteins (unpublished data). Plasma membranes from *Arabidopsis thaliana* were analysed and the results demonstrated a very high relative amount of plasma membrane proteins, compared with soluble proteins and membrane proteins originating from other membranes. The LPI™ FlowCell has also been used for studying bacterial membrane proteomics of anammox (anaerobic ammonium oxidation) bacteria residing in oceans and waste-water treatment plants (Karlsson *et al.*, 2009). Anammox bacteria are very important in the global nitrogen cycle and are estimated to contribute between 25% and 50% of the marine production of N_2 (Devol, 2003; Arrigo, 2005; Kuypers *et al.*, 2005). Analysis of the membrane proteome of the organelle-like compartment of the anammox bacteria revealed key proteins of the reaction, some of which were previously unknown.

Here, the outer membrane proteomes of the two pathogens, *H. pylori* and *S.* Typhimurium are reported as examples to demonstrate the potential of the technology. In the case of *H. pylori*, we demonstrate the use of the LPI™ FlowCell in conjunction with intact bacterial cells while for *S.* Typhimurium outer membrane vesicles were prepared prior to analysis.

Figure 8.3 *Schematic image showing the membrane embedded part of the ompA protein. (a) The ompA β-barrel contains eight transmembrane β-sheets with short loops on both sides of the membrane. (Left) Snakeplot of the membrane embedded part. (Right) Three-dimensional structure of the membrane embedded part. (b) Outer membrane vesicles of S. typhimurium were digested in the LPI™ FlowCell using a multi-step procedure with one short 30 min trypsin digestion followed by another trypsin digestion for 2 h. The first digestion step enabled the discovery of a certain set of peptides (shown in red). (c) The second digestion step with trypsin also contained an acid-labile detergent that enabled another set of peptides to be discovered (shown in red). (d) Summary of all peptides that were found using this multi-step procedure (shown in red). The remaining fragment does not have any trypsin-cleavable sites and is too large to detect in the mass spectrometer. Hence, all fragments that were possible to find of the membrane embedded ompA β-barrel were detected in this study*

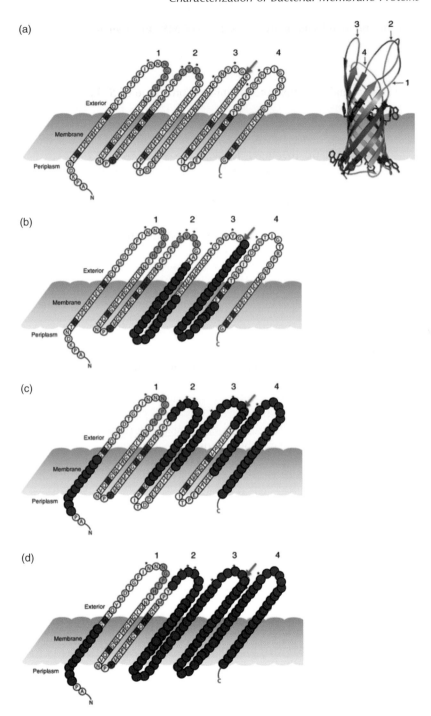

8.5   *Salmonella* Typhimurium – Disease Mechanism and Outer Membrane Proteins

Salmonella is classified under the large family Enterobacteriaceae, which includes other pathogens such as *Escherichia* and *Shigella* (Bopp *et al.*, 2003). Organisms that belong to this family are Gram-negative, straight rods and can either be nonmotile or motile via flagella (Prescott *et al.*, 2002). The *Salmonella* genus is composed of two species, *Salmonella bongori* and *Salmonella enterica* (Popoff *et al.*, 2003). *Salmonella enterica* comprises six subspecies which are frequently subdivided further based on the expression and detection of three cell surface antigens, the somatic 'O' antigen, the flagella 'H1' and 'H2' antigen, and the capsular virulence antigen 'Vi' (Popoff *et al.*, 2003). Infection caused by *Salmonella* is referred to as 'Salmonellosis', with symptoms that include enteric fever (typhoidal *Salmonella*) and gastroenteritis (nontyphoidal *Salmonella*). Currently, there are more than 2500 different *Salmonella* serotypes that have been identified, with each having the potential to cause illness in humans. Even though these serotypes are closely related, they all have a variable, and in some cases limited, host spectrum. For example, the serotypes Typhimurium and Enteriditis have the potential to infect and cause disease in a variety of hosts such as humans, poultry, pigs, rodents and cattle and thus jump the species barrier with ease. Others such as Typhi, Dublin, Choleraesuis and Pullorum are more host-specific but can still cause disease in other animal hosts (Baumler *et al.*, 1998). Salmonellosis is one of the most frequent and widespread food-borne infections and is contracted by the consumption of contaminated food and water. The bacteria gain entry into the host by invading the lymphoid nodules located in the walls of the small intestine. In nontyphoidal Salmonellosis, the bacterium invades the intestinal mucosa and infects the phagocytes within. This leads to the localization of the infection and inflammation of the small intestine resulting in gastroenteritis (Baumler *et al.*, 1998; Monack *et al.*, 2004). However, in typhoidal Salmonellosis, *Salmonella* Typhi infected phagocytes gain access to the lymphatic and circulatory systems. The organism can thereby spread and invade other parts of the body such as the liver, spleen, gall bladder and bone marrow, which may lead to further complications such as septicaemia, encephalitis, endocarditis and even death (Monack *et al.*, 2004). *Salmonella* Typhi, it is responsible for causing between 13 and 33 million cases of typhoid fever, leading to 200 000–600 000 deaths annually worldwide (Huang and DuPont, 2005).

The pathogenicity of a particular strain of *Salmonella* is determined by its virulence factors. Regions of the chromosome that code for these virulence factors are known as pathogenicity islands. *Salmonella enterica* contains two pathogenicity islands, encoding for two different types of secretion systems that are responsible for enabling pathogenic *Salmonella* to transfer virulence factors into the host, allowing it to invade and use the host cellular processes (Gophna *et al.*, 2003; Kuhle and Hensel, 2004). One secretion system is responsible for the invasion of the host's intestinal cells, while the other is responsible for the survival and proliferation of the bacteria within the host macrophages (Fardini *et al.*, 2007). Overall, the secretion system consists of more than 20 proteins including soluble cytoplasmic proteins, integral membrane proteins and OMPs (Gophna *et al.*, 2003). Despite widespread and thorough studies of *Salmonella*, there are important areas where crucial information is lacking.. A key to understanding host–pathogen interaction, including host invasion and colonization is the outer membrane proteome. Application of LPI technology should help to unravel further details of the components of the outer membrane complex and provide further insight into understanding the pathogensis of this species.

NCBI database searches of *S.* Typhimurium for proteins annotated as OMPs, including homologues of *E. coli* OMPs found in the *S.* Typhimurium genome, revealed at least 131 proteins as potential OMPs (McClelland *et al.*, 2001; Coldham and Woodward, 2004). The list includes, for example, integral OMPs and porins, lipoproteins and receptors. Many of the predicted surface proteins are very likely key components for the pathogenicity of *S.* Typhimurium. For example, the TolC and Tsx channels are probably important in the antibiotic resistance of *S.* Typhimurium (Coldham and Woodward, 2004), as might also the outer membrane porin OmpD, while the receptor BtuB is essential for acquiring vitamin B_{12} nutrition into the *Salmonella* cells.

8.6 Outer Membrane Proteins of *S.* Typhimurium

The lipid-based immobilization technique was used in order to elucidate the outer membrane proteome of *S.* Typhimurium. First, outer membrane vesicles were prepared through chemical and enzymatic treatments. The vesicles formed on the surface of the bacteria were budded off from the bacteria through the addition of energy in the form of shaking (Figure 8.4). The vesicles were loaded onto the surfaces of the LPI™ FlowCell and the membrane associated proteins were digested through a multi-step trypsin digestion. This multi-step procedure was designed to optimize the number of peptides and proteins found in the MS/MS analysis, and consisted of a short 30 min trypsin digestion followed by a 2 h trypsin digestion. The eluted peptides were subsequently analysed by LC-MS/MS.

Among the identified proteins, at least 71 were known to belong to the class of OMPs. Different functions were represented by, for example, transporter proteins such as BtuB (vitamin B_{12}) and FadL (fatty acids), OMPs, for example OmpA, channels such as TolC and Tsx as well as a large number of lipoproteins (Table 8.3). LPI technology was therefore shown to be a marked improvement over previous methods for elucidating the OMPs of *S.* Typhimurium.

In order to obtain more information about the outer membrane proteome of *Salmonella*, protocols including the use of multi-step, multi-protease digestion schemes will be further utilized. Membrane proteins buried deep in the lipid bilayer membrane are normally not digested by trypsin alone. For this reason, other proteases will be included in the sequential digestion scheme. This strategy will most definitely elucidate more porins, a group of OMPs consisting of ß-barrel structures, which mainly reside in the hydrophobic core of the lipid bilayer. In a continuation of the project, quantitative measurements of OMPs will be performed as it is well known that *S.* Typhimurium changes its expression profile upon invading a host. For example, proteins weakly expressed in a cell culture may be more prominent once the bacteria are proliferating inside the host. Comparative studies on levels of protein expression between cultured cells and cells extracted from an infected host or tissue culture will illuminate key proteins involved in host colonization as well as increase the number of suitable candidates for vaccine development. Additional areas of interest include comparisons of the protein expression of *S.* Typhimurium and *S.* Typhi. As previously mentioned these two serovars are genetically very similar but display very different

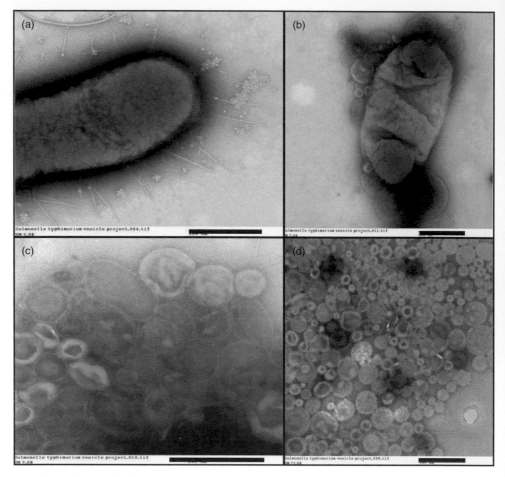

Figure 8.4 *Electron microscopy images showing outer membrane vesicles from Salmonella typhimurium cells. (a) An intact Salmonella cell prior to vesiculation treatment. (b) A Salmonella cell during vesiculation of the outer membrane. The cell is imploding and vesicles are budded from the cell by shaking. (c and d) Outer membrane vesicles. Scale bar, 500 nm*

pathogenicity. Quantitative measurements of the protein expression levels may give an indication of what proteins are responsible, thus increasing the knowledge of the mechanism of disease.

8.7   *Helicobacter pylori* – Disease Mechanism and Outer Membrane Proteins

Helicobacter pylori is a Gram-negative bacterium that infects various areas of the stomach and duodenum (Brown, 2000; Yamaoka, 2008), and is known to cause peptic ulcers,

Table 8.3   *A list of selected OMPs found in the MS/MS analysis of the outer membrane vesicles enzymatically digested by trypsin in the LPI™ FlowCell. From the data analysis 71 OMPs were identified. Transporters, outer membrane proteins, channels and lipoproteins are all represented*

Gene	Name (description)	No. of peptides
btuB	Vitamin B12 transporter btuB precursor	14
fadL	Long-chain fatty acid transport protein	6
lamB	Maltoporin	2
ompA	Outer membrane protein A	26
ompD	Outer membrane porin protein ompD precursor	6
ompW	Outer membrane protein W precursor	6
ompX	Outer membrane protease, receptor for phage OX2	3
tolC	Outer membrane channel	10
tsx	Nucleoside channel; receptor of phage T6 and colicin K	11
lolB	Outer-membrane lipoprotein lolB precursor	6
lpp1	Major outer membrane lipoprotein 1 precursor	8
lpp2	Major outer membrane lipoprotein 2 precursor	2
nlpB	Lipoprotein-34	16
pal	Peptidoglycan-associated lipoprotein	8
rlpA	Minor lipoprotein	11
rlpB	LPS-assembly lipoprotein rlpB precursor	9
slp	Putative lipoprotein	6
slyB	Outer membrane lipoprotein SlyB	7
vacJ	Lipoprotein	7

gastritis, duodenitis, and gastrointestinal cancers (Suerbaum and Michetti, 2002) (see Chapter 17). The bacteria express adhesins which bind to membrane-associated lipids and carbohydrates to facilitate adhesion to epithelial cells (Kusters *et al.*, 2006 Liu *et al.*, 2006). Once attached, the *H. pylori* cells produce ammonia and secrete proteases, catalase and phospholipases, which all cause damage to epithelial cells in the host (Yamaoka, 2008). It is estimated that up to 70% of infection is asymptomatic and that about two-thirds of the world's population is infected by the bacterium, making it the most widespread infection in the world (Sipponen, 1997). Unfortunately, an increasing number of infected individuals are found to harbour antibiotic-resistant bacteria. Attempts are being made to understand the pathogenesis of *H. pylori* and the ability of this organism to cause disease. From the genome, 64 proteins have been predicted to be localized on the surface (Sabarth *et al.*, 2002b). However, it is not always well known whether these candidates are actually expressed by the bacteria, since their presence has so far not been verified experimentally. These include surface proteins, such as urease, catalase, Hsp60 and Hsp70 while flagellar sheath proteins are excluded by theoretical predictions since they do not possess the characteristics of a surface protein (Sabarth *et al.*, 2002b). Several studies using different strategies and methods have been used in order to identify surface proteins of *H. pylori* (Jungblut *et al.*, 2000; Buman *et al.*, 2002a, b; Kim *et al.*, 2002; Sabarth *et al.*, 2002a, b; Yamoaka *et al.*, 2002; Krah *et al.*, 2003; Baik *et al.*, 2004; de Jonge *et al.*, 2004; Dossumbekova *et al.*, 2006; Carlsohn *et al.*, 2006b; Timothy, 2006). Many of the identified proteins are well-known virulence factors (Jungblut *et al.*, 2000; Krah *et al.*, 2003). These include urease B which is involved in acid neutralization; proteins involved in adhesion such as BabA and the outer membrane proteins AlpA and AlpB (Carlsohn *et al.*, 2006a) as well as the sialic acid lectin HpaA (HP0410) (Carlsohn *et al.*, 2006b); proteins involved in motility such as flagellar proteins; and proteins linked to the pathogenicity, such as cag pathogenicity island encoded proteins, for example CagA (Cag26), and the vacuolating toxin VacA (Jungblut *et al.*, 2000; Krah *et al.*, 2003). Interestingly, recent studies have shown that certain proteins are secreted into the extracellular space and then attached to the outer membrane, where they are believed to play an important role in the pathogenicity of *H. pylori* (Buman *et al.*, 2002b).

8.8   Surface Proteins of Intact *H. pylori*

Intact bacteria of *H. pylori* were allowed to attach to the membrane-attracting surfaces of the LPI™ FlowCell. Following trypsin digestion, peptides from the proteins exposed on the cell surface of the bacteria were eluted and analysed by LC-MS/MS. Preliminary results confirm many of the predicted OMPs. Several of the proteins secreted to the extra-cellular space were also detected. Ten of the 13 proteins classified as protective *H. pylori* antigens are currently being used in preclinical immunization trials and were found in this first preliminary trial run without any optimization (Table 8.4). These include urease B (Smoot, 1997), the vacuolating cytotoxin VacA (Smoot, 1997), the sialic acid lectin HpaA (Carlsohn *et al.*, 2006b) and CagA, from the pathogenicity island gene sequence, a protein which is injected into the stomach's epithelial cells, where it disrupts the cytoskeleton, adherence to adjacent cells, intracellular signalling, and other cellular activities (Backert and Selbach, 2008).

Protocols will be further optimized for efficient enzymatic digestion of surface-exposed proteins of intact *H. pylori* cells in the LPI™ FlowCell. Membrane preparation protocols will also be used for the production of purified outer membrane fractions of the bacteria. The results arising from intact bacterial cells will then be compared with the results from

Table 8.4   *Protective H. pylori antigens detected by proteomics approaches and tested in preclinical immunization trials. Previous investigations have been inconclusive in determining surface exposure. Using the LPI™ FlowCell, 10 of the 13 proteins were identified*

Protein	Surface exposed	Detected by LPI technology
Urease B	Data not conclusive	Yes
Catalase	Yes	Yes
HspA	Data not conclusive	Yes
HspB	Data not conclusive	Yes
VacA	Yes	Yes
L7/L12 ribosomal protein	Not detected so far	Yes
Hemolysin secretion protein	Possible	Yes
Citrate synthase	Not detected so far	Yes
CagA	Yes	Yes
HP0410 HpaA	Yes	Yes
Urease A	Yes	Not yet
NapA	Yes	Not yet
HP0231	Yes	Not yet

the outer membrane vesicles to help confirm the presence of the OMPs, whilst surface-shaving of intact bacteria will also show proteins secreted (and attached) to the surface of the *H. pylori* cells.

References

Arrigo, K.R. Marine microorganisms and global nutrient cycles. Nature 2005, 437: 349–355.

Backert, S. and Selbach, M. Role of type IV secretion in *Helicobacter pylori* pathogenesis Cell. Microbiol. 2008, 10(8), 1573–1581.

Baik, S.C., Kim, K.M., Song, S.M., Kim, D.S., Jun, J.S., Lee, S.G., Song, J.Y., Park, J.U., Kang, H.L., Lee, W.K., Cho, M.J., Youn, H.S., Ko, G.H. and Rhee, K.H. Proteomic analysis of the sarcosine-insoluble outer membrane fraction of *Helicobacter pylori* strain 26695. J. Bacteriol. 2004, 186(4), 949–955.

Baker, S. and Dougan, G. The genome of Salmonella *enterica* serovar Typhi. Clin. Infect. Dis. 2007, 45, 29–33.

Baumler, A.J., Tsolis, R.M., Ficht, T.A. and Adams, L.G. Evolution of host adaptation in *Salmonella enterica*. Infect. Immun. 1998, 66(10), 4579–4587.

Bergey, D.H., Holt, J.G., Krieg, N.R. and Sneath, P.H.A. Bergey's Manual of Determinative Bacteriology, 9th edn. Baltimore: Lippincott Williams & Wilkins, 1994.

Beveridge, T.J. Use of the gram stain in microbiology. Biotech. Histochem. 2001, 76(3), 111–118.

Bopp, C.A., Brenner, F.W., Wells, J.G. and Strockbine, N.A. In: Manual of Clinical Microbiology, Murray, P.R., Baron, E.J., Jorgensen, J.H., Pfaller, M.A. and Yolken, R.H., editors. Washington, DC: ASM Press, 2003, 654–671.

Braun, R.J., Kinkl, N., Beer, M. and Ueffing, M. Two-dimensional electrophoresis of membrane proteins. Anal. Bioanal. Chem. 2007, 389(4), 1033–1045.

Brown, L.M. *Helicobacter pylori*: epidemiology and routes of transmission. Epidemiol. Rev. 2000, 22(2), 283–297.

Bumann, D., Aksu, S., Wendland, M., Janek, K., Zimny-Arndt, U., Sabarth, N., Meyer, T.F. and Jungblut, P.R. Protcome analysis of sccrctcd protcins of thc gastric pathogen *Helicobacter pylori*. Infect. Immun. 2002a, 70(7), 3396–3403.

Buman, D., Holland, P., Siejak, F., Koesling, J.,Sabarth, N., Lamer, S., Zimny-Arndt, U., Jungblut, P.R. and Meyer, T.F. A comparison of murina and human immunoproteomes of *Helicobacter pylori* validates the preclinical murine infection model for antigen screening. Infect. Immun. 2002b, 70(11), 6494–6498.

Carlsohn, E., Nyström, J., Karlsson, H., Svennerholm, A.-M. and Nilsson, C.L. Characterization of the outer membrane protein profile from disease-related *Helicobacter pylori* isolates by subcellular fractionation and nano-LC FT-ICR MS analysis. J. Proteome Res. 2006a, 5, 3197–3204.

Carlsohn, E., Nyström, J., Bölin, I., Nilsson, C.L. and Svennerholm, A.-M. HpaA is essential for *Helicobacter pylori* colonization in mice. Infect. Immun. 2006b, 74(2), 920–926.

Coldham, N.G. and Woodward, M.J. Characterization of the *Salmonella* Typhimurium proteome by semi-automated two-dimensional HPLC-mass spectrometry: Detection of proteins implicated in multiple antibiotic resistance. J. Proteome Res. 2004, 3, 595–603.

Cole, J.N., Ramirez, R.D., Currie, B.J., Cordwell, S.J., Djordjevic, S.P. and Walker, M.J. Surface analyses and immune reactivities of major cell wall-associated proteins of group a streptococcus. Infect. Immun. 2005, 73(5), 3137–3146.

Cordwell, S.J. Sequential extraction of proteins by chemical reagents. Methods Mol. Biol. 2008, 424, 139–146.

de Jonge, R., Durrani, Z., Rijpkema, S.G., Kuipers, E.J., van Vliet, A.H.M. and Kusters, J.G. Role of the *Helicobacter pylori* outer-membrane proteins AlpA and AlpB in colonization of the guinea pig stomach. J. Med. Microbiol. 2004, 53, 375–379.

Delcour, A.H. Structure and function of pore-forming b-barrels from bacteria. J. Mol. Microbiol. Biotechnol. 2002, 4(1), 1–10.

Devol, A.H. Nitrogen cycle: Solution to a marine mystery. Nature 2003, 422, 575–576.

Dossumbekova, A., Prinz, C., Mages, J., Lang, R., Kusters, J.G., van Vliet, A.H.M., Reindl, W., Backert, S., Saur, D., Schmid, R.M. and Rad, R. *Helicobacter pylori* HopH (OipA) and bacterial pathogenicity: genetic and functional genomic analysis of hopH gene polymorphisms. J. Infect. Dis. 2006, 194, 1346–1355.

Edwards, R.A., Olsen, G.J. and Maloy, S.R. Comparative genomics of closely related salmonellae. Trends Microbiol. 2002, 10, 94–99.

Fardini, Y., Chettab, K., Grepinet, O., Rochereau, S., Trotereau, J., Harvey, P., Amy, M., Bottreau, E., Bumstead, N., Barrow, P.A. and Virlogeux-Payant, I. The YfgL lipoprotein is essential for type III secretion system expression and virulence of *Salmonella enterica* serovar Enteritidis. Infect. Immun. 2007, 75(1), 358–370.

Ge, Y. and Rikihisa, Y. Surface-exposed proteins of *Ehrlichia chaffeensis*. Infect. Immun. 2007, 75(8), 3833–3841.

Gophna, U., Ron, E.Z. and Graur, D. Bacterial type III secretion systems are ancient and evolved by multiple horizontal-transfer events. Gene 2003, 312, 151–163.

Gram, H.C. Über die isolierte Färbung der Schizomyceten in Schnitt- und Trockenpräparaten. Fortschr. Medizin 1884, 2, 185–189.

Harding, S.V., Sarkar-Tyson, M., Smither, S.J., Atkins, T.P., Oyston, P.C., Brown, K.A., Liu, Y., Wait, R. and Titball, R.W. The identification of surface proteins of *Burkholderia pseudomallei*. Vaccine 2007, 25(14), 2664–2672.

Huang, D.B. and DuPont, H.L. Problem pathogens: extra-intestinal complications of *Salmonella enterica* serotype Typhi infection. Lancet Infect. Dis. 2005, 5(6), 341–348.

Jungblut, P.R., Bumann, D., Haas, G., Zimny-Arndt, U., Holland, P., Lamer, S., Siejak, F., Aebizher, A. and Meyer, T.F. Comparative proteome analysis of *Helicobacter pylori*. Mol. Microbiol. 2000, 36(3), 710–725.

Karlsson, R., Karlsson, A., Bäckman, O., Johansson, B.R. and Hulth, S. Identification of key proteins involved in the anammox reaction. FEMS Microbiol. Lett. 2009, 297(1), 87–94. Epub 2009 Jun3.

Kim, N., Weeks, D.L., Moo Shin, J., Scott, D.R., Young, M.K. and Sachs, G. Proteins released by *Helicobacter pylori in vitro*. J. Bacteriol. 2002, 184(22), 6155–6162.

Koebnik, R., Locher, K.P. and van Gelders, P. Structure and function of bacterial outer membrane proteins: barrels in a nutshell. Mol. Microbiol. 2000, 37(2), 239–253.

Krah, A., Schmidt, F., Becher, D., Schmid, M., Albrecht, D., Rack, A., Buttner, K. and Jungblut, P.R. Analysis of automatically generated peptide mass fingerprints of cellular proteins and antigens from *Helicobacter pylori 26695* separated by two-dimensional electrophoresis. Mol. Cell. Proteomics 2003, 2(12), 1271–1283.

Kuhle, V. and Hensel, M. Cellular microbiology of intracellular *Salmonella enterica*: functions of the type III secretion system encoded by Salmonella pathogenicity island 2. Cell. Mol. Life Sci. 2004, 61(22), 2812–2826.

Kusters, J.G., van Vliet, A.H. and Kuipers, E.J. Pathogenesis of *Helicobacter pylori* infection. Clin. Microbiol. Rev. 2006, 19(3), 449–490.

Kuypers, M.M.M., Lavik, G., Woebken, D., Schmid, M., Fuchs, B.M., Amann, R., Jørgensen, B.B. and Jetten, M.S. Massive nitrogen loss from the Benguela upwelling system through anaerobic ammonium oxidation. Proc. Natl Acad. Sci. USA 2005, 102, 6478–6483.

Liu, G.Y., Nie, P., Zhang, J. and Li, N. Proteomic analysis of the sarcosine-insoluble outer membrane fraction of *Flavobacterium columnare*. J. Fish Dis. 2008, 31(4), 269–276.

Liu, Z.F., Chen, C.Y., Tang, W., Zhang, J.Y., Gong, Y.Q. and Jia, J.H. Gene-expression profiles in gastric epithelial cells stimulated with spiral and coccoid *Helicobacter pylori*. J. Med. Microbiol. 2006, 55(Pt 8), 1009–1015.

Macher, B.A. and Yen, T.-Y. Proteins at membrane surfaces – a review of approaches. Mol. Biosyst. 2007, 3, 705–713.

McClelland, M., Sanderson, K.E., Spieth, J., Clifton, S.W., Latreille, P., Courtney, L., Porwollik, S., Ali, J., Dante, M., Feiyu, D., Shunfang, H., Layman, D., Leonard, S., Nguyen, C., Scott, K., Holmes, A., Grewal, N., Mulvaney, E., Ryan, E., Sun, H., Florea, L., Miller, W., Stoneking, T., Nhan, M., Waterston, R. and Wilson, R.K. Complete genome sequence of *Salmonella enterica* serovar Typhimurium LT2. Nature 2001, 413, 852–856.

Molloy, M.P. Isolation of bacterial cell membranes proteins using carbonate extraction. Methods Mol. Biol. 2008, 424, 397–401.

Molloy, M.P., Herbert, B.R., Slade, M.B., Rabilloud, T., Nouwens, A.S., Williams, K.L. and Gooley, A.A. Proteomic analysis of the *Escherichia coli* outer membrane. Eur. J. Biochem. 2000, 267, 2871–2881.

Monack, D.M., Mueller, A. and Falkow, S. Persistent bacterial infections: the interface of the pathogen and the host immune system. Nat. Rev. Microbiol. 2004, 2(9), 747–765.

Mujahid, S., Pechan, T. and Wang, C. Improved solubilization of surface proteins from Listeria monocytogenes for 2-DE. Electrophoresis 2007, 28(21), 3998–4007.

Nandakumar, R., Nandakumar, M.P., Marten, M.R. and Ross, J.M. Proteome analysis of membrane and cell wall associated proteins from *Staphylococcus aureus*. J. Proteome Res. 2005, 4(2), 250–257.

Nyblom, M., Öberg, F., Lindkvist-Petersson, K., Hallgren, K., Findlay, H., Wikström, J., Karlsson, A., Hansson, Ö., Booth, P.J., Bill, R.M., Neutze, R. and Hedfalk, K. Exceptional overproduction of a functional human membrane protein. Prot. Exp. Purif. 2007, 56(1), 110–120.

Parkhill, J., Dougan, G., James, K.D., Thomson, N.R., Pickard, D., Wain, J., Churcher, C., Mungall, K.L., Bentley, S.D., Holden, M.T.G., Sebaihia, M., Baker, S., Basham, D., Brooks, K., Chillingworth, T., Connerton, P., Cronin, A., Davis, P., Davies, R.M., Dowd, L., White, N., Farrar, J., Feltwell, T., Hamlin, N., Haque, A., Hien, T.T., Holroyd, S., Jagels, K., Kroghk, A., Larsenk, T.S., Leather, S., Moule, S., O'Gaora, P., Parry, C., Quail, M., Rutherford, K., Simmonds, M., Skelton, J., Stevens, K., Whitehead, S. and Barrell, B.G. Complete genome sequence of a multiple drug resistant *Salmonella enterica* serovar Typhi CT18. Nature 2001, 413, 848–852.

Peng, X., Xu, C., Ren, H., Lin, X.; Wu, L. and Wang, S. Proteomic analysis of the sarcosine-insoluble outer membrane fraction of *Pseudomonas aeruginosa* responding to ampicillin, kanamysin, and tetracycline resistance. J. Proteome Res. 2005, 4(6), 2257–2265.

Popoff, M.Y., Bockemuhl, J. and Gheesling, L.L. Supplement 2001 (no. 45) to the Kauffmann-White scheme. Res. Microbiol. 2003, 154, 173–174.

Prescott, L., Harley, J. and Klein D. Microbiology, 5th edn. New York: McGraw-Hill, 2002.

Rhomberg, T.A., Karlsberg, O., Mini, T., Zimny-Arndt, U., Wickenberg, U., Röttgen, M., Jungblut, P.R., Jenö, P., Andersson, S.G. and Dehio, C. Proteomic analysis of the sarcosine-insoluble outer membrane fraction of the bacterial pathogen *Bartonella henselae*. Proteomics 2004, 4(10), 3021–3033.

Rivas, L., Fegan, N. and Dykes, G.A. Expression and putative roles in attachment of outer membrane proteins of *Escherichia coli* O157 from planktonic and sessile culture. Foodborne Pathog. Dis. 2008, 5(2), 155–164.

Sabarth, N., Hurwitz, R., Meyer, T.F. and Bumann, D. Multiparameter selection of *Helicobacter pylori* antigens identifies two novel antigens with high protective efficacy. Infect. Immun. 2002a, 70(11), 6499–6503.

Sabarth, N., Lamer, S., Zimny-Arndt, U., Jungblut, P.R., Meyer, T.F. and Bumann, D. Identification of surface proteins of *Helicobacter pylori* by selective biotinylation, affinity purification, and two-dimensional gel electrophoresis. J. Biol. Chem. 2002b, 277(31), 27896–27902.

Schindler, J., Jung, S., Niedner-Schatteburg, G., Friauf, E. and Nothwang, H.G. Enrichment of integral membrane proteins from small amounts of brain tissue. J. Neural Transm. 2006, 113(8), 995–1013.

Schulz, G.E. The structure of bacterial outer membrane proteins. Biochim. Biophys. Acta 2002, 1565, 308–317.

Sipponen, P. *Helicobacter pylori* gastritis-epidemiology. J Gastroenterol. 1997, 32(2), 273–277.

Smither, S.J., Hill, J., van Baar, B.L., Hulst, A.G., de Jong, A.L. and Titball, R.W. Identification of outer membrane proteins of *Yersinia pestis* through biotinylation. J. Microbiol. Methods 2007, 68(1), 26–31.

Smoot, D.T. How does *Helicobacter pylori* cause mucosal damage? Direct mechanisms. Gastroenterology 1997, 113(6 Suppl.), S31–4, discussion S50.

Suerbaum, S. and Michetti, P. *Helicobacter pylori* infection. N. Engl. J. Med. 2002, 347(15), 1175–1186.

Timothy, L. Role of *Helicobacter pylori* outer membrane proteins in gastroduodenal disease. J. Infect. Dis. 2006, 194, 1343–1345.

Washburn, M.P. and Yates, J.R. Analysis of the microbial proteome. Curr. Opin. Microbiol. 2000, 3, 292–297.

Wehmhoner, D., Dietrich, G., Fisher, E., Baumgartner, M., Wehland, J. and Jansch, L. 'Lanespector', a tool for membrane protein profiling based on sodium dodecyl sulphate-polyacrylamide get electrophoresis/liquid chromatography-tandem mass spectrometry analysis: application to Listeria monocytogenes membrane proteins. Electrophoresis 2005, 26, 2450–2460.

Xiong, Y., Chalmers, M.J., Gao, F.P., Cross, T.A. and Marshall, A.G. Identification of Mycobacterium tuberculosis H37Rv integral membrane proteins by one-dimensional gel electrophoresis and liquid chromatography electrospray ionization tandem mass spectrometry. J. Proteome Res. 2005, 4, 855–861.

Yamaoka, Y. *Helicobacter pylori*: Molecular Genetics and Cellular Biology. Norwich: Caister Academic, 2008.

Yamaoka, Y., Kita, M., Kodama, T., Imamura, S., Ohno, T., Sawai, N., Ishimaru, A., Imanishi, J. and Graham, D.Y. *Helicobacter pylori* infection in mice: role of outer membrane proteins in colonization and inflammation. Gastroenterology 2002, 123, 1992–2004.

Zahedi, R.P., Meisinger, C. and Sickmann, A. Two-dimensional benzyldimethyl-n-hexadecylammonium chloride/SDS-PAGE for membrane proteomics. Proteomics 2005, 5(14), 3581–3588.

Zoubi-Hasona, K. and Brady, L.J. Isolation and solubilization of cellular membrane proteins from bacteria. Methods Mol. Biol. 2008, 425, 287–293.

Zuobi-Hasona, K., Crowley, P.J., Hasona, A., Bleiweis, A.S. and Brady, L.J. Solubilization of cellular membrane proteins from Streptococcus mutants for two-dimensional gel electrophoresis. Electrophoresis 2005, 26(6), 1200–1205.

9

Wider Protein Detection from Biological Extracts by the Reduction of the Dynamic Concentration Range

Luc Guerrier[1], Pier Giorgio Righetti[2] and Egisto Boschetti[1]
[1] *Bio-Rad Laboratories, Marnes-la-Coquette, France*
[2] *Department of Chemistry, Materials and Chemical Engineering 'Giulio Natta',*
Politecnico di Milano, Milan, Italy

9.1 Introduction

In the history of technological developments, there were always periods of rapid progress and quiescence for thought on finding new solutions. Life sciences also follow this rule as demonstrated in the last few decades with the deciphering of the fundamental mechanisms of life driven by advances in genomics and subsequently proteomics. Impressive, sound understanding and applications of genomic discoveries did not, however, stimulate the development of proteomics under the same dynamics. In reality proteomics, as a reflection of genomics, is far more complicated, not only due to the transcription and translation processes, but also because genes are not all expressed into proteins at a similar rate, but also because many proteins are modified post-translationally, often truncated or might be misfolded and degraded into a multitude of small and large peptides.

Initially, research in proteomics showed rapid progress because the field was widely open and the complex issues to be addressed were still a distant goal. Applications focused on areas that were likely to be more fruitful such as 'differential expression' as for example in pathology to search for specific biological markers for diagnostic and biomedical applications. Unfortunately, the drive to progress as fast as possible has been slowed down because many discoveries were based around the same relatively small group of proteins and particularly those that are present in high or intermediate abundance [1]. In human

Mass Spectrometry for Microbial Proteomics Edited by Haroun N. Shah and Saheer E. Gharbia
© 2010 John Wiley & Sons, Ltd

blood plasma, for instance, extremely large numbers of proteins are present. If one assumes that most genes are expressed at different levels and under different circumstances and each one has a dozen splice variants and each gene product is represented by a few dozen glycosylated forms, this alone would yield several hundred thousand species. The concentration difference between the most abundant and the most dilute proteins spans over 10 orders of magnitude and will further add to this complexity. Although serum might be an exception, any other cell lysate might not fare much better, since the cytoskeletal and housekeeping proteins are the most abundant ones compared with enzymes, hormones, and other catalytic and membrane-involved proteins. As a general picture of protein abundance, it can be estimated that high-abundance proteins represent as little as 1% of the total while moderately abundant proteins cover ca. 10% of the total diversity. Low abundance proteins account for 90% of the overall protein content.

This complexity therefore presents a huge technical challenge. Thus, bringing low and very low abundance proteins to a level of detection is an extremely important task because (i) this could represent a way to decipher the composition of various proteomes (serum, cells, subcellular structures, membranes, urine, etc.) and (ii) because pathological conditions may be related to mis-regulated mechanisms where only very low abundance proteins would be involved especially in the initiation of a disease.

The search for such species involves treatment of the initial sample by different types of fractionation or modification steps depending on the composition or relative abundance. In the first case, extensive fractionation methods have been described alone and under complex conditions; in the second case, the apparent complexity of the resulting 'manipulated' proteome, results in even higher complexity, necessitating additional fractionation steps to yield better analysis of single components. However, the development of proteomic technology is not only restricted to sample treatment but also to improvements in instrumentation especially in resolution and sensitivity to allow a better detectability of species that were impossible to identify a decade ago.

Overall prefractionation became a necessity to reduce the complexity of the proteomic samples prior to two-dimensional electrophoresis (2DE) or two-dimensional liquid chromatography-mass spectrometry (2D LC-MS) or liquid chromatography-tandem mass spectrometry (LC-MS/MS). Very elaborate sample preparation approaches were suggested as described below, with the aim to identify and also quantify low-abundance proteins.

The following sections give a general review of the most modern methods of protein fractionation and investigation, focusing mainly on sample treatment. Fractionation itself, depletion of high abundance species and reduction of dynamic concentration range, will be reviewed and evaluated.

9.2   Fractionation as a Means to Decipher Proteome Complexity

The term complexity covers here two distinct features and generally refers to the analysis of protein constituents of the initial biological extract.

The first feature is related to the very large number of different proteins regardless of their concentration. There are numerous gene products depending on the type of extract and theoretically this should be equivalent to the number of expressed genes. In addition, there are proteins in different isoforms resulting from a large number of post-translational modifications, while in biological fluids such as serum the complexity is increased by the large number of different antibodies. The number of total proteins is so large that at present it is impossible to state a precise value. The second feature is the large difference in concentration of expressed components that can reach more than ten orders of magnitude. Proteins of massive abundance include albumin in serum, haemoglobin in erythrocytes, actin inside certain cells, or ribulose-1,5-biphosphate carboxylase/oxygenase (Rubisco) in plant leaf extracts. Their presence is so disproportionate that they mask the signals of very low abundance proteins.

To try to unravel the complexity, in such samples four main approaches are suggested: (i) extensively fractionate the initial sample into partial portions and apply current analytical procedures; (ii) focus the detection to partial areas and then reconstruct the entire picture; (iii) remove the high abundance species so that other species are better represented; and (iv) reduce the dynamic range by decreasing the concentration of high abundance proteins while concomitantly increasing the concentration of low abundance species. In the latter case the 'visible' complexity becomes even higher, and, for easier handling, in many cases the resulting processed sample should be fractionated by means of current tools.

9.2.1   Subcellular Fractionation

One of the most commonly used traditional approaches to fractionate proteomes is to separate cell substructures by centrifugation. Subcellular structures are very specialized in their function and hence their protein composition is compartmented. In the early 1960s [2–4], this procedure allowed, via a series of runs at different centrifugal speeds, isolation in a reasonably pure form subfractionation such as mitochondria, lysosomes, Golgi, nuclei, peroxisomes, synaptosomes and various microbodies. Analysing proteomes of such enriched substructures has been the hallmark of separation technologies and enabled the discovery of proteins and their relationship to their cell local function [5–7]. The mitochondrial proteome of rat liver has been investigated by Lopez [8] using differential centrifugation in sucrose-mannitol based buffers. Solubilization of proteins was obtained using a solution comprising urea, thiourea, CHAPS and dithiothreitol and their analysis directly performed by 2DE. As a second fractionation method, different chromatographic approaches were also used including affinity and hydrophobic interactions. In this study, authors identified among them hundreds of soluble glycoproteins belonging to membranes as well as calcium-binding species.

In a more recent study, Sun *et al.* [9] described the isolation of mitochondria from liver by density gradient centrifugation with the view to investigate their soluble proteins and elucidate the role of some of these in liver regeneration following partial hepatectomy. They discovered more than 20 differentially expressed species.

The potential of mitochondrial proteomics for toxicological studies and therapeutic purposes was also investigated [10].

Today mitochondrial separation by gradient centrifugation is challenged by free-flow electrophoresis separation as described by Zischka *et al.* [11]. Free-flow electrophoresis could also constitute a complementary mode of purification after density centrifugation. The inner mitochondrial membranes of yeast were successfully studied using first differential centrifugation, followed by a free-flow electrophoresis purification step [12].

The nucleus of the cell comprises important proteins that are at the centre of gene regulation and expression. They are classified as 'low abundance' and are difficult to analyse without preliminary separation of the nucleus from other cell organelles. At present this is achieved by centrifugation using a density gradient [13]. In human cancer, understanding the process is dependent on knowing the involvement and interaction of relevant proteins *in situ*. Proteins that are restricted in this organelle are mostly involved in the strong interaction with nucleic acid and are thus difficult to isolate; moreover, they are poorly soluble in physiological conditions. Nevertheless the preparation of nuclear proteins is first highly dependent on the proper isolation of this subcellular organelle. Density centrifugation, using one or more density systems, is necessary to obtain purified material as described by Komatsu [14]. This author reported detailed methods and ascertained the purity of nuclear proteins by specific immunoblots after solubilization with dedicated buffers of sodium dodecyl sulfate. Mass spectrometry was also used by other authors for assessing the nature and identity of nuclear proteins following isolation of cell nuclei by centrifugation [15].

Centrifugation, although commonly used for cells and subcellular particle separation, is labour intensive and needs experienced scientists to obtain highly purified biological material from which suborgan-selective proteins can be extracted.

9.2.2   Protein Precipitation

Protein precipitation, used extensively in the past, is not common at present because it generates overlapping proteins between fractions; however, it deserves to be mentioned. Three chemical agents are commonly used for protein precipitation: (i) ammonium sulfate at different degrees of saturation; (ii) polyethylene glycol; and (iii) organic solvents [16]. Precipitation of proteins in the presence of these agents is based on the decreased solubility and hence aggregation brought about mostly by hydrophobic interactions with lyotropic salts or hydrophilic interactions with organic solvents. Ammonium sulfate precipitation is perhaps the most popular. It is used when undesired substances are present in the initial cell or tissue extract. This is the case with plant tissues where major contaminants such as polyphenols and polysaccharides need to be removed [17]. It is used also when dealing with prokaryotic cell extracts to eliminate nucleic acids. The protein pellets obtained are then resolubilized in appropriate buffers for proteomics investigations [18]. However, while precipitation processes resolve some specific issues, they generate some drawbacks such as the loss of peptides and small proteins that do not precipitate. Additionally, the presence of a large amount of ammonium sulfate needs to be removed, often by dialysis or desalting.

Organic solvents such as acetone alone or as a mixture with other solvents (e.g. trichloroacetic acid) are also used for protein precipitation for similar purposes as ammonium sulfate [19]. This is often referred to as 'protein clean-up' and is performed before proteome analysis. This method has similar drawbacks to ammonium sulfate precipitation in causing some protein denaturation.

9.2.3   Immunoprecipitation

Immunoprecipitation is used frequently in proteomic investigations prior to protein analysis using mass spectrometry or 2DE analysis. The principle is based on the use of antibodies that are selective for one or a group of proteins sharing a similar epitope as in the case of phosphorylated proteins or protein isoforms. In practice the specific antibody is mixed with the protein extract and incubated for a period necessary to form an immunocomplex after which it is separated using a Protein A column. Isolation of isoforms of tumor necrosis factor was described by Watts *et al.* [20], where the separated proteins were then resolved by 2DE.

Immunoprecipitation includes also the selective adsorption of proteins on solid phases where specific antibodies are grafted. The investigation domains of phosphoproteins and serine/threonine-phosphorylated peptides were significantly improved by using specific antibodies [21]. Immunoprecipitation approaches, used alone or in association with other separation methods, were extensively reported as probes for elucidating aspects of signal transduction [22] as well as for the analysis of phosphotyrosyl proteins in cerebrospinal fluid [23] and the mapping of phosphorylation sites from human T-cells [24]. The immunoprecipitation principle is also helpful for investigating the formation of protein-protein complexes [25–27], and therefore contributing to the elucidation of some pathways.

9.2.4   Chromatographic Fractionation

Liquid chromatography techniques are most commonly used in proteome prefractionation prior to in-depth analysis. They are based on the use of solid-phase adsorbents of a large variety of selectivity. Low, as well as very high, interaction specificity can be operated by selecting appropriate sorbents and conditions of adsorption-elution. They are generally used as a single column with single or stepwise elution; however, a second column may be used in a sequence mode, i.e. an orthogonal configuration with complementary adsorption properties. Moreover they can be used under cascade mode with the same adsorption buffer for all selected sorbents (see Section 9.3.3.7).

Excellent reviews on protein fractionation have been published [28–32]. The following is a concise overview of chromatographic fractionation illustrating the vast number of permutations possible.

The most popular, liquid chromatography fractionation, is based upon ion-exchange chromatography. Proteins are captured by cation or anion exchange according to their net charge depending on the environmental pH. Thus acidic proteins are mostly fractionated by using anion exchange chromatography while basic proteins are fractionated by cation exchange chromatography. Adsorbed proteins on ion exchangers are desorbed under a discontinuous or a linear elution gradient involving salt or pH gradients. Using these approaches, proteomes are simplified and thus more easily analysed by mass spectrometry and 2DE. Investigations on human serum for the detection of biomarkers of diagnostic interest were successfully achieved by using anion exchange chromatography prefractionation [33, 34]. Weak cation exchange chromatographic resins were used for instance for the study of kinase families [35].

Hydrophobic interaction chromatography (HIC) acts through molecular association involving two hydrophobic moieties in the presence of high concentrations of lyotropic salts exposing hydrophobic parts of proteins towards hydrophobic patches of solid phase sorbents. This lyotropic effect is the basis of hydrophobic chromatography separation of proteins. Ammonium sulfate is frequently used as a lyotropic salt to promote hydrophobic interaction. Desorption is promoted by a simple progressive decrease of the lyotropic salts concentration. The least hydrophobic proteins are collected first while the most hydrophobic proteins are desorbed when the concentration of the salt is reduced to zero. In some cases, the hydrophobic interaction is so strong that it necessitates the use of hydro-organic mixtures.

HIC was used for the prefractionation of cytosolic soluble proteins of *Haemophilus influenzae* [36] on a phenyl column for further proteomics studies.

A particular aspect of HIC is reverse phase chromatography (RPC) where the degree of hydrophobicity is higher due to longer hydrocarbon chains grafted at very high density on the surface of solid phase sorbents. RPC is extensively used in proteomic investigations (see for example ref. [37] and Chapter 4) not only in protein prefractionation but also as peptide fractionation prior to mass spectrometry analysis.

To target a higher level of specificity as for instance in the separation of protein groups having common predefined properties, other types of chromatographic fractionation have been extensively described. For example, heparin chromatography has been used for enrichment purposes [38, 39] or for more specific applications as in the case of separation of proteins involved in transcription processes [40] or even for depletion operations [41].

Lectins, as proteins capable of interacting with the glycan moiety of glycoproteins, have been used extensively for the separation of post-translational modified proteins [42–44]. Various types of lectins can be used for the subfractionation of glycoproteins and glyco-peptides as a function of the glycan composition and structure [45–47].

Phosphorylated proteins and peptides are additional examples of species having common phosphate esters as a result of post-translational modifications. The separation of this group of proteins or even subgroups with different phosphorylation sites is also important to deconvolute signal transduction pathways. This category of proteins is separated using specifically designed antibodies [48] or solid phase chelating metal ions such as Fe^{3+}, Ga^{3+}, Ti^{4+} and Zr^{4+}, having the possibility to interact selectively with phosphate groups [49–52].

9.2.5   Electrokinetic Methods in Proteome Fractionation

A number of preparative electrophoretic methodologies have been developed for a diverse range of proteomic investigations [53]. Although initially relatively difficult to use as a practical approach, these methods were progressively improved to become one of the most commonly used in the field, particularly with the introduction of 'mining tools' for proteome analysis. Some recent examples are given. A novel device is also described in Chapter 10.

In a recent review, Nissum and Foucher [54] described the analysis of human plasma proteome after fractionation by free-flow electrophoresis where the separation principles are described along with selected examples. Free-flow electrophoresis is also a useful technique when included in multidimensional separation strategies and in the definition of various workflows.

Perhaps the most developed electrokinetic separation methods are based on the isoelectric focusing (IEF) principle, where a gradient of pH is created with the migration of protein species to reach the pH corresponding to its isoelectric point. Popular devices include Rotofor, multicompartment devices and Off-Gel.

Rotofor is a device assembled from sample chambers, separated by liquid-permeable nylon screens, except at the extremities, where cation- and anion-exchange membranes are placed against the anodic and cathodic compartments, respectively. At the end of protein migration, protein fractions are collected [55]. This methodology can also be taken as a first dimension separation, each Rotofor fraction being then subfractionated by HIC, using nonporous reversed-phase HPLC [56]. Each HIC collected peak is then treated with trypsin, and the peptides subjected to mass spectrometry analysis. By this approach, Davidsson *et al.* [57] have subfractionated human cerebrospinal fluid and brain tissue with good results. The major drawback is the presence of carrier ampholytes that create perturbations during proteome analysis. However, when Rotofor is used for peptide digests of an entire proteome, the migration can be performed in an ampholyte-free environment. The peptides themselves act as carrier ampholyte-buffers to create a pH gradient.

Modern multicompartment electrolyzers are a class of devices based on conventional IEF separation in the presence of a low amount of carrier ampholytes, where discrete chambers are created with isoelectric membranes. Described for the first time by Righetti *et al.* [58, 59], the advantages of such devices are the possible collection of proteins within a predetermined isoelectric points (between the value of two adjacent isoelectric membranes) and the full compatibility of protein fractions with the subsequent first dimension separation in 2DE maps. A known representative of such devices is 'Zoom' which is capable of fractionating small volumes of proteins into several different fractions. This device was used for instance for the prefractionation of a mouse brain extract on a narrow pH range with successful identification of a large number of proteins including about 10% hydrophobic species.

An offshoot of the multicompartment pI separation technique is the so-called 'Off-Gel IEF' [60]. Briefly, the sample is placed in a series of compartments of small volume positioned on top of an immobilized pH gradient (IPG) gel strip. At the end of the separation process, protein groups are found in the compartments by ranges of isoelectric points depending on their positioning over the IEF gel strip. Theoretical calculations and modelling have shown that the protonation of an ampholyte occurs in the thin layer of solvation close to the IPG gel/solution interface [61]. Upon application of a voltage gradient, perpendicular to the liquid chamber, the electric field penetrates into the channel and extracts all charged species (those having pI values above and below the pH of the IPG gel), thus removing them from the sample cup. This initial system was improved and adapted to a multiwell device, composed of a series of compartments of small volume (100–300 μL) and compatible with current instruments for separation [62].

Proteome analysis of human plasma alone [63] and both human plasma and amniotic fluid by Off-Gel IEF followed by nano-LC-MS/MS have been reported [64]. Here the Off-Gel principle was used in two distinct steps: first for the fractionation of proteins and then each fraction, after digestion with trypsin, was subjected to a second fractionation of the resulting peptides. However, the initial material was treated with immunosorbents to remove major proteins such as albumin, immunoglobulins and four other proteins that would interfere with isoelectric migration due to their over abundance. Analysis of fractions by LC-MS/MS revealed a number of differential proteins, between plasma and amniotic fluid.

Although this method is very attractive because it allows analysis of a very low volume of biological material, protein separation suffers from two distinct issues: (i) the possible precipitation of proteins at the vicinity of their isoelectric point; and (ii) difficulties in the extraction of proteins from the gel slab especially when the mass is relatively large. It is in this context that the method is mostly devoted to peptide fractionation as described first by Heller *et al.* [65]. The method was then extended to peptide models for technology evaluation [66] and then a number of peptide fractionation applications were described which included cerebrospinal fluid [67]. In spite of some limitations, this method appears promising for selected orthogonal separation combinations.

9.3   Dealing with Low Abundance Proteins

Conventional methods for proteome analysis such as 2DE and mass spectrometry associated or not associated with liquid chromatography (LC-MS/MS) can explore quite a large dynamic range of about four orders of magnitude in protein concentration differences from biological material. However, due to a much larger dynamic concentration range of most protein extracts, that can reach values larger than ten orders of magnitude, the exploration of deep proteomes remains challenging. In spite of sophisticated methodologies devised in the last few years, the identification of traces of gene products is hindered by their very low abundance level.

The extreme complexity of sample composition, comprising not only many proteins encoded by the genome, but also splicing variants and numerous post-translational modifications of large complexity, adds to the large difference in protein concentration from the most abundant species to the proteins that are present only in a few copies. A possible way would be to concentrate biological samples; however, in practice, a simple concentration operation is not workable because the concentration would be performed indiscriminately, leaving proteins that are already present in high abundance, without any change in the dynamic range of concentration between species.

Possible strategies to improve the detectability of proteins from a complex proteome could be as follows:

- Use of more sensitive detection methods (fluorescent staining for 2DE; enhanced mass spectrometry detectors, etc.).
- Adsorbing protein groups on solid phases (selective concentration by chromatography) or capturing specifically undetectable species.
- Removing species that mask minor species, via for instance depletion approaches.
- Zooming in on species that are close to each other in their physico-chemical properties such as isoelectric point. This approach will impact also on resolution.
- Reducing the dynamic concentration range by concentrating on low abundance proteins while concomitantly and proportionally reducing the concentration of high abundance species.

In this section, three approaches will be reviewed: (i) depletion methods; (ii) zooming in with narrow IPG; and (iii) reducing the dynamic concentration range. The principles and applications of the latter will then be discussed in detail.

9.3.1   Depletion of a Few High Abundance Proteins

The presence of a few very high abundance proteins in a number of biological fluids and cell extracts is one of the major obstacles to detecting rare species that are present in trace amounts. In two-dimensional separation, major proteins with spots that are proportional to their concentration mask the rare and diluted species of similar mass and similar isoelectric point. Thus many species that are in the surroundings are difficult, if not impossible, to identify. This is further exacerbated during subsequent mass spectrometry when the ionization energy is for example mostly taken by the most concentrated proteins, leaving the rare species nonionized and thus not detected. As a possible solution to this problem, the selective removal of high abundance proteins has been pursued by a variety of strategies. These included dye ligands, protein A and G, lectins, affibodies and specific anti-bodies [68]. Initially, the depletion addressed two major human serum proteins, albumin and immunoglobulins. Later, this was considerably extended.

Affinity depletion ligands may be categorized into large groups: (i) nonantibody deple-tion ligands; and (ii) antibodies for immunodepletion, both grafted on beads similar to those used for affinity chromatography purposes.

Among the nonantibody-based immobilized ligands are dye-based methods. There are numerous dyes that could be used for protein depletion but due to relatively poor specifi-city, only Cibacron Blue is frequently used for the removal of serum albumin [69–71]. Cibacron Blue adsorbs a number of other polypeptides nonspecifically, removing them from the original proteome where they might never be detected. This dye has some spe-cificity for NAD-dependent enzymes, and a large number of other enzymes and cytokines. Its structure of aromatic mobile electrons, sulfate anionic groups and primary amines is ideal for promoting molecular interactions, hydrophobic associations, ion exchange effects and hydrogen bonding. The depletion fraction (proteins that are removed by the dye), could eventually be considered as the fraction of interest where minor proteins are also concentrated as a result of nonspecific binding. However, such an approach complicates the task and goes against the purpose of the original depletion process.

Proteins A and G have been proposed as capturing ligands to remove immunoglobulin G (IgG) from serum and plasma as a consequence of quite specific interaction on the Fc region of the antibody. Both have been well known microbial ligands for the purification of antibodies for several decades [72, 73]. Because of its specific binding to the Fc-region of IgG, Protein G has been extensively used alone or in association with Protein A, to deplete serum IgG [74, 75]. Unfortunately these two bacterial proteins also display non-specific binding for a number of polypeptides and render the depletion significantly less specific than required. Additionally both Proteins A and G do not recognize certain immu-noglobulin classes, which will not be removed from the sample. Finally with nonantibody-based depletion, lectins have been used with other methods; for example to identify plasma proteins (at concentrations of $10–100\,ng\,ml^{-1}$) that are of particular interest for studying a variety of disease conditions [76].

Affibody-mediated depletion, as an alternative to those described above, has been suc-cessfully evaluated recently for albumin, immunoglobulins and transferrin in human cer-ebrospinal fluid and serum [77] but was not very efficient for the former. Clearly, affibody-based depletion, while attractive because of higher binding capacity and specifi-city, needs additional development work.

To date the most effective depletion processes are based on specific antibodies. In 2005, a comparative study using different depletion technologies concluded that multiple immuno-affinity sorbents offered the most promising depletion approaches [78, 79].

Monoclonal antibodies with high affinity to a unique epitope on the macromolecule are required to completely deplete the targeted proteins [80]. However, monoclonal antibodies may not recognize truncated proteins or differences in post-translational modification of the same gene product or even misfolded epitopes for which they are designed [71]. Using polyclonal antibodies will in principle result in better depletion since they recognize many regions on the target surface.

However, even though the masking effect of high abundance proteins may be removed by some of these methods, several authors reported specific drawbacks that might limit the use of depletion methods depending on the objectives. Granger *et al.* [81] and Shen *et al.* [82] demonstrated co-depletion of many low abundance proteins during immunodepletion. They included proteins associated with albumin and IgG as well as co-depletion due to nonspecific binding. Thus, in spite of better visibility of the remaining depleted proteome, the number of new spots detected on 2DE analysis was modest, and most of the newly visualized spots were minor isoforms of relatively abundant proteins [69].

Because immunoglobulin Y (IgY) polyclonal hen's egg yolk antibodies have less cross-reactivity and are evolutionarily distant from mammal IgG-based antibodies, they have been used for high abundance protein depletion [83, 84]. Their advantage extends also to the fact that the Fc-region does not bind to other proteins as is the case for mammal antibodies.

In summary, the immunodepletion method described above is by definition highly specific and usable for only one protein species. However, immunodepletion has been described for the removal of other high abundance proteins such as haemoglobin from red blood cell lysate [85, 86] and Rubisco from plant leaf extracts with quite a high degree of success [87].

9.3.2   Narrow IPG

One- and two-dimensional IPG provide a possible means to detect low abundance proteins. As described a few years ago [88, 89], this approach is based on the use of a series of narrow or very narrow range IPG strips (covering no more than 1 pH unit). Under a relatively large protein load and under these pI ranges, rare proteins appear as a result of better spot resolution. However, this method provides only a focused vision around the zone covered by the selected pI range. After the separation in a number of narrow pI ranges all close to each other, the entire map can then be reconstructed electronically to provide an image of the coverage of the proteome. Although this approach largely overcomes the problem of spot overlapping, it does not resolve other difficulties. For example, when using very narrow IPG strips, the entire nonfractionated sample needs to be loaded often causing precipitation of proteins at the extremities of the pI gradient. Once precipitation occurs, it entraps species that are no longer able to migrate and therefore escape detection. Protein precipitation is progressively enhanced as larger and larger loadings are necessary to render visible species that are much diluted. In practice this method is cumbersome and infrequently used.

9.3.3   Reduction of the Dynamic Concentration Range with Combinatorial Ligand Libraries

Reduction of the dynamic concentration range aims to reduce the concentration of high abundance proteins while concomitantly increasing proportionally the concentration of low abundance species. The process does not seek to remove species as described above (Section 9.3.2) but rather sets out to minimize the difference in concentration between the two extremes. Conceptually, the process is built on a multitude of affinity moieties that are all treated in large overloading conditions after which captured species are eluted quantitatively. Theoretically, it is an interaction process involving a solid phase and proteins in solution where concentrations and dissociation constants of each individual solute are at the heart of the competition system.

9.3.3.1 Principle

The solid phase used is a combinatorial peptide ligand library where each individual peptide is grafted onto a bead in a multitude of copies, each bead containing a unique peptide structure. Hexapeptides are synthesized via a short spacer according to a modified Merrifield approach [90], by using the split, couple and recombine method [91–93]. The ligands are represented throughout the bead's porous structure at a density of ca. 40–60 μmol ml^{-1} of settled beads. Each single bead has millions of copies of a single, unique ligand structure and each bead, potentially, has a different ligand from every other bead. Depending on the number of starting amino acids used, a hexapeptide library contains a population of dozens of millions of different ligands (e.g. 11 million for 15 different amino acids or 24 million for 17 different amino acids or even 64 million if the number of starting amino acids is 20).

The preparation of the entire combination of beads is relatively simple; diversity is dependent on the number of amino acids and the length of the peptide chain. The process is extensively described by Lam *et al.* [91].

The vast population of solid phase baits means that, in principle, an appropriate volume of beads could contain enough partners able to interact with each of all the proteins present in a complex proteome (be it a biological fluid or a tissue or cell lysate of any origin).

If one assumes that there existed affinity columns for each single gene product and all these affinity systems could work with the same adsorption-elution conditions, it would be possible to mix all affinity sorbents together and have a means to capture all proteins. In the present case, the affinity system is a bead carrying one single affinity bait (the hexapeptide) and the adsorption bed comprises statistically the same amount of beads of different specificity. The number of beads with similar affinity being limited in number, they display a limited protein binding capacity. When a complex protein extract such as serum is exposed to such a ligand library, in large overloading conditions each bead with affinity to an abundant protein will rapidly become saturated and the vast majority of the same protein will remain unbound. In contrast, trace proteins will not be able to saturate all the beads that have an affinity for them unless the volume of the biological sample is very large. Thus, on the basis of ability to capture each single gene product present and of the saturation-overloading chromatographic principle, a solid phase ligand library enriches for trace proteins while concomitantly reducing the concentration of abundant species (Figure 9.1). Limitations of this process are essentially inherent to the value of the dissociation constants between bead baits and their protein partner; temperature, ionic strength and pH play an important role in the modulation of affinity constants and on protein competition. Other perturbations could come from possible protein-protein interactions; moreover, the presence of chemicals that prevent protein docking on their respective peptides and possible nonspecific binding are also considered possible threats.

This principle along with the first experimental data are extensively explained in the seminal paper of Thulasiraman *et al.* [94] and then discussed in a few other published papers [95–97].

In spite of the above-described principle, individual proteins alone or from a mixture do not form a single partnership with a given ligand but rather with a number of other peptides of similar or different composition according to a variety of similar or dissimilar affinity constants.

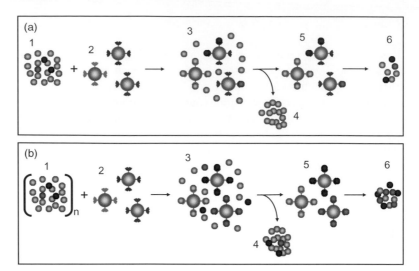

Figure 9.1 *Schematic representation of partial depletion of high abundance proteins (a) versus reduction of the dynamic concentration range (b) with concomitant increase in low abundance species. In (a) the sample load was insufficient to saturate all the beads of the combinatorial library while under overloading conditions (b), the reduction of high abundance species is accompanied by an increase in the rare species (blue and red circles). The numbers 1–6 represent the sequence of steps to get the final result as explained in the text: 1, sample loading on combinatorial library '2'; 3, washing away the excess of proteins ('4'): 5, the partially (a) or fully (b) saturated library is then treated to elute captured species ('6')*

In addition, a single bead carrying a given peptide does not capture only one protein, but several polypeptidic structures depending on the ability to form more or less stable complexes determined by the internal competition between these proteins for the same ligand [96].

The mass action law which is at the basis of the described protein-hexapeptide formation implies that interactions between partners are an equilibrium of forces governed by an association and dissociation constant. With complex mixtures of millions of hexapeptides on the one hand and thousands of gene products on the other hand, a large number of interaction possibilities takes place. Affinity constants between hexapeptides and proteins can range between almost zero and very large numbers so that it can be hypothesized that all proteins have the potential to interact with one bead and all beads to interact with one protein. As a result of this intricate situation, it can be stated that the mechanism of reduction of the protein concentration range is more complex than can be anticipated and is fully dependent on a number of parameters including the number of hexapeptide baits.

The number of beads used for the treatment of a biological sample should also be included in the interaction consideration. Since the beads have a given particle diameter, it is possible to calculate the number of beads and thus the number of diverse baits in a given volume of beads. It becomes apparent that this number is lower than the possible number of hexapeptides initially present. First of all, generally the volume of beads used for the sample treatment ranges between $100\,\mu l$ and $1000\,\mu l$ corresponding to 350000 and

3 500 000 compared with several millions of hexapeptide combinations. Secondly, even if the volume is large enough to comprise the required number of beads, the representation of the entire library would not be complete. Fortunately in the large majority of cases it is not necessary to have the entire library to obtain consistent and reproducible results because (i) there are many similar peptide combinations (e.g. those sharing all the structure but one single amino acid), (ii) penta- and even tetrapeptides show similar capabilities compared with hexapeptides and (iii) the most distal three amino acids comprising the peptide are effective for the interaction with the protein to capture.

Thus, after treatment with a combinatorial solid phase peptide library, not all proteins are equally concentrated; however, a strong reduction in the protein concentration range occurs, rendering detectable proteins that are normally at very low concentration and undetectable under current analytical approaches. It has been reported that the process has remarkable reproducibility in spite of the complex mechanism of action. Reproducibility has been demonstrated [95–97] not only qualitatively by 2DE and mass spectrometry experiments but also by quantitation measurements of low abundance species enhanced as a consequence of library treatment. The library dedicated to the described sample treatment is commercially available under the trade name of ProteoMiner (Bio-Rad Laboratories).

It is important to note that using hexapeptide ligands for establishing an affinity interaction might be considered to represent a rather weak binding event; however, experience has demonstrated that the complexes obtained can be of high affinity requiring strong elution conditions. Table 9.1 gives a summary of what could be used as a desorption agent as a function of molecular interaction and application. Further details have been reported elsewhere [95–97].

Table 9.1 *Elution protocols for proteins captured by solid phase peptide libraries as a function of molecular interaction between prey proteins and bait peptides*

Bonds involved	Examples of dissociation solutions
Ionic interactions	1 M sodium chloride Other salts Changes in pH below 3 or above 11
Hydrogen bonding	4–8 M urea
Mildly hydrophobic associations	50% ethylene glycol in water
Strong hydrophobic associations	Acetonitrile (6.6)-isopropanol (33.3)-trifluoroacetic acid (0.5)-water(49.5) Acetonitrile (6.6)-isopropanol (33.3)-ammonia(0.5)-water(49.5)
Mixed mode, hydrophobic associations, hydrogen bonding	2 M thiourea-7 M urea-4% CHAPS
Combined hydrogen bonding, ionic interactions	9 M urea, 2% CHAPS, citric acid or acetic acid to pH 3.0–3.5 9 M urea, 2% CHAPS, ammonia to pH 11 2 M thiourea-7 M urea-4% CHAPS, cysteic acid or hydrochloric acid to pH 2–3
All types of interactions	6 M Guanidine-HCl. pH 6 10% SDS-2% dithiothreitol
Spatial conformation docking	200 mM Glycine-HCl, pH 2.5
Molecular recognition	Specific displacers

9.3.3.2   Impact on 2DE Analysis

Two-dimensional electrophoresis is one of the most popular analytical methods and illustrates the complexity of protein extracts by positioning all proteins into a map of isoelectric points and molecular masses. The use of combinatorial ligand libraries increases the number of protein spots throughout the surface of the map. What is remarkable is that high abundance proteins, that prevent observation of other proteins that are in the surroundings, are reduced, rendering the map more readable and spots better resolved. Many other undetectable proteins appear as a consequence of the concentration increase of low abundance species, thus enhancing the detectability of many new polypeptides that are 'amplified' at a higher level than medium–large proteins. The lower parts of the map where small masses are located are largely populated suggesting that probably even lower masses are present but migrated from the plate because of their small mass. For better detection of masses below 20 kDa it is therefore suggested to use steeper gel gradients capable of retaining relatively small proteins.

As far as the isoelectric point positioning is concerned, the largest number of proteins ranges between about 4.5 and 8.5; this is generally the pI zone where most of the proteins are present with a preferred pI range of between 4.5 and 6.5. The use of the hexapeptide combinatorial ligand library does not appear to change the picture. With libraries that comprise many primary amine terminations, the average pI is a little displaced towards alkaline pHs, whereas with carboxylated libraries the protein population displaces towards acidic pHs. This situation suggests that there are complementary possibilities to make protein captures at different pHs with larger coverage compared with regular current use. Recently published data demonstrated this assumption when using a human serum proteins sample; compared with the capture performed under physiological conditions, acidic and alkaline sample treatment allowed the detection of additional spots by 2DE [98].

Protein treatment with the described libraries implies that there is a total desorption of captured proteins by appropriate eluting agents. Due to some extremely strong interactions, denaturing elution agents are used such as concentrated urea acidified with either acetic acid or citric acid or even using concentrated solutions of guanidine hydrochloride. Recovered proteins here are incompatible with direct 2DE unless dialysed against appropriate aqueous solutions and also lyophilized. To restore full compatibility between protein elution and 2D analysis, selected eluting agents have been reported [99] such as thiourea-urea-chaps (TUC) solution added with acids [Figure 9.2(a)]. While regular TUC solution allows elution between 60% and 80% of proteins according to the protein extract, acidified TUC increases the elution capability to more than 95–97%. Acidification with acetic acid or citric acid engenders focusing problems due to their relatively high pKs, while with other acids, such as formic and cysteic acid, the drawback is circumvented. Adding 40 mM formic acid, all 2DE maps exhibit more even distribution of spots in the entire pH 3–10 interval with no distortion of zones being apparent all along the stated pH interval. This might not only be due to formic acid condensing around pH 3, but also due to its fully protonated state that will allow it to slowly evaporate thus substantially diminishing its concentration in the gel phase. Nevertheless it has been reported that formic acid could induce formylation (esterification) at serine and threonine residues as well as N-formylation (amidation) at ε-amino groups of lysine side chains, and the terminal amines [100]. These potential reactions induce little changes in both protein mass and isoelectric point thus displacing their position throughout the 2D map.

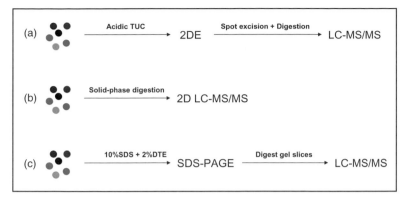

Figure 9.2 *Recommended protein elution approaches from peptide libraries as a function of proteome analysis. Each proposed elution allows a direct interphasing with the analysis to follow without any preliminary treatment (concentration, dialysis, desalting, precipitation, etc.)*

Instead, cysteic acid is not reactive towards chemical residues from protein structures. It exhibits an isoelectric point (pI) of 1.80 (pK of the carboxyl group close to 2.1 and pK of the sulfate group close to 1.5). It provides an extraordinary buffering power and as a direct consequence, it could be used at only 30 mM concentration in the presence of TUC; the pH decreases and stabilizes to about 3, ensuring complete desorption of captured proteins. At this pH value all free carboxyls of hexapeptides are fully protonated annihilating ionic interactions. Other components of TUC then contribute to weaken hydrogen bonding and hydrophobic associations.

9.3.3.3 Impact on Mass Spectrometry Analysis (Top-Down and Bottom-Up)

The impact of proteins collected out of peptide libraries depends on the type of mass spectrometry analysis.

If LC-MS/MS is considered for the identification of proteins comprising the proteome, two different situations are encountered: (i) full digestion of the protein mixture followed by LC-MS/MS analysis with one or two types of LC steps; and (ii) separation of the proteome by sodium dodecyl sulfate polyacrylamide gel electrophoresis (SDS-PAGE) followed by lane slicing and trypsin digestion. Each digest is then loaded onto a LC-MS/MS system for full analysis of peptides. For both applications, eluates from combinatorial libraries obtained using acidic urea or even acidic TUC are not ideal; in fact trypsin enzymatic activity is reduced or totally denatured while SDS-PAGE migration would not be as good. In such cases alternative procedures should be adopted. Full digestion of proteins can be operated in the solid phase without the need for elution of proteins as described [101]. Peptides desorbed by volatile organic solvents/acids are then usable directly for bottom-up analysis [Figure 9.2(b)].

When proteins from the ProteoMiner are to be separated first by SDS-PAGE, elution can be achieved directly with a solution of 10% SDS containing 2% dithiothreitol (DTE) [Figure 9.2(c)] at room temperature or even better at boiling temperature. This elution agent has been reported to be highly successful for urine protein elution with a high recovery of 97% [102]. Classical elution using concentrated solutions of urea acidified with acetic or citric acid are compatible with surface enhanced laser desorption/ionization (SELDI) MS analysis because the proteins are first diluted in the buffer for the intended chip surface and undesirable chemical agents are then eliminated by the washing step of the mass spectrometry chips prior to matrix assisted laser desorption/ionization (MALDI) analysis [103, 104].

9.3.3.4   Applications to Biological Fluids

The majority of data related to biological samples treated with combinatorial ligand libraries are biological fluids. These include blood serum, urine, cerebrospinal fluid, bile fluid all from human origin, milk whey from bovine and even snake venom, the latter being the only one not of mammalian origin. All these studies have in common the unambiguous enhancement of many rare species that would otherwise be impossible to detect or identify by current analytical methods such as 2DE and even advanced mass spectral techniques. Another common result between the treated samples is that polypeptides below 10 kDa are significantly increased compared with those greater than 40 kDa. When making a comparative gene ontology classification of proteins before and after treatment, it has been found also that there is a proportional increase in the presence of protein categories that have properties to interact with others in their biological function, such as transport proteins and signalling species.

As a consequence of the large increase in the number of rare species that are normally undetectable, losses of proteins that were present in the crude sample were below the detectability level and were therefore 'lost'. The number of 'lost' proteins is variable between samples, but some studies report the loss of a few per cent while in others this can be greater than 30% as for instance in bile fluid. Specific reasons for this have not been demonstrated experimentally but may be due to the absence of peptide structures with affinity for lost species, or repulsion conditions of the lost species due to environmental conditions such as temperature, ionic strength and pH. Although unlikely, incomplete elution of captured proteins has also been suggested as a possible reason for protein losses. In fact, protein traces remaining on the beads after stringent elution have not revealed different protein patterns. When using complementary libraries such as primary amine terminal and carboxylated libraries, the number of 'lost' species significantly diminished confirming the repulsion hypothesis mentioned above which results probably from very strong competition of species for their common peptide structure. Since pH modulates the protein net charge and hence the affinity between peptide and protein partnerships, trials of protein capture at different pHs were reported; thus different SDS-PAGE patterns were found [105] with the presence of complementary proteins. Two-dimensional analysis of these fractions (Figure 9.3) confirmed that additional proteins were collected when acidic and alkaline pH were used confirming the drastic reduction of protein losses (unpublished data).

Using the peptide library treatment the number of detected spots on a 2DE map of human serum was increased more than six times (from 115 to 790). A significant increase was obtained when using a large sample and three distinct elutions from a peptide library with evidence of many very low abundance gene products. Masses and isoelectric points were extended throughout the map. The analysis of the same fractions by LC-MS/MS after SDS-PAGE separation and slicing of migration lanes into 15 parts, confirmed the perform-

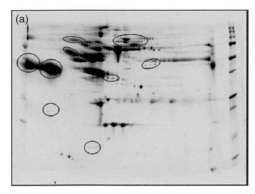 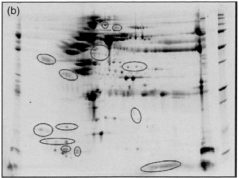

Figure 9.3 *Two-dimensional analysis of proteins captured by the same peptide library as a function of pH. (a) The capture was performed at pH 4; (b) the capture was performed at pH 9.3. Both captured proteins were eluted by a solution composed of 9 M urea containing 2% CHAPS and 100 mM acetic acid. As shown, a large number of proteins are common; however, a significant number is dependent on the pH at which the capture was performed. Blue rings indicate exclusive proteins captured at acidic pH; red rings indicate proteins captured at alkaline pH*

ance of the study, with the identification either of 1559 or 3869 gene products, depending on the level of confidence (99 or 95%) [106]. Out of the protein species 75 belong to immunoglobulins, a group that is normally absent in depletion-based sample treatment. Some unexpected proteins came out from the analysis of peptide library eluates, such as nuclear proteins which are not known to be present in serum and may be the result of erythrocyte cell lysis during the preparation of serum.

Compared with previously published protein lists of the serum proteome, the relatively low overlap observed reinforces the notion that the peptide ligand library retains proteins that are mostly complementary to what is normally found using conventional methods. Another observation by the authors was that alkaline proteins were over-represented after serum treatment with the library.

In a number of published reports for example with human urinary proteome, two libraries were used in series and the discovered proteins summed up. Here the number of gene products found in the control sample was 134 while the sum of proteins of two sequential peptide ligand bead eluates comprised 249 nonredundant unique proteins [107]. This study contributed to an increase in the known list of urine proteins with 251 new gene products. Among them were proteins of larger masses than expected (more than 200 kDa), suggesting that the enhancement effect was also operating for proteins that were presumably from cells of the urinary ducts. Among them were membrane, receptor and binding proteins and cell adhesion molecules. They are probably proteins that have not filtered though kidney glomeruli but rather collected after filtration by duct wall cell desquamation.

Cerebrospinal fluid, a low concentration protein fluid, was also treated with peptide ligand libraries. Under conditions of very large overloading the number of gene products found using nano-LC-MS/MS after SDS-PAGE separation of obtained fractions, increased

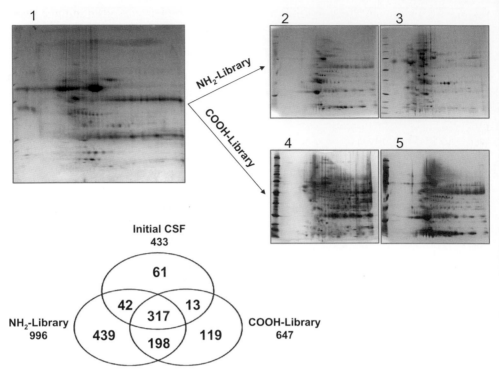

Figure 9.4 *Proteome analysis of human cerebrospinal fluid: Two-dimensional electrophoresis (upper) and overlap diagram (lower) of gene products found by nano-LC-MS/MS.*
1, 2D map of control cerebrospinal fluid; 2 and 3, 2D maps of eluates from a primary amine peptide library desorbed, respectively, by thiourea-urea-CHAPS and acidic urea-CHAPS solutions; 4 and 5, 2D maps of eluates from a carboxylated peptide library desorbed, respectively, by thiourea-urea-CHAPS and acidic urea-CHAPS solutions. Coomassie staining. The numbers inside the circles of the Venn diagram represent the gene product found in the cerebrospinal fluid prior to treatment and after contact with the above-mentioned libraries

by a factor of 2.6 (from 433 to 1128) (Figure 9.4). This included nonredundant species that were from the two libraries positioned in series as mentioned above. The first library alone captured 996 unique proteins while the second library added another 132 proteins with a common pool of 515 gene products. The analysis of the same fractions using 2DE showed many more spots compared with an untreated sample. Eluates from the first library showed some more acidic proteins while from the carboxylated library more alkaline proteins were found. These observations are common with a number of other biological samples. Analysis of the proteins following peptide library treatment revealed the presence of a large number of polypeptides related to the brain which could be expected from such a biological fluid.

The analysis of human bile fluid following treatment with peptide libraries has also been reported [108]. Since bile contains a large amount of lipid-like products that tend to interfere with docking mechanisms between proteins and their respective peptide bait and produce bead aggregates, the first operation was to undertake a clean-up step. The bile was treated with two libraries in series and the captured proteins desorbed in three steps. The analysis of the material by SDS-PAGE and LC-MS/MS showed that the initial material comprised 141 gene products whereas the sample treated with the peptide ligand libraries allowed detection of 197 species, 116 of them being redundant with the initial untreated sample. From the difference, 81 new gene products were identified, most of them previously unreported in bile protein fluids. As stated, about 11% of the high abundance proteins were not captured. The large majority of products (about 75%) were binding proteins and those with transport activities. Among the newly identified proteins in bile were the ERM family, i.e. radixin and/or ezrin that are both known to be expressed at significant levels in hepatocytes and in biliary epithelial cells. Moesin, another protein detected after peptide library treatment, is known to be expressed at very low levels in hepatocytes. Detoxifying enzymes, derived from hepato-biliary lining cells were also found such as glutathione *S*-transferase of three types, microsomal, α and π, the latter being specific for the biliary epithelium, while the others are generally expressed both in hepatocytes and in biliary epithelial cells. Other proteins related to defence system were found such as azurocidin and defensins along with proteins reflecting the presence of cell debris from the bile duct.

In the study of the proteome of bovine milk whey proteome two complementary peptide libraries were used. The first, a single elution of all proteins allowed discovery and identification of a large number of previously unreported proteins [109]. A total of 149 unique protein species were found, of which 100 were hitherto not reported previously. From the carboxylated peptide library, polymorphic alkaline protein signals were observed with strong positive signals when blotted with sera of allergic patients.

Many allergens were detectable from this study; many of which were not part of the lists of known allergens. Among them immunoglobulins were formally identified after having been largely amplified. This is a very minor allergen generally undetectable unless after fractionation of milk whey. Other allergens located in the acidic part of IEF analysis of protein fractions have not been identified yet. These findings could be extended to a number of other protein extracts from various sources and used as a very sensitive method to detect minor milk allergens with potential applications in pharmaceutical and diagnostic fields.

The analysis of snake venom (*Crotalus atrox*) fluid after peptide library treatment [110] was achieved using two peptide libraries and the captured proteins collected as a single fraction. The latter was then analysed by 2DE and differential spots excised and identified by LC-MS/MS after trypsin digestion. This first in-depth venomic study showed about 24 proteins belonged to two distinct families of proteases, metalloproteinases and serine proteinases, both representing almost 70% of total venom proteome. Additionally four medium abundance protein groups were found: cysteine-rich secretory protein, medium-size disintegrin, PLA_2, and L-amino acid oxidase. Among the low abundance proteins (less than 5% of venom proteome), three groups of proteins were found: endogenous inhibitors of proteinases, C-type lectins and vasoactive peptides.

Globally, the amplified low-copy proteins from *C. atrox* venom comprised a C-type lectin-like protein, several PLA_2 molecules, PIII-SVMP isoforms, glutaminyl cyclase, and a 2-cys peroxiredoxin highly conserved across the animal kingdom. Moreover, the authors reported proteins involved in redox processes leading to the structural/functional diversification of toxins, and in the N-terminal pyrrolidone carboxylic acid formation required in the maturation of bioactive peptides such as bradykinin-potentiating peptides and endogenous inhibitors of metalloproteases. This pioneering study may also be of interest in the search for novel medical applications since snake venom is a huge and largely unexplored reservoir of bioactive components.

Apart from full biological fluids other studies have demonstrated the capability of the peptide ligand library in enhancing rare proteins in a number of other biological extracts such as egg white [111] and egg yolk [112]. These studies not only allowed discovery of new species and thus contribute to enlarge the protein components of these proteomes, but could also open up possible applications for the pharmaceutical and food industries. In vaccines produced in eggs for example, it may be possible to find protein impurities that could normally be excluded for injections.

9.3.3.5   Applications to Cell Extracts

The effect of hexapeptide ligand libraries has also been reported for cell extracts in physiological saline conditions. Red blood cells, platelets and yeast cells were also investigated with the aim of discovering proteins involved in oxidative stress. The exploration of the platelets' proteome after treatment by the peptide ligand libraries, followed by 2DE and mass spectrometry analysis, allowed expansion of the list of known proteins with 147 newly described gene products [113]. Using two different libraries, it was also possible to distinguish proteins captured by an amino terminal peptide library from those that were captured by a carboxylated library. The complementary effect of these two libraries was thus demonstrated with their tendency to enhance, respectively, mostly acidic proteins, on the one hand, and alkaline proteins, on the other hand, although a lot of proteins were found in common.

Several very interesting proteins were reported such as PMKase, signalosome subunit 7a, LMW-PT, reticulon-4 and the nucleosome-assembly protein-2. Other proteins that are either functionally expected in platelet lysates but never reported or known by just their biological activity were Rho-GTPase-activating protein 18, PAF-acetyl hydrolase and Galpha 13 together with several proteins from the same signalling pathway. Actin, which was highly represented in the initial platelet extract, was significantly reduced after peptide library treatment.

The red blood cell lysate is probably one of the most representative examples of enhancement of low abundance proteins following reduction of its major protein, haemoglobin which represents more than 97% of the overall protein mass. Starting from a large volume of cell extract and using two libraries in series, 1578 gene products were identified [114]. With a more modest load of the same lysate and just a single elution instead of three, the number of proteins identified was smaller (525 gene products) [115] but still twice the number of proteins reported recently [116]. Two-dimensional analysis of the sample after treatment showed a large number of spots populating the entire surface of molecular masses and isoelectric points. Of the proteins, 72% were cytoplasmic; however, a relatively large number of proteins were also classified as nuclear in spite of the anuclear character of red blood cells. This might suggest that a number of proteins that normally stay in the nucleus of erythroblasts are still present in the mature erythrocytes. Among other interesting discoveries, eight different haemoglobin chains were detected; including not only the well known α, β, γ and δ chains, but also two embryonic ε and ζ chains from genes that are incompletely silenced during development, and for the first time the μ and θ chains [117].

By contrast, in the study on yeast there was no dominant protein but rather a continuum of the concentration of proteins even when the dynamic concentration range was very large. So when treating such extracts, the goal was mostly to enhance the presence of very low abundance proteins that are otherwise difficult to detect. This was performed with yeast with the view of detecting the maximum number of proteins that are involved in oxidative stress. To that end, *Saccharomyces cerevisiae* cells were collected after culture and lysed in the presence of antiproteases and after removal of the pellet the proteins were treated with peptide ligand libraries in large overloading conditions. Eluted proteins from beads using acidic concentrated urea were then submitted to 2D analysis and compared with the same proteins prior to treatment. The difference was immediately apparent since major groups of isoforms were significantly reduced and many more spots became detectable. Using a well established procedure, proteins involved in oxido-reductive processes were isolated and identified by LC-MS/MS. The result was a significant increase of more than 200 undetectable proteins out of a total of about 550 species (unpublished data).

An *E. coli* whole extract was also investigated [94] and resulted in the detection of many more proteins that are normally difficult to detect from a normal crude extract. The identification was performed by first separating eluted proteins by SDS-PAGE, slicing out bands of migration lane and digesting proteins by in-gel procedure followed by MS/MS analysis of peptides. Among the identified proteins was ADP-L-glycero-B-manno-heptose-6-epimerase that is generally present in *E. coli* cells in around 220 copies only; there were enzymes not previously detected by 2DE and a putative tagatose 6-phosphate kinase gatZ that was previously reported only by DNA sequencing.

9.3.3.6   Application to Plant Proteomes

As with animal protein extracts or biological fluids, the difficulty of detecting low abundance proteins in plant extracts is also frequently impeded by the concentration of some abundant species. As an example, the most abundant protein in plant leaves and thus in the biosphere, rubisco (ribulose-1,5-biphosphate carboxylase/oxygenase) accounts for more than 40% of the total leaf protein [118] and is the most frequently identified by 2DE. In a study by Giavalisco *et al.* [119], rubisco alone represented a total of 366 spots, corresponding to 12.5% of those assigned.

However, within the field of proteomics, plant proteomes are the least reported. Moreover, published papers deal with about three to four dozen species, most focus either on the primary dicot *Arabidopsis thaliana* [120] or on the monocot rice (*Oryza sativa*), both of which have had their full genome sequenced [121]. With few exceptions, protein concentrations in plant extracts are very dilute and their analysis is therefore difficult. It is anticipated that this situation could be largely improved with the use of the combinatorial peptide ligand libraries for mining the low abundance proteome, as outlined above. To add to this challenge, plant extracts are composed of many substances [122] that can interfere with separation techniques. There are effectively substances that interfere with the protein capture by peptide libraries by competing with proteins for the same binding sites. Among these are the presence of powerful proteases, which require specific precautions and elaborate protease inhibitor cocktails during extraction procedures [123] and various plant-specific cellular components, such as polysaccharides, lipids, polyphenols and secondary metabolites, which also interfere with protein separation and subsequent analyses [124]. Complex assemblies of polysaccharides with large amounts of anionic pectins contribute to prevent affinity protein capture using large overloading conditions. Further complications arise from the nature of plant proteins which are heavily glycosylated comprising proteoglycans that interfere with separation and analytical determinations. To improve the treatment of crude plant extracts with combinatorial peptide libraries, pretreatments should be considered. A recent published protocol suggests several approaches prior to treatment of plant extract by peptide libraries [125]. Depending on the type of plant tissue, a lipid removal step may be justified especially when dealing with seeds such as corn [126], sunflower and soya beans. Plant proteins can easily be precipitated with trichloroacetic acid (TCA) or ammonium sulfate to collect proteins that are free or almost free of interfering substances. While ammonium sulfate does not denature proteins, the use of TCA may induce denaturation and therefore the loss of biological properties.

It has been observed that carboxyl-terminal peptide libraries perform better compared with the -NH$_2$-terminus peptide libraries. This observation could be explained by the residual presence of negatively charged peptide baits as opposed to positively charged peptides. Whereas interfering polyanionic substances would surely bind to the -NH$_2$-terminus resin, thus hampering the binding of proteins, they would be repelled by the carboxyl-terminus beads, allowing unhindered binding of the proteinaceous ligands.

To achieve maximum performance of the peptide library of a corn extract, lipid extraction with a chloroform-methanol mixture was first performed followed by extraction under physiological conditions. From the collected clear solution, the proteins were recovered by precipitation with ammonium sulfate at 80% saturation. This ensured that virtually no nucleic acids were present and most of the polysaccharides were eliminated. The protein extracts captured by peptide libraries did not show any improvement in protein detection if the extract had not been treated to remove interfering substances. However, after proper pretreatment, a number of newer 2DE spots could be visualized in the various eluates. This was also clearly visible in the SELDI-MS profiling on the Q10 ProteinChip array of the six extracts, as compared with the control (Figure 9.5).

The latex from *Hevea brasiliensis* was also analysed before and after treatment with peptide libraries. Centrifugation of the clear aqueous phase corresponding to the cytosol was treated with 80% saturation of ammonium sulfate and protein pellets redissolved in physiological saline before contact with the library. Again, not much was visible in untreated cytosolic fractions. However, when 900–1200 mg protein harvested from 100 ml of latex cytosol were additionally treated by precipitation of proteins in 80% ammonium sulfate, excellent results were obtained [127] with the recovery of all bands along with a large number of new ones. This was also clearly visible by mass spectrometry profiling using SELDI-MS.

9.3.3.7   Fractionation Requirements as a Consequence of Dynamic Range Reduction

Sample treatment with a combinatorial hexapeptide ligand library increases the apparent complexity of the resulting sample. When using current analytical methods as for example one- or two-dimensional electrophoresis, many more spots are visible as additional gene products and novel isoforms. The complexity of the treated sample could be so high that the probability of species co-migrating or with similar or even the same isoelectric point and same or similar mass is high. Thus overlapping signals render the detection almost impossible and create confusion in the identification of species. In this situation, fractionation of the sample except subcellular fractionation remains one of the preferred methods to prevent signal overlapping. The selection of a fractionation method is based on the properties that best reduce protein redundancy from fraction to fraction. It is preferable to produce a limited number of fractions with low level of redundancy than many large overlapping fractions.

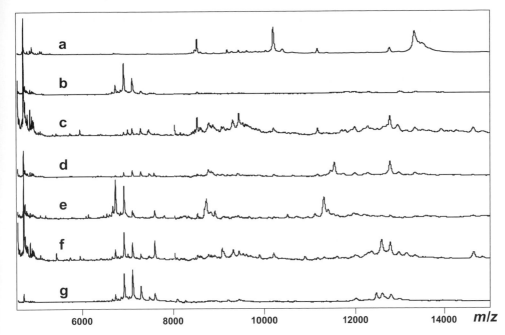

Figure 9.5 *SELDI-MS analysis of* Hevea brasilensis *protein extract (a) compared with fractions obtained from the treatment of the initial sample with two sequential libraries (primary amino terminal, see b, c and d and carboxylated, see e, f and g) eluted using 1 M sodium chloride (b and e), thiourea-urea-CHAPS (c and f) and urea-CHAPS-acetic acid (d and g). Complementary signals can be identified easily along with a significantly larger number of species. The analysis was carried out at masses ranging from 5 to 15 kDa; m/z, mass to charge ratio*

In general, liquid chromatography with its related gradient elution (linear or stepwise) produces overlapping fractions since a protein peak is never really symmetric and forms tails that interfere with the following elution of species. Additionally, desorbed species cross the column from the top to the bottom while other proteins are still adsorbed creating carryover with unavoidable cross contamination. To prevent or reduce this phenomenon, sectional columns were described a few years ago where proteins interacted on their selective section and separated during a single binding step. These sections were then separated and the captured proteins were individually desorbed [128]. Using this principle a positive reduction of overlapping was achieved and a large number of species identified.

Recent experiments were reported that demonstrated a very effective approach for polypeptide separation by pI ranges in combination with sectional columns [129]. Based on the use of solid-state buffers associated with ion exchangers, the method does not use carrier ampholytes and the separation is rapid, making the method attractive for the treatment of several samples within a short period of time. Overlapping risks related to column-based separation is thus prevented using the principle of sectional columns mentioned above. It was shown that a significantly larger number of proteins could be identified by this method when compared with conventional ion exchange elution methods [130]. Thus, whatever method was selected for fractionation, the number of proteins found was always larger than analysing the sample just after treatment with peptide libraries.

9.4   Conclusions and Envisioned Outcome

Major technical progress has been made in the last decade to improve proteomic analysis especially relating to the discovery of low and very low abundance gene products. However, in spite of the depletion of major proteins, the reduction of dynamic concentration range and the instrument sensitivity and resolution, further improvements are still needed to enlarge protein coverage. Although the use of combinatorial peptide libraries represents a breakthrough in rare protein detection, additional progress is expected. The future will see the introduction of specialized libraries designed for the selective reduction of the dynamic range of protein categories such as phosphoproteins or other post-translational modifications.

Continued research should also focus on quantitative aspects of rare species, since their regulation may be of importance in diseases as well as in drug interaction studies. As discussed above, fractionation remains a necessary means to simplify downstream analysis; therefore designing strategies for fractionation capable of reducing protein overlapping between fractions is a key step to its success. Attention will need to be given to better exploit the separation possibilities of available technologies as well their direct use after the treatment of the initial sample. Fractionations targeting predefined categories of gene products are also expected. However, progress will only be possible after there is a sound understanding of the detailed mechanisms of interactions between protein and peptides. In this way a more rational approach would become possible for the optimization of available technologies.

Acknowledgements

PGR is supported by Fondazione Cariplo (Milano).

References

1. Righetti, P.G., Castagna, A., Antonucci, F., Piubelli, C., Cecconi, D., Campostrini, N., Rustichelli, C., Antonioli, P., Zanusso, G., Monaco, S., Lomas, L., Boschetti, E. Proteome analysis in the clinical chemistry laboratory: myth or reality? *Clin. Chim. Acta* 2005, **357**, 123–139.
2. de Duve, C., Pressman, B.C., Gianetto, R., Wattiaux, R., Appelmans, F. Tissue fractionation studies. 6. Intracellular distribution patterns of enzymes in rat-liver tissue. *Biochem. J.* 1955, **60**, 604–612.
3. Beaufay, H., Bendall, D.S., Baudhun, P., Wattiaux, R., de Duve, C. Tissue fractionation studies. 13. Analysis of mitochondrial fractions from rat liver by density-gradient centrifuging. *Biochem J.* 1959, **73**, 628–637.
4. de Duve, C. The separation and characterization of subcellular particles. *Harvey Lectures* 1965, **59**, 49–59.
5. Börnig, H. Separation and characterization of subcellular organoids by tissue fractionation. *Acta Histochem. Suppl.* 1975, **15**, 13–34.
6. Pertoft, H. Fractionation of cells and subcellular particles with Percoll. *J Biochem. Biophys. Methods* 2000, **44**, 1–30.

7. Gauthier, D.J., Lazure, C. Complementary methods to assist subcellular fractionation in organellar proteomics. *Expert Rev. Proteomics* 2008, **5**, 603–617.
8. Lopez, M.F. Better approaches to finding the needle in a haystack: optimizing proteome analysis through automation. *Electrophoresis* 2000, **21**, 1082–1093.
9. Sun, Q., Miao, M., Jia, X., Guo, W., Wang, L., Yao, Z., Liu, C., Jiao, B. Subproteomic analysis of the mitochondrial proteins in rats 24 h after partial hepatectomy. *J. Cell. Biochem.* 2008, **105**, 176–184.
10. McDonald, T.G., Van Eyk, J.E. Mitochondrial proteomics. Undercover in the lipid bilayer. *Basic Res. Cardiol.* 2003, **98**, 219–227.
11. Zischka, H., Kinkl, N., Braun, R.J., Ueffing, M. Purification of *Saccharomyces cerevisiae* mitochondria by zone electrophoresis in a free flow device. *Methods Mol. Biol.* 2008, **432**, 51–64.
12. Braun, R.J., Kinkl, N., Zischka, H., Ueffing, M. 16-BAC/SDS-PAGE analysis of membrane proteins of yeast mitochondria purified by free flow electrophoresis. *Methods Mol. Biol.* 2009, **528**, 83–107.
13. Henrich, S., Cordwell, S.J., Crossett, B., Baker, M.S., Christopherson, R.I. The nuclear proteome and DNA-binding fraction of human Raji lymphoma cells. *Biochim Biophys Acta* 2007, **1774**, 413–432.
14. Komatsu, S. Extraction of nuclear proteins. *Methods Mol. Biol.* 2007, **355**, 73–77.
15. Cox, B., Emili, A. Tissue subcellular fractionation and protein extraction for use in mass-spectrometry-based proteomics. *Nat. Protoc.* 2006, **1**, 1872–1878.
16. Jiang, L., He, L., Fountoulakis, M. Comparison of protein precipitation methods for sample preparation prior to proteomic analysis. *J Chromatogr., A* 2004, **1023**, 317–320.
17. Maldonado, A.M., Echevarría-Zomeño, S., Jean-Baptiste, S., Hernández, M., Jorrín-Novo, J.V. Evaluation of three different protocols of protein extraction for *Arabidopsis thaliana* leaf proteome analysis by two-dimensional electrophoresis. *J Proteomics* 2008, **71**, 461–472.
18. Park, J.W., Lee, S.G., Song, J.Y., Joo, J.S., Chung, M.J., Kim, S.C., Youn, H.S., Kang, H.L., Baik, S.C., Lee, W.K., Cho, M.J., Rhee, K.H. Proteomic analysis of *Helicobacter pylori* cellular proteins fractionated by ammonium sulfate precipitation. *Electrophoresis* 2008, **29**, 2891–2903.
19. Méchin, V., Damerval, C., Zivy, M. Total protein extraction with TCA-acetone. *Methods Mol Biol.* 2007, **355**, 1–8.
20. Watts, A.D., Hunt, N.H., Hambly, B.D., Chaudhri, G. Separation of tumor necrosis factor alpha isoforms by two-dimensional polyacrylamide gel electrophoresis. *Electrophoresis* 1997, **18**, 1806–1091.
21. Gronborg, M., Kristiansen, T.Z., Stensballe, A., Andersen, J.S., Ohara, O., Mann, M., Jensen, O.N., Pandey, A. A mass spectrometry-based proteomic approach for identification of serine/threonine-phosphorylated proteins by enrichment with phospho-specific antibodies: identification of a novel protein, Frigg, as a protein kinase A substrate. *Mol. Cell. Proteomics* 2002, **1**, 517–527.
22. Sheffield, L.G., Gavinski, J.J. Proteomics methods for probing molecular mechanisms in signal transduction. *J. Anim. Sci.* 2003, **81**, 48–57.
23. Yuan, X., Desiderio, D.M. Proteomics analysis of phosphotyrosyl-proteins in human lumbar cerebrospinal fluid. *J. Proteome Res.* 2003, **2**, 476–487.
24. Brill, L.M., Salomon, A.R., Ficarro, S.B., Mukherji, M., Stettler-Grill, M., Peters, E.C. Robust phosphoproteomic profiling of tyrosine phosphorylation sites from human T cells using immobilized metal affinity chromatography and tandem mass spectrometry. *Anal. Chem.* 2004, **76**, 2763–2772.
25. Figeys, D., McBroom, L.D., Moran, M.F. Mass spectrometry for the study of protein-protein interactions. *Methods* 2001, **24**, 230–239.
26. Ren, L., Emery, D., Kaboord, B., Chang, E., Qoronfleh, M.W. Improved immunomatrix methods to detect protein:protein interactions. *J. Biochem. Biophys. Methods* 2003, **57**, 143–157.
27. Schulze, W.X., Mann, M. A novel proteomic screen for peptide-protein interactions. *J. Biol. Chem.* 2004, **279**, 10756–10764.
28. Stasyk, T., Huber, L.A. Zooming in: fractionation strategies in proteomics. *Proteomics.* 2004, **4**, 3704–3716.

29. Righetti, P.G., Castagna, A., Antonioli, P., Boschetti, E. Prefractionation techniques in proteome analysis: the mining tools of the third millennium. *Electrophoresis.* 2005, **26**, 297–319.
30. Lee, H.J., Lee, E.Y., Kwon, M.S., Paik, Y.K. Biomarker discovery from the plasma proteome using multidimensional fractionation proteomics. *Curr. Opin. Chem. Biol.* 2006, **10**, 42–49.
31. Matt, P., Fu, Z., Fu, Q., Van Eyk, J.E. Biomarker discovery: proteome fractionation and separation in biological samples. *Physiol Genomics* 2008, **33**, 12–17.
32. Jmeian, Y., El Rassi, Z. Liquid-phase-based separation systems for depletion, prefractionation and enrichment of proteins in biological fluids for in-depth proteomics analysis. *Electrophoresis* 2009, **30**, 249–261.
33. Goncalves, A., Esterni, B., Bertucci, F., Sauvan, R., Chabannon, C., Cubizolles, M., Bardou, V.J., Houvenaegel, G., Jacquemier, J., Granjeaud, S., Meng, X.-Y., Fung, E.T., Birnbaum, D., Maraninchi, D., Viens, P., Borg, J.-P. Postoperative serum proteomic profiles may predict metastatic relapse in high-risk primary breast cancer patients receiving adjuvant chemotherapy. *Oncogene* 2005, **25**, 981–989.
34. Ehmann, M., Felix, K., Hartmann, D., Schno, M., Nees, M., Vorderwu, S., Bogumil, R., Buchler, M.W., Friess, H. Identification of potential markers for the detection of pancreatic cancer through comparative serum protein expression profiling. *Pancreas* 2007, **34**, 205–214.
35. Wissing, J., Jänsch, L., Nimtz, M., Dieterich, G., Hornberger, R., Kéri, G., Wehland, J., Daub, H., Proteomics analysis of protein kinases by target class-selective prefractionation and tandem mass spectrometry. *Mol. Cell. Proteomics* 2007, **6**, 537–547.
36. Fountoulakis, M., Takacs, M.F., Takacs, B. Enrichment of low-copy-number gene products by hydrophobic interaction chromatography. *J. Chromatogr., A* 1999, **833**, 157–168.
37. Moritz, R.L., Ji, H., Schütz, F., Connolly, L.M., Kapp, E.A., Speed, T.P., Simpson, R.J. A proteome strategy for fractionating proteins and peptides using continuous free-flow electrophoresis coupled off-line to reversed-phase high-performance liquid chromatography. *Anal Chem.* 2004, **76**, 4811–4824.
38. Fountoulakis, M., Takács, B. Enrichment and proteomic analysis of low-abundance bacterial proteins. *Methods Enzymol.* 2002, **358**, 288–306.
39. Xiong, S., Zhang, L., He, Q.Y. Fractionation of proteins by heparin chromatography. *Methods Mol. Biol.* 2008, **424**, 213–221.
40. Möncke-Buchner, E., Mackeldanz, P., Krüger, D.H., Reuter, M. Overexpression and affinity chromatography purification of the Type III restriction endonuclease EcoP15I for use in transcriptome analysis. *J. Biotechnol.* 2004, **114**, 99–106.
41. Lei, T., He, Q.Y., Wang, Y.L., Si, L.S., Chiu, J.F. Heparin chromatography to deplete high-abundance proteins for serum proteomics. *Clin. Chim. Acta* 2008, **388**, 173–178.
42. Geng, M., Zhang, X., Bina, M., Regnier, F. Proteomics of glycoproteins based on affinity selection of glycopeptides from tryptic digests. *J. Chromatogr.* 2001, **752**, 293–306.
43. Lopez, M.F., Kristal, B.S., Chernokalskaya, E., Lazarev, A., Shestopalov, A.I., Bogdanova, A., Robinson, M. High-throughput profiling of the mitochondrial proteome using affinity fractionation and automation. *Electrophoresis* 2000, **21**, 3227–3440.
44. Ghosh, D., Krokhin, O., Antonovici, M., Ens, W., Standing, K.G., Beavis, R.C., Wilkins, J.A. Lectin affinity as an approach to the proteomic analysis of membrane glycoproteins. *J. Proteome Res.* 2004, **3**, 841–850.
45. Jiang, X., Ye, M., Zou, H. Technologies and methods for sample pretreatment in efficient proteome and peptidome analysis. *Proteomics* 2008, **8**, 686–705.
46. Dayarathna, M.K., Hancock, W.S., Hincapie, M. A two step fractionation approach for plasma proteomics using immunodepletion of abundant proteins and multi-lectin affinity chromatography: Application to the analysis of obesity, diabetes, and hypertension diseases. *J. Sep. Sci.* 2008, **31**, 1156–1166.
47. Madera, M., Mann, B., Mechref, Y., Novotny, M.V. Efficacy of glycoprotein enrichment by microscale lectin affinity chromatography. *J. Sep. Sci.* 2008, **31**, 2722–2732.
48. Rush, J., Moritz, A., Lee, K.A., Guo, A., Goss, V.L., Spek, E.J., Zhang, H., Zha, X.M., Polakiewicz, R.D., Comb, M.J. Immuno-affinity profiling of tyrosine phosphorylation in cancer cells. *Nature Biotech.* 2005, **23**, 94–101.
49. Sun, X., Chiu, J.F., He, Q.Y. Fractionation of proteins by immobilized metal affinity chromatography. *Methods Mol. Biol.* 2008, **424**, 205–212.

50. Li, Y., Leng, T., Lin, H., Deng, C., Xu, X., Yao, N., Yang, P., Zhang, X. Preparation of Fe_3O_4-ZrO_2 core-shell microspheres as affinity probes for selective enrichment and direct determination of phosphopeptides using matrix-assisted laser desorption ionization mass spectrometry. *J. Proteome Res.* 2007, **6**, 4498–4510.

51. Feng, S., Ye, M., Zhou, H., Jiang, X., Zou, H., Gong, B. Immobilized zirconium ion affinity chromatography for specific enrichment of phosphopeptides in phosphoproteome analysis. *Mol. Cell. Proteomics* 2007, **6**, 1656–1665.

52. Li, Y., Xu, X., Qi, D., Deng, C., Yang, P., Zhang, X. Novel $Fe_3O_4.@TiO_2$ core-shell microspheres for selective enrichment of phosphopeptides in phosphoproteome analysis. *J. Proteome Res.* 2008, **7**, 2526–2538.

53. Righetti, P.G., Faupel, M., Wenisch, E., in: Chrambach, A., Dunn, M.J., Radola, B.J. (Eds), *Advances in Electrophoresis*, Vol. **5**, VCH, Weinheim, 1992, pp. 159–200.

54. Nissum, M., Foucher, A.L. Analysis of human plasma proteins: a focus on sample collection and separation using free-flow electrophoresis. *Expert Rev Proteomics* 2008, **5**, 571–587.

55. Brobey, R.K., Soong, L. Establishing a liquid-phase IEF in combination with 2-DE for the analysis of Leishmania proteins. *Proteomics* 2007, **7**, 116–120.

56. Yamauchi, K., Tata, J.R. Purification and characterization of a cytosolic thyroid-hormone-binding protein (CTBP) in *Xenopus* liver. *Eur. J. Biochem.* 2005, **225**, 1105–1112.

57. Davidsson, P., Paulson, L., Hesse, C., Blennow, K., Nilsson, C.L. Proteome studies of human cerebrospinal fluid and brain tissue using a preparative two-dimensional electrophoresis approach prior to mass spectrometry. *Proteomics* 2001, **1**, 444–452.

58. Herbert, B.R., Righetti, P.G. A turning point in proteome analysis: sample prefractionation via multicompartment electrolyzers with isoelectric membranes. *Electrophoresis* 2000, **21**, 3639–3648.

59. Righetti, P.G., Castagna, A., Herbert, B. Prefractionation techniques in proteome analysis. *Anal. Chem.* 2001, **73**, 320–326.

60. Ros, A., Faupel, M., Mees, H., Oostrum, J.V., Ferrigno, R., Reymond, F., Michel, P., Rossier, J.S., Girault, H.H. Protein purification by Off-Gel electrophoresis. *Proteomics* 2002, **2**, 151–156.

61. Arnaud, I.L., Josserand, J., Rossier, J.S., Girault, H.H. Finite element simulation of Off-Gel trade mark buffering. *Electrophoresis* 2002, **23**, 3253–3261.

62. Michel, P.E., Reymond, P., Arnaud, I.L., Josserand, J., Girault, H.H., Rossier, J.S. Protein fractionation in a multicompartment device using Off-Gel isoelectric focusing. *Electrophoresis* 2003, **24**, 3–11.

63. Michel, P.E., Crettaz, D., Morier, P., Heller, M., Gallot, D., Tissot, J.D., Reymond, F., Rossier, J.S. Two-stage Off-Gel isoelectric focusing: protein followed by peptide fractionation and application to proteome analysis of human plasma. *Electrophoresis.* 2005, **26**, 1174–1188.

64. Michel, P.E., Crettaz, D., Morier, P., Heller, M., Gallot, D., Tissot, J.D., Reymond, F., Rossier, J.S. Proteome analysis of human plasma and amniotic fluid by Off-Gel isoelectric focusing followed by nano-LC-MS/MS. *Electrophoresis* 2006, **27**, 1169–1181.

65. Heller, M., Ye, M., Michel, P.E., Morier, P., Stalder, D., Jünger, M.A., Aebersold, R., Reymond, F., Rossier, J.S. Added value for tandem mass spectrometry shotgun proteomics data validation through isoelectric focusing of peptides. *J. Proteome Res.* 2005, **4**, 2273–2282.

66. Busnel, J.M., Lion, N., Girault, H.H. Capillary electrophoresis as a second dimension to isoelectric focusing for peptide separation. *Anal. Chem.* 2007, **79**, 5949–5955.

67. Waller, L.N., Shores, K., Knapp, D.R. Shotgun proteomic analysis of cerebrospinal fluid using off-gel electrophoresis as the first-dimension separation. *J. Proteome Res.* 2008, **7**, 4577–4584.

68. Govorukhina, N.I., Keizer-Gunnink, A., van der Zee, A.G., de Jong, S., de Bruijn, H.W., Bischoff, R., Sample preparation of human serum for the analysis of tumor markers. Comparison of different approaches for albumin and gamma-globulin depletion. *J. Chromatogr., A* 2003, **1009**, 171–178.

69. Echan, L.A., Tang, H.Y., Ali-Khan, N., Lee, K., Speicher, D.W., Depletion of multiple high-abundance proteins improves protein profiling capacities of human serum and plasma. *Proteomics* 2005, **5**, 3292–3303.

70. Lopez, M.F., Mikulskis, A., Kuzdzal, S., Bennett, D.A. Kelly, J., Golenko, E., DiCesare, J., Denoyer, E., Patton, W.F., Ediger, R., Sapp, L., Ziegert, T., Lynch, C., Kramer, S., Whiteley,

G.R., Wall, M.R., Mannion, D.P., Della Cioppa, G., Rakitan, J.S., Wolfe, G.M. High-resolution serum proteomic profiling of Alzheimer disease samples reveals disease-specific, carrier-protein-bound mass signatures. *Clin. Chem.* 2005, **51**, 1946–1954.

71. Wang, Y., Seneviratne, C.J. Biomarker discovery in clinical proteomics: Strategies for exposing low abundant proteins. *J. Curr. Proteomics* 2008, **5**, 104–114.

72. Eliasson, M., Olsson, A., Palmcrantz, E., Wiberg, K., Inganäs, M., Guss, B., Lindberg, M., Uhlén, M. Chimeric IgG-binding receptors engineered from staphylococcal protein A and streptococcal protein G. *J. Biol. Chem.* 1988, **263**, 4323–4327.

73. Ma, Z., Ramakrishna, S. Electrospun regenerated cellulose nanofiber affinity membrane functionalized with protein A/G for IgG purification S. *J. Membr. Sci.* 2008, **319**, 23–28.

74. Colantonio, D.A., Dunkinson, C., Bovenkamp, D.E., van Eyk, J.E. Effective removal of albumin from serum. *Proteomics* 2005, **5**, 3831–3835.

75. Vestergaard, M.D., Tamiya, E. A rapid sample pretreatment protocol: improved sensitivity in the detection of a low-abundant serum biomarker for prostate cancer. *Anal. Sci.* 2007, **23**, 1443–1446.

76. Plavina, T., Wakshull, E., Hancock, W.S., Hincapie, M. Combination of abundant protein depletion and multi-lectin affinity chromatography (M-LAC) for plasma protein biomarker discovery. *J. Proteome Res.* 2007, **6**, 662–671.

77. Gronwall, C., Sjoberg, A., Ramstrom, M., Hoiden-Guthenberg, I., Hober S., Jonasson P., Ståhl S. Affibody-mediated transferrin depletion for proteomics applications. *Biotechnol. J.* 2007, **2**, 1389–1398.

78. Zolotarjova, N., Martosella, J., Nicol, G., Bailey, J., Boyes, B.E., Barrett, W.C. Differences among techniques for high-abundant protein depletion. *Proteomics* 2005, **5**, 3304–3313.

79. Björhall, K., Miliotis, T., Davidsson, P. Comparison of different depletion strategies for improved resolution in proteomic analysis of human serum samples. *Proteomics* 2005, **5**, 307–317.

80. Steel, L.F., Trotter, M.G., Nakajima, P.B., Mattu, T.S., Gonye, G., Block, T. Efficient and specific removal of albumin from human serum samples. *Mol. Cell. Proteomics* 2003, **2**, 262–270.

81. Granger, J., Siddiqui, J., Copeland, S., Remick, D., Albumin depletion of human plasma also removes low abundance proteins including the cytokines. *Proteomics* 2005, **5**, 4713–4718.

82. Shen, Y., Kim, J., Strittmatter, E.F., Jacobs, J.M., Camp, D.G., 2nd, Fang, R., Tolié, N., Moore, R.J., Smith, R.D. Characterization of the human blood plasma proteome. *Proteomics* 2005, **5**, 4034–4045.

83. Huang, L., Harvie, G., Feitelson, J.S., Gramatikoff, K. Herold, D.A., Allen, D.L., Amunngama, R., Hagler, R.A., Pisano, M.R., Zhang, W.W., Fang, X. Immunoaffinity separation of plasma proteins by IgY microbeads: meeting the needs of proteomic sample preparation and analysis. *Proteomics* 2005, **5**, 3314–3328.

84. Hinerfeld, D., Innamorati, D., Pirro, J., Tam Sun, W. Serum/plasma depletion with chicken immunoglobulin Y antibodies for proteomic analysis from multiple mammalian species. *J. Biomol. Tech.* 2004, **15**, 184–190.

85. Ringrose, J.H., van Solinge, W.W., Mohammed, S., O'Flaherty, M.C., van Wijk, R., Heck, A.J., Slijper, M. Highly efficient depletion strategy for the two most abundant erythrocyte soluble proteins improves proteome coverage dramatically. *J. Proteome Res.* 2008, **7**, 3060–3063.

86. Bhattacharya, D., Mukhopadhyay, D., Chakrabarti, A. Hemoglobin depletion from red blood cell cytosol reveals new proteins in 2-D gel-based proteomics study. *Proteomics Clin. Appl.* 2007, **1**, 561–564.

87. Cellar, N.A., Kuppannan, K., Langhorst, M.L., Ni, W., Xu, P., Young, S.A. Cross species applicability of abundant protein depletion columns for ribulose-1,5-bisphosphate carboxylase/ oxygenase. *J. Chromatogr., B* 2008, **861**, 29–39.

88. Hoving, S., Voshol, H., van Oostrum, J., Towards high performance two-dimensional gel electrophoresis using ultrazoom gels. *Electrophoresis* 2000, **21**, 2617–2621.

89. Westbrook, J.A., Yan, J.X., Wait, R., Welson, S.Y., Dunn, M.J., Zooming-in on the proteome: very narrow-range immobilized pH gradients reveal more protein species and isoforms. *Electrophoresis* 2001, **22**, 2865–2871.

90. Merrifield, R.B. Automated synthesis of peptides. *Science* 1965, **150**, 178–185.

91. Lam, K.S., Salmon, S.E., Hersh, E.M., Hruby, V.J., Kazmierski, W.M., Knapp, R.J. A new type of synthetic peptide library for identifying ligand-binding activity. *Nature* 1991, **354**, 82–84.
92. Furka, A., Sebestyen, F., Asgedom, M., Dibo, G. General method for rapid synthesis of multicomponent peptide mixtures. *Int. J. Pept. Protein Res.* 1991, **37**, 487–493.
93. Watts, A.D., Hunt, N.H., Hambly, B.D., Chaudhri, G. Separation of tumor necrosis factor alpha isoforms by two-dimensional polyacrylamide gel electrophoresis. *Electrophoresis* 1997, **18**, 1086–1091.
94. Thulasiraman, V., Lin, S., Gheorghiu, L., Lathrop, J., Lomas, L., Hammond, D., Boschetti, E. Reduction of the concentration difference of proteins in biological liquids using a library of combinatorial ligands. *Electrophoresis* 2005, **26**, 3561–3571.
95. Righetti, P.G., Boschetti, E., Lomas, L., Citterio, A. Protein Equalizer technology: The quest for a 'democratic proteome'. *Proteomics* 2006, **6**, 3980–3992.
96. Boschetti, E., Lomas, L., Citterio, A., Righetti, P.G. Romancing the 'hidden proteome', Anno Domini two zero zero six. *J. Chromatogr., A* 2007, **1153**, 277–290.
97. Righetti, P.G., Boschetti, E., Sherlock Holmes and the proteome: a detective story. *FEBS J.* 2007, **274**, 897–905.
98. Fasoli, E., Farinazzo, A., Sun, C.J., Kravchuck, A.V., Guerrier, L., Fortis, F., Boschetti, E., Righetti, P.G. Interaction between proteins and peptide libraries in proteome analysis: pH involvement for a larger capture of species. *J. Proteomics* 2010, **73**, 733–742.
99. Farinazzo, A., Fasoli, E., Kravchuk, A., Candiano, G., Aldini, G., Regazzoni, L., Righetti P.G. En bloc elution of proteomes from combinatorial peptide ligand libraries. *J. Proteomics* 2009, **72**, 725–730.
100. Hamdan, M., Galvani, M., Righetti, P.G. Monitoring 2-D gel-induced modifications of proteins by MALDI-TOF mass spectrometry. *Mass Spectrom. Rev.* 2001, **20**, 121–141.
101. Fortis, F., Guerrier, G., Areces, L., Antonioli, P., Hayes, T., Carrick, K., Hammond, D., Boschetti, E., Righetti, P.G. A new approach for the amplification, and identification of protein impurities from purified biopharmaceuticals using combinatorial solid phase ligand libraries. *J. Proteome Res.* 2006, **5**, 2577–2585.
102. Candiano, G., Dimuccio, V., Bruschi, M., Santucci, L., Gusmano, R., Boschetti, E., Righetti, P.G., Ghiggeri, G.-M. Combinatorial peptide ligand libraries for urine proteome analysis: investigation of different elution systems. *Electrophoresis* 2009, **30**, 2405–2411.
103. Merchant, M., Weinberger, S.M. Recent advancements in surface-enhanced laser desorption/ionisation-time of flight-mass spectrometry. *Electrophoresis* 2000, **21**, 1164–1167.
104. Seibert, V., Ebert, M.P., Buschmann, T. Advances in clinical cancer proteomics: SELDI-ToF-mass spectrometry and biomarker discovery. *Brief Funct. Genomic Proteomic* 2005, **4**, 16–26.
105. Guerrier, L., Thulasiraman, V., Castagna, A., Fortis, F., Lin, S., Lomas, L., Righetti, P.G., Boschetti, E. Reducing protein concentration range of biological samples using solid-phase ligand libraries. *J. Chromatogr., B* 2006, **833**, 33–40.
106. Sennels, L., Salek, M., Lomas, L., Boschetti, E., Righetti, P.G., Rappsilber, J. Proteomic analysis of human blood serum using peptide library beads. *J. Proteome Res.* 2007, **6**, 4055–4062.
107. Castagna, A., Cecconi, D., Sennels, L., Rappsilber, J., Guerrier, L., Fortis, F., Boschetti, E., Lomas, L., Righetti, P.G. Exploring the hidden human urinary proteome via ligand library beads. *J. Proteome Res.* 2005, **4**, 1917–1930.
108. Guerrier, L., Claverol, S., Finzi, L., Paye, F., Fortis, F., Boschetti, E., Housset, C. Contribution of solid-phase hexapeptide ligand libraries to the repertoire of human bile proteins. *J. Chromatogr., A* 2007, **1176**, 192–205.
109. Bachi, A., Restuccia, U., Fasoli, E., Boschetti, E., Peltre, G., Sénéchal, H., Righetti, P.G. In-depth exploration of cow's whey proteome and 'hidden allergens' via combinatorial peptide ligand libraries. *J. Prot. Res.* 2009, **8**, 3925–3936.
110. Calvete, J.J., Fasoli, E., Sanz, L., Boschetti, E., Righetti, P.G. Exploring the venom proteome of the western diamondback rattlesnake, *Crotalus atrox*, via snake venomics and combinatorial peptide ligand library approaches. *J. Prot. Res.* 2009, **8**, 3055–3067.
111. D'Ambrosio, C., Arena, S., Scaloni, A., Fortis, F., Boschetti, E., Mendieta, M.E., Citterio, A., Righetti, P.G. Exploring the chicken egg white proteome with combinatorial peptide ligand libraries. *J. Prot Res.* 2008, **7**, 3461–3474.

112. Farinazzo, A., Restuccia, U., Bachi, A., Guerrier, L., Boschetti, E., Fasoli, E., Citterio, A., Righetti, P.G. The chicken egg yolk cytoplasmic proteome, mined via combinatorial peptide ligand libraries. *J. Chromatogr.* 2009, **1216**, 1241–1252.

113. Guerrier, L., Claverol, S., Fortis, F., Rinalducci, S., Timperio, A.M., Antonioli, P., Jandrot-Perrus, M., Boschetti, E., Righetti, P.G., Exploring the platelet proteome via combinatorial, hexapeptide ligand libraries. *J. Proteome Res.* 2007, **6**, 4290–4303.

114. Roux-Dalvai, F., Gonzalez de Peredo, A., Simò, C., Guerrier, L., Bouyssie, D., Zanella, A., Citterio, A., Burlet-Schiltz, O., Boschetti, E., Righetti, P.G., Monsarrat, B. Extensive analysis of the cytoplasmic proteome of human erythrocytes using the peptide ligand library technology and advanced mass spectrometry. *Mol. Cell Proteomics* 2008, **7**, 2254–2269.

115. Simó, C., Bachi, A., Cattaneo, A., Guerrier, L., Fortis, F., Boschetti, E., Podtelejnikov, A., Righetti, P.G. Performance of combinatorial peptide libraries in capturing the low-abundance proteome of red blood cells.I. Behavior of mono- to hexa-peptides. *Anal. Chem.* 2008, **80**, 3547–3556.

116. Pasini, E.M., Kirkegaard, M., Mortensen, P., Lutz, H.U., Thomas, A.W., Mann, M. In-depth analysis of the membrane and cytosolic proteome of red blood cells. *Blood* 2006, **108**, 791–801.

117. Clegg, J.B. Can the product of the theta gene be a real globin? *Nature* 1987, **329**, 465–466.

118. McCabe, M.S., Garratt, L.C., Schepers, F., Jordi, W.J., Stoopen, G.M., Davelaar, E., van Rhijn, J.H., Power, J.B., Davey, M.R. Effects of P(SAG12)-IPT gene expression on development and senescence in transgenic lettuce. *Plant Physiol.* 2001, **127**, 505–516.

119. Giavalisco, P., Nordhoff, E., Kreitler, T., Klöppel, K.D., Lehrach, H., Klose, J., Gobom, J. Proteome analysis of *Arabidopsis thaliana* by two-dimensional gel electrophoresis and matrix-assisted laser desorption/ionisation-time of flight mass spectrometry. *Proteomics* 2005, **5**, 1902–1913.

120. The Arabidopsis Genome Initiative. Analysis of the genome sequence of the flowering plant *Arabidopsis thaliana*. *Nature* 2000, **408**, 796–815.

121. International Rice Genome Sequencing Project. The map-based sequence of the rice genome. *Nature* 2005, **436**, 793–800.

122. Rabilloud, T. Solubilization of proteins for electrophoretic analyses. *Electrophoresis* 1996, **17**, 813–829.

123. des Francs, C.C., Thiellement, H., de Vienne, D. Analysis of leaf proteins by two-dimensional gel electrophoresis: protease action as exemplified by ribulose bisphosphate carboxylase/oxygenase degradation and procedure to avoid proteolysis during extraction. *Plant Physiol.* 1985, **78**, 178–182.

124. Gengenheimer, P. Preparation of extracts from plants. *Methods Enzymol.* 1990, **182**, 174–193.

125. Boschetti, E., Bindschedler, L.V., Tang, C., Fasoli, E., Righetti, P.G. Combinatorial peptide ligand libraries and plant proteomics: A winning strategy at a price. *J. Chromatogr., A* 2009, **1216**, 1215–1222.

126. Fasoli, E., Pastorello, E.A., Farioli, L., Scibilia, J., Aldini, G., Carini, M., Marocco, A., Boschetti, E., Righetti, P.G. Searching for allergens in maize kernels via proteomic tools. *J. Proteomics* 2009, **72**, 501–510.

127. Sookmark, U., Pujade-Renaud, V., Chrestin, H., Lacote, R., Naiyanetr, C., Seguin, M., Rom-ruensukharom, P., Narangajavana, J. Characterization of polypeptides accumulated in the latex cytosol of rubber trees affected by the tapping panel dryness syndrome. *Plant Cell Physiol.* 2002, **43**, 1323–1333.

128. Guerrier, L., Lomas, L., Boschetti, E. A simplified monobuffer multidimensional chromatography for high-throughput proteome fractionation. *J. Chromatogr.* 2005, **1073**, 25–33.

129. Fortis, F., Guerrier, L., Girot, P., Fasoli, E., Righetti, P.G., Boschetti, E. A pI-based protein fractionation method using Solid-State Buffers. *J. Proteomics* 2007, **71**, 379–389.

130. Restuccia, U., Boschetti, E., Fasoli, E., Fortis, F., Guerrier, G., Bachi, A., Kravchuk, A., Righetti, P.G. pI-based fractionation of serum proteomes versus anion exchange after enhancement of low-abundance proteins by means of peptide library. *J. Proteomics* 2009, **72**, 1061–1070.

10

3D-Gel Electrophoresis – A New Development in Protein Analysis

Robert Ventzki[1] and Josef Stegemann[2]
[1] Am Neuberg 2, Dossenheim, Germany
[2] Kühler Grund 50, Heidelberg, Germany

10.1 Introduction

10.1.1 State of the Art in Protein Analysis

Despite the substantial advances of gel-free liquid chromatography-mass spectrometry (LC-MS)-based techniques in recent years, two-dimensional gel electrophoresis (2DE) still remains the most widely accepted, reliable and comprehensive analytical method for the evaluation of complex protein mixtures extracted from cells, tissues, or other biological samples. Moreover, optimal proteome coverage demands the complementation of LC-MS analysis by 2DE studies due to the incomplete overlap of accessible protein size ranges of both methods. By combining isoelectric focusing (IEF) as a first separation step with sodium dodecyl sulfate polyacrylamide gel electrophoresis (SDS-PAGE) as a second step, 2DE allows the analysis of protein extracts according to two independent parameters – pI and molecular mass (MM). With the second separation step being carried out in slab gels, 2DE is commonly employed prior to the identification of proteins by mass spectrometry, or for comparative expression studies of specific sets of previously identified proteins. Since its introduction by O'Farrell in 1975 [1], the power of 2DE as a biochemical separation technique has been widely recognized and the method has been constantly refined with respect to resolution, sensitivity, and handling. However, with conventional SDS-PAGE performed in slab gels remaining a time- and labour-intensive procedure, 2DE analysis has never been considered suitable for large-scale proteomics studies involving hundreds of samples. In addition, the reproducibility of the results obtained with 2DE

Mass Spectrometry for Microbial Proteomics Edited by Haroun N. Shah and Saheer E. Gharbia
© 2010 John Wiley & Sons, Ltd

remains a key issue that initiated considerable efforts for the standardization of protocols and the normalization of proteomic data [2]. In particular, controlling the technical gel-to-gel variations in conventional 2DE is an indispensable prerequisite for quantitative results [3]. Moreover, the mobility shifts inherent to 2DE slab gels require elaborate software corrections (image warping) of the gel scans and large numbers of gel runs to obtain reliable results in differential protein analysis.

10.1.2   Innovations by 3D-Gel Electrophoresis

Here, we describe a new method and instrument for high-throughput analysis of proteins, based on electrophoretic separation in a three-dimensional (3D) geometry gel combined with online detection of laser-induced fluorescence (LIF) [4]. The 3D-gel technique overcomes the above-mentioned limitations of standard slab gel protein analysis, introducing several distinct features: It virtually eliminates the gel-to-gel variations and mobility shifts by analysing a series of samples in the same 3D-gel block under identical electrophoretic and thermal conditions. With its online detection system, the 3D-gel instrument provides a high degree of automatization and improves the efficiency and throughput of gel-based protein analysis considerably. A proprietary loading device allows the simultaneous manual transfer of up to 1536 samples out of a microtitre plate (MTP) directly on the 3D-gel.

With the 3D-gel instrument in its last phase of development, we are currently investigating and initiating applications in various fields and disciplines. We foresee a wide range of applications in molecular discovery, clinical diagnosis, pharmacology, and toxicology. Examples include high-throughput differential protein expression studies, diagnostic protein level monitoring during the onset and progression of diseases, as well as the screening of chemical compounds for their effect on the proteome or macromolecular assemblies in drug discovery programmes.

10.1.3   Concept of 3D-Gel Electrophoresis

Standard electrophoretic separation methods are limited to systems with a one- or two-dimensional geometry, where the separation medium has no significant spatial extension in the third dimension. Examples for a one-dimensional geometry are capillary [5, 6] and microchannel [7, 8] electrophoresis, where separation occurs linear along one dimension. Typical examples for a two-dimensional geometry are slab gels. With these, the second geometric dimension can either be used to increase the number of samples run in parallel lanes, or to analyse a single sample according to two independent separation parameters (e.g. size fractionation of proteins after IEF). The difference in meaning of the common wording two-dimensional electrophoresis (2DE) referring to two separation parameters, and two-dimensional geometry referring to spatial dimensions should be noted here.

In the 3D-gel electrophoresis system, the separation medium extends in all three spatial dimensions, with each one providing functionality [4]. The samples are applied in a two-dimensional planar array to the top surface of a 3D geometry gel body [Figure 10.1(a)]. The migration and electrophoretic separation occur along the third spatial dimension, in a direction perpendicular to the sample loading surface. During electrophoresis, the sample fractions are detected by LIF, similar to the operation principle of automated DNA sequencers [9]. A digital camera located underneath the 3D-gel takes images at regular

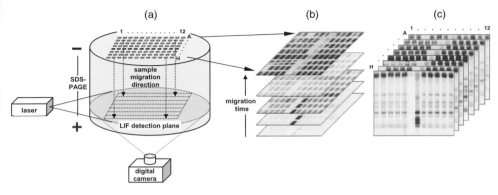

Figure 10.1 *Principle of 3D-gel electrophoresis. (a) Samples are arrayed in a two-dimensional loading surface on the top of a 3D geometry gel body. Separation occurs along the third (z) coordinate, perpendicular to the loading surface. Fluorescently labelled samples are detected online while passing a laser-illuminated plane. (b) Electronically recorded 3D image stack acquired by a digital camera at regular time intervals during electrophoresis (five of ~2000 images shown). Original Cy3 orange colour inverted for better visualization. (c) Computer-generated series of vertical sections of the 3D image stack. Each section represents a conventional slab gel scan. A regular sample migration without 'smiling' effects can be observed, independent of the position in the 3D-gel. Reproduced from Ventzki et al., Electrophoresis 2006, 27, 3338–3348. Copyright Wiley-Blackwell. Reproduced with permission*

time intervals as the fluorescently labelled sample fractions pass a laser-illuminated detection plane within the gel. These images are electronically read out in real time and stored as a 3D image stack on a computer [Figure 10.1(b)]. To visualize the separation patterns, image processing software generates a series of adjacent vertical sections out of the 3D image stack, with each section representing a conventional slab gel scan [Figure 10.1(c)]. The separation patterns can then conveniently be compared and quantitatively evaluated using publicly available software packages.

The concept of 3D-gel electrophoresis was developed and published independently by two groups in 2003 [4, 10]. The team of Bao-Shiang Lee proposed a 3D-gel cube to improve the MM resolution of SDS-PAGE protein analysis [10]. We published concurrently a feasibility study, introducing a specific thermal management of the 3D-gel to solve the problem of irregular sample migration ('smiling') and gel-to-gel variations [4]. In addition, our first 3D-gel instrument provided a system for online LIF detection of the samples. We proposed and tested several applications, such as DNA sequencing, DNA fragment analysis, and protein separation by SDS-PAGE. These early studies demonstrated the suitability of 3D-gel electrophoresis for the analysis of biomolecules according to one, two or three independent separation parameters.

Examples for a separation according to one parameter (i.e. MM) are the SDS-PAGE analysis of high-performance liquid chromatography (HPLC)-prefractionated proteins or the quality validation of purified proteins prior to their crystallization [11]. The loading in a two-dimensional array onto the 3D-gel allows for a very high sample capacity, so that all samples of a 1536-well MTP can be separated in parallel.

An example for separation according to two parameters is the analysis of protein aggregates by blue native electrophoresis (BNE), followed by SDS-PAGE in the 3D-gel for separation of the aggregates' components [12]. A comparative 2DE protein analysis of a series of 36 samples in one electrophoresis run can also be carried out very efficiently with the 3D-gel method [13]. By providing an immediate comparability of all separation patterns of a sample series, the 3D-gel method inherently supports high-throughput differential 2DE protein analysis. Together with the online LIF detection this makes the overall analysis process considerably more efficient.

Although not yet implemented, a separation by three independent parameters is technologically feasible. An example would be the analysis of protein complexes by BNE, followed by IEF of the complexes' subunits in a slab gel, and SDS-PAGE in the 3D-gel. This sequence of separation parameters was reported with the use of conventional slab gels [14].

In the case of two- or three-parameter analysis, the 3D-gel is used only for the last SDS-PAGE separation step. The preceding separation steps are carried out using conventional methods (i.e. BNE or IEF in slab gels or strips, respectively).

10.2   Methods

10.2.1   The 3D-Gel Instrument

The current 3D-gel electrophoresis instrument (Figure 10.2) is an improved version based on the prototypes described previously [4, 12]. It was designed to accomodate a gel body with an upper surface of 12 cm by 8 cm to achieve compatibility with the standard MTP format. The gel chamber is removable from the lower part of the 3D-gel instrument for easy gel casting and cleaning. A sheet of platinum wire mesh, placed in the upper buffer tank at a distance of 1 cm above the gel surface serves as cathode. To acquire images through the anode buffer tank, the anode wire and the bubbles created during electrophoresis had to be kept out of the optical path. This was achieved by placing an annular platinum anode wire in a channel encircling the lower buffer tank. A tube connecting the anode channel with the upper buffer tank allows the bubbles to escape. Due to the relatively large distance between anode and gel bottom (~20 cm), the distortion of field lines in the gel is negligible and thus the electric field homogeneous. Electrophoresis is performed in the 3D-gel with a maximum voltage of 300 V and current of 1.5 A, resulting in a field strength of 10 V cm^{-1} and a dissipated power of 450 W.

10.2.2   Thermal Management of the 3D-Gel

To track the large number of samples that are applied to the 3D-gel, and particularly to assign separation patterns to individual samples, the migration lanes must be easily discernible and definable. This obviously is greatly facilitated if the sample migration paths are perfectly straight. However, all electrophoretic separation methods suffer to a variable extent from Joule's heat, which is a result of the dissipation of electric power throughout the separation medium. As sample mobility and electrolytic conductivity depend heavily on temperature, the sample migration is influenced by this effect in three ways: (1) There is the well known tendency of samples to migrate faster in regions of higher temperature

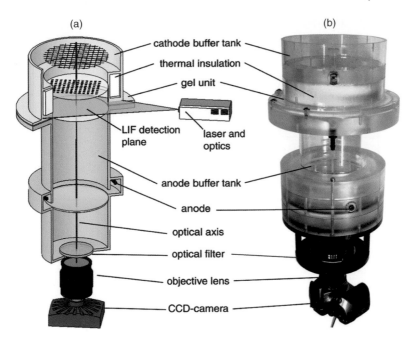

(a)   (b)

cathode buffer tank

thermal insulation

gel unit

LIF detection   laser and
plane   optics

anode buffer tank

anode

optical axis

optical filter

objective lens

CCD-camera

Figure 10.2 *3D-gel electrophoresis instrument. Cutaway view of the 3D-gel set-up (a) and the instrument prototype (b). During operation, the instrument is contained in a light-tight housing. An external buffer circulation and cooling system, an electrophoresis power supply, and a computer for data recording and evaluation complement it.*
© 2008 BioTechniques. Used by Permission

due to their lower gel viscosity; this phenomenon is commonly referred to as 'smiling' due to the band front appearance. (2) There is focusing of electric field lines towards regions of higher gel temperature due to their higher electric conductivity, resulting in the bending of migration lanes towards those regions. (3) Most disruptive to sample resolution however are intra-band temperature gradients that result from heat flow perpendicular to the sample migration direction. If large amounts of heat are dissipated, or if heat transport is inefficient, the migration speed within a band is not homogeneous, leading to band widening and degraded resolution. Furthermore, irregular sample migration causes gel-to-gel variations that impair the comparability between a series of 2DE slab gels in differential protein expression studies.

So far, countermeasures have focused on optimizing heat transport perpendicular to the sample migration direction, as is the case with ultrathin slab gels [15, 16], capillaries [5, 6], and microchannels [7, 8]. However, perhaps unexpectedly, radial temperature gradients can also be avoided by inhibiting radial heat transfer [4]. To this end, the inner gel chamber of the 3D-gel instrument is sleeved with styrofoam insulation, blocking heat transfer in the radial (horizontal) direction (Figure 10.3). This way, the heat flow is forced up- and downwards in the 3D-gel, i.e. towards the upper and lower buffer tank. The electrode buffer that is in contact with both top and bottom surfaces of the 3D-gel body serves as a cooling medium. The buffer is pumped through a heat exchanger in a waterbath held at

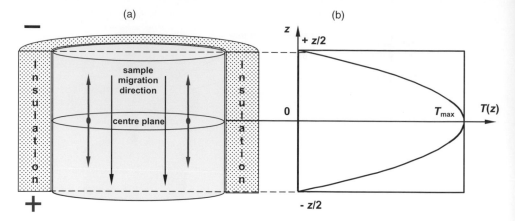

Figure 10.3   *Thermal management of the 3D-gel. (a) A thermal insulation encircling the 3D-gel body inhibits heat transfer in the radial direction, thereby forcing the heat flow in the vertical (z) direction towards the top and bottom surfaces of the gel, as indicated by the red arrows. As the heat flow is parallel to the sample migration direction (black arrows), distortion of the sample lanes and 'smiling' effects are avoided. (b) The temperature profile prevailing in the 3D-gel along the sample migration axis z is given by $T(z) = T_{max} - kz^2$, where k is a constant. The maximum temperature T_{max} (~40 °C) is reached in the centre plane of the gel. The temperature drops towards the top and bottom surfaces of the 3D-gel, which are cooled to ~5 °C by the electrode buffer. The total dissipated power is ~450 W*

5 °C. Under these conditions, temperature gradients can occur only vertically, i.e. parallel to the sample migration direction.

As the radial temperature gradient is negligible, the gel viscosity and ion mobility are radially isotropic, which ensures an even and straight sample migration, whether in the centre or in the perimeter of the 3D-gel body. In conclusion, this novel temperature management significantly reduces band widening, smiling and bending of the migration lanes in the 3D-gel. This results in more accurately identified and matched spots than conventional 2DE, thus rendering the comparison of a series of separation patterns obtained with the 3D-gel less ambiguous. Together, this makes the instrument an ideal tool for large-scale differential protein analysis.

10.2.3   Online Detection of Laser-Induced Fluorescence

For fluorescence excitation, an air-cooled argon laser (λ = 488 nm) is used. The laser light is projected into the 3D-gel through a glass window at the side of the gel chamber. The beam is shaped by a specifically designed optical system to evenly illuminate a plane of ~100 μm thickness within the transparent 3D-gel body. This sheet of light, oriented perpendicularly to the sample migration direction and located 5 mm above the bottom of the 3D-gel, serves as a LIF detection plane [Figure 10.1(a)]. While the sample fractions pass through the detection plane, their fluorescent labels are detected by a digital camera mounted below the 3D-gel. Specifically designed optics, consisting of an objective lens and additional lenses ensure an efficient fluorescence light collection for high detection sensitivity. The laser excitation light is suppressed by an optical long-pass filter. The camera is connected to a computer via a universal serial bus (USB) for exposure control and electronic read-out of the images at regular time intervals during the electrophoresis run.

10.2.4 Casting of the 3D-Gel

For casting of the 3D-gel, the gel chamber is removed from the lower part of the 3D-gel instrument and sealed with a plate on the bottom. As electrode and gel buffers, Tris-Borate (TB; 0.1 M), pH 8.8, 0.1% SDS is used. The separation gels are prepared in the concentration range of 6–9% with a 30%, 37.5 : 1 acrylamide/Bis stock solution (Bio-Rad Laboratories, Hercules, CA, USA) to final volumes of up to 900 ml, resulting in a gel thickness of up to 9 cm. After addition of the catalysts APS and TEMED, the gel solution is poured into the gel chamber and covered with a floating plastic sheet to prevent the access of ambient air during polymerization [4]. The plastic sheet remains on top of the 3D-gel until use, and, after removal, leaves a smooth gel surface ready for sample loading. In our experience, the casting of a 3D-gel is much easier than of a standard slab gel and results in higher quality gels, as bubbles and gel disruptions that frequently cause artifacts in spot detection with conventional slab gels are much less likely to occur. As the 3D-gel chamber is made of sturdy acrylic plastic (perspex) without fragile glass parts, its handling during gel casting and cleaning after electrophoresis is very easy.

10.2.5 Sample Preparation and Fluorescent Labelling

The online LIF detection of samples in the 3D-gel obviates the need for post-electrophoresis gel staining with several time-consuming incubation steps. Instead, samples are labelled with covalently binding fluorescent dyes in a few simple steps prior to their separation. With a limit of detection (LOD) of ~1 ng, depending on the fluorophore the sensitivity of the online LIF detection is comparable with that of standard slab gels that are fluorescently stained and scanned after electrophoresis [11].

For separation according to one parameter by SDS-PAGE, an optimum detection sensitivity and dynamic range is attained by saturation labelling, where multiple dye molecules bind to the N-terminus and the ε-amino group of lysine residues in the protein molecules. To this end, total lysates, purified proteins and markers are labelled with equal amounts of Cy3 monoreactive dye (GE Healthcare, Freiburg, Germany) according to the manufacturer's instructions. The unbound dye is removed either using a NAP10 column (GE Healthcare), or with the supernatant after precipitation of proteins in the presence of 10% trichloroacetic acid (TCA). The natural fluorescence of green fluorescent protein (GFP)-fused proteins proved also sufficient for their direct detection with a LOD of tens of nanograms [11].

For 2DE analysis in the 3D-gel, protein samples are fluorescently labelled prior to IEF separation according to protocols optimized for differential in-gel electrophoresis (DIGE) found in the literature [17, 18]. However, as samples are not pooled in the 3D-gel, only a single dye label (Cy3) is used. Alternatively, any other fluorescent dye that matches the specific laser excitation wavelength is suitable for labelling. By changing the laser and optical filter, the 3D-gel instrument can be adapted for detection of a broad range of fluorescent dyes. A 3D-gel instrument suitable for the simultaneous detection of multiple dyes is planned.

10.2.6   Sample Loading

In contrast to slab gels where samples are loaded in a linear row on one edge of the gel, for 3D-gel electrophoresis samples are applied as a two-dimensional array on top of the 3D-gel body. This surface loading offers a very high sample capacity in which the density is only limited by lateral diffusion. Over a separation distance of 7 cm, a 0.5 mm wide sample spot is broadened by horizontal diffusion to an area of less than 2 mm diameter, hence a distance of 2 mm in between samples suffices to prevent crosstalk [4]. The resulting large number of samples requires adequate loading techniques for both separation modes of the 3D-gel.

For analysis according to one parameter, the 3D-gel instrument is completely scalable and can accommodate all samples of a standard 96-, 384- or 1536-well MTP for simultaneous separation. We developed a convenient method for facile manual transfer of large numbers of samples to the 3D-gel: all samples are picked up simultaneously from the MTP wells by capillary force into an array of hollow stainless steel tips (e.g. machined syringe needles), similar to a replicator (Figure 10.4). The capillary array is placed on top of the 3D-gel body with the tips touching the smooth loading surface. The aluminium back plate

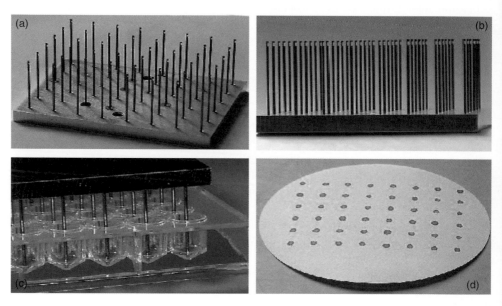

Figure 10.4   *Loading of individual samples to the 3D-gel. (a) Prototype of the loading tool, consisting of an array of stainless steel capillaries inserted in an aluminium base plate. (b) A cut-out is machined at the side of each capillary close to its tip. The cut-out position defines the volume picked up by capillary action. (c) By manually dipping the tips into the wells of a MTP, all samples are picked up simultaneously. (d) A test print on absorbent paper shows the reproducibility of the deposited volumes. By positioning the tips on top of the 3D-gel and connecting the array to the cathode, all samples are electrokinetically injected into the smooth gel surface in a few seconds. After sample loading, the capillary array is cleaned in an ultrasonic waterbath*

of the array is connected to the cathode. By applying an electric current for about 10s, all samples are injected electrokinetically into the 3D-gel. After loading, the array is removed, buffer is filled into the upper gel chamber and electrophoresis is started. The capillary array can be easily cleaned in an ultrasonic waterbath for reuse. This novel loading technique avoids the strenuous manual application of samples with a (multi-)pipette and reduces cost by obviating the pipette tips. Moreover, the forming of sample wells in the gel surface is not necessary.

For 2DE analysis of a series of proteins, up to 36 samples are separated by IEF following standard protocols [19, 20] in a first step (Figure 10.5, Step 1). The samples are transferred from the immobilized pH gradient (IPG) strips to the 3D-gel using a specific tool consisting of a plastic grid (1.5 mm thickness) the size of a MTP with 36 parallel slots (11 cm long, 0.5 mm wide) cut with 2 mm separation. Up to 36 IPG strips containing pre-separated samples are inserted into the slots, such that the edges of the (3 mm wide) strips protrude out of the grid on both sides. The grid is positioned on top of the 3D-gel with the lower edges of the strips touching the smooth gel surface (Figure 10.5, Step 2).

A sheet of absorbent paper soaked with buffer is placed on top of the grid to cover the strips' upper edges. The paper sheet provides electrical contact to a stainless steel plate on top that serves as the cathode. For electrokinetic transfer of the samples to the 3D-gel, a current of 5–10 mA per strip (~100 V) is applied. After 45 min, the grid and the IPG

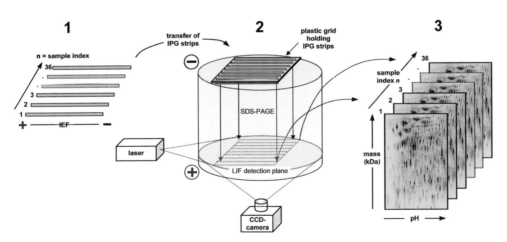

Figure 10.5 *Workflow of comparative 2DE protein analysis in a 3D-gel. Step 1: A series of fluorescently labelled protein samples, indexed n = 1, … , 36, is separated by conventional IEF in 11 cm long IPG strips. Step 2: IPG strips are arrayed in a slotted plastic grid that is placed on top of the 3D-gel and covered with a buffer-soaked absorbent paper and an electrode plate. Electrophoresis is switched on, and the samples are drawn into the 3D-gel. As the samples undergo SDS-PAGE, they pass a laser-illuminated plane located at the lower part of the 3D-gel. Their fluorescence is recorded as an image series, taken in regular time intervals by a digital camera mounted below the gel. Step 3: Computer-generated vertical sections of the 3D image stack, representing a series of 36 conventional 2DE slab gels. The sample index n can denote a third experimental parameter (e.g. time, concentration or dose regimen), depending on the application.* © 2008 BioTechniques. Used by Permission

Figure 10.6 *Data acquisition and evaluation. (a) Left panel: Original camera image, showing the fluorescence signals of a series of IPG strips loaded in parallel to the same 3D-gel. All protein fractions in the image are of the same molecular mass. The dotted line connects spots of the same proteins through all IPG strips. The strip-to-strip variations in pH gradient can be clearly observed as horizontal shifts in spot positions. Right panel: Plot of the fluorescence intensity (arbitrary units) as a function of the sample index n, obtained from the spots along the dotted line. By taking the samples as a series with increments in time, dose or concentration with the sample index denoting that variable, the protein abundance as a function of that parameter can be displayed. (b) Overlay of two digitally recoloured vertical sections (simulated DIGE display), acquired in the same run at different positions in the 3D-gel. The images were not warped before overlay, yet show a remarkable congruence with only minute differences in spot positions. Some horizontally shifted red and green spots (indicated by white arrows) can be attributed to pH-gradient variations of the IPG strips. The lack of vertically adjacent red and green spots demonstrates the absence of mobility shifts during SDS-PAGE in the 3D-gel. (c) Comparison of a 2DE protein separation carried out in a standard slab gel and in a 3D-gel. Left panel: Scan of a pH 4–7, 12% SDS gel, showing protein fractions in a size range of 10–120 kDa (Haemophilus influenzae control sample). Right panel: Computer-generated vertical section of the 3D image stack, showing the same sample. The software (Delta2D; Decodon, Greifswald, Germany) was able to match over 95% of all protein fractions in both separation patterns*

---

strips are removed, the remaining volume of the gel chamber is filled with buffer, and electrophoresis is started. To facilitate handling, we plan to replace the individual strips by a single IPG slab gel, containing all samples of a series pre-separated according to pI in 36 parallel lanes. As first tests with IPG gel strips placed flat on the top surface of the 3D-gel after removal of the plastic backing have shown, this will also improve the resolution for MM as well as pI. In addition, the use of an IPG slab gel will eliminate the strip-to-strip variations that occur with individual IPG strips [Figure 10.6(a)], thereby improving the homogeneity of the pI separation within a sample series. As backing of the IPG slab gel we intend to use a porous plastic sheet (e.g. Cyclopore track-etched membrane; Whatman, Maidstone, UK). Such material, when soaked with buffer, is electrically conductive and therefore would not need to be removed before placing the IPG gel slab (gel side facing downwards) on top of the 3D-gel.

10.2.7 Image Processing and Data Evaluation

The online data collection during electrophoresis makes the results of a 3D-gel run immediately available without subsequent gel scanning. Analysis and evaluation of the separation patterns is done after the electrophoresis run using common image processing software. Depending on the application, the software (ImageJ) [21] allows the display of the results in different formats. To visualize the separation patterns, vertical sections representing a series of 36 standard slab gel scans [Figures 10.1(c) and 10.5] are digitally generated from the 3D image stack using the software plug-in TransformJ [22].

For the evaluation of one-parameter separations, plots of the fluorescence intensity of individual samples as a function of migration time are generated. The data can be exported

(a)

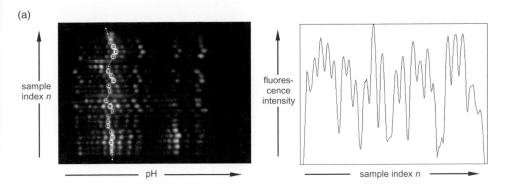

(b)

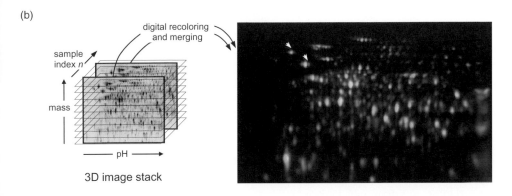

(c)

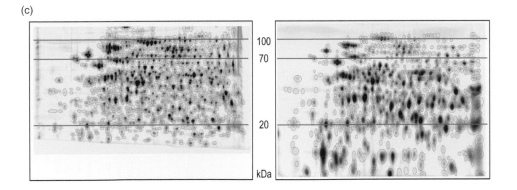

in different formats into Microsoft Excel for a convenient MM determination and quantification of the protein fractions [11].

The evaluation of 2DE separations is done using commercial 2DE analysis software (e.g. Delta2D; Decodon, Greifswald, Germany). For comparing the spot intensities over a whole experiment, each spot on a certain separation pattern has to be mapped to the corresponding spots on the other separation patterns ('spot matching'). It is generally recognized that the quality of the raw data in 2DE image analysis has a significant impact on the final result, i.e. the reliability of the spot matching depends heavily on the quality and reproducibility of the electrophoretic separation. As the separation patterns of a sample series analysed with the 3D-gel are highly congruent, only minute software corrections are required for their comparison and image overlay. Due to the particular thermal management of the 3D-gel, shifts in vertical direction caused by mobility differences are negligible. The only observed differences in spot position from sample to sample are owed to strip-to-strip variations in the pH gradient of the IPG strips [Figure 10.6(a)] and cannot be attributed to the 3D-gel instrument.

To compensate for these strip-to-strip variations, the software automatically aligns the images horizontally in only a minimal computer processing time. The time-consuming manual editing of detected spots and computer-generated matches, as it is still required for the comparison of standard 2DE gel scans [23], is virtually obviated with the 3D-gel method. The user can navigate through the resulting image series to select separation patterns of individual samples (Figure 10.5, Step 3). Two separation patterns can be displayed simultaneously by digitally recolouring the images of choice in different colours (i.e. red and green) and then superimposing those images [Figure 10.6(b)]. Emulating a colour display obtained with the DIGE method [17, 18], the resulting false-colour overlay image is useful to visually inspect differences in protein expression. Furthermore, the commercial software provides all state-of-the-art analyses, like the comparison with standard 2DE gel scans [Figure 10.6(c)] and the statistical evaluation of protein expression levels. A comprehensive review of 2DE gel image analysis methods and software products can be found in the literature [24–26].

In addition to commonly known standard analysis methods, data obtained with the 3D-gel can be displayed and evaluated in multiple ways: As an example, fluorescence intensity plots can be generated as a function of the sample index n, displaying a histogram of protein abundance in the different samples of the series [Figure 10.6(a)]. By extracting the protein samples from the cells as a series with increments in time, dosage regimen or concentration, with the sample index n denoting that variable, protein abundance is displayed as a function of the respective parameter. The large image series resulting from such a time course or dose–response experiment can also be displayed as an animation, visualizing the variations in expression levels as changing spot intensities [27].

10.3   Results and Discussion

10.3.1   Comparison of 3D-Gel with Standard Slab Gel Separation

By conventional electrophoresis with scanned slab gels, the migration *distance* of the protein fractions is determined as a measure of their MM. In contrast, the 3D-gel instrument with its online detection system measures the migration *time* of the fractions from the loading surface to the LIF detection plane. Hence, in separation patterns obtained with the 3D-gel, the small MM fragments appear condensed, whereas the large MM compounds appear as vertically expanded spots as compared with standard 2-DE gel scans. In order to compare the separation patterns obtained with the two techniques, the vertical (z) axis of the 3D-gel images needs to be rescaled inverse to migration time (i.e. by a factor of 1/time). To this end, a specific software routine bulk-processes all images of a series accordingly. After this conversion step, the 3D-gel separation patterns are exported to commercially available software for comparison with standard 2DE gel scans [Figure 10.6(c)].

In conventional slab gels, the large-mass proteins have moved only a small fraction of the total gel length when the migration front reaches the bottom and electrophoresis is stopped for scanning of the gel. Therefore, the resolution is higher for smaller proteins as they travel longer separation distances than larger proteins. In the 3D-gel with its online LIF detection, all sample fractions are intercepted by the laser beam after having travelled the same distance, independently of their MM. This leads to an inferior resolution for small proteins as compared with conventional electrophoresis, whereas high MM proteins travel relatively longer distances and therefore are better resolved in the 3D-gel. The effect described here relatively can be clearly seen in the separation patterns obtained with both techniques [Figure 10.6(c)].

10.3.2 Applications of 3D-Gel Electrophoresis

Any application involving high-throughput SDS-PAGE protein analysis can benefit from the advantages of the 3D-gel electrophoresis technique presented here. The innovative aspects and advantages of the 3D-gel method as compared with conventional slab gel electrophoresis are compiled in Table 10.1.

With its two operation modes (one- or two-parameter separation), the 3D-gel lends itself to a variety of applications in different fields and disciplines. In structural and functional proteomics, there is a large demand for instruments and protocols to evaluate the output of protein expression in order to find suitable conditions for expressing and purifying soluble and correctly folded proteins [28]. Conventional SDS-PAGE remains the only comprehensive and reliable analysis method for this purpose. Here, the 3D-gel method with its high throughput one-parameter separation mode could help overcome the bottleneck represented by the protein quality evaluation in large screening experiments.

A very promising new application for this separation mode is the profiling of binding profiles of small compounds in affinity-based screens. The possibility to separate and identify affinity-captured GFP-tagged fusion proteins by their mass in 3D-gel SDS-PAGE increases the throughput (up to 1536 extracts pools per gel) such that the screening of an entire proteome in only a few gels becomes feasible.

However, among the possible applications of the 3D-gel electrophoresis method, we consider comparative IEF/SDS-PAGE protein analysis to be the most promising. A single 3D-gel separates the equivalent of 36 standard 2DE slab gels under identical electrophoretic and thermal conditions, thereby avoiding the gel-to-gel variations pertaining to conventional 2DE. This allows a quick, straightforward and unambiguous spot matching and comparison of the separation patterns with reliable results. The immediate comparability allows to precisely judge variations of protein samples obtained as a series with increments in time, concentration or treatment dosage. With the sample index n ($n = 1, \ldots, 36$) denoting this variable, the 3D-gel method quantitatively measures the protein abundance as a function of the different parameters.

To our knowledge, the 3D-gel instrument provides the only method for high-throughput 2DE protein analysis with online LIF detection, making results immediately available

Table 10.1 *Comparison of conventional slab gel and 3D-gel electrophoresis. The difference in MM resolution is caused by the separation distance in the 3D-gel being equal for all compounds (high and low MM) due to the online detection, while the separation distance in conventional slab gels varies with the MM of the compounds*

	Conventional slab gel	3D-gel
Performance		
Molecular mass resolution	• Moderate for high MM, good for low MM proteins	• Good for high MM, moderate for low MM proteins
Detection sensitivity	• High, LOD 1 ng	• High, LOD 1 ng
Dynamic range	• ~10 000	• 3500 at present
Electrophoresis run time	• ~9 h for 20 cm long gel	• ~6 h
Handling		
Gel casting and cleanup	• Laborious and time consuming	• Very easy
Post-electrophoresis gel processing, documentation	• Requires fixing, staining, de-staining, scanning	• Not required, results are immediately accessible
One-parameter separation		
Sample capacity per gel	• Max. ~50 samples	• Up to 1536 samples of a MTP
Gel loading	• Manual pipetting	• Convenient simultaneous loading of all samples of a MTP
Two-parameter separation		
Sample capacity per gel	• 1 IPG strip	• 36 IPG strips
pH resolution	• Up to 24 cm IPG strips	• Limited to 11 cm IPG strips
Multi-dye (DIGE) capability	• Yes	• Yes
Sample recovery for mass spectrometry identification	• Directly out of the 2DE gel	• Out of a conventional 2DE gel containing pooled samples
Irregular sample migration, gel-to-gel variations	• Software correction needed for comparison of gel scans	• Negligible, allows direct comparison of images

without any further gel processing. However, due to the online detection principle, the sample fractions are lost in the lower buffer tank as they travel through the laser-illuminated detection plane at the bottom of the gel. Therefore, the current version of the 3D-gel instrument does not allow their collection for identification by mass spectrometry. Hence, we propose its use for comparative expression studies of specific sets of previously identified proteins under different applied treatments or experimental conditions. Alternatively, the protein abundance can be quantitated in the 3D-gel as the first step of the analysis. Only those proteins of interest that show significant over- or underexpression are then later collected for mass spectrometry identification from a conventional 2-DE slab gel.

For protein expression analysis as a function of dose or concentration, we foresee many interesting applications in pharmacology and molecular toxicology. Examples are dose–response assessments, drug absorption, distribution, metabolism, toxicity (ADMeTox) and interaction studies, as well as the screening of drug candidates. Also, the proteome signatures of organisms exposed to environmental stress (toxicological fingerprint) could be established very efficiently with the help of the 3D-gel method [29]. With the European Union directive for the registration, evaluation and authorization of chemicals (REACH) that was introduced in 2006, a thorough investigation of chemicals for their potentially hazardous effect on organisms becomes mandatory for their producers [30]. Here, the 3D-gel method could be used for studying changes in protein expression levels in response to those substances [31].

For protein expression analysis as a function of time, we predict applications in developmental and cell biology, such as expression studies during (embryonal) development or the cell cycle [27]. For instance, the appearance of differentiation-specific marker proteins is highly diagnostic for tissue/cell differentiation (e.g. in stem cell research) and de-differentiation (e.g. during cancer). In the fields of healthcare and pathology, the changes in protein expression over the progression of diseases can be investigated for diagnostic purposes, e.g. for the detection of disease markers or therapy monitoring. In biotechnology, the monitoring of bacterial metabolic activity during growth is highly relevant for efficient process engineering, e.g. for bioreactor condition optimization [32].

10.3.3 Future Developments

As an emerging technology, the 3D-gel electrophoresis method is subject to continuous improvement. (Regularly updated information is available at www.3D-gel.com). We are currently working toward further increasing the resolution, sensitivity and dynamic range of the 3D-gel instrument. This work entails an optimization of the optical and photodetection system to attain maximum fluorescence detection efficiency. In the course of this improvement, we will incorporate additional lasers and matching optical filters for simultaneous excitation and detection of fluorescence at different wavelengths. This will enable the 3D-gel instrument to multiplex assays by detecting differently labelled samples that co-migrate in the same vertical gel layers and will increase its throughput accordingly. Moreover, the multidye detection capability will give the 3D-gel instrument full compatibility with the DIGE method [17, 18]. Protein extracts, for example control and treated, can be labelled with different fluorescent dyes (e.g. Cy2, Cy3 and Cy5), then pooled and co-separated by IEF, followed by SDS-PAGE in the 3D-gel. In addition, dye multiplexing allows for a quantitative normalization over several separations by using an internal standard, i.e. a mixture of equal aliquots of every sample under analysis [33]. The internal standard is separated together with sample pairs and serves as a quantitative reference for an exact calibration. This way, system variations can be controlled, allowing the identification of biological variations and changes in protein expression with statistical confidence. In combination with its high throughput, this will make the 3D-gel instrument a valuable tool for large-scale, high-precision differential protein analysis.

Acknowledgements

The authors are grateful to the following people for their support and encouragement: Jörg Bernhardt, Christiane Fertig, Jane Lefken, Andreas Schleifenbaum and Fred Wouters.

References

1. O'Farrell, P.H., High resolution two-dimensional electrophoresis of proteins; J. Biol. Chem. 1975, 250, 4007–4021
2. http://www.fixingproteomics.org
3. Fuxius, S.; Eravci, M.; Broedel, O.; Weist, S.; Mansmann, U; Eravci, S.; Baumgartner, A., Technical strategies to reduce the amount of 'false significant' results in quantitative proteomics; Proteomics 2008, 8, 1780–1784
4. Ventzki, R.; Stegemann, J., High-throughput separation of DNA and proteins by 3D geometry gel electrophoresis: Feasibility studies; Electrophoresis 2003, 24, 4153–4160
5. Kambara, H.; Takahashi, S., Multiple-sheathflow capillary array DNA analyser; Nature 1993, 361, 565–566
6. Marsh, M.; Tu, O.; Dolnik, V.; Roach, D.; Solomon, N.; Bechtol, K.; Smietana, P.; Wang, L.; Li, X.; Cartwright, P.; Marks, A.; Barker, D.; Harris, D.; Bashkin, J., High-throughput DNA sequencing on a capillary array electrophoresis system; J. Capillary Electrophor. 1997, 4, 83–89
7. Medintz, I.L.; Paegel, B.M.; Blazej, R.G.; Emrich, C.A.; Berti, L.; Scherer, J.R.; Mathies, R.A., High-performance genetic analysis using microfabricated capillary array electrophoresis microplates; Electrophoresis 2001, 22, 3845–3856
8. Paegel, B.M.; Emrich, C.A.; Wedemayer, G.J.; Scherer, J.R.; Mathies, R.A., High throughput DNA sequencing with a microfabricated 96-lane capillary array electrophoresis bioprocessor; Proc. Natl Acad. Sci. USA 2002, 99, 574–579
9. Ansorge, W.; Sproat, B.; Stegemann, J.; Schwager, C.; Zenke, M., Automated DNA sequencing: Ultrasensitive detection of fluorescent bands during electrophoresis; Nucleic Acids Res. 1987, 15, 4593–4602
10. Lee, B.-S.; Gupta, S.; Morozova, I., High-resolution separation of proteins by a three-dimensional sodium dodecyl sulfate polyacrylamide cube gel electrophoresis; Anal. Biochem. 2003, 317, 271–275
11. Ventzki, R.; Stegemann, J.; Martinez, L.; de Marco, A., Automated protein analysis by online detection of laser-induced fluorescence in slab gels and 3D geometry gels; Electrophoresis 2006, 27, 3338–3348
12. Stegemann, J.; Ventzki, R., Schrödel, A.; de Marco, A., Comparative analysis of protein aggregates by blue native electrophoresis and subsequent SDS-PAGE in a 3D geometry gel; Proteomics 2005, 5, 2002–2009
13. Ventzki, R.; Rüggeberg, S.; Leicht, S.; Franz, T.; Stegemann, J., Comparative 2-DE protein analysis in a 3D geometry gel; Biotechniques 2007, 42, 271–279
14. Werhahn, W.; Braun, H.-P., Biochemical dissection of the mitochondrial proteome from *Arabidopsis thaliana* by three-dimensional gel electrophoresis; Electrophoresis 2002, 23, 640–646
15. Garoff, H.; Ansorge, W., Improvements of DNA sequencing gels; Anal. Biochem. 1981, 115, 450–457
16. Ansorge, W.; Barker, R., System for DNA sequencing with resolution of up to 600 base pairs; J. Biochem. Biophys. Methods 1984, 9, 33–47
17. Ünlü, M.; Morgan, M.E.; Minden, J.S., Difference gel electrophoresis: a single gel method for detecting changes in protein extracts; Electrophoresis 1997, 18, 2071–2077
18. Tonge, R.; Shaw, J.; Middleton, B.; Rowlinson, R.; Rayner, S.; Young, J.; Pognan, F.; Hawkins, E.; Currie, I.; Davison, M., Validation and development of fluorescence two-dimensional differential gel electrophoresis proteomics technology; Proteomics 2001, 1, 377–396
19. Bjellqvist, B.; Ek, K.; Righetti, P.G.; Gianazza, E.; Görg, A.; Westermeier, R.; Postel, W., Isoelectric focusing in immobilized pH gradients: Principle, methodology and some applications; J. Biochem. Biophys. Methods 1982, 6, 317–339
20. Görg, A.; Obermaier, C.; Boguth, G.; Harder, A.; Scheibe, B.; Wildgruber, R.; Weiss, W., The current state of two-dimensional electrophoresis with immobilized pH gradients; Electrophoresis 2000, 21, 1037–1053
21. Rasband, W.S., Image J., National Institutes of Health, Bethesda, MD, USA, 1997, http://rsb.info.nih.gov/ij/

22. Meijering, E., Transform J., University Medical Center, Rotterdam, The Netherlands, 2001, http://www.imagescience.org/meijering/software/transformj/

23. Clark, B.N.; Gutstein, H.B., The myth of automated, high-throughput two-dimensional gel analysis; Proteomics 2008, 8, 1197–1203

24. Aittokallio, T.; Salmi, J.; Nyman, N.A.; Nevalainen, O.S., Geometrical distortions in two-dimensional gels: Applicable correction methods; J. Chromatogr., B 2005, 815, 25–37

25. Berth, M.; Moser, F.M.; Kolbe, M.; Bernhardt, J., The state of the art in the analysis of two-dimensional gel electrophoresis images; Appl. Microbiol. Biotechnol. 2007, 76, 1223–1243

26. Bandow, J.E.; Baker, J.D.; Berth, M.; Painter, C.; Sepulveda, O.J.; Clark, K.A.; Kilty, I.; Van-Bogelen, R.A., Improved image analysis workflow for 2-D gels enables large-scale 2-D gel-based proteomics studies – COPD biomarker discovery study; Proteomics 2008, 8, 3030–3041

27. Bernhardt, J.; Weibezahn, J.; Scharf, C.; Hecker, M., *Bacillus subtilis* during feast and famine: visualization of the overall regulation of protein synthesis during glucose starvation by proteome analysis; Genome Res. 2003, 13, 224–237, http://microbio1.biologie.uni-greifswald.de/starv/movie.htm

28. Cornvik, T.; Dahlroth, S.-L.; Magnusdottir, A.; Herman, M.D.; Knaust, R.; Ekberg, M.; Nordlund, P., Colony filtration blot: a new screening method for soluble protein expression in *Escherichia coli*; Nat. Methods 2005, 2, 507–509

29. Oberemm, A.; Meckert, C.; Brandenburger, L.; Herzig, A.; Lindner, Y.; Kalenberg, K.; Krause, E.; Ittrich, C.; Kopp-Schneider, A.; Stahlmann, R.; Richter-Reichhelm, H.B.; Gundert-Remy, U., Differential signatures of protein expression in marmoset liver and thymus induced by single-dose TCDD treatment; Toxicology 2005, 206, 33–48

30. European Union directive for the registration, evaluation and authorization of chemicals (REACH), 2006, http://ecb.jrc.it/reach

31. Oberemm, A.; Onyon, L.; Gundert-Remy, U., How can toxicogenomics inform risk assessment? Toxicol. Appl. Pharmacol. 2005, 207 (2 Suppl.), 592–598

32. Hecker, M.; Volker, U., Towards a comprehensive understanding of *Bacillus subtilis* cell physiology by physiological proteomics; Proteomics 2004, 4, 3727–3750

33. Alban, A.; Davis, S.O.; Bjorkesten, L.; Andersson, C.; Sloge, E.; Lewis, S.; Currie, I., A novel experimental design for comparative two-dimensional gel analysis: two dimensional difference gel electrophoresis incorporating a pooled internal standard; Proteomics 2003, 3, 36–44

Part IV

Characterisation of Microorganisms by Pattern Matching of Mass Spectral Profiles and Biomarker Approaches Requiring Minimal Sample Preparation

11

Microbial Disease Biomarkers Using ProteinChip Arrays

Shea Hamilton, Michael Levin, J. Simon Kroll and Paul R. Langford
Department of Paediatrics, Imperial College London, London, UK

11.1 Introduction

Surface enhanced laser desorption/ionization time-of-flight mass spectrometry (SELDI, also called SELDI-TOF or SELDI-TOF-MS) has been widely used for the discovery of biomarkers. There has been a rapid explosion in SELDI papers with only 13 PubMed citations in 2000 compared with 342 in 2007 with those in the field of oncology being predominant. In contrast, there are relatively few studies in the area of infectious diseases, although they have increased over the last few years. Possible reasons include the proven track record of identifying biomarkers, especially in the cancer field [1–3], and clear application of the technique, with an increasing number of successful examples to infectious disease. Additionally, it is possible to use SELDI to measure biomarkers where conventional methods are technically demanding or not possible [4–6]. One of the early criticisms of SELDI was the sensitivity of the instrumentation and the lack of reproducibility. However, the newly available PC4000 instrument has excellent sensitivity and has been designed to achieve high levels of reproducibility. The latter is facilitated by quality control of the ProteinChip arrays. Batch to batch variation of ProteinChips was unquestionably a problem during the early use of SELDI but this is no longer the case.

How does SELDI differ from conventional matrix assisted laser desorption/ionization time-of-flight mass spectrometry (MALDI-TOF)? With the latter, the surfaces, such as stainless steel, are designed so that all peptides and proteins present in a sample can be deposited and analysed. With SELDI, however, the sample is placed on a ProteinChip array which has a particular surface, each one having different peptide and protein binding

Mass Spectrometry for Microbial Proteomics   Edited by Haroun N. Shah and Saheer E. Gharbia
© 2010 John Wiley & Sons, Ltd

characteristics. The ProteinChip surfaces available include WCX2 or CM10 (weak cationic exchange), SAX2 or Q10 (strong anionic exchange), IMAC (immobilized metal affinity chromatography) and H4 or H50 (hydrophobic). If required NP20 or gold Protein-Chips can be used to analyse all of the proteins in a sample similar to MALDI-TOF. It should be noted that, in both SELDI and MALDI-TOF, not all proteins in a sample may be detected, as some proteins and peptides do not ionize and the signal obtained for some proteins can suppress that from others. Protein samples may be prefractionated (e.g. with Q-Hyper D anion-exchange resin, see Table 11.1) or directly placed on the ProteinChip surface and unbound proteins and interfering substances removed by washing, an energy-absorbing matrix applied and the ProteinChip inserted into the mass spectrometer. By varying ProteinChip surface and wash conditions (e.g. pH and organic solvents), the proteins that remain bound to the ProteinChip vary depending on their physical characteristics (e.g. isoelectric point). Different energy-absorbing matrices can also affect the results. Typically, α-cyano-4-hydroxy cinnamic acid or sinapinic acid are used dependent on the

Table 11.1 *Summary of the patient/control populations investigated, sample preparation and data analysis methods used in the referenced studies. See main text for more detailed information and discussion of these studies*

Disease or aetiological agent	Type and number of samples analysed	Sample preparation	ProteinChip surface(s) used
Hepatitis B	Chronic hepatitis B = 46	Serum denatured with CHAPS/urea	CM10
Hepatitis B	Liver cirrhosis = 25 Liver cancer = 20 Healthy = 25	Serum denatured with CHAPS/urea	WCX2
Hepatitis B	HCC = 67 HBV = 50 Normal = 44	Serum denatured with CHAPS/urea	IMAC (Cu)
Hepatitis B	HCC = 81 Cirrhosis = 36 Liver cirrhosis = 43	Serum denatured with CHAPS/urea	WCX2
Hepatitis B	HCC = 81 Controls = 33	Serum	WCX2
Hepatitis B	HCC = 26 Cirrhosis = 18	Liver biopsy	CM10

mass range of interest. A different SELDI proteomic profile can be obtained for the same sample over the same mass range using different energy-absorbing matrices, e.g. sinapinic acid versus fusidic acid [7]. Each ProteinChip and wash condition will select for a subset of the proteins present in a sample (sub-proteome). There are two big advantages of SELDI over conventional protein analysis techniques such as two-dimensional polyacrylamide gel electrophoresis (2D-PAGE). First, the amount of sample that is required for analysis is very small e.g. 1–4 μl of serum is sufficient to obtain a spectrum, 100 ng–1 μg of protein generally being sufficient for a good profile. In addition to serum, sample types that have been analysed by SELDI include plasma, bronchial lavage fluid, induced sputum, nipple aspirates, tears, urine, peritoneal dialysate and biopsy material from many tissues. Although not used in the infectious disease arena so far, it is also possible to analyse samples from archived fixed tissue, including capture by laser dissection. Secondly, the method can be high throughput – hundreds or thousands of samples can be analysed in a comparatively short space of time, although typically this would involve the use of robots.

Data analysis	Addition of clinical data	Validation	Peak ID	Reference
Significance analysis of microarrays Artificial neural network	Yes	Cross-validation	No	[24]
Decision tree	No	Independent validation set Liver cirrhosis = 15 Liver cancer = 10 Healthy = 12	No	[28]
Hierarchical clustering Logistic regression	No	Independent validation set HCC = 17 HBV = 8 Normal = 14 Biomarker identified: SAA	Yes	[32]
Direct comparison of normalized peak intensities	No	No	No	[29]
Decision tree	No	Cross-validation Independent validation set HCC = 48 Controls = 33 Biomarker identified: neutrophil activating protein-2	Yes	[31]
Direct comparison of normalized peak intensities	No	No Biomarker inferred from mass: Bax (expression confirmed by immunohistochemistry)	No	[16]

(*continued overleaf*)

Table 11.1   (continued)

Disease or aetiological agent	Type and number of samples analysed	Sample preparation	ProteinChip surface(s) used
Hepatitis B	HCC = 29 Liver cirrhosis = 30	Serum denatured with CHAPS/urea	IMAC (Cu)
Hepatitis C	HCC = 57 Cirrhosis = 38 Noncirrhotic liver disease = 36 No liver disease = 39	Serum denatured with CHAPS/urea	CM10 Q10 IMAC (Zn)
Hepatitis C	Responders = 68 Nonresponders = 28	Serum denatured with CHAPS/urea	CM10 IMAC (Zn)
Hepatitis C	HCC = 55 Chronic hepatitis C = 48 Normal = 9	Serum denatured with CHAPS/urea	WCX2
Hepatitis C	HCC = 60 Non-HCC = 84	Serum denatured with CHAPS/urea	IMAC (Cu)
Hepatitis C	F1/F2 fibrosis = 39 F4 fibrosis = 44 HCC = 34	Serum Q-Hyper D anion-exchange resin fractionation	CM10
Hepatitis C	Liver cirrhosis-HCV-HCC = 77 Liver cirrhosis-HCV = 76	Serum denatured with CHAPS/urea	CM10
SARS	SARS = 8 Non-SARS = 15	Serum denatured with CHAPS/urea	CM10
SARS	Acute SARS = 37 Non-SARS = 74	Serum denatured with CHAPS/urea	WCX2
SARS	SARS = 28 Non-SARS = 72 Healthy = 10	Serum Q-Hyper D anion-exchange resin fractionation	CM10 IMAC (Cu)

Data analysis	Addition of clinical data	Validation	Peak ID	Reference
Discriminant analysis	Yes	Cross-validation	No	[15]
Decision tree	Yes	Cross-validation Independent validation set HCC = 56 Healthy = 42	No	[25]
Logistic regression	No	Cross-validation Independent validation set Responders = 38 Nonresponders = 13	No	[22]
Direct comparison of normalized peak intensities	No	Biomarker identified: complement C3a Immunoassay and semi-quantitative western blot on samples used in initial study	Yes	[21]
Artificial neural network	No	Cross-validation Independent validation set HCC = 17 Non-HCC = 21 Biomarkers identified: complement C3a, κ and λ immunoglobulin light chains	Yes	[26]
Decision tree	Yes	Cross-validation Biomarker identified: apolipoprotein C-1 Confirmed by ELISA	Yes	[27]
Decision tree	No	Cross-validation Independent validation sets Set 1 Liver cirrhosis-HCV-HCC = 35 Liver cirrhosis-HCV = 44 Set 2 Samples from patients prior to developing HCC Subsequently developed HCC = 7 Did not develop HCC = 5	No	[30]
Direct comparison of normalized peak intensities	No	Independent validation set SARS = 7	No	[23]
Decision tree	No	Independent validation set Acute SARS = 37 Non-SARS = 993	No	[33]
Direct comparison of normalized peak intensities	No	Biomarker identified: SAA Longitudinal follow-up tracked correlation between SAA levels and severity of pneumonia	Yes	[34]

(*continued overleaf*)

Table 11.1   (continued)

Disease or aetiological agent	Type and number of samples analysed	Sample preparation	ProteinChip surface(s) used
SARS	SARS = 39 Non-SARS = 39	Serum denatured with CHAPS/urea	CM10
HIV-1	HIV seropositive = 31 HIV seronegative = 10	Cellular lysates of cultured monocyte derived macrophages Triton-x 100	WCX2
HIV-1	HIV-associated dementia = 11 HIV seropositive no dementia = 13 HIV seronegative = 9	Secreted proteins from cultured macrophages/monocytes CHAPS/urea	WCX2
HIV-1	HIV-infected no dementia = 15 HIV-infected with dementia = 15 Matched HIV-negative controls = 15	Secreted proteins from cultured macrophages/monocytes CHAPS/urea	Gold
HTLV-1	ATL = 32 HAM/TSP = 40 Healthy = 28	Serum denatured with CHAPS/urea	IMAC (Cu)
BK virus	Stable graft = 29 BKV nephropathy = 21 Acute allograft rejection = 28	Urine denatured with CHAPS/urea	CM10 IMAC (Cu)
CMV	CMV infected with hepatitis = 20 CMV infected without hepatitis = 5 Hepatitis but no CMV infection = 10 Healthy = 10	Serum	WCX2
PRRS	Controls = 8 Infected = 8	Serum	WCX IMAC (Cu) H50

Data analysis	Addition of clinical data	Validation	Peak ID	Reference
Significance analysis of microarrays Hierarchical clustering	No	No Biomarkers identified: fragment of fibrinogen and an N-terminal fragment of complement C3c α-chain	Yes	[13]
Decision tree	No	Cross-validation	No	[36]
Direct comparison of normalized peak intensities	No	Biomarker identified: lysozyme ELISA of lysates and supernatants for tumour necrosis factor α and lysozyme	Yes	[38]
Direct comparison of normalized peak intensities	No	No	No	[39]
Decision tree	No	Independent validation set ATL = 10 HAM/TSP = 10 Healthy = 10 Biomarkers identified: a fragment of α trypsin inhibitor and two contiguous fragments of haptoglobin-2	Yes	[41]
Support vector machines Random forest Classification and regression tree (decision tree)	No	No	No	[42]
Decision tree	No	No	No	[45]
Direct comparison of normalized peak intensities	No	Cross-validation Controls = 16 Infected = 16	No	[47]

(continued overleaf)

Table 11.1   (continued)

Disease or aetiological agent	Type and number of samples analysed	Sample preparation	ProteinChip surface(s) used
IAI Rhesus monkeys + humans *Monkeys* Group B *Streptococcus* *Ureaplasma* *parvum* *U. urealyticum* *U. hominis*	*Monkeys* Pre- and post-IAI = 19 *Humans* Preterm delivery <35 weeks associated with clinical IAI = 11 Preterm delivery <35 weeks without clinical IAI = 11 Preterm controls and term delivery >35 weeks = 11	Amniotic fluid	NP20 H4 IMAC (Ni)
IAI	PTL + WBC + AFC = 21 PTL + WBC − AFC = 7 PTL − WBC + AFC = 8 PTL − WBC − AFC = 24 Normal = 17	Amniotic fluid	H4
IAI (Twin pregnancies)	Three mothers with twins = 6	Amniotic fluid	H4
IAI	Intact membranes = 27 PPROM = 21 Paired abdominal vs vaginal amniotic fluid	Amniotic fluid	H4
IAI	PTL IAI = 7 PTL = 7 PPROM IAI = 7 PPROM = 6	Amniotic fluid Cervical fluid	Q10
IAI	PPROM = 70 Intact membranes = 99	Amniotic fluid	H4
IAI	Second trimester control (2T-CRL) = 31 2T-CRL without AFC = 27 2T-CRL with AFC = 27 Third trimester control = 28	Amniotic fluid	H4

Data analysis	Addition of clinical data	Validation	Peak ID	Reference
Direct comparison of normalized peak intensities	No	Cross-validation Biomarkers identified: calgranulin B, vitamin D binding protein, IGFBP-1 (intact and proteolytic fraction) Western blot analysis of calgranulin B and IGFBP-1	Yes	[61]
Direct comparison of normalized peak intensities	No	Independent validation set = 24 Biomarkers identified: neutrophil defensin-1 neutrophil defensin-2 calgranulin A calgranulin C Used as the basis for MR score Validation by western blotting	Yes	[59]
Direct comparison of normalized peak intensities - MR score of 4 proteins	No	No	No	[60]
	No	No Calgranulin A poorly detectable in ELISAs (4/48 amniotic fluids) compared with 19/48 amniotic fluids with SELDI	No	[4]
Direct comparison of normalized peak intensities	No	No Biomarker identified: neutrophil protein 1–3 calgranulins A and B Western blot of calgranulin A–C ELISA of human neutrophil protein in amniotic fluid	Yes	[62]
Direct comparison of normalized peak intensities - MR score of 4 proteins	No	No	No	[58]
Direct comparison of normalized peak intensities - MR score of 4 proteins	No	No Measured S100A12/ENRAGE levels using SELDI	No	[64]

(continued overleaf)

Table 11.1   (continued)

Disease or aetiological agent	Type and number of samples analysed	Sample preparation	ProteinChip surface(s) used
IAI Histologic chorioamniotis	PPROM + PTL = 158 Newborns = 125	Amniotic fluid	H4
IAI	Singletons with symptoms of preterm labour No inflammation and no bleeding = 193 Inflammation = 71 Bleeding = 6 Inflammation and bleeding = 15	Amniotic fluid	H4
IAI	Consecutive mothers = 132 MR score 0 = 26 MR score 1–2 = 39 MR score 3–4 = 67 Umbilical cord blood	Amniotic fluid	H4
IAI	Term = 59 PTL + IAI = 60	Amniotic fluid	CM10 H50
Tuberculosis	Tuberculosis = 102 Controls = 79 Healthy = 12	Serum denatured with CHAPS/urea	CM10
Bacterial endocarditis	Endocarditis = 23 Controls = 36	Serum Q-Hyper D anion-exchange resin fractionation	CM10
CF	CF = 39 Other respiratory diseases = 38	BALF	CM10 Q10 IMAC (Cu) IMAC (Ni)
COPD	COPD = 30 Age sex, smoking history matched controls = 30	Plasma Q-Hyper D anion-exchange resin fractionation Serum denatured with CHAPS/urea	Q10 IMAC (Cu)

Data analysis	Addition of clinical data	Validation	Peak ID	Reference
Direct comparison of normalized peak intensities - MR score of 4 proteins	No	No	No	[63]
Direct comparison of normalized peak intensities - MR score of 4 proteins Q-profile (5 peaks in 10–12.5 kDa) range	No	No Description of Q-profile Biomarkers identified: numerous including apolipoprotein A-I IFGBP-1 apolipoprotein A-IV lumican 1 IgM cold agglutinin α-1-antitrypsin	Yes	[67]
Direct comparison of normalized peak intensities - MR score of 4 proteins	No	No Correlation of MR score with clinical parameters	No	[65]
Binary fingerprinting and pattern discovery	No	Cross-validation Independent validation set Term = 15 PTL + IAI = 16	No	[68]
Single and multiple layer perception Tree classifiers Support vector machines	No	Cross-validation Independent validation set Tuberculosis = 77 Controls = 70 Healthy = 9 Biomarkers identified: SAA des Arg and transthyretin	Yes	[12]
Direct comparison of normalized peak intensities Partial least squares Logistic regression	No	Iterative random re-sampling Independent validation set Endocarditis = 11 Controls = 18	No	[53]
Direct comparison of normalized peak intensities	No	No Biomarker identified: S100 A8 (calgranulin A) Correlation of two biomarkers with % bronchoalveolar neutrophil numbers	Yes	[56]
Classification and regression tree analysis (decision tree)	No	Cross-validation	No	[54]

(continued overleaf)

Table 11.1   *(continued)*

Disease or aetiological agent	Type and number of samples analysed	Sample preparation	ProteinChip surface(s) used
CF	Controls chronic cough and/or stridor = 27 CF = 24	BALF	H4
CF + COPD	Healthy controls = 20 Asthma = 24 COPD = 24 CF = 28 Bronchiectasis = 19 CF patients sampled before and after antibiotic therapy for an infective exacerbation = 12	Induced sputum	CM10 IMAC (Ni) Q10
Peritonitis	Infected = 5 Controls = 11	Peritoneal dialysis fluid	CM10 IMAC
Trypanosomiasis	Trypanosomiasis = 45 Control = 77	Serum denatured with CHAPS/urea	CM10
Fasciolosis	Infected = 8 Serial bleeds 0–12 weeks	Plasma Q-Hyper D anion-exchange resin fractionation Serum denatured with CHAPS/urea	CM10

HCC, Hepatocellular carcinoma; CHAPS, 3-[(3-cholamidopropyl)dimethylammonio]-1-propanesulfonate hydrate; SARS, severe acute respiratory syndrome; CM10, weak cationic exchange array; Q10, strong anionic exchange array; IMAC, immobilized metal affinity array; ATL, adult T-cell leukaemia; HAM/TSP, human T-cell leukaemia virus type 1-associated myelopathy/tropical spastic paraparesis; CMV, cytomegalovirus; PTL, preterm labour; WBC, white blood cell; AFC, amniotic fluid culture positive; MR, mass restricted; IAI, intra-amniotic inflammation; PPROM, preterm premature rupture of membranes; ELISA, enzyme linked immunosorbent assay; SAA, serum amyloid A; CF, cystic fibrosis; COPD, chronic obstructive pulmonary disease; BALF, bronchial lavage fluid: IGFBP-1, insulin-like growth factor binding protein-1.

In recent years, it has been recognized that a number of variables can affect the SELDI mass spectral profiles. This has been extensively reviewed elsewhere [8, 9] and only a few salient points will be mentioned here. Broadly these can be put into two categories: (1) sample, machine and method of analysis related; and (2) study design. In brief, it is important to have standardized protocols for sample collection and storage prior to analysis.

Data analysis	Addition of clinical data	Validation	Peak ID	Reference
Direct comparison of normalized peak intensities	No	No Biomarkers identified: defensin 1 and 2, S100A8, S100A9, and S100A12, novel forms of S100A8 and S100A12 with equivalent C-terminal deletions Some confirmation by ELISA and western blotting	Yes	[55]
Direct comparison of normalized peak intensities	No	No Biomarkers identified: calgranulin A–C, Clara cell secretory protein, proline rich salivary peptide, lysozyme C and cystatin S There were decreases in the levels of calgranulins A and B (measured by SELDI) and calprotectin (heterodimer of calgranulins A and B) measured by ELISA	Yes	[57]
Direct comparison of normalized peak intensities	No	No Biomarker identified: SAA	Yes	[69]
Artificial neural network Genetic algorithm Decision tree	No	Cross-validation Independent validation set Trypanosomiasis = 40 Control = 69	No	[14]
Direct comparison of normalized peak intensities	No	No Biomarkers identified: transferrin and apolipoprotein A-IV	Yes	[49]

Storage at −80 °C rather than −20 °C is generally preferable. For example, breakdown of cystatin B was found at −20 °C [10, 11]. Such proteolytic breakdown can potentially result in the identification of erroneous biomarkers. This is not a trivial issue since many of the biomarkers that have been discovered are peptide fragments that have arisen because of proteolytic activity of endogenous proteases whose regulation has been altered either directly or indirectly as a result of the disease process. It may be possible to circumvent this problem by adding protease inhibitors when the sample is collected, although this may lead to the disappearance of relevant biomarkers.

Study design is crucial in biomarker discovery. Typically, representative spectra from known diseased and control patients are obtained – the training set. This is followed by a second blind study with additional samples from the same cohort – the testing set. Ideally, a validation study with a separate cohort, either collected at the same or another site, would be undertaken and this should be done blind. The numbers of samples that are tested in each of the phases varies but 20–50 are typically used in the training and testing sets. Clearly this will be influenced by the nature of the infectious disease and the number of samples available, e.g. biologically relevant biopsy samples are often hard to obtain. Analytical methods will also influence the results. These vary from the simple analysis of relative normalized peak intensities (available in the standard Ciphergen Express software) to the more complex artificial neural networks, support vector machines, genetic algorithms, principal component analysis and decision trees. More commonly, investigators are using more than one analytical method and/or combining existing clinical parameters since such approaches can result in increased sensitivity and specificity [12–15].

Once biomarkers of interest [mass/charge (*m/z*) ratios on the SELDI spectra] have been discovered, the next stage is to determine their identity at the molecular level. It is possible to search for proteins and/or peptides with a particular mass using programs such as TagIdent (ExPASy). However, identification from *m/z* is rarely possible directly because of post-translational modification, e.g. glycosylation. Nonetheless, some of the smaller molecular mass proteins such as defensins have been identified by this route. Additionally, taking into account the biological context may be of value in directing resources [16]. Identification can be achieved by optimizing conditions to separate peaks of interest on a small scale and then a larger scale with resins compatible with the appropriate ProteinChip surfaces. Proteins can be separated further by 1D- or 2D-PAGE or liquid chromatography, and identified through peptide mass fingerprinting and/or tandem mass spectrometry. Thereafter, peak identification may be unequivocally determined by immunoprecipitation using specific antibodies reactive with the protein of interest. Examples of all these approaches have been used to identify infectious disease biomarkers.

Knowing the identity of the biomarker can be used in a number of ways and depends on the rationale of the study in the first place. There are difficulties in using conventional immunological methods to screen for some of the biomarkers, so the fact that SELDI can identify the relevant proteins on one ProteinChip is advantageous. A typical scenario would be to identify biomarkers with a view to using a simpler cheaper test in the clinic e.g. ELISA. Other secondary methods e.g. western blotting or a functional assay may also be used for verification of a marker. Problems arise when the biomarker is a peptide fragment as off-the-shelf tests may not be available and therefore need to be developed. In some cases, e.g. tuberculosis and trypanosomiasis, the rationale behind studies is to use powerful proteomic methods to identify biomarkers with a view to formulation of simple rapid affordable point-of-care tests, which are urgently needed in the developing world.

This chapter will be restricted to those studies that involve SELDI analysis of body fluids or specific cell types from human or animal test and control subjects where infectious disease is the primary reason for investigation. Table 11.1 summarizes important features of all the relevant infectious disease-based SELDI studies to date.

11.2 Biomarker Studies Involving Patients Infected with Viruses

11.2.1 Hepatitis B and C

Hepatocellular carcinoma (HCC) is the fifth most common and the third leading cause of death from cancer worldwide, with hepatitis B virus (HBV) or hepatitis C virus (HCV) being major risk factors. As with many of the diseases where SELDI is being employed to search for biomarkers, current diagnostic methods are inadequate. A lack of both specificity and sensitivity is associated with α-fetoprotein levels in serum and liver biopsy, which while definitive, is expensive and potentially risky. Thus, there are clear advantages to identifying biomarkers from readily available body fluids and especially for early diagnosis and/or prediction of those patients that are most at risk from HCC. There are a number of comprehensive reviews of HCC biomarkers discovered through proteomic approaches [3, 17–20]. Only SELDI-based papers published since our previous review [20] will be considered, although salient features of studies prior to this [21–28] can be found in Table 11.1. The majority (5/6) of the studies reviewed here involve the search for serum biomarkers [15, 29–32] as opposed to those present in liver biopsy samples [16]. HBV is the focus of all of the studies reviewed except that of Kanmura *et al.* [30] where it is HCV.

Zhang *et al.* [16] found 16 peaks in fresh liver biopsies (7 up-regulated, 9 down-regulated) that could separate HCC (HBV related) and cirrhotic (non-HBV related) patients. Discrimination between moderately or well-differentiated and poorly differentiated HCC (as determined by histology) was not possible. There was no correlation between α-fetoprotein staining and biomarker intensity. Putative protein identification was made after interrogation of the ExPASy database using the parameters *m/z* +/– 0.1% and assuming the isoelectric point was 9 +/– 5. A 4674 Da peak was putatively identified as Bax – a protein which activates caspase 3 resulting in nuclear fragmentation (apoptosis). Subsequent immunohistochemistry found a correlation between Bax expression levels and 4674 Da peak intensity. Whilst the experimental procedure did not unequivocally identify the 4674 Da peak as Bax, it was highly suggestive. Relatively few investigators have used this approach as identification from *m/z* can be problematic due to post-translational modifications such as glycosylation and methylation. Nevertheless, as an initial screen and taking into account the nature of the disease under investigation, the approach may have some merits.

Forty-five serum protein/peptide peaks separated HBV-related HCC from HBV-related liver cirrhosis (LC) patients [15]. Twenty-six and 16 peaks were up-regulated in the HCC and LC groups, respectively. Disappointingly, the diagnostic value of the most significant peak (3892 Da) was only as good as α-fetoprotein. However, the combination of two peaks and AFP resulted in a sensitivity and specificity of 82.3 and 93.3%, respectively. Studies by Cui *et al.* [29] found protein peaks that could separate the serum profiles of HBV-related HCC, LC and chronic hepatitis patients. Of particular interest were five peaks that were down-regulated in serum from HBV-related HVC patients and two that were up-regulated in both HCC and LC patients, as it was suggested that they could prove useful HCC pathology markers.

A combination of SELDI and 2D-PAGE was used to identify serum biomarkers for the prediction of HBV-related HCC [32]. The groups investigated were HBV-related HCC,

HBV-infected and normal controls. Three peaks were identified that could discriminate HBV-related HCC patients from the other two groups when a predictive model was used. The predictive model had a 100% sensitivity and specificity at identifying HBV-related HCC subjects and those with HBV infection. Immunodepletion experiments were used to confirm the identity of this protein as serum amyloid A (SAA). The increase in intensity for the SAA peak from normal controls to HBV-infected to HBV-related HCC observed in the SELDI proteomic profiles was mirrored in western blotting experiments. SAA is an acute phase reactant and increases have been observed in other diseases, such as tuberculosis [12]. The other two peaks proved refractory to identification. HBV-related HCC and noninfected normal controls could be separated with a sensitivity of 100% and a specificity of 97% upon decision tree analysis when the six most discriminatory peaks were used [31]. Neutrophil-activating peptide 2 (NAP-2) was identified as one of the six peaks and was found to be expressed in HCC cells but not in normal liver tissues. NAP-2 had previously been described as a biomarker for HCC [31].

Serum samples from HCC-associated chronic liver disease with or without HCV involvement were analysed by Kanmura *et al.* [30]. Similar to He *et al.* [32], a decision tree using six biomarkers – although none of them were the same – was highly discriminatory on a test group. A final third phase of the study prospectively investigated serum from five patients free of HCC for 3 years and seven patients prior to diagnosis of HCC. The decision tree predicted HCC diagnosis in six out of seven patients before clinically apparent disease. Although the numbers were small, the results suggest that SELDI may be useful for both early detection and prediction of HCV-associated HCC.

11.2.2   Severe Acute Respiratory Syndrome (SARS)

Fortunately SARS did not develop into the worldwide pandemic as was originally feared at the time of the first outbreak. However, it is important that the world remains vigilant. Diagnosis of SARS is based primarily on clinical definition and the similarity of symptoms with other respiratory infections can make diagnosis difficult. Early detection is vital given the high transmissibility of the virus and high mortality associated with infection. Other methods used to detect the virus include quantitative real-time SARS coronavirus (SARS-CoV) RNA detection, which suffers from poor sensitivity and specificity, and immunological tests, which are only reliable weeks after onset of disease. SELDI was investigated as a diagnostic tool because rapid analysis is possible. Poon *et al.* [23] analysed SELDI spectra from plasma of a small number of paediatric patients (8 SARS and 15 controls). Fifteen peaks had the desired characteristics, i.e. were positively correlated with the concentration of SARS-CoV in plasma and increased in SARS patients at the onset of fever and decreased at the time of recovery. Unusually for SELDI studies, only one peak in the <20 kDa range (where most peaks are found upon SELDI analysis of serum or plasma) fulfilled the three criteria as did 14 peaks >75 kDa. Whilst no protein identification work was carried out and the sample numbers used were low, the results indicated that SELDI analysis of readily obtainable body fluid was a promising area of research. Two further serum-based studies quickly followed [33, 34]. Impressively, in what is the largest single study so far reported, 2000 samples were analysed for SARS biomarkers [33]. It was possible to identify acute SARS and non-SARS controls with an accuracy of 100 and 97.3%, respectively.

Yip *et al.* [34] found 12 peaks (9 up-regulated, 3 down-regulated) that could differentiate SARS from non-SARS patients. One of the up-regulated peaks was presumed, on the basis of *m/z* value, to be derived from SAA and this was subsequently investigated further [35]. Using optimal SELDI analysis conditions for SAA, three peaks presumed to be full length SAA, des Arg SAA and des Arg/des Ser variants at the amino terminus were analysed. Although there was correlation between peak intensities and serum concentrations of SAA, these were not specific to SARS patients. The high peak intensities and serum SAA concentrations were also found in non-SARS patients exhibiting similar symptoms. A further study identified 20 peaks (15 up-regulated, 5 down-regulated) that differentiated SARS patients from those displaying similar symptoms [13]. The most discriminating biomarkers (sensitivities 97–100%, specificities 95–97%) were identified as an internal fragment of fibrinogen and an N-terminal fragment of complement C3c α-chain.

11.2.3   Human Immunodeficiency Virus (HIV)

One of the potential complications with HIV infection is cognitive impairment ranging from acute neurological disease to dementia. It was hypothesized by Luo *et al.* [36] that there was an HIV dementia-associated monocyte derived macrophage (MDM) proteomic signature. Two peaks from MDMs (whole cells), obtained from patients and cultured in vitro, could differentiate HIV-1 infected from HIV-1 seronegative subjects with a sensitivity of 100% and a specificity of 80%. A further single peak separated HIV-1 infected patients with and without cognitive impairment with a sensitivity and specificity of 100 and 75%, respectively. The result thus appeared to support their hypothesis. However, the significance of the results was questioned [37], mainly due to the low patient numbers and restricted nature of the cohort (Hispanic women aged 21–45 years of age).

Sun *et al.* [38] carried out a similar study except MDM secreted proteins were used instead of whole cells. Monocytes/macrophages were obtained from patients with HIV-1 associated dementia, HIV-1 seropositive (no dementia) and HIV-1 seronegative subjects, cultured *in vitro* and the supernatants analysed. A 14.6 kDa peak had consistently lower intensities in the supernatants obtained from HIV-1 associated dementia patients compared with those from HIV-1 seropositive (no dementia) and HIV-1 seronegative groups. There was not a correlation between the 14.6 kDa peak intensity and lysozyme, viral load, CD4 cell count or antiretroviral therapy.

That ethnicity is an important factor in host response is illustrated by the results of Ratto-Kim *et al.* [39] who investigated a predominantly Thai cohort, although it also contained some individuals from Hawaii. A CD14CD16HLADR monocyte phenotype was significantly higher in healthy HIV-negative Thai individuals compared with controls from Hawaii. In the Thai cohort, no peaks derived from monocyte supernatants differentiated HIV-positive patients with or without dementia from HIV-negative matched controls. All samples from HIV-positive patients were obtained prior to the initiation of antiretroviral therapy. Finally, no peaks were found that separated HIV-positive patients with and without dementia when cerebrospinal fluid was analysed [40].

11.2.4   Human T-Cell Leukaemia Virus Type-1 (HTLV-1)

HTLV-1 is, like HIV, a retrovirus and infection can result in adult T-cell leukaemia (ATL), a common form of non-Hodgkins lymphoma, and a progressive neurological disorder called HTLV-1 associated myelopathy/tropical spastic paraparesis (HAM/TSP). Only small numbers of individuals get overt disease, but it is estimated that 10–20 million people worldwide are infected via spread from person to person through infected cells in semen, blood, breast milk and sharing needles. Detection of early infection and accurate disease status is problematic. Hence, Semmes *et al.* [41] compared SELDI profiles of serum obtained from ATL, HAM/TSP and 38 normal controls. Three peaks, identified as a fragment of α-trypsin inhibitor and two contiguous fragments of haptoglobin-2, were up-regulated in ATL patients. It was hypothesized that a primary response to HLTV-1 infection induces the expression of a yet unidentified endogenous protease, and subsequent secondary events (cleavage of haptoglobin-2) result in the formation of informative biomarkers. It was claimed that this was the first description of protein profiling being able to distinguish between disease states resulting from infection with a single pathogen.

11.2.5   BK Virus (BKV)

In immunocompromised patients, such as those that have had a bone marrow transplant or with a solid organ-allograft, infection with BKV virus can cause BKV renal allograft nephropathy (BKVAN). Clinically and histologically, BKVAN can resemble acute allograft rejection (AR). Distinguishing BKVAN from AR is important as the treatment regimes are radically different with immunosupression being decreased with the former and increased with the latter. Current noninvasive tests are inadequate. Jahnukainen *et al.* [42] compared the SELDI proteomic profiles of urine from BKVAN, AR and stable transplant groups. Five peaks were significantly higher in BKVAN patients. Some of the samples were positive for cytomegalovirus (CMV), HHV-6 or adenoviruses. Schaub *et al.* [43] included five patients with unspecified urinary tract infections in an AR urine biomarker study and reported a difference between the two groups. There were not any additional peaks in urine samples obtained from patients who had CMV-viraemia. Differences in peaks were found in some patients with urinary tract infection compared with controls and AR [44]. The results from the BKVAN study show promise but it has been argued that, in general, the application of proteomic approaches in the renal transplantation field has failed to live up to expectations [43].

11.2.6   Cytomegalovirus (CMV)

Liu *et al.* [45] searched for serum biomarkers for CMV, a major cause of congenital viral infection, with a view to greater understanding of the disease. Serum was obtained from infants with congenital CMV and hepatitis, with asymptomatic congenital CMV infection (control group 1), with hepatitis but without CMV infection (control group 2), and age-matched healthy infants (control group 3). Four protein peaks significantly differentiated the congenital CMV group from controls. A presumptive identification, based on searching the Swiss-Prot database, was that one peak was derived from β-defensin 8 and the other a macrophage-derived chemokine. In an analogous manner, two of the five peaks that differentiated those with CMV infection from those without were presumptively identified as platelet factor 4/CXCL4 and IL-25. Biologically plausible roles for some of these proteins in the disease process were made, although further work is required to confirm identity.

11.2.7   Porcine Reproductive and Respiratory Disease Syndrome (PRRS)

PRSS is responsible for a substantial amount of morbidity and mortality in the worldwide pig industry. The causative agent is a single stranded RNA virus (PRRSV) and infection with the virus can result in failure of reproduction, respiratory distress and influenza-like symptoms. PRRSV is of huge economic significance due to reduced daily weight gain and increased use of medication required for control of secondary pathogens [46]. In a pilot study, SELDI was used to identify serum biomarkers that differentiate weaning asymptomatic piglets positive for PRRSV viraemia (as adjudged by PCR) from negative controls [47]. Eighteen highly significant ($P < 0.01$) and 34 significant ($P = 0.01$–0.05) peaks were capable of differentiating infected from control animals in phase 1 of the study. However, although the peaks were present in an independent validation set of sera, none of the 18 peaks was statistically different between infected and control animals. Reasons put forward for the disparity between phase 1 and the validation arms of the study were differences in protein concentration and the masking of potential early disease biomarkers by more abundant proteins. Of positive note, IMAC and WCX ProteinChips gave excellent porcine serum SELDI profiles.

11.3   Biomarker Studies Involving Patients Infected with Parasites

11.3.1   Trypanosomiasis

The worldwide burden of human African trypanosomiasis (sleeping sickness) is immense. There is a real need for a simple diagnostic test capable of detecting early disease that will enable those that need treatment to get it promptly. Hazardous drugs are used in treatment, so early diagnosis will also reduce the amount of inappropriate therapy. SELDI serum profiles could separate trypanosomiasis infected patients from those with other infectious disease or inflammatory conditions with a sensitivity and specificity of 100 and 98.6%, respectively, provided that the majority opinion of decision tree, artificial neural networks and genetic algorithms were considered [14]. No peak identification was carried out and it was not possible to distinguish patients from controls with a single biomarker.

11.3.2   Fasciolosis

The trematode *Fasciola hepatica* (common liver fluke) can cause Fasciolosis in many herbivorous mammals (including sheep and cattle) and humans. It is found mainly in Europe, America and Oceania. The parasite is responsible for large economic losses in livestock and human zoonotic infection is an increasing concern. Early on in the disease process, there is encystment in the small intestine and passage to the liver and subsequent tissue inflammation and necrosis [48]. Rioux *et al.* [49] evaluated SELDI to identify serum biomarkers associated with *F. hepatica* infection of Corriedale sheep. In particular, the authors were interested in host molecules that may be involved in the establishment or suppression of host immunity. Sera were collected at weekly intervals from eight experimentally infected sheep. At each respective time point sera were pooled, fractionated and analysed. For the validation phase, sera from individual sheep (3 and 9 weeks post infection) were used. Six biomarkers were up-regulated at weeks 3 and 9 post infection, 16 were up-regulated only at week 9 post infection and four biomarkers were down-regulated at week 9 post infection. Two of the biomarkers that were up-regulated at 9 weeks post infection were identified as transferrin and apolipoprotein A-IV and these were verified by western blot analysis. Overall the results demonstrated that many complex changes were occurring in the serum with progression of the disease. The use of pooled serum may have been suboptimal. For example, pooling of serum samples from patients with invasive aspergillosis and controls resulted in 50% loss of peak clusters detected in individual samples [50]. Nevertheless the approach taken, i.e. pooling in phase 1 and using individual sera in the validation phase, did result in identification of useful biomarkers.

11.4   Biomarker Studies Involving Patients Infected with Bacteria

11.4.1   Tuberculosis

Tuberculosis is a major global concern, especially the increasing appearance of multiple drug resistant strains. Diagnosis can be problematic especially for latent tuberculosis and early disease. A positive bacterial culture – typically from induced sputum – is the gold standard but results are not obtainable for 2–6 weeks. Additionally, where there is a paucity of bacterial numbers, as occurs in infants, culture has low sensitivity. Exposure to environmental mycobacteria can confound tuberculin skin test results, as can BCG immunization [51]. One possible approach to diagnose tuberculosis is to identify serum biomarkers and this has been investigated in adults [12]. SELDI spectra were obtained from patients with culture proven tuberculosis (Gambia and Uganda) and controls (Gambia, Uganda, Angola and the UK). The initial phase of the study identified 20 peaks that could differentiate the groups with an impressive sensitivity of 94.4% and specificity of 91.8%. Two peaks were identified: transthyretin and SAA des Arg. Results from immunological assays of these two proteins were combined with that from measurements of neopterin and C-reactive protein, which had previously been used in tuberculosis diagnosis. When all four markers were considered in the analysis, sensitivities and specificities of >80% were obtained. In a second exclusively UK cohort, the four biomarkers could differentiate tuberculosis from controls with a sensitivity of 89% and a specificity of 74%.

11.4.2   Infectious Endocarditis

Infectious endocarditis is a serious disease which can be caused by a number of organisms, although staphylococci and oral streptococci predominate. Current diagnosis relies mainly on cardiac imaging and blood isolation of causative organisms. Diagnosis can be especially challenging when the fastidious nature of the causative organism or prior antibiotic therapy makes blood isolation difficult [52]. The question asked by Fenollar *et al.* [53] was: is a specific serum proteomic signature associated with infectious endocarditis? In a retrospective study it was possible to identify 66 peaks that differentiated infectious endocarditis patients from controls with a sensitivity and specificity of 81 and 94%, respectively. This study has parallels with that of intra-amniotic inflammation and preterm birth in that it is possible to identify a signature associated with, in infection terms, polymicrobial causes.

11.4.3   Respiratory Diseases

Chronic obstructive pulmonary disease (COPD) is responsible for substantial morbidity and mortality worldwide. The disease is characterized by progressive and irreversible decline in lung function resulting in reduced airflow in the lungs. COPD is an umbrella term covering two main diseases: bronchitis and emphysema. The biggest risk factor for COPD is smoking, although it is also associated with occupational exposure to fumes, chemicals and dusts. Disease progression is characterized by intermittent worsening of symptoms called acute exacerbations, which are typically associated with infections of either viral or bacterial origin, in particular nontypable *Haemophilus influenzae*. There is only one blood test that is used to determine whether a smoker is at risk for COPD, measurement of α-1 antitrypsin and phenotype. However, only a few patients are α-1 antitrypsin-deficient. Bowler *et al.* [54], therefore, evaluated whether SELDI could identify serum biomarkers that separate COPD and age, sex and smoking history controls. Five biomarkers could distinguish COPD patients from controls with a sensitivity and specificity of 91.7 and 88.3%, respectively. No biomarkers were identified using the Q10 ProteinChip. Ten-fold cross-validation resulted in both sensitivity and specificity of 81.7%. The authors concluded that substantial work was required for SELDI to become a clinical test. None of the peaks were identified.

Cystic fibrosis (CF) is a hereditary disease caused by a mutation(s) in a gene encoding the cystic fibrosis trans-membrane conductance regulator (CFTR), which is a chloride ion channel. Hallmarks of CF disease include recurrent and persistent bacterial infection. Early infections may be caused by a variety of bacterial pathogens such as *Staphylococcus aureus*, *H. influenzae* and nonmucoid *Pseudomonas aeruginosa*. However, mucoid *P. aeruginosa* infections predominate amongst older patients and these are associated with a substantial decline in lung function. Airway inflammation is predominantly associated with neutrophils. To further define the immune response, McMorran *et al.* [55] compared the SELDI profiles of induced sputum on H4 ProteinChips of children less than 5 years of age with and without CF experiencing endobronchial infection. Multiple samples were obtained from those children with CF and one from those without. Samples were assigned into four groups dependent on their CF status and the bacterial/fungal load of respiratory pathogens in their bronchial lavage fluid. Complete protease inhibitor cocktail (Roche) was added to bronchial lavage fluid samples prior to analysis. Twenty-four unique peak intensities were associated with CF, 11 of which were identified at the molecular level. These include the neutrophil-derived proteins DEFA1, DEFA2 and DEFA3. The latter was absent in a subset of patients consistent with a known genetic polymorphism that suppresses gene expression. Six were members of the S100A protein family, again neutrophil derived, with a number having known or unknown modifications. For example, two modified forms of S100A9 (calgranulin B) were found. One was missing the N-terminal methionine and had added acetylation at threonine 2, and the other was missing the first five N-terminal amino acids and had an added acetylation at serine 6. Novel truncated forms, missing their last two C-terminal amino acids, of S100A8 (calgranulin A) and S100A12 were detected. The final markers to be identified with CF patients were lysozyme and β-globin. ELISA (DEFA1-3) and western blotting (S100A8 and S100A9) were used to validate specific biomarkers. In the case of DEFA1-DEFA2, S100A8 and S100A12 there was a correlation between presence and neutrophil numbers. A biologically plausible contribution to CF for all of the biomarkers was hypothesized.

In a similar study, MacGregor *et al.* [56] compared the SELDI proteomic profiles of bronchial lavage fluid from children with CF or other respiratory disease (i.e. lower respiratory tract infection, chronic cough, primary ciliary dyskinesia, croup). Twelve different binding surfaces and wash conditions were used, and 202 proteins/peptides were differentially expressed in the CF samples. The most abundant biomarker was identified as S100A8 (calgranulin A), which was also found in induced sputum of CF patients [55]. The most discriminatory biomarker was at 5163 Da although this was not identified at the molecular level. There was no correlation between the 5163 Da peak and the neutrophil count in BALF, although there was a negative correlation with calgranulin A. Three markers at 10 186 (S100A12), 6214 and 21 066 Da, were positively correlated with neutrophil counts.

Another SELDI study involved comparison of the proteomic profiles of induced sputum from both COPD and CF patients, 18/28 of the latter were chronically infected with *P. aeruginosa* [57]. Additional groups were bronchiectasis, asthma and healthy controls. Longitudinal samples were taken from 12 CF patients during an infective exacerbation, including at onset and after the completion of intravenous antibiotic therapy. Samples were analysed on CM10, Q10 and IMAC ProteinChips to maximize the discovery of new biomarkers. There were a substantial number of differential peaks associated with all disease conditions compared with healthy controls. There was more similarity between SELDI profiles between the two obstructive diseases and between the two supparative diseases. Only 16 proteins (from the three different ProteinChip surfaces) were significantly different between the asthma and COPD patient groups. Biomarker peaks that differentiated each disease group from healthy controls were found at 105, 113, 381 and 377 Da for asthma, COPD, CF and bronchiectasis, respectively. A number of the peaks were identified, including calgranulin A–C, Clara cell secretory protein, proline rich salivary peptide, lysozyme C and cystatin S. CF and bronchiectasis were associated with high levels of calgranulins and confirmed by western blotting, including a truncated form of calgranulin A. No correlation between the amount of calgranulin A and neutrophil number was observed. Clara cell secretory protein was lower in abundance in all disease groups and this was confirmed by western blotting. Thus, SELDI shows promise in identifying novel biomarkers in respiratory disease that may lead to advances in our understanding in pathophysiology and may be useful for monitoring disease progression or treatment.

11.4.4   Intra-Amniotic Infection

Preterm birth is a major cause of morbidity and mortality worldwide. In the United States, one in eight births in 2003 were preterm and accounted for 70% of neonatal deaths and 75% of neonatal morbidity. Long-term handicaps include blindness, deafness, developmental delay, chronic lung disease and an increased risk of cerebral palsy. A number of factors have been associated with preterm birth, with intra-amniotic inflammation being of high significance. Infection and/or sepsis are major causes of intra-amniotic inflammation [58]. Organisms including *Fusobacterium nucleatum*, *Mycoplasma hominis* and *Ureaplasma* species have come under particular spotlight but many others have been implicated. There have been a number of studies, in particular from Buhimschi and colleagues, with amniotic fluid that aimed at identifying those at risk of preterm birth and understanding the pathophysiology involved. The first study [59] involving single pregnancies identified 13 markers of interest, four of which were identified at the molecular level: defensin-1, defensin-2, calgranulin A and calgranulin C. The four markers were used to formulate a mass-restricted (MR) score which ranged from 0 to 4 depending on the presence or absence of particular peaks. If a peak is present, it is given a value of 1 and if absent a value of 0. Values of 3–4 and 0–2 indicate that intra-amniotic inflammation is present and absent, respectively. There was a 100% correlation between a MR score of 3–4 and the presence of intra-amniotic inflammation. A subsequent twin study [60] showed that it was possible to predict the presence or absence of intra-amniotic inflammation in each of the independent amniotic sacs. Two biomarkers (calgranulin B and insulin-like growth factor binding protein 1) were identified as being up-regulated in both the amniotic fluid and maternal serum biomarkers in a rhesus monkey model of infection and also patients [61]. It should be noted that the latter study has not been without its critics, specifically regarding the identification of calgranulin B and the use of pooled maternal serum. An invasive and not without risk procedure is required to obtain amniotic fluid. This prompted the concurrent analysis of amniotic and cervical fluids for biomarkers [62]. Masses indicative of calgranulins A and C and defensins 1–3 were among the 17 peaks that were significantly raised in the amniotic fluid of patients with intra-amniotic inflammation. All 17 peaks were present in cognate cervical fluid samples but there was no significant increase in signal intensities.

A major paper in the field is that of Buhimschi and colleagues [58] who did a prospective blinded study to determine the reproducibility and accuracy of the MR score when compared with previously established or proposed clinical markers of intra-amniotic inflammation or infection. The clinical markers included glucose, white blood cell counts, lactate dehydrogenase, gram stain, IL-6 and matrix metalloprotease 8. Additionally, the relationship between the SELDI proteomic results and pregnancy outcome, pathologic evaluation of the placenta and early-onset neonatal sepsis were determined. The study involved investigation of 169 consecutive women with singleton pregnancies admitted with preterm labour or premature rupture of membranes. Shorter amniocentesis-to-delivery intervals were found for women who had an MR score of 3–4 compared with 0–2, although some inflammation was associated with preterm birth independent of membrane status. There were significant associations between the MR score and histopathological chorioamnionitis, haematology and early-onset sepsis. In comparison with standard biological tests, they found that the MR score was the most accurate in detecting inflammation (i.e. where the white cell blood count was >100 cells mm^{-3}). A combination of MR score and gram stain proved to be the best predictor of amniotic infection. A follow up study of 158 patients investigated the relationship between MR score and histologic chorioamniotis [63]. There was a significant correlation between the stage of histologic chorioamnionitis, the grade of choriodecidutis and amnionitis and MR score, with African-American women being overrepresented in the inflammation group. Calgranulin C was the most informative marker with regard to stage III chorioamnionitis, time to delivery and gestational age and independent of mode of delivery and exposure to antibiotics or steroids. Calgranulin C is part of the S100 protein family (S100A12) and a ligand for the receptor for advanced glycation end-products (RAGE) and is associated with neutrophils. The study provided further evidence that the S100A12/extracellularly newly identified

RAGE-binding protein (ENRAGE) system is important in intra-amniotic infection and inflammation as had previously been reported [64]. The relationship between MR score and foetal inflammatory status has also been assessed in a prospective observational cohort of 132 patients [65]. The ratio of IL-6 level in cord blood compared with amniotic fluid was used as an indicator of the differential inflammatory response in each compartment. Higher IL-6 levels were found in the amniotic fluid of women and in the cord blood of neonates at birth. Early onset neonatal sepsis was also associated with increased levels of IL-6 in cord blood. There was a correlation between the amniotic fluid–cord blood ratio and neutrophil influx in the chorionic plate, choriodecidua and umbilicus, but not the amnion. It was not possible to isolate bacteria from the amniotic fluid of approximately 30% of the women where the MR score suggested severe intra-amniotic inflammation. While infection is a major cause of intra-amniotic inflammation other causes such as bleeding are known. Another reason for the lack of positive bacterial culture, the gold standard for detecting intra-amniotic infection, is the presence of unculturable or difficult to cultivate bacteria. It should be noted that a wide variety of pathogens have been associated with intra-amniotic infection whether present singularly or in multiples. To address this question further, Han and colleagues [66] determined the presence of bacteria in amniotic fluid by culture and sequencing of 16S rRNA and this was related to the MR score. Amongst the preterm birth samples, bacterial DNA was amplified from all culture positive samples and 17% of culture negative samples. 16S rRNA analysis detected further species in the culture positive samples and there was evidence of intra-amniotic inflammation, as adjudged by the MR score, in all five samples which were culture negative but PCR positive.

Buhimschi *et al.* [67] have more recently devised a new ranking system based on the proteomic signature of amniotic fluid – the Q-profile – that is complementary to the MR score. The profile was based on five peaks in the 10–12.5 kDa mass range. The aim was to identify pathways related to preterm birth in the absence of inflammation or bleeding. Additional 2D-PAGE electrophoresis experiments and protein analysis through evolutionary relationships (PANTHER) analysis identified differentially expressed proteins not associated with inflammation pathways. The results suggest that in some patients there is a novel intra-amniotic inflammation-independent pathway that results in preterm birth. The study illustrates the powerful combination of proteomics and pathway analysis.

In a retrospective cross-sectional SELDI study, Romero *et al.* [68] investigated patients with spontaneous preterm labour who delivered at term and those that delivered preterm with intra-amniotic inflammation. A novel computational method which involved analysis of the combined data from two ProteinChip surfaces (CM10 and H50), two energy absorbing matrices and two laser intensities resulted in 69 spectral features corresponding to 39 *m/z* values. It was possible to predict preterm labour with a sensitivity and specificity of 91.7 and 91.5%, respectively. Some of the *m/z* values corresponded to those found in previous studies and used in the MR score.

11.4.5   Bacterial Peritonitis

Lin *et al.* [69] sought biomarkers of infection in peritoneal fluid of patients undergoing continuous ambulatory peritoneal dialysis. *Staphylococcus epidermidis*, *S. aureus* or *Klebsiella pneumoniae* were isolated from the peritoneal fluid of infected patients. A combination of 2D-PAGE and SELDI found an 11 117.4 Da peak that could differentiate infected patients from controls with an accuracy of 95%. The peak was identified as a peptide fragment derived from β-2-microglobulin (full length 11 731 Da) and confirmed by SELDI-based immunodepletion experiments. 2-microglobulin is an established peritoneal fluid biomarker for infection.

11.5   Other Diseases of Possible Infectious Origin

11.5.1   Kawasaki Disease

Kawasaki disease is characterized by inflammation of the blood vessels and predominantly affects children under 4 years of age, particularly those of Asian origin. The cause is unknown but is thought to either result from an infection or be an autoimmune disease stimulated by exposure to an environmental toxin or infection. It is the most common cause of acquired heart disease in children in developed countries. Carlson *et al.* [70] evaluated a pre-processing algorithm to analyse SELDI data derived from analysis of plasma samples obtained from either patients with Kawasaki disease or those with other febrile diseases. Three significant clusters were identified. The two most significant individual features of the most significant cluster had m/z values of 7586 and 7665 Da. It was hypothesized that the m/z 7665 Da peak was a post-translational modification of m/z 7586 Da. While the number of samples that was investigated was small, it suggests that SELDI may be useful in identifying biomarkers for Kawasaki disease.

11.6   Conclusions

The use of SELDI to identify biomarkers of infectious disease is growing as investigators exploit the small sample volume required, its high throughput capability, increased sensitivity and reproducibility of the new instruments – all facilitated by a greater understanding of the variables that can affect the quality of the proteomic signatures. Has SELDI delivered in the area of infectious disease? The answer is arguably not but SELDI shows great promise. Important advances have been made in diagnostic fields, e.g. for intra-amniotic inflammation a number of candidates have been identified that may increase our understanding of the pathophysiology of various infectious diseases. However, it is unlikely that SELDI will become the gold standard method for the diagnosis of particular infectious diseases worldwide on availability and cost grounds, particularly given the burden of infectious disease in developing countries. The exception may be where the diagnosis of a particular infectious disease is challenging in a developed world setting. SELDI is unquestionably useful in identifying biomarkers for inclusion in cheaper tests such as are required for some of the major infectious diseases in developing countries. There are still criticisms that the biomarkers identified are relatively high abundant proteins. However, immunological reagents that deplete the most abundant proteins in serum are readily commercially available and are useful in allowing less abundant proteins to be analysed. The use of ProteoMiner beads (BioRad) can also assist in this regard [71] (see Chapter 9). Certainly SELDI is capable of detecting small alterations in proteins that may be of diagnostic significance, e.g. SAA des Arg for tuberculosis diagnosis. Sample pretreatment methods allowing the 'sub-proteome' to be explored in more detail are rapidly advancing and we envisage that they will increase the usefulness of SELDI even further.

References

[1] Engwegen JY, Gast MC, Schellens JH, Beijnen JH. Clinical proteomics: searching for better tumour markers with SELDI-TOF mass spectrometry. Trends Pharmacol Sci 2006, 27, 251–9

[2] Seibert V, Ebert MP, Buschmann T. Advances in clinical cancer proteomics: SELDI-ToF-mass spectrometry and biomarker discovery. Brief Funct Genomic Proteomic 2005, 4, 16–26

[3] Sun S, Lee NP, Poon RT, Fan ST, He QY, Lau GK, Luk JM. Oncoproteomics of hepatocellular carcinoma: from cancer markers' discovery to functional pathways. Liver Int 2007, 27, 1021–38

[4] Buhimschi IA, Buhimschi CS, Weiner CP, Kimura T, Hamar BD, Sfakianaki AK, Norwitz ER, Funai EF, Ratner E. Proteomic but not enzyme-linked immunosorbent assay technology detects amniotic fluid monomeric calgranulins from their complexed calprotectin form. Clin Diagn Lab Immunol 2005, 12, 837–44

[5] Howard CT, McKakpo US, Quakyi IA, Bosompem KM, Addison EA, Sun K, Sullivan D, Semba RD. Relationship of hepcidin with parasitemia and anemia among patients with uncomplicated *Plasmodium falciparum* malaria in Ghana. Am J Trop Med Hyg 2007, 77, 623–6

[6] Fujita N, Sugimoto R, Motonishi S, Tomosugi N, Tanaka H, Takeo M, Iwasa M, Kobayashi Y, Hayashi H, Kaito M, Takei Y. Patients with chronic hepatitis C achieving a sustained virological response to peginterferon and ribavirin therapy recover from impaired hepcidin secretion. J Hepatol 2008, 49, 702–10

[7] Hodgetts A, Bossé JT, Kroll JS, Langford PR. Analysis of differential protein expression in *Actinobacillus pleuropneumoniae* by Surface Enhanced Laser Desorption Ionisation-ProteinChip (SELDI) technology. Vet Microbiol 2004, 99, 215–25

[8] Kiehntopf M, Siegmund R, Deufel T. Use of SELDI-TOF mass spectrometry for identification of new biomarkers: potential and limitations. Clin Chem Lab Med 2007, 45, 1435–49

[9] Villar-Garea A, Griese M, Imhof A. Biomarker discovery from body fluids using mass spectrometry. J Chromatogr B 2007, 849, 105–14

[10] Carrette O, Burkhard PR, Hughes S, Hochstrasser DF, Sanchez JC. Truncated cystatin C in cerebrospiral fluid: Technical artefact or biological process? Proteomics 2005, 5, 3060–5

[11] Del Boccio P, Pieragostino D, Lugaresi A, Di Ioia M, Pavone B, Travaglini D, D'Aguanno S, Bernardini S, Sacchetta P, Federici G, Di Ilio C, Gambi D, Urbani A. Cleavage of cystatin C is not associated with multiple sclerosis. Ann Neurol 2007, 62, 201–4

[12] Agranoff D, Fernandez-Reyes D, Papadopoulos MC, Rojas SA, Herbster M, Loosemore A, Tarelli E, Sheldon J, Schwenk A, Pollok R, Rayner CF, Krishna S. Identification of diagnostic markers for tuberculosis by proteomic fingerprinting of serum. Lancet 2006, 368, 1012–21

[13] Pang RT, Poon TC, Chan KC, Lee NL, Chiu RW, Tong YK, Wong RM, Chim SS, Ngai SM, Sung JJ, Lo YM. Serum proteomic fingerprints of adult patients with severe acute respiratory syndrome. Clin Chem 2006, 52, 421–9

[14] Papadopoulos MC, Abel PM, Agranoff D, Stich A, Tarelli E, Bell BA, Planche T, Loosemore A, Saadoun S, Wilkins P, Krishna S. A novel and accurate diagnostic test for human African trypanosomiasis. Lancet 2004, 363, 1358–63

[15] Wu C, Wang Z, Liu L, Zhao P, Wang W, Yao D, Shi B, Lu J, Liao P, Yang Y, Zhu L. Surface enhanced laser desorption/ionization profiling: New diagnostic method of HBV-related hepatocellular carcinoma. J Gastroenterol Hepatol 2009, 24, 55–62

[16] Zhang J, Li D, Zheng Y, Cui Y, Feng K, Zhou J, Wu J. Proteomic profiling of hepatitis B virus-related hepatocellular carcinoma in China: a SELDI-TOF-MS study. Int J Clin Exp Pathol 2008, 1, 352–61

[17] El-Aneed A, Banoub J. Proteomics in the diagnosis of hepatocellular carcinoma: focus on high risk hepatitis B and C patients. Anticancer Res 2006, 26, 3293–300

[18] Vivekanandan P, Singh OV. High-dimensional biology to comprehend hepatocellular carcinoma. Expert Rev Proteomics 2008, 5, 45–60

[19] Wright LM, Kreikemeier JT, Fimmel CJ. A concise review of serum markers for hepatocellular cancer. Cancer Detect Prev 2007, 31, 35–44

[20] Hodgetts A, Levin M, Kroll JS, Langford PR. Biomarker discovery in infectious diseases using SELDI. Future Microbiol 2007, 2, 35–49

[21] Lee IN, Chen CH, Sheu JC, Lee HS, Huang GT, Chen DS, Yu CY, Wen CL, Lu FJ, Chow LP. Identification of complement C3a as a candidate biomarker in human chronic hepatitis C and HCV-related hepatocellular carcinoma using a proteomics approach. Proteomics 2006, 6, 2865–73

[22] Paradis V, Asselah T, Dargere D, Ripault MP, Martinot M, Boyer N, Valla D, Marcellin P, Bedossa P. Serum proteome to predict virologic response in patients with hepatitis C treated by pegylated interferon plus ribavirin. Gastroenterology 2006, 130, 2189–97

[23] Poon TC, Chan KC, Ng PC, Chiu RW, Ang IL, Tong YK, Ng EK, Cheng FW, Li AM, Hon EK, Fok TF, Lo YM. Serial analysis of plasma proteomic signatures in pediatric patients with severe acute respiratory syndrome and correlation with viral load. Clin Chem 2004, 50, 1452–5

[24] Poon TC, Hui AY, Chan HL, Ang IL, Chow SM, Wong N, Sung JJ. Prediction of liver fibrosis and cirrhosis in chronic hepatitis B infection by serum proteomic fingerprinting: a pilot study. Clin Chem 2005, 51, 328–35

[25] Schwegler EE, Cazares L, Steel LF, Adam BL, Johnson DA, Semmes OJ, Block TM, Marrero JA, Drake RR. SELDI-TOF MS profiling of serum for detection of the progression of chronic hepatitis C to hepatocellular carcinoma. Hepatology 2005, 41, 634–42

[26] Ward DG, Cheng Y, N'Kontchou G, Thar TT, Barget N, Wei W, Billingham LJ, Martin A, Beaugrand M, Johnson PJ. Changes in the serum proteome associated with the development of hepatocellular carcinoma in hepatitis C-related cirrhosis. Br J Cancer 2006, 94, 287–92

[27] Ward DG, Cheng Y, N'Kontchou G, Thar TT, Barget N, Wei W, Martin A, Beaugrand M, Johnson PJ. Preclinical and post-treatment changes in the HCC-associated serum proteome. Br J Cancer 2006, 95, 1379–83

[28] Zhu XD, Zhang WH, Li CL, Xu Y, Liang WJ, Tien P. New serum biomarkers for detection of HBV-induced liver cirrhosis using SELDI protein chip technology. World J Gastroenterol 2004, 10, 2327–9

[29] Cui JF, Liu YK, Zhou HJ, Kang XN, Huang C, He YF, Tang ZY, Uemura T. Screening serum hepatocellular carcinoma-associated proteins by SELDI-based protein spectrum analysis. World J Gastroenterol 2008, 14, 1257–62

[30] Kanmura S, Uto H, Kusumoto K, Ishida Y, Hasuike S, Nagata K, Hayashi K, Ido A, Stuver SO, Tsubouchi H. Early diagnostic potential for hepatocellular carcinoma using the SELDI ProteinChip system. Hepatology 2007, 45, 948–56

[31] He M, Qin J, Zhai R, Wei X, Wang Q, Rong M, Jiang Z, Huang Y, Zhang Z. Detection and identification of NAP-2 as a biomarker in hepatitis B-related hepatocellular carcinoma by proteomic approach. Proteome Sci 2008, 6, 10

[32] He QY, Zhu R, Lei T, Ng MY, Luk JM, Sham P, Lau GK, Chiu JF. Toward the proteomic identification of biomarkers for the prediction of HBV related hepatocellular carcinoma. J Cell Biochem 2008, 103, 740–52

[33] Kang X, Xu Y, Wu X, Liang Y, Wang C, Guo J, Wang Y, Chen M, Wu D, Bi S, Qiu Y, Lu P, Cheng J, Xiao B, Hu L, Gao X, Liu J, Song Y, Zhang L, Suo F, Chen T, Huang Z, Zhao Y, Lu H, Pan C, Tang H. Proteomic fingerprints for potential application to early diagnosis of severe acute respiratory syndrome. Clin Chem 2005, 51, 56–64

[34] Yip TT, Chan JW, Cho WC, Wang Z, Kwan TL, Law SC, Tsang DN, Chan JK, Lee KC, Cheng WW, Ma VW, Yip C, Lim CK, Ngan RK, Au JS, Chan A, Lim WW. Protein chip array profiling analysis in patients with severe acute respiratory syndrome identified serum amyloid A protein as a biomarker potentially useful in monitoring the extent of pneumonia. Clin Chem 2005, 51, 47–55

[35] Pang RT, Poon TC, Chan KC, Lee NL, Chiu RW, Tong YK, Chim SS, Sung JJ, Lo YM. Serum amyloid A is not useful in the diagnosis of severe acute respiratory syndrome. Clin Chem 2006, 52, 1202–4

[36] Luo X, Carlson KA, Wojna V, Mayo R, Biskup TM, Stoner J, Anderson J, Gendelman HE, Melendez LM. Macrophage proteomic fingerprinting predicts HIV-1-associated cognitive impairment. Neurology 2003, 60, 1931–7

[37] Everall I, Grant I. Proteomic pointers in HIV neurocognitive disorder. Lancet 2004, 363, 1091–2

[38] Sun B, Rempel HC, Pulliam L. Loss of macrophage-secreted lysozyme in HIV-1-associated dementia detected by SELDI-TOF mass spectrometry. AIDS 2004, 18, 1009–12

[39] Ratto-Kim S, Chuenchitra T, Pulliam L, Paris R, Sukwit S, Gongwon S, Sithinamsuwan P, Nidhinandana S, Thitivichianlert S, Shiramizu BT, de Souza MS, Chitpatima ST, Sun B, Rempel H, Nitayaphan S, Williams K, Kim JH, Shikuma CM, Valcour VG. Expression of monocyte markers in HIV-1 infected individuals with or without HIV associated dementia and normal controls in Bangkok Thailand. J Neuroimmunol 2008, 195, 100–7

[40] Irani DN, Anderson C, Gundry R, Cotter R, Moore S, Kerr DA, McArthur JC, Sacktor N, Pardo CA, Jones M, Calabresi PA, Nath A. Cleavage of cystatin C in the cerebrospinal fluid of patients with multiple sclerosis. Ann Neurol 2006, 59, 237–47

[41] Semmes OJ, Cazares LH, Ward MD, Qi L, Moody M, Maloney E, Morris J, Trosset MW, Hisada M, Gygi S, Jacobson S. Discrete serum protein signatures discriminate between human retrovirus-associated hematologic and neurologic disease. Leukemia 2005, 19, 1229–38

[42] Jahnukainen T, Malehorn D, Sun M, Lyons-Weiler J, Bigbee W, Gupta G, Shapiro R, Randhawa PS, Pelikan R, Hauskrecht M, Vats A. Proteomic analysis of urine in kidney transplant patients with BK virus nephropathy. J Am Soc Nephrol 2006, 17, 3248–56

[43] Schaub S, Wilkins JA, Nickerson P. Proteomics and renal transplantation: searching for novel biomarkers and therapeutic targets. Contrib Nephrol 2008, 160, 65–75

[44] Wittke S, Haubitz M, Walden M, Rohde F, Schwarz A, Mengel M, Mischak H, Haller H, Gwinner W. Detection of acute tubulointerstitial rejection by proteomic analysis of urinary samples in renal transplant recipients. Am J Transplant 2005, 5, 2479–88

[45] Liu Z, Tian Y, Wang B, Yan Z, Qian D, Ding S, Song X, Bai Z, Li L. Serum proteomics with SELDI-TOF-MS in congenital human cytomegalovirus hepatitis. J Med Virol 2007, 79, 1500–5

[46] Cho JG, Dee SA. Porcine reproductive and respiratory syndrome virus. Theriogenology 2006, 66, 655–62

[47] Genini S, Cantu M, Botti S, Malinverni R, Costa A, Marras D, Giuffra E. Diagnostic markers for diseases: SELDI-TOF profiling of pig sera for PRRS. Dev Biol (Basel) 2008, 132, 399–403

[48] Marcos LA, Terashima A, Gotuzzo E. Update on hepatobiliary flukes: fascioliasis, opisthorchiasis and clonorchiasis. Curr Opin Infect Dis 2008, 21, 523–30

[49] Rioux MC, Carmona C, Acosta D, Ward B, Ndao M, Gibbs BF, Bennett HP, Spithill TW. Discovery and validation of serum biomarkers expressed over the first twelve weeks of *Fasciola hepatica* infection in sheep. Int J Parasitol 2008, 38, 123–36

[50] Sadiq ST, Agranoff D. Pooling serum samples may lead to loss of potential biomarkers in SELDI-ToF MS proteomic profiling. Proteome Sci 2008, 6, 16

[51] Newton SM, Brent AJ, Anderson S, Whittaker E, Kampmann B. Paediatric tuberculosis. Lancet Infect Dis 2008, 8, 498–510

[52] Lang S. Getting to the heart of the problem: serological and molecular techniques in the diagnosis of infective endocarditis. Future Microbiol 2008, 3, 341–9

[53] Fenollar F, Goncalves A, Esterni B, Azza S, Habib G, Borg JP, Raoult D. A serum protein signature with high diagnostic value in bacterial endocarditis: results from a study based on surface-enhanced laser desorption/ionization time-of-flight mass spectrometry. J Infect Dis 2006, 194, 1356–66

[54] Bowler RP, Canham ME, Ellison MC. Surface enhanced laser desorption/ionization (SELDI) time-of-flight mass spectrometry to identify patients with chronic obstructive pulmonary disease. COPD 2006, 3, 41–50

[55] McMorran BJ, Patat SA, Carlin JB, Grimwood K, Jones A, Armstrong DS, Galati JC, Cooper PJ, Byrnes CA, Francis PW, Robertson CF, Hume DA, Borchers CH, Wainwright CE, Wainwright BJ. Novel neutrophil-derived proteins in bronchoalveolar lavage fluid indicate an exaggerated inflammatory response in pediatric cystic fibrosis patients. Clin Chem 2007, 53, 1782–91

[56] MacGregor G, Gray RD, Hilliard TN, Imrie M, Boyd AC, Alton EW, Bush A, Davies JC, Innes JA, Porteous DJ, Greening AP. Biomarkers for cystic fibrosis lung disease: application of SELDI-TOF mass spectrometry to BAL fluid. J Cyst Fibros 2008, 7, 352–8

[57] Gray RD, MacGregor G, Noble D, Imrie M, Dewar M, Boyd AC, Innes JA, Porteous DJ, Greening AP. Sputum proteomics in inflammatory and suppurative respiratory diseases. Am J Respir Crit Care Med 2008, 178, 444–52

[58] Buhimschi CS, Bhandari V, Hamar BD, Bahtiyar MO, Zhao G, Sfakianaki AK, Pettker CM, Magloire L, Funai E, Norwitz ER, Paidas M, Copel JA, Weiner CP, Lockwood CJ, Buhimschi IA. Proteomic profiling of the amniotic fluid to detect inflammation, infection, and neonatal sepsis. PLoS Med 2007, 4, e18

[59] Buhimschi IA, Christner R, Buhimschi CS. Proteomic biomarker analysis of amniotic fluid for identification of intra-amniotic inflammation. BJOG 2005, 112, 173–81

[60] Buhimschi IA, Buhimschi CS, Christner R, Weiner CP. Proteomics technology for the accurate diagnosis of inflammation in twin pregnancies. BJOG 2005, 112, 250–5

[61] Gravett MG, Novy MJ, Rosenfeld RG, Reddy AP, Jacob T, Turner M, McCormack A, Lapidus JA, Hitti J, Eschenbach DA, Roberts CT, Jr., Nagalla SR. Diagnosis of intra-amniotic infection by proteomic profiling and identification of novel biomarkers. JAMA 2004, 292, 462–9

[62] Ruetschi U, Rosen A, Karlsson G, Zetterberg H, Rymo L, Hagberg H, Jacobsson B. Proteomic analysis using protein chips to detect biomarkers in cervical and amniotic fluid in women with intra-amniotic inflammation. J Proteome Res 2005, 4, 2236–42

[63] Buhimschi IA, Zambrano E, Pettker CM, Bahtiyar MO, Paidas M, Rosenberg VA, Thung S, Salafia CM, Buhimschi CS. Using proteomic analysis of the human amniotic fluid to identify histologic chorioamnionitis. Obstet Gynecol 2008, 111, 403–12

[64] Buhimschi IA, Zhao G, Pettker CM, Bahtiyar MO, Magloire LK, Thung S, Fairchild T, Buhimschi CS. The receptor for advanced glycation end products (RAGE) system in women with intraamniotic infection and inflammation. Am J Obstet Gynecol 2007, 196, 181 e1–13

[65] Buhimschi CS, Dulay AT, Abdel-Razeq S, Zhao G, Lee S, Hodgson EJ, Bhandari V, Buhimschi IA. Fetal inflammatory response in women with proteomic biomarkers character-istic of intra-amniotic inflammation and preterm birth. BJOG 2009, 116, 257–67

[66] Han YW, Shen T, Chung P, Buhimschi IA, Buhimschi CS. Uncultivated bacteria as etio-logic agents of intra-amniotic inflammation leading to preterm birth. J Clin Microbiol 2009, 47, 38–47

[67] Buhimschi IA, Zhao G, Rosenberg VA, Abdel-Razeq S, Thung S, Buhimschi CS. Multidimen-sional proteomics analysis of amniotic fluid to provide insight into the mechanisms of idio-pathic preterm birth. PLoS ONE 2008, 3, e2049

[68] Romero R, Espinoza J, Rogers WT, Moser A, Nien JK, Kusanovic JP, Gotsch F, Erez O, Gomez R, Edwin S, Hassan SS. Proteomic analysis of amniotic fluid to identify women with preterm labor and intra-amniotic inflammation/infection: the use of a novel computational method to analyze mass spectrometric profiling. J Matern Fetal Neonatal Med 2008, 21, 367–88

[69] Lin WT, Tsai CC, Chen CY, Lee WJ, Su CC, Wu YJ. Proteomic analysis of peritoneal dialysate fluid in patients with dialysis-related peritonitis. Ren Fail 2008, 30, 772–7

[70] Carlson SM, Najmi A, Cohen HJ. Biomarker clustering to address correlations in proteomic data. Proteomics 2007, 7, 1037–46

[71] Sihlbom C, Kanmert I, Bahr H, Davidsson P. Evaluation of the combination of bead technology with SELDI-TOF-MS and 2-D DIGE for detection of plasma proteins. J Proteome Res 2008, 7, 4191–8

12

MALDI-TOF MS for Microbial Identification: Years of Experimental Development to an Established Protocol

Wibke Kallow[1], Marcel Erhard[1], Haroun N. Shah[2], Emmanuel Raptakis[3] and Martin Welker[1]

[1]*AnagnosTec GmbH, Potsdam, Germany*
[2]*Department for Bioanalysis and Horizon Technologies, Health Protection Agency Centre for Infections, London, UK*
[3]*Shimadzu Biotech-Kratos Analytical, Manchester, UK*

12.1 Identification of Microorganisms in Clinical Routine

A crucial step in the epidemiology and the successful therapy of any infectious disease is the identification of the causative microbe. For more than a century, clinical microbiology has relied on the isolation of the suspected pathogen from various samples such as stools, throat swabs, blood, or urine on selective growth media and an identification procedure that is based on the metabolic capacities of the isolate. An array of carbohydrate fermentation and enzymatic reactions are tested that generally involve a colour change of an indicator when a particular substrate is catabolised. The profile of positive and negative reactions is assumed to be characteristic for a bacterial taxon and is consequently used for identification. Modern microbial identification systems are miniaturised, combining some tens of reactions into a single strip or card to allow for high throughput analysis. The major shortcomings of these systems are the need to incubate isolates for several hours to obtain pure cultures, and a required pre-selection of tests. Although this

Mass Spectrometry for Microbial Proteomics   Edited by Haroun N. Shah and Saheer E. Gharbia
© 2010 John Wiley & Sons, Ltd

is still the most commonly used method in clinical diagnostic laboratories, microbiologists have been seeking alternative methods for the identification of pathogens for decades.

A new era has dawned with the arrival of molecular methods such as the polymerase chain reaction (PCR) and nucleotide sequence analysis. In diagnostic and systematic microbiology today, analysis of genomic sequences is rapidly displacing biochemical tests for the provision of new characters for the circumscription of taxa. For example, a prerequisite for the description of a new species is the inclusion of the sequence of the 16S rRNA gene which now plays a pivotal role in microbial phylogeny. However, despite the widespread use of PCR and sequencing in all fields of microbiology, the technology is still lagging behind in clinical microbiology and is largely restricted to research applications. On the other hand, the high sensitivity and specificity of molecular methods make them indispensable in modern microbiological laboratories, as for example in the detection of methicillin resistant *Staphylococcus aureus* (MRSA) by real-time PCR assays, or the identification of atypical or very rare pathogens.

12.2   Mass Spectrometry and Microbiology

The application of chemical analyses (referred to as chemotaxonomy) for the identification and classification of microorganisms has been explored extensively prior to molecular analysis. These were based on the characterisation of polar (e.g. phospholipids) and non-polar lipids such as respiratory quinones (e.g. ubiquinones and menaquinones) and long-chained cellular fatty acids.[1] The structure of these lipids were challenging and ushered in a period of intense mass spectral analysis to characterise the vast array of lipids present in the microbial kingdom. While these methods provided characters at the genus and species levels, pyrolysis mass spectrometry was introduced as a means of typing bacterial isolates.[2] Such approaches were motivated by the need for a rapid method to identify pathogens in only a fraction of the time required for biochemical tests. However, because of the limitations of the technology at that time, mass spectral approaches were confined to the detection of organic molecules in a mass range up to 1500 Da[3] (see also Chapter 1). Detection of larger molecules was hitherto only possible with techniques such as plasma desorption mass spectrometry.[4] This was about to change dramatically within a few years with the invention of matrix assisted laser desorption/ionisation time-of-flight mass spectrometry (MALDI-TOF MS) and its success in many fields of life sciences has been phenomenal.

The laser desorption and mass spectral analysis of large biomolecules was developed simultaneously by two research groups in Japan and Germany. While the group of Tanaka could successfully detect proteins up to a mass to charge ratio (m/z) of 100000 Da by direct laser ionisation,[5] Karas and Hillenkamp[6] relied on a light absorbing matrix, and achieved similar results by this method. A matrix effect on the desorption rate was observed earlier for smaller molecules[7] and further studies quickly directed the search for matrix candidates to a few small organic molecules that are still the primarily used ones in MALDI-TOF MS applications today, namely cinnamic acid derivatives[8] and 2,5-dihydrobenzoic acid.[9] Accuracy and resolution could be significantly improved by the introduction of delayed ion extraction,[10] which compensates for the variability in initial ion velocity.[11] The sensitivity of MALDI-TOF MS for the detection of large proteins was rapidly increased to the femtomolar range[12] and further, aided by sophisticated handling procedures to the zeptomolar range.[13] Within less than a decade, MALDI-TOF MS developed into a widely applied methodology in diverse fields of life sciences,[14] promoted by a number of major advantages it had compared with other mass spectral technologies. These included the possibility to detect unfragmented large molecules, the speed with which a full scan over a wide mass range can be achieved, and the simplicity of sample preparation. For the analysis of whole cells and crude extracts, an outstanding advantage of MALDI compared with other ionisation techniques such as electrospray ionisation (ESI) or fast atom bombardment (FAB), is the fact that in MALDI-TOF MS predominantly singly charged ions are detected[15] which simplified the interpretation and treatment of mass spectral data considerably.[16]

12.3　Mass Spectral 'Fingerprints' of Whole Cells

The possibility of introducing whole cells into a mass spectrometer and detecting biomolecules in a mass range extending to several kilodaltons was immediately recognised by microbiologists. Shortly after the first MALDI-TOF mass spectrometer was commercially available (by Vestec and Kratos) the first reports on intact cell mass spectrometry (ICMS) of bacteria were published, some of which highlighted its potential for microbial diagnostics.[17–21] Essentially these studies showed that mass spectra of whole cells of bacterial strains revealed patterns of mass signals that were reproducible and specific for strains or species. Because cells could be analysed after minimal processing and preparation, required only minute biomass, either as a cell suspension or cells placed directly on a target plate, the implications for diagnostic microbiology were immediately evident and numerous studies ensued.

The mass ranges that were selected in early studies varied from m/z 500–2200[18] to 2000–20000.[22] In the lower mass range, constituents of the cell wall are detected and also, but to a lesser degree, protein components.[23] The lower mass range was also used for the typing of potentially toxic cyanobacteria[24] and proved to be suitable for the metabolic typing of sub-specific taxonomic units in natural populations.[25]

The possibility that mass fingerprints could be used to classify and eventually identify unknown microbial isolates was pursued in the following years by a number of groups. Most studies focused on the detection of proteins in mass ranges spanning from above m/z 2000 to below m/z 25000. The identity of proteins detected at this time was not very clear because only a few genomic sequences of microorganisms were available. It was generally assumed that the proteins desorbed from whole cells, that were not subjected to typical mechanical or chemical lysis, were attached to the cell surface.[23] However, studies on membrane-associated proteins of *Escherichia coli* K12, revealed that the molecular mass of most of these proteins exceeds 20 kDa[26].

Comparative MALDI-TOF mass spectral studies of isolated ribosomal subunits with those of whole cell mass spectra of *E. coli* K12 revealed that they shared a large number of mass signals that are commensurate with ribosomal proteins[27] (Figure 12.1). This large number of ionised ribosomal proteins is in accord with the high level of these proteins (ca. 30% of total proteins) in a cell in exponential growth phase. Furthermore, studies on isolated ribosomes also revealed that many proteins were modified posttranslationally, and that the observed mass signals were in agreement with the general rules for methionine cleavage.[28] Such studies emphasised that direct 'translation' of genomic data in mass patterns is not straightforward. With the rapidly increasing number of available complete

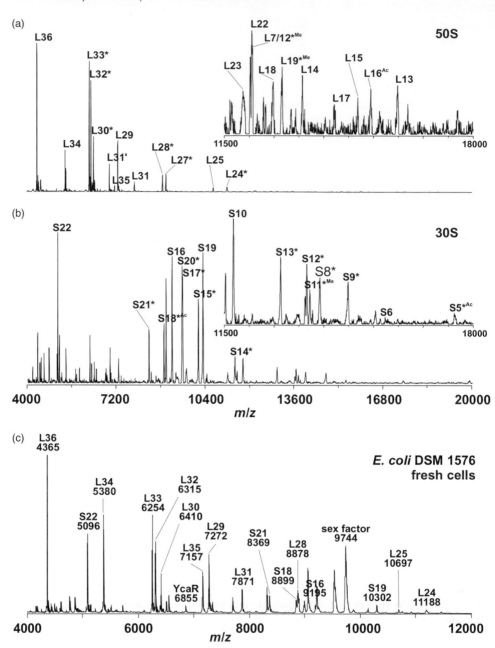

Figure 12.1   *MALDI-TOF mass spectra of isolated ribosomes of Escherichia coli K12 and of fresh cells directly prepared on the sample target. Mass spectra of 50S (a) and 30S (b) ribosomal subunits were recorded on a Voyager DE Pro system in 2000 while fresh cells (c) were analysed on a Shimadzu AXIMA CFRplus in 2007. Ribosomal proteins are indicated by L (large 50S subunit) or S (small 30S subunit) and numbers. In (a) and (b) the inserts show an enlarged mass range as indicated. Asterisks indicate an N-terminal methionine cleavage, superscript 'Me' a single methylation and superscript 'Ac' an acetylation.[28] YcaR refers to a noncharacterised protein and 'sex factor' to an F-plasmid related protein[35]*

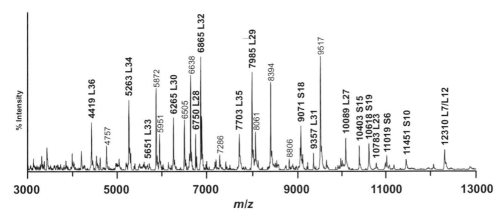

Figure 12.2   *Whole cell MALDI-TOF mass spectrum of a clinical isolate of* Streptococcus pneumoniae. *Ribosomal proteins are indicated by their numbers referring to the small 30S subunit (S) or the large 50S subunit (L) proteins. Assignment was done based on comparison of protein masses obtained from translated genomic sequences of multiple* Str. pneumoniae *strains available from the SwissProt website and taking into consideration possible posttranslational modifications by N-terminal methionine cleavage and/or monomethylation*

genomes the identification of ribosomal proteins in mass spectra of many bacterial species becomes feasible[29] and in combination with bioinformatical tools can be expanded even to unsequenced species.[30]

Further microbial taxa have been studied in detail to reveal the identity of proteins detected by whole cell mass spectrometry. A high level of observed peaks in mass spectra could be assigned to ribosomal proteins in *Helicobacter pylori*,[29] *Saccharomyces cerevisiae* (together with mitochondrion related proteins),[31] *Campylobacter jejuni*,[32] *Lactobacillus plantarum*[33] and *Pseudomonas putida*.[34] With the growing number of completed genomes in databases, the assignment of mass signals to proteins will be extended to further species. In Figure 12.2, a mass spectrum of a clinical isolate of *Streptococcus pneumoniae* is shown with the mass signals matching to theoretical masses of ribosomal proteins obtained from the SwissProt database. Other proteins identified in whole cell mass spectra include for example, DNA-binding proteins[32] and a protein in *E. coli* that correlates with the possession of an F-plasmid (*m/z* 9743).[35] The latter example also underlines the limits of *in silico* studies, since no candidate among some 30 proteins encoded on the F-plasmid had a sequence matching the observed mass signal when considering only modifications such as methionine cleavage or methylation. The *in vivo* protein could be further modified for example by cyclisation or formation of disulfide bridges, or cleaved from a larger premature protein.

The detection of ribosomal and other small structural proteins (i.e. proteins without a direct metabolic function) in whole cell MALDI-TOF mass spectra of various microbial taxa is very promising for the development of identification systems for clinical applications. To this end, two major objectives need to be addressed. Firstly, the reproducibility of mass fingerprints acquired under various culture conditions, on different instruments, or from multiple strains of a single species and, secondly, the discriminating power of mass fingerprints among recognised species.

12.4   Reproducibility of Mass Spectral Fingerprints

A principle problem for the development of a standard protocol using mass spectrometry is the fact that even replicate analyses of a single clone do not result in identical mass fingerprints. This applies to the individual peak's height as well as to the presence and absence of peaks, primarily of those with low intensities. The reproducibility of mass spectral fingerprints therefore has been the subject of a number of studies assessing biological, physical, and chemical effects on whole cell mass spectra. Wang *et al.*[36] demonstrated that variations in the sample preparation procedures and in the experimental conditions resulted in a significant change in the observed spectra, especially in relative peak intensities. Thus, a number of variable peaks were observed in the spectra from the same bacterial sample obtained under different conditions, especially when acidic extraction was employed. Similar observations were made by Williams *et al.*[37] who also reported that acidification of the matrix solution with 1–2.5% trifluoroacetic acid (TFA) yielded the best results in terms of peak number and reproducibility with different matrices. The shape of recorded peaks is itself dependent on a number of factors such as sample concentration and laser intensity, and consequently these factors need to be optimised for a given instrument.[38]

Since peak recognition partly depends on the software settings rather than on the hardware components, a sample preparation procedure that yields reproducible results on one mass spectrometer is likely to be suitable for others. This was assessed in an interlaboratory study where a single sample was analysed in three laboratories on different mass spectrometers.[39] Again, a number of mass signals were invariably recorded in all fingerprint spectra, while others were recorded with differing frequency in different laboratories. Thus, to a certain degree differences in whole cell mass spectra of a single bacterial sample can be attributed to physical and chemical factors. This is important to note since spectrum-to-spectrum variability could be mistakenly taken to indicate biological differences.

Biological variability is expected in whole cell mass spectral fingerprints due to the dynamic regulation of protein expression in response to growth conditions and culture age. Thus, for *E. coli* it was shown that mass spectral profiles changed with cultivation time.[40] Assuming that the mass ions of the ribosomal proteins are recorded in mass spectral fingerprints of many species, it is expected that the signal intensity of these proteins will be highest in the exponential growth phase. Figure 12.3 shows the mass spectra of whole cells of *Proteus mirabilis* recorded after cultivation times varying between 16 h and 120 h. Although the quality of the mass fingerprint declines considerably after 48 h (in terms of peak number and resolution) the major mass signals were still recorded and allowed automated identification (see below). However, our own observations have shown that the effect of culture conditions on mass spectral fingerprints is less pronounced than would be expected considering the complex system of protein expression in living cells. This has been also confirmed by Valentine *et al.*[41] who examined the mass spectra of several species using different culture media and Wunschel *et al.*[42] who additionally analysed cells grown at different temperatures, pH and growth rates.

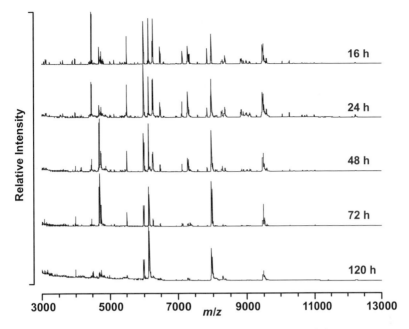

Figure 12.3 *Whole cell MALDI-TOF mass spectra of* Proteus mirabilis *DSM 4479 acquired after the cultivation times indicated. The strain was grown on Columbia blood agar at 37 °C and shows a remarkably stable profile in spite of gross changes taking place within the cell during various phases of growth*

12.5 Species and Strain Discrimination by Mass Spectrometry

Early studies that explored the use of mass spectral fingerprints tested the discriminating power on rather distantly related taxa, however, as confidence grew in the potential of the technique, more closely related taxa were studied (Table 12.1). *Bacillus* spores and cells have been studied intensively because of their potential use as bioterrorism agents and the need for a fast, safe and reproducible method to identify them in suspicious samples. It proved that with MALDI-TOF MS (and other mass spectral techniques) cells and spores of one of the most dangerous species, *B. anthracis* and the generally inseparable species, *B. cereus*, could be differentiated.[43–45] Reliable biomarkers were reported as 'small acid soluble proteins' (SASPs) that were characteristic of a number of strains[46] and had the potential for identification of cells and spores.[47,48]

Lynn *et al.*[49] studied a number of Enterobacteriaceae of different genera and showed that a small number of mass signals were recorded for all strains, e.g. *m/z* 4364 (ribosomal protein L36), 5380 (ribosomal protein L34) and 6856 (YcaR). For particular genera, species and strains, specific mass signals could be identified that allowed the discrimination of *E. coli, Klebsiella pneumoniae, Providencia rettgeri*, and *Salmonella typhimurium*. Such studies illustrated that mass spectral fingerprints can potentially be used for the classification of bacterial strains. Further studies showed that environmental *E. coli* isolates

Table 12.1 Compilation of studies on mass spectral analysis of microbial taxa and the potential to differentiate closely related species based on mass spectral fingerprints

Taxon	Comment	Ref.
Prokaryotes		
Archeae	Characterisation of archeae and extremophile bacteria	95
Eubacteria		
Aeromonas sp.	Genus-specific mass signals and discrimination of species	96
Enterobacteriaceae	Family- and genus-specific biomarkers	49
Enterobacteriaceae	Analysis after on-slide treatment with trypsin	97
E. coli	Discrimination of ampicillin-resistant and -susceptible strains	98
E. coli	Differentiation of environmental isolates by host origin	50
Salmonella sp.	Subspecies differentiation	99
Erwinia sp.	Typing of subspecies	81
Yersinia sp.	Mass spectral typing after inactivation with 80% TFA	62
Vibrio sp.	Differentiation of clinical and veterinary species	100
Legionella sp.	Species identification	101
Coxiella burnetii	Identification in different growth phases	102
Moraxella catarrhalis	Mass spectral typing consistent with genotyping	103
Francisella tularensis	SELDI-TOF MS discrimination of subspecies	60
Pseudomonas putida	Mass spectral classification consistent with genotyping	34
Haemophilus sp.	Discrimination of pathogenic and nonpathogenic species	104
Burkholderia sp.	Discrimination of *B. cepacia*-complex species	70, 91
Campylobacter sp.	Characterisation of protein biomarkers in three species	105
Campylobacter sp.	Discrimination of six species; identification of biomarkers	106
Campylobacter jejuni	Subspeciation; identification of biomarkers	107
Helicobacter pylori	Species- and strain-specific biomarkers	108
Arthrobacter sp.	Strain level differentiation	109
Mycobacterium sp.	Characterisation and identification of species and strains	61
Mycobacterium sp.	Discrimination of 13 species	110
Bacillus anthracis	Discrimination from other species of *B. cereus* group	48
Bacillus anthracis	Biomarkers in spores	45
Bacillus anthracis	Characterisation of biomarkers by IT MS/MS	111
Bacillus sp.	Biomarkers in spores; strain-specific in *B. cereus*	112
Bacillus sp.	Identification of SASPs in spores as biomarkers	43
Bacillus sp.	Mass spectral typing after inactivation of spores with TFA	62
Bacillus cereus	Pretreatment with corona plasma discharge	113
Listeria sp.	Differentiation of clonal lineages	114
Staphylococcus	Discrimination of 23 species and subspecies	54
Staphylococcus	discrimination of MRSA and MSSA	57
Staphylococcus	Discrimination of MRSA and MSSA	56
Staphylococcus	High throughput identification of hospital isolates	55
Staphylococcus	MRSA differentiation with respect to teicoplanin resistance	115
Streptococcus viridans gr.	Identification of isolates (excluding *Str. pneumoniae*)	53
Streptococcus mutans gr.	Determination of species of clinical isolates	51
Streptococcus pyogenes	Differentiation of invasive and noninvasive isolates	52
Enterococcus sp.	Source tracking of environmental isolates	116
Lactobacillus plantarum	Identification of ribosomal proteins as biomarkers	33
Bacteroides sp.	Differentiation of clinical species	117
Nonfermenter	Identification of clinical isolates	90
Eukaryota		
Fungi		
Fusarium sp.	Discrimination of *Fusarium* species	65, 66
Serpula sp.	Identification of indoor wood-decay fungi	63

Table 12.1   *(continued)*

Taxon	Comment	Ref.
Dermatophytes	Identifcation of clinical *Trichophyton* isolates	64
Fungal spores	Reproducibility of mass fingerprints	118
Protists		
Plasmodium sp.	Detection in processed blood samples	119
Cryptosporidia	Specific biomarkers for two species	69
Giardia lamblia	Discrimination of species	68
Dinoflagellates	Identification of HAB species	120
Metazoa		
Nematoda	Identification of plant parasites	121
Aphids insecta	Discrimination of species	122

could be classified by MALDI-TOF MS according to their origin. Thus, avian *E. coli* isolates could be distinguished from those of bovine origin.[50]

In other studies, clinically relevant *Streptococcus* species were analysed by MALDI-TOF MS. Identification and differentiation of *Streptococcus* species is generally challenging by routine clinical methods while some groups of species share more than 99% nucleotide sequence similarity in the 16S rRNA gene. However, using MALDI-TOF MS mutans streptococci could be discriminated down to the subspecies level[51] permitting the reclassification of some strains. Similarly, *Streptococcus pyogenes* strains could be identified correctly and differentiated into invasive and noninvasive isolates.[52] *Streptococcus* strains belonging to the viridans group were analysed by Friedrichs *et al.*[53] and it was shown that closely related species such as *Str. mitis* and *Str. oralis* could be reproducibly separated.

Some twenty clinically relevant species and subspecies of *Staphylococcus* were studied by Carbonelle *et al.*[54] who reported that each species had a distinct mass spectral fingerprint that allowed its discrimination for the high throughput identification of *Staph. aureus*.[55] Discrimination between methicillin-resistant (MRSA) and methicillin-susceptible (MSSA) *Staph. aureus* strains by MALDI-TOF MS was reported by Edwards-Jones *et al.*[56] and Du *et al.*[57] Both types of strains differed in mass signals around *m/z* 2500 and their assignment was in accordance with the results of a *mecA*-specific PCR assay. A few exceptions, however, were observed and the discrimination of MRSA and MSSA by MALDI-TOF MS could not be confirmed by the analysis of further strains (AnagnosTec, unpublished data).[58] Nonetheless, considering the high costs of MRSA identification by real-time PCR assays, an alternative methodology would be advantageous. Since MRSA are not a monophyletic unit but have arisen from different MSSA lineages independently[59] new approaches will require a different strategy. Direct detection from whole cells of gene products of the staphylococcal chromosomal cassette (SCC*mec*) responsible for the β-lactam resistance is, however, unlikely to be feasible. The penicillin binding protein 2a, for example, has a mass of 78 kDa, far beyond the mass range generally used for mass fingerprinting. Methods involving surface enhanced laser desorption/ionisation (SELDI) show promise.[55] This technique has also been used to delineate subspecies of *Francisella tularensis*, the causative organism of tularaemia.[60]

A number of *Mycobacterium* species were found to be distinguishable by mass spectrometry, and strains could even be discriminated when mass signal intensity was accounted for in addition to simple presence/absence data.[61]

As agents of bioterrorism usually do not emerge in laboratory settings, a procedure is required to make any potentially hazardous substance innocuous immediately, without disabling its biomarkers for subsequent identification. A number of *Bacillus* (spores and cells) and *Yersinia* species were studied for the applicability of a thorough inactivation protocol prior to mass spectral characterisation.[62] Cells and spores were treated with 80% TFA resulting in complete inactivation that did not significantly deteriorate the quality of mass spectral fingerprints, which could still be used for identification.

Compared with bacteria, less attention has been given to the mass spectral analysis of fungi. However, the few studies undertaken so far indicate that pathogenic fungi can also be identified by whole cell MALDI-TOF MS. Closely related wood decay fungi could be discriminated and identified.[63] Clinical isolates of dematophytes could be automatically identified by matching the isolates' spectra to a spectral database.[64] An infection with *Fusarium proliferatum* was confirmed by mass spectral typing and comparison with reference strains[65] A major obstacle in the mass spectral identification of fungi is the variability in a strain's mass fingerprint due to cell differentiation and spore formation. With well established reference databases, however, the routine identification of clinical filamentous fungi and yeasts is possible.[66,67]

Among the protists, some notorious pathogens are relevant for public health surveillance such as *Giardia lamblia* or Cryptosporidia. Both taxa have been analysed by MALDI-TOF MS and the results suggest that mass spectral fingerprints can also be applied for their correct identification.[68,69]

Proof of principle has been demonstrated for several species (see Table 12.1) but the short-comings of some studies prevent their direct transfer to a diagnostic laboratory. First, in many studies only a limited number of strains were tested, which is unlikely to fully represent the intra-specific diversity of a given taxon, and secondly, these identification principles have only rarely been tested for unknown samples.

For mass fingerprints to be used for the routine identification of unknown samples, methods have to be developed that treat the mass spectral information in a reproducible and standardised manner, which can then be compared with reference mass spectra of well characterised strains.

12.6   Pattern Matching Approaches for Automated Identification

In general, the comparison of mass spectral fingerprints is done by mathematical approaches based on various algorithms for cluster analyses, or multivariate approaches such as principal component analysis. The key issue for a successful method is not the particular algorithm that is applied to the processed data since different algorithms may lead to very similar results.[70] However, the processing of the raw data and the identification of mass signals is of paramount importance.

Different approaches have been proposed for pattern matching procedures. Arnold and Reilly[71] applied a cross-correlation analysis to mass spectra of 25 *E. coli* strains by dividing the full spectrum into a varying number of intervals. By this procedure, individual strains could be discriminated even when the corresponding mass spectra appeared very similar. Demirev *et al.*[72] proposed a combination of mass spectrometry and protein database search to assign signals in the mass spectrum of an unknown bacterial sample to a protein sequence translated from internet-accessible genomic sequences. Jarman *et al.*[73] developed an algorithm for the extraction of fingerprints from mass spectra (m/z 1000–10000) of bacteria considering the variability in intensity, frequency, and accuracy of mass signals in replicate analyses of an individual strain. By doing so, a fingerprint was calculated that accounts for differences in replicate mass spectra caused by physiological regulation, analytical error and detection limit. The extracted fingerprint data were applied to the comparison of blind samples with library reference fingerprints of five species by calculating similarities between an unknown and a reference fingerprint based on the absence and presence of individual biomarkers.[74] For the blind test samples, the same strains and mixtures thereof with *Shewanella algae* cells were used, and fingerprints calculated from repeated analyses. For all species except *Bacillus cereus,* a correct identification was achieved when these were part of the mixture, and for all strains correct negative results were returned in the case of their absence. Bright *et al.*[75] applied a pattern recognition algorithm to mass spectra in a range of m/z 500–10000. The software tool MUSE translated each quality-controlled mass spectrum into a single point vector in an n-dimensional space. The spectra of replicate samples of a strain had a location in an n-dimensional space close to, but not identical to, each other. As a reference library the mass spectral data of 35 strains from 20 species were included. When the same strains were analysed as blind samples, the correct match was achieved for more than 95% on the species level. A hierarchical clustering algorithm was applied by Hsieh *et al.*[76] in combination with analysis of variance (ANOVA) to extract biomarkers from multiple isolates belonging to six human pathogens. The set of specific markers was then reduced to 2–4, which was sufficient to identify isolates of the same species correctly.

12.7   Mass Spectral Identification of Microorganisms – Requirements for Routine Diagnostics

Proof of principle of mass spectral identification of microorganisms has been shown for various taxa (see Table 12.1). This is an essential prerequisite for diagnostics in clinical microbiology but for daily routine use there are further demands. Thus, mass spectral identification needs to be:

- simple and fast in handling;
- robust to account for variations and variability in culture conditions;
- reproducible to allow identifications at different locations;
- applicable to the majority of clinically relevant microorganisms;
- economic to allow identifications at competitive costs.

The measuring process itself, as described in most publications, is straightforward. Several extraction solvents and matrix compounds have been tested for their effect on resulting mass fingerprints.[77,78] Despite various suggestions for extraction procedures, cells of most taxa can generally be applied directly to sample targets, either from solid media or from suspensions. One particular exception, yeasts, has been shown to result in considerably improved mass fingerprints after an additional extraction step with formic acid[31] but for most bacteria, extraction performed on the sample spot by adding matrix solution can be considered as sufficient. Extraction procedures involving more steps may lead to increased reproducibility, but at the expense of a significant increase in handling time.

Once raw mass spectra have been obtained, mass signals need to be processed by peak recognition algorithms to extract the relevant information for further analysis. Although this is performed by all software packages supplied with a mass spectrometer, differences in peak recognition algorithms can have a considerable effect on the processing and need to be adjusted with care.[79] Raw data could be used for spectral comparison directly, e.g. by the BioNumerics software package as it has been applied to mass spectra of *Burkholderia cepacia* complex species by Vanlaere *et al.*[70] A major disadvantage of this approach is the extended computing time for large datasets. For practical reasons therefore, the extraction of peak data from raw mass spectra substantially facilitates the downstream data handling, and can be generally performed with the mass spectrometry software with appropriate settings for baseline correction, noise filtering, etc.

Finally, the identifying step involves the comparison of the mass spectral fingerprint of an unknown sample with the mass spectral fingerprints of well characterised strains. The principles applied to the identification of microorganisms based on mass spectral fingerprints are essentially the same as for similar problems where pattern recognition is required, for example, restriction fragment length polymorphism analysis. A major obstacle in this process is the fact that every mass signal in a spectrum has an analytical error. This is in contrast to the comparison of (nucleotide) sequence data where only discrete values are compared. In MALDI-TOF MS several factors contribute to the analytical error, e.g. small variability of acceleration voltage, shape of matrix crystals, and peak recognition by the processing software. For most linear MALDI-TOF mass spectrometers an error of some 500 ppm is considered as acceptable, i.e. a deviation of 5 Da for m/z 10 000. A higher precision can be achieved by internal standards but for ICMS this is not applicable because the calibrants potentially interfere with biomarkers. Consequently, a particular biomarker is detected with a variability in m/z values that has to be taken into consideration when pattern matching algorithms are applied.

The most important part of a system for automated mass spectral identification of microorganisms is a validated reference database.[58,80,81] The reference database should contain not only the mass spectral data of pathogens but also of related, nonpathogenic species that are frequently found in a clinical environment to avoid the danger of false positive identifications. Further, intra-specific variability in mass fingerprints can be considerable, demanding a number of individual strains to represent a single species in a database.[82] From these considerations, it is evident that a database needs to contain reference spectra of multiple strains of hundreds of microbial species.

Such a database has been established by AnagnosTec since 1998 and is constantly amended for rare and newly emerging pathogens, microbes relevant in veterinary medicine, and environmental species. In the next section the principle and workflow for automated mass spectral identification of clinical microorganisms using SARAMIS (spectral archiving and microbial identification system) is briefly explained and examples of application in routine laboratories are given.

12.8 Automated Mass Spectral Analysis of Microorganisms in Clinical Routine Diagnostics

The identification system SARAMIS consists of a database containing reference spectra, and software that allows the comparison of the mass fingerprint of an unknown sample with the reference spectra and was developed by AnagnosTec.[83] A similar principle is applied in the MALDI BioTyper software package developed by Bruker.[84]

As reference spectra, SARAMIS uses so-called 'SuperSpectra' that have been calculated from replicate mass spectra of multiple isolates of a single species or subspecies. To generate SuperSpectra, high quality mass spectra of a species of interest are collected from multiple strains grown under varying conditions and in different laboratories. Raw mass spectra are then processed to yield peak lists that are subjected to cluster analysis by applying a single link agglomerative clustering algorithm that allows a pre-set analytical error, generally 500 ppm. Each cluster represents a mass fingerprint type of a species as shown in Figure 12.4. When numerous strains of a single species have been analysed, cluster analysis may reveal sub-specific units and in that case, multiple SuperSpectra can be calculated. In the next step, a consensus spectrum is calculated containing only these mass signals that have been recorded in a frequency of 50–100% to exclude unspecific signals. This threshold frequency needs to be set specifically with a dataset at hand that allows estimation of the intraspecific variability. Consensus spectra inevitably contain peaks that are specific for higher taxonomic units such as genera or families which can give mass spectral fingerprints of members of respective taxa a high background similarity (see, for example, Enterobacteriaceae;[49] Figure 12.4). For species identification respective mass signals cannot be reclaimed and are given a low weighting. For the calculation of

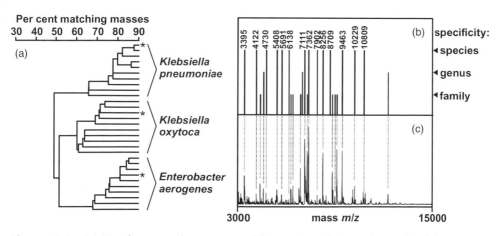

Figure 12.4 *(a) Dendrogram of mass spectral fingerprints of 10 strains each of three closely related species within the diverse family Enterobacteriaceae. (b) Illustration of a SARAMIS SuperSpectrum for* Klebsiella oxytoca *with taxonomic specificities of mass signals indicated by their height. (c) Mass spectrum of a clinical isolate of K. oxytoca with mass signals matching to the SuperSpectrum indicated by dotted lines. Asterisks indicate DSM reference strains of respective species. For the calculation of SuperSpectra see text*

SARAMIS SuperSpectra such nonspecific mass signals are identified by comparing each m/z value with m/z values in the entire database of reference spectra (>100 000; 2009). Before SuperSpectra are released for automated identification, each mass signal contained therein is weighted according to its specificity for the taxon it represents, e.g. a mass signal that has been recorded exclusively with high frequency in one species receives a high value while one that has been recorded in multiple species of the same genus or family, respectively receives a lower value.

When a mass spectral fingerprint is matched to the SuperSpectra in the database, weighting values for each matching mass signal are summed up and as a result those SuperSpectra for which the highest specific concordance was established are listed. The sum of the mass signal specific match values can then be translated into a confidence value for identification.

The performance of SARAMIS was evaluated in several clinical diagnostic laboratories[85] and the results of one study are presented here. In 2007, routine identification of isolates from human urine samples was performed with two systems in parallel in the laboratory Dr Stein & Kollegen (Mönchengladbach, Germany). Samples were analysed by the VITEK 2 system already established in the laboratory as well as by MALDI-TOF MS/SARAMIS that was set up in the laboratory in 2007. A total of 569 bacterial isolates were analysed by both methods representing the commonly encountered, clinically relevant taxa, including diverse species of Enterobacteriaceae, gram-positive cocci, and non-fermentative species. Both methods yielded consistent results for 96.6% of all isolates, including 6.3% for which the SARAMIS identification was correct but only with a confidence level of 80–90% (Table 12.2). The best hit rates were achieved for Enterobacteriaceae and Nonfermenters (98 and 100%, respectively) while for staphylococci and enterococci/streptococci only 92.4 and 82.8% correct identifications were attained, respectively. For these two groups, the highest number of unidentified isolates or isolates for which automatic data acquisition did not yield a mass fingerprint was encountered. Thus, although the mass spectrum had passed the quality control, when a match to SuperSpectra

Table 12.2 *Summary of the evaluation of MALDI-TOF MS/SARAMIS identification of isolates from urine samples compared with identification obtained using VITEK II. (The studies were carried out in the laboratory Dr Stein & Kollegen, Mönchengladbach, Germany.) Numbers for correct identifications with different confidence levels (conf.) are given separately. Numbers in parentheses are percentages. For details see text*

	n	>99–90% conf.	90–80% conf.	No ID	False ID	No spectrum
Enterobacteriaceae	338	319 (94.4)	12 (3.6)	6 (1.8)	0 (0)	1 (0.3)
Enterococci and streptococci	84	70 (83.3)	8 (9.5)	3 (3.6)	0 (0)	3 (3.6)
Staphylococci	79	61 (77.2)	12 (15.2)	4 (5.1)	0 (0)	2 (2.5)
Nonfermenter	68	64 (94.1)	4 (5.9)	0 (0)	0 (0)	0 (0)
Total	569	514 (90.3)	36 (6.3)	13 (2.3)	0 (0)	6 (1.1)

accounts for <70% confidence, SARAMIS issues a nonidentification result. The main reason for a negative result is generally gaps in the database, either encompassing rare species for which sufficient data have not yet been collected or subtypes of well-characterised species for which the mass fingerprint deviates considerably from typical mass fingerprints of that species. For example, in *Streptococcus* sp. the transition between species is frequently accompanied by new mass fingerprint types (own, unpublished data). For 1.1% of the isolates, no mass fingerprint could be generated by automated spectrum acquisition, but in the manual data acquisition mode for part of these samples a mass spectrum could be obtained. One of the most important outcomes of the evaluation was that not a single sample was mis-identified. These results clearly underline that a mass spectrometry based identification system is now readily available as a diagnostic tool for the clinical laboratory and can be implemented after a short familiarisation phase for the routine identification of clinical samples.

Nonetheless, it has to be emphasised that mass spectrometry based identification of microorganisms has its limits. This is partly due to the fact that bacterial identification is based indirectly on a genomospecies concept (i.e. genes translated into proteins that were detected) while classical clinical microbiology is based on phenotypic characters derived from metabolic profiles.[86] A typical example is the *Shigella* species that are recognised by taxonomists as polyphyletic lineages within the species *E. coli*.[87] The main phenotypic characteristics of *Shigella* sp. are that they are nonmotile and cannot metabolise lactose; characteristics that also occur, however, in enteroinvasive *E. coli* strains[88] and which cannot be determined directly by ICMS. By a liquid chromatography/mass spectrometry approach of crude cell extracts, however, *Shigella* species could be discriminated from *E. coli*.[89]

On the other hand, species discrimination by mass fingerprinting can be straightforward for taxa that are poorly discriminated by biochemical methods,[90] for example for species of the *Burkholderia cepacia* complex.[70,91] Consequently, the prospects and capacities of MALDI-TOF MS based methods and biochemical microbial identification procedures are not completely congruent and the application of the former in routine diagnostics will require some training of the medical staff to interpret the data. However, traditional bio-chemical tests are now giving way to genomics approaches for species identification especially since the introduction of 16S rRNA gene sequencing. Throughout the entire microbial kingdom, many groups of biochemically similar species have been further resolved and clarified based on 16S rRNA and other genomic sequences and the current edition of *Bergey's Manual of Systematic Bacteriology*[92] has adopted this approach for a 'road map' of the microbial kingdom. There is therefore a great opportunity to interphase a MALDI-TOF MS based method with 16S rRNA sequencing to provide a modern and reliable approach to microbial identification. The simplicity, speed, and reliability of mass based identification techniques are indisputable advantages[93,94] and it is expected that the methodology will be strengthened by the large number of studies in progress and in the future will become part of the work of routine diagnostics laboratories.

Acknowledgements

We wish to thank Dr Lothar Kruska for making the data of the SARAMIS evaluation available.

References

1. J.P. Anhalt and C. Fenselau, Identification of bacteria using mass spectrometry, *Anal. Chem.* **47**, 219–225 (1975).
2. H.R. Schulten, H.D. Beckey, H.L.C. Meuzelaar and A.J.H. Boerboom, High resolution field ionization mass spectrometry of bacterial pyrolysis products, *Anal. Chem.* **45**, 191–195 (1973).
3. D.N. Heller, R.J. Cotter, C. Fenselau and O.M. Uy, Profiling of bacteria by Fast Atom Bombardment Mass Spectrometry, *Anal. Chem.* **59**, 2806–2809 (1987).
4. B. Sundqvist, A. Hedin, P. Hakansson, I. Kamensky, M. Salehpour and G. Säwe, Plasma desorption mass spectrometry (PDMS): limitations and possibilities, *Int. J. Mass Spectrom. Ion Processes* **65**, 69–89 (1985).
5. K. Tanaka, H. Waki, Y. Ido, S. Akita, Y. Yoshida and T. Yoshida, Protein and polymer analyses up to m/z 100000 by laser ionization time-of-flight mass spectrometry, *Rapid Commun. Mass Spectrom.* **2**, 151–153 (1988).
6. M. Karas and F. Hillenkamp, Laser desorption ionization of proteins with masses exceeding 10000 Da, *Anal. Chem.* **60**, 2299–2303 (1988).
7. M. Karas, D. Bachmann, U. Bahr and F. Hillenkamp, Matrix-assisted ultraviolet laser desorption of non-volatile compounds, *Int. J. Mass Spectrom. Ion Processes* **78**, 53–68 (1987).
8. R.C. Beavis and B.T. Chait, Cinnamic acid derivatives as matrices for ultraviolet laser desorption mass spectrometry, *Rapid Commun. Mass Spectrom.* **3**, 432–435 (1989).
9. K. Strupat, M. Karas and F. Hillenkamp, 2,5-Dihydroxybenzoic acid: a new matrix for laser desorption-ionization mass spectrometry, *Int. J. Mass Spectrom. Ion Processes* **111**, 89–102 (1991).
10. U. Bahr, J. Stahl-Zeng, E. Gleitsmann and M. Karas, Delayed extraction time-of-flight MALDI mass spectrometry of proteins above 25000 Da, *J. Mass Spectrom.* **32**, 1111–1116 (1997).
11. M. Glückmann and M. Karas, The initial ion velocity and its dependence on matrix, analyte and preparation method in ultraviolet matrix-assisted laser desorption/ionization, *J. Mass Spectrom.* **34**, 467–477 (1999).
12. M. Karas, A. Ingendoh, U. Bahr and F. Hillenkamp, Ultraviolet-laser desorption/ionization mass spectrometry of femtomolar amounts of large proteins, *Biol. Mass Spectrom.* **18**, 841–843 (1989).
13. S. Jespersen, W.M.A. Niessen, U.R. Tjaden, J. van der Greef, E. Litborn, U. Lindberg, J. Roeraade and F. Hillenkamp, Attomole detection of proteins by matrix-assisted laser desorption/ionization mass spectrometry with the use of picolitre vials, *Rapid Commun. Mass Spectrom.* **8**, 581–584 (1994).
14. F. Hillenkamp and M. Karas, Matrix-assisted laser desorption/ionization, an experience, *Int. J. Mass Spectrom.* **200**, 71–77 (2000).
15. M. Karas, M. Glückmann and J. Schäfer, Ionization in matrix-assisted laser desorption/ionization: singly charged molecular ions are the lucky survivors, *J. Mass Spectrom.* **35**, 1–12 (2000).
16. J.G. Wilkes, D.A. Buzatu, D.J. Dare, Y.P. Dragan, M.P. Chiarelli, R.D. Holland, M. Beaudoin, T.M. Heinze, R. Nayak and A.A. Shvartsburg, Improved cell typing by charge-state deconvolution of matrix-assisted laser desorption/ionization mass spectra, *Rapid Commun. Mass Spectrom.* **20**, 1595–1603 (2006).
17. T. Cain, D.M. Lubman and W.J. Weber Jr, Differentiation of bacteria using protein profiles from matrix-assisted laser desorption/ionization time-of-flight mass spectrometry, *Rapid Commun. Mass Spectrom.* **8**, 1026–1030 (1996).
18. M.A. Claydon, S.N. Davey, V. Edwards-Jones and D.B. Gordon, The rapid identification of intact microorganisms using mass spectrometry, *Nat. Biotechnol.* **14**, 1584–1586 (1996).
19. J.G.W. Holland, Rapid identification of intact whole bacteria based on spectral patterns using matrix-assisted laser desorption/ionization with time-of-flight mass spectrometry, *Rapid Commun. Mass Spectrom.* **10**, 1227–1232 (1996).
20. T. Krishnamurthy, P.L. Ross and U. Rajamani, Detection of pathogenic and non-pathogenic bacteria by matrix assisted laser desorption/ionization time of flight mass spectrometry, *Rapid Commun. Mass Spectrom.* **10**, 883–888 (1996).

21. X. Liang, K. Zheng, M.G. Qian and D.M. Lubman, Determination of bacterial protein profiles by matrix-assisted laser desorption/ionization mass spectrometry with high-performance liquid chromatography, *Rapid Commun. Mass Spectrom.* **10**, 1219–1226 (1996).

22. T. Krishnamurthy and P.L. Ross, Rapid identification of bacteria by direct matrix-assisted laser desorption/ionization mass spectrometric analysis of whole cells, *Rapid Commun. Mass Spectrom.* **10**, 1992–1996 (1996).

23. D.J. Evason, M.A. Claydon and D.B. Gordon, Exploring the limits of bacterial identification by intact cell-mass spectrometry, *J. Am. Soc. Mass Spectrom.* **12**, 49–54 (2001).

24. M. Erhard, H. von Döhren and P. Jungblut, Rapid typing and elucidation of new secondary metabolites of intact cyanobacteria using MALDI-TOF mass spectrometry, *Nat. Biotechnol.* **15**, 906–909 (1997).

25. M. Welker and M. Erhard, Consistency between chemotyping of single filaments of *Planktothrix rubescens* (cyanobacteria) by MALDI-TOF and the peptide patterns of strains determined by HPLC-MS, *J. Mass Spectrom.* **42**, 1062–1068 (2007).

26. C. Cirulli, G. Marino and A. Amoresano, Membrane proteome in *Escherichia coli* probed by MS3 mass spectrometry: a preliminary report, *Rapid Commun. Mass Spectrom.* **21**, 2389–2397 (2007).

27. V. Ryzhov and C. Fenselau, Characterization of the protein subset desorbed by MALDI from whole bacterial cells, *Anal. Chem.* **73**, 746–750 (2001).

28. R.J. Arnold and J.P. Reilly, Observation of *Escherichia coli* ribosomal proteins and their post-translational modifications by mass spectrometry, *Anal. Biochem.* **269**, 105–112 (1999).

29. P.A. Demirev, J.S. Lin, F.J. Pineda and C. Fenselau, Bioinformatics and mass spectrometry for microorganism identification: proteome-wide post-translational modifications and database search algorithms for characterization of intact *H. pylori*, *Anal. Chem.* **73**, 4566–4573 (2001).

30. C. Wynne, C. Fenselau, P.A. Demirev and N. Edwards, Top-down identification of protein biomarkers in bacteria with unsequenced genomes, *Anal. Chem.* **81**, 9633–9642 (2009).

31. B. Arniri-Eliasi and C. Fenselau, Characterization of protein biomarkers desorbed by MALDI from whole fungal cells, *Anal. Chem.* **73**, 5228–5231 (2001).

32. C.K. Fagerquist, W.G. Miller, L.A. Harden, A.H. Bates, W.H. Vensel, G. Wang and R.E. Mandrell, Genomic and proteomic identification of a DNA-binding protein used in the 'fingerprinting' of *Campylobacter* species and strains by MALDI-TOF-MS protein biomarker analysis, *Anal. Chem.* **77**, 4897–4907 (2005).

33. L. Sun, K. Teramoto, H. Sato, M. Torimura, H. Tao and T. Shintani, Characterization of ribosomal proteins as biomarkers for matrix-assisted laser desorption/ionization mass spectral identification of *Lactobacillus plantarum*, *Rapid Commun. Mass Spectrom.* **20**, 3789–3798 (2006).

34. K. Teramoto, H. Sato, L. Sun, M. Torimura, H. Tao, H. Yoshikawa, Y. Hotta, A. Hosoda and H. Tamura, Phylogenetic classification of *Pseudomonas putida* strains by MALDI-MS using ribosomal subunit proteins as biomarkers, *Anal. Chem.* **79**, 8712–8719 (2007).

35. J.A. Karty, S. Lato and J.P. Reilly, Detection of the bacteriological sex factor in *E. coli* by matrix-assisted laser desorption/ionization time-of-flight mass spectrometry, *Rapid Commun. Mass Spectrom.* **12**, 625–629 (1998).

36. Z. Wang, L. Russon, L. Li, D.C. Roser and S.R. Long, Investigation of spectral reproducibility in direct analysis of bacteria proteins by matrix-assisted laser desorption/ionization time-of-flight mass spectrometry, *Rapid Commun. Mass Spectrom.* **12**, 456–464 (1998).

37. T.L. Williams, D. Andrzejewski, J.O. Lay and S.M. Musser, Experimental factors affecting the quality and reproducibility of MALDI TOF mass spectra obtained from whole bacteria cells, *J. Am. Soc. Mass Spectrom.* **14**, 342–351 (2003).

38. J. Ramirez and C. Fenselau, Factors contributing to peak broadening and mass accuracy in the characterization of intact spores using matrix-assisted laser desorption/ionization coupled with time-of-flight mass spectrometry, *J. Mass Spectrom.* **36**, 929–936 (2001).

39. S.C. Wunschel, K.H. Jarman, C.E. Petersen, N.B. Valentine, K.L. Wahl, D. Schauki, J. Jackman, C.P. Nelson and E. White, Bacterial analysis by MALDI-TOF mass spectrometry: an inter-laboratory comparison, *J. Am. Soc. Mass Spectrom.* **16**, 456–462 (2005).

40. R.J. Arnold, J.A. Karty, A.D. Ellington and J.P. Reilly, Monitoring the growth of a bacteria culture by MALDI-MS of whole cells, *Anal. Chem.* **71**, 1990–1996 (1999).
41. N.B. Valentine, S.C. Wunschel, D.S. Wunschel, C.E. Petersen and K.L. Wahl, Effect of culture conditions on microorganism identification by matrix-assisted laser desorption ionization mass spectrometry, *Appl. Environ. Microbiol.* **71**, 58–64 (2005).
42. D.S. Wunschel, E.A. Hill, J.S. McLean, K.H. Jarman, Y.A. Gorby, N.B. Valentine and K.L. Wahl, Effects of varied pH, growth rate and temperature using controlled fermentation and batch culture on Matrix Assisted Laser Desorption/Ionization whole cell protein fingerprints, *J. Microb. Meth.* **62**, 259–271 (2005).
43. Y. Hathout, B. Setlow, R.M. Cabrera-Martinez, C. Fenselau and P. Setlow, Small, acid-soluble proteins as biomarkers in mass spectrometry analysis of *Bacillus* spores, *Appl. Environ. Microbiol.* **69**, 1100–1107 (2003).
44. P.A. Demirev, J. Ramirez and C. Fenselau, Tandem mass spectrometry of intact proteins for characterization of biomarkers from *Bacillus cereus* T spores, *Anal. Chem.* **73**, 5725–5731 (2001).
45. E. Elhanany, R. Barak, M. Fisher, D. Kobiler and Z. Altboum, Detection of specific *Bacillus anthracis* spore biomarkers by matrix-assisted laser desorption/ionization time-of-flight mass spectrometry, *Rapid Commun. Mass Spectrom.* **15**, 2110–2116 (2001).
46. J.R. Whiteaker, B. Warscheid, P. Pribil, Y. Hathout and C. Fenselau, Complete sequences of small acid-soluble proteins from *Bacillus globigii*, *J. Mass Spectrom.* **39**, 1113–1121 (2004).
47. P.A. Pribil, E. Patton, G. Black, V. Doroshenko and C. Fenselau, Rapid characterization of *Bacillus* spores targeting species-unique peptides produced with an atmospheric pressure matrix-assisted laser desorption/ionization source, *J. Mass Spectrom.* **40**, 464–474 (2005).
48. E.R. Castanha, A. Fox and K.F. Fox, Rapid discrimination of *Bacillus anthracis* from other members of the *B. cereus* group by mass and sequence of 'intact' small acid soluble proteins (SASPs) using mass spectrometry, *J. Microbiol. Meth.* **67**, 230–240 (2006).
49. E.C. Lynn, M.-C. Chung, W.-C. Tsai and C.-C. Han, Identification of Enterobacteriaceae bacteria by direct matrix-assisted laser desorptiom/ionization mass spectrometric analysis of whole cells, *Rapid Commun. Mass Spectrom.* **13**, 2022–2027 (1999).
50. T.J. Siegrist, P.D. Anderson, W.H. Huen, G.T. Kleinheinz, C.M. McDermott and T.R. Sandrin, Discrimination and characterization of environmental strains of *Escherichia coli* by matrix-assisted laser desorption/ionization time-of-flight mass spectrometry (MALDI-TOF-MS), *J. Microbiol. Meth.* **68**, 554–562 (2007).
51. S. Rupf, K. Breitung, W. Schellenberger, K. Merte, S. Kneist and K. Eschrich, Differentiation of mutans streptococci by intact cell matrix-assisted laser desorption/ionization time-of-flight mass spectrometry, *Oral Microbiol. Immunol.* **20**, 267–273 (2005).
52. H. Moura, A.R. Woolfitt, M.G. Carvalho, A. Pavlopoulos, L.M. Teixeira, G.A. Satten and J.R. Barr, MALDI-TOF mass spectrometry as a tool for differentiation of invasive and noninvasive *Streptococcus pyogenes* isolates, *FEMS Immun. Med. Microbiol.* **53**, 333–342 (2008).
53. C. Friedrichs, A.C. Rodloff, G.S. Chhatwal, W. Schellenberger and K. Eschrich, Rapid identification of viridans streptococci by mass spectrometric discrimination, *J. Clin. Microbiol.* **45**, 2392–2397 (2007).
54. E. Carbonnelle, J.L. Beretti, S. Cottyn, G. Quesne, P. Berche, X. Nassif and A. Ferroni, Rapid identification of staphylococci isolated in clinical microbiology laboratories by matrix-assisted laser desorption ionization – Time of flight mass spectrometry, *J. Clin. Microbiol.* **45**, 2156–2161 (2007).
55. L. Rajakaruna, G. Hallas, L. Molenaar, D. Dare, H. Sutton, V. Encheva, R. Culak, I. Innes, G. Ball, A.M. Sefton, M. Eydmann, A.M. Kearns and H.N. Shah, High throughput identification of clinical isolates of *Staphylococcus aureus* using MALDI-TOF MS of intact cells, *Infect. Gen. Evol.* **9**, 507–513 (2009).
56. V. Edwards-Jones, M.A. Claydon, D.J. Evason, J. Walker, A.J. Fox and D.B. Gordon, Rapid discrimination between methicillin-sensitive and methicillin-resistant *Staphylococcus aureus* by intact cell mass spectrometry, *J. Med. Microbiol.* **49**, 295–300 (2000).
57. Z. Du, R. Yang, Z. Guo, Y. Song and J. Wang, Identification of *Staphylococcus aureus* and determination of its methicillin resistance by matrix-assisted laser desorption/ionization time-of-flight mass spectrometry, *Anal. Chem.* **74**, 5487–5491 (2002).

58. C.J. Keys, D.J. Dare, H. Sutton, G. Wells, M. Lunt, T. McKenna, M. McDowall and H.N. Shah, Compilation of a MALDI-TOF mass spectral database for the rapid screening and characterisation of bacteria implicated in human infectious diseases, *Infect. Genet. Evol.* **4**, 221–242 (2004).

59. M.C. Enright, D.A. Robinson, G. Randle, E.J. Feil, H. Grundmann and B.G. Spratt, The evolutionary history of methicillin-resistant Staphylococcus aureus (MRSA), *Proc. Natl. Acad. Sci. USA* **99**, 7687–7692 (2002).

60. E. Seibold, R. Bogumil, S. Vorderwulbecke, S. Al Dahouk, A. Buckendahl, H. Tomaso and W. Splettstoesser, Optimized application of surface-enhanced laser desorption/ionization time-of-flight MS to differentiate *Francisella tularensis* at the level of subspecies and individual strains, *FEMS Immunol. Med. Microbiol.* **49**, 364–373 (2007).

61. J.M. Hettick, M.L. Kashon, J.E. Slaven, Y. Ma, J.P. Simpson, P.D. Siegel, G.N. Mazurek and D.N. Weissman, Discrimination of intact mycobacteria at the strain level: A combined MALDI-TOF MS and biostatistical analysis, *Proteomics* **6**, 6416–6425 (2006).

62. P. Lasch, H. Nattermann, M. Erhard, M. Stämmler, R. Grunow, N. Bannert, B. Appel and D. Naumann, A MALDI-ToF mass spectrometry compatible inactivation method for highly pathogenic microbial cells and spores, *Anal. Chem.* **80**, 2026–2034 (2008).

63. O. Schmidt and W. Kallow, Differentiation of indoor wood decay fungi with MALDI-TOF mass spectrometry, *Holzforschung* **59**, 374–377 (2005).

64. M. Erhard, U.C. Hipler, A. Burmester, A.A. Brakhage and J. Wöstemeyer, Identification of dermatophyte species causing onychomycosis and tinea pedis by MALDI-TOF mass spectrometry, *Exp. Dermatol.* **17**, 365–371 (2008).

65. F. Seyfarth, M. Ziemer, M. Kaatz, H. Sayer, A. Burmester, M. Erhard, M. Welker, S. Schliemann, E. Strabe and U.C. Hipler, Case report: The use of ITS DNA sequence analysis and MALDI-TOF mass spectrometry in diagnosing an infection with *Fusarium proliferatum*, *Exp. Dermatol.* **17**, 965–971 (2008).

66. C. Marinach-Patrice, A. Lethuillier, A. Marly, J.Y. Brossas, J. Gene, F. Symoens, A. Datry, J. Guarro, D. Mazier and C. Hennequin, Use of mass spectrometry to identify clinical *Fusarium* isolates, *Clin. Microbiol. Infect.* **15**, 634–642 (2009).

67. G. Marklein, M. Josten, U. Klanke, E. Muller, R. Horre, T. Maier, T. Wenzel, M. Kostrzewa, G. Bierbaum, A. Hoerauf and H.G. Sahl, Matrix-assisted laser desorption ionization-time of flight mass spectrometry for fast and reliable identification of clinical yeast isolates, *J. Clin. Microbiol.* **47**, 2912–2917 (2009).

68. E.N. Villegas, S.T. Glassmeyer, M.W. Ware, S.L. Hayes and F.W. Schaefer, Matrix-assisted laser desorption/ionization time-of-flight mass spectrometry-based analysis of *Giardia lamblia* and *Giardia muris*, *J. Eukaryotic Microbiol.* **53**, S179-S181 (2006).

69. M.L. Magnuson, J.H. Owens and C.A. Kelty, Characterization of *Cryptosporidium parvum* by matrix-assisted laser desorptio/ionisation time-of-flight mass spectrometry, *Appl. Environ. Microbiol.* **66**, 4720–4724 (2000).

70. E. Vanlaere, K. Sergeant, P. Dawyndt, W. Kallow, M. Erhard, H. Sutton, D. Dare, B. Samyn, B. Devreese and P. Vandamme, Matrix assisted laser desorption ionization-time of flight mass spectrometry of intact cells allows rapid identification of *Burkholderia cepacia* complex species, *J. Microbiol. Meth.* **75**, 279–286 (2008).

71. R.J. Arnold and J.P. Reilly, Fingerprint matching of *E. coli* strains with matrix-assisted laser desorption/ionization time-of-flight mass spectrometry of whole cells using a modified correlation approach, *Rapid Commun. Mass Spectrom.* **12**, 630–636 (1998).

72. P.A. Demirev, Y.P. Ho, V. Ryzhov and C. Fenselau, Microorganism identification by mass spetrometry and protein database searches, *Anal. Chem.* **71**, 2732–2738 (1999).

73. K.H. Jarman, D.S. Daly, C.E. Peterson, A.J. Saenz, N.B. Valentine and K.L. Wahl, Extracting and visualizing matrix-assisted laser desorption/ionization time-of-flight mass spectral fingerprints, *Rapid Commun. Mass Spectrom.* **13**, 1586–1594 (1999).

74. K.H. Jarman, S.T. Cebula, A.J. Saenz, C.E. Petersen, N.B. Valentine, M.T. Kingsley and K.L. Wahl, An algorithm for automated bacterial identification using matrix-assisted laser desorption/ionization mass spectrometry, *Anal. Chem.* **72**, 1217–1223 (2000).

75. J.J. Bright, M.A. Claydon, M. Soufian and D.B. Gordon, Rapid typing of bacteria using matrix-assisted laser desorption ionisation time-of-flight mass spectrometry and pattern recognition software, *J. Microbiol. Meth.* **48**, 127–138 (2002).

76. S.-Y. Hsieh, C.-L. Tseng, Y.-S. Lee, A.-J. Kuo, C.-F. Sun, Y.-H. Lin and J.-K. Chen, Highly efficient classification and identification of human pathogenic bacteria by MALDI-TOF MS, *Mol. Cell. Proteomics* **7**, 448–456 (2008).
77. M.A. Domin, K.J. Welham and D.S. Ashton, The effect of solvent and matrix combinations on the analysis of bacteria by matrix-assisted laser desorption/ionisation time-of-flight mass spectrometry, *Rapid Commun. Mass Spectrom.* **13**, 222–226 (1999).
78. D.J. Evason, M.A. Claydon and D.B. Gordon, Effects of ion mode and matrix additives in the identification of bacteria by intact cell mass spectrometry, *Rapid Commun. Mass Spectrom.* **14**, 669–672 (2000).
79. K.H. Jarman, D.S. Daly, K.K. Anderson and K.L. Wahl, A new approach to automated peak detection, *Chemomet. Intell. Lab. Systems* **69**, 61–76 (2003).
80. Z. Wang, K. Dunlop, S.R. Long and L. Li, Mass spectrometric methods for generation of protein mass database used for bacterial identification, *Anal. Chem.* **74**, 3174–3182 (2002).
81. S. Sauer, A. Freiwald, T. Maier, M. Kube, R. Reinhardt, M. Kostrzewa and K. Geider, Classifiction and identification of bacteria by mass spectrometry and computational analysis, *PLoS One* **3**, e2843 (2008).
82. M.F. Lartigue, G. Hery-Arnaud, E. Haguenoer, A.S. Domelier, P.O. Schmit, N. van der Mee-Marquet, P. Lanotte, L. Mereghetti, M. Kostrzewa and R. Quentin, Identification of *Streptococcus agalactiae* isolates from various phylogenetic lineages by matrix-assisted laser desorption ionization-time of flight mass spectrometry, *J. Clin. Microbiol.* **47**, 2284–2287 (2009).
83. W. Kallow, R. Dieckmann, N. Kleinkauf, M. Erhard and T. Neuhof, *Method of identifying microorganisms using MALDI-TOF-MS*, European Patent (2000).
84. W. Pusch, Bruker Daltronics: leading the way from basic research to mass-spectrometry-based clinical applications, *Pharmacogenomics* **8**, 663–668 (2007).
85. J. Gielen, M. Erhard, W. Kallow, M. Krönke and O. Krut. Rapid pathogen identification by MALDI-TOF mass spectrometry/SARAMIS database in clinical microbiological routine diagnostics, 17th ECCMID Conference, Munich, Germany, abstract (2007). Munich, Germany.
86. E. Stackebrandt, W. Frederiksen, G.M. Garrity, P.A.D. Grimont, P. Kämpfer, M.C.J. Maiden, X. Nesme, R. Rosselló-Mora, J. Swings, H.G. Trüper, L. Vauterin, A.C. Ward and W.B. Whiotman, Report of the *ad hoc* committee for the re-evaluation of the species definition in bacteriology, *Int. J. Syst. Evol. Microbiol.* **52**, 1043–1047 (2002).
87. G.M. Pupo, R. Lan and P.R. Reeves, Multiple independent origins of *Shigella* clones of *Escherichia coli* and convergent evolution of many of their characteristics, *Proc. Natl. Acad. Sci. USA* **97**, 10567–10572 (2000).
88. R. Lan, M.C. Alles, K. Donohhoe, M.B. Martinez and P.R. Reeves, Molecular evolutionary relationships of enteroinvasive *Escherichia coli* and *Shigella* ssp., *Infect. Immun.* **72**, 5080–5088 (2004).
89. R.A. Everley, T.M. Mott, S.A. Wyatt, D.M. Toney and T.R. Croley, Liquid chromatography/mass spectrometry characterization of *Escherichia coli* and *Shigella* species, *J. Am. Soc. Mass Spectrom.* **19**, 1621–1628 (2008).
90. A. Mellmann, J. Cloud, T. Maier, U. Keckevoet, I. Rarnminger, P. Iwen, J. Dunn, G. Hall, D. Wilson, P.R. Lasala, M. Kostrzewa and D. Harmsen, Evaluation of Matrix-Assisted Laser Desorption Ionization-Time-of-Flight mass spectrometry in comparison to 16S rRNA gene sequencing for species identification of nonfermenting bacteria, *J. Clin. Microbiol.* **46**, 1946–1954 (2008).
91. A. Minan, A. Bosch, P. Lasch, M. Stammler, D.O. Serra, J. Degrossi, B. Gatti, C. Vay, M. D'aquino, O. Yantorno and D. Naumann, Rapid identification of *Burkholderia cepacia* complex species including strains of the novel Taxon K, recovered from cystic fibrosis patients by intact cell MALDI-ToF mass spectrometry, *Analyst* **134**, 1138–1148 (2009).
92. Multiple authors, *Bergey's Manual of Systematic Bacteriology, Vols 1–5*, Springer, New York (2001).
93. P. Seng, M. Drancourt, F. Gouriet, B. La Scola, P.E. Fournier, J.M. Rolain and D. Raoult, Ongoing revolution in bacteriology: routine identification of bacteria by matrix-assisted laser desorption ionization time-of-flight mass spectrometry, *Clin. Infect. Dis.* **49**, 543–551 (2009).

94. S. Sauer and M. Kliem, Mass spectrometry tools for the classification and identification of bacteria, *Nature Rev. Microbiol.* **8**, 74–82 (2010).

95. P. Krader and D. Emerson, Identification of archaea and some extremophilic bacteria using matrix-assisted laser desorption/ionization time-of-flight (MALDI-TOF) mass spectrometry, *Extremophiles* **8**, 259–268 (2004).

96. M.J. Donohue, A.W. Smallwood, S. Pfaller, M. Rodgers and J.A. Shoemaker, The development of a matrix-assisted laser desorption/ionization mass spectrometry-based method for the protein fingerprinting and identification of *Aeromonas* species using whole cells, *J. Microbiol. Meth.* **65**, 380–389 (2006).

97. P.A. Pribil and C. Fenselau, Characterization of Enterobacteria using MALDI-TOF mass spectrometry, *Anal. Chem.* **77**, 6092–6095 (2005).

98. J.E. Camara and F.A. Hays, Discrimination between wild-type and ampicillin-resistant *Escherichia coli* by matrix-assisted laser desorption/ionization time-of-flight mass spectrometry, *Anal. Bioanal. Chem.* **389**, 1633–1638 (2007).

99. R. Dieckmann, R. Helmuth, M. Erhard and B. Malorny, Rapid classification and identification of Salmonellae at the species and subspecies levels by whole-cell matrix-assisted laser desorption ionization-time of flight mass spectrometry, *Appl. Environ. Microbiol.* **74**, 7767–7778 (2008).

100. R. Dieckmann, E. Strauch and T. Alter, Rapid identification and characterization of *Vibrio* species using whole-cell MALDI-TOF mass spectrometry, *J. Appl. Microbiol.* in press.

101. C. Moliner, C. Ginevra, S. Jarraud, C. Flaudrops, C. Bedotto, C. Couderc, J. Etienne and P.E. Fournier, Rapid identification of Legionella species by mass spectrometry, *J. Med. Microbiol.* in press.

102. C.Y. Pierce, J.R. Barr, A.R. Woolfitt, H. Moura, E.I. Shaw, H.A. Thompson, R.F. Massung and F.M. Fernandez, Strain and phase identification of the US category B agent *Coxiella burnetii* by matrix assisted laser desorption/ionization time-of-flight mass spectrometry and multivariate pattern recognition, *Anal. Chim. Acta* **583**, 23–31 (2007).

103. A. Schaller, R. Troller, D. Molina, S. Gallati, C. Aebi and P.S. Meier, Rapid typing of *Moraxelia catarrhalis* subpopulations based on outer membrane proteins using mass spectrometry, *Proteomics* **6**, 172–180 (2006).

104. A.M. Haag, S.N. Taylor, K.H. Johnston and R.B. Cole, Rapid identification and speciation of *Haemophilus* bacteria by matrix-assisted laser desorption/ionization time-of-flight mass spectrometry, *J. Mass Spectrom.* **33**, 750–756 (1998).

105. C.K. Fagerquist, E. Yee and W.G. Miller, Composite sequence proteomic analysis of protein biomarkers of *Campylobacter coli*, *C. lari* and *C. concisus* for bacterial identification, *Analyst* **132**, 1010–1023 (2007).

106. R.E. Mandrell, L.A. Harden, A. Bates, W.G. Miller, W.F. Haddon and C.K. Fagerquist, Speciation of *Campylobacter coli*, *C. jejuni*, *C. helveticus*, *C. lari*, *C. sputorum*, and *C. upsaliensis* by matrix-assisted laser desorption/ionization time-of-flight mass spectrometry, *Appl. Environ. Microbiol.* **71**, 6292–6307 (2005).

107. C.K. Fagerquist, A.H. Bates, S. Heath, B.C. King, B.R. Garbus, L.A. Harden and W.G. Miller, Sub-speciating *Campylobacter jejuni* by proteomic analysis of its protein biomarkers and their post-translational modifications, *J. Proteome Res.* **5**, 2527–2538 (2006).

108. C.L. Nilsson, Fingerprinting of *Helicobacter pylori* strains by matrix-assisted laser desorption/ionization mass spectrometric analysis, *Rapid Commun. Mass Spectrom.* **13**, 1067–1071 (1999).

109. M. Vargha, Z. Takats, A. Konopka and C.H. Nakatsu, Optimization of MALDI-TOF MS for strain level differentiation of *Arthrobacter* isolates, *J. Microbiol. Meth.* **66**, 399–409 (2006).

110. M. Pignone, K.M. Greth, J. Cooper, D. Emerson and J. Tang, Identification of mycobacteria by matrix-assisted laser desorption/ionization time-of-flight mass spectrometry, *J. Clin. Microbiol.* **44**, 1963–1970 (2006).

111. M.J. Stump, G. Black, A. Fox, K.F. Fox, C.E. Turick and M. Matthews, Identification of marker proteins for *Bacillus anthracis* using MALDI-TOF IVIS and ion trap MS/MS after direct extraction or electrophoretic separation, *J. Sep. Sci.* **28**, 1642–1647 (2005).

112. Y. Hathout, P.A. Demirev, Y.P. Ho, J.L. Bundy, V. Rhyzov, L. Sapp, J. Stutler, J. Jackman and C. Fenselau, Identification of *Bacillus* spores by matrix-assisted laser desorption ionization-mass spectrometry, *Appl. Environ. Microbiol.* **65**, 4313–4319 (1999).

113. V. Ryzhov, Y. Hathout and C. Fenselau, Rapid characterization of spores of *Bacillus cereus* group bacteria by matrix-assisted laser desorption-ionization time-of-flight mass spectrometry, *Appl. Environ. Microbiol.* **66**, 3828–3834 (2000).

114. S. Barbuddhe, T. Maier, G. Schwarz, M. Kostrzewa, E. Domann, T. Chakraborty and T. Hain, Rapid identification of *Listeria* species using MALDI-TOF MS fingerprinting with MALDI BioTyper, *Int. J. Med. Microbiol.* **297**, 18–19 (2007).

115. P.A. Majcherczyk, T. McKenna, P. Moreillon and P. Vaudaux, The discriminatory power of MALDI-TOF mass spectrometry to differentiate between isogenic teicoplanin-susceptible and teicoplanin-resistant strains of methicillin-resistant *Staphylococcus aureus*, *FEMS Microbiol. Lett.* **255**, 233–239 (2006).

116. R.A. Giebel, W. Fredenberg and T.R. Sandrin, Characterization of environmental isolates of *Enterococcus* spp. by matrix-assisted laser desorption/ionization time-of-flight mass spectrometry, *Water Res.* **42**, 931–940 (2008).

117. H.N. Shah, R.C. Jacinto, N. Ahmod, S. Langham, S.E. Gharbia, W. Kallow and M. Welker, The genus *Bacteroides*, in *Encyclopedia of Life Sciences*, John Wiley & Sons, Ltd, Chichester (2008).

118. K.J. Welham, M.A. Domin, K. Johnson, L. Jones and D.S. Ashton, Characterization of fungal spores by laser desorption/ionization time-of-flight mass spectrometry, *Rapid Commun. Mass Spectrom.* **14**, 307–310 (2000).

119. P.A. Demirev, A.B. Feldman, D. Kongkasuriyachai, P. Scholl, D. Sullivan and N. Kumar, Detection of malaria parasites in blood by laser desorption mass spectrometry, *Anal. Chem.* **74**, 3262–3266 (2002).

120. F.W.F. Lee, K.C. Ho and S.C.L. Lo, Rapid identification of dinoflagellates using protein profiling with matrix-assisted laser desorption/ionization mass spectrometry, *Harmful Algae* **7**, 551–559 (2008).

121. M.R. Perera, V.A. Vanstone and M.G.K. Jones, A novel approach to identify plant parasitic nematodes using matrix-assisted laser desorption/ionization time-of-flight mass spectrometry, *Rapid Commun. Mass Spectrom.* **19**, 1454–1460 (2005).

122. M.R. Perera, R.D.F. Vargas and M.G.K. Jones, Identification of aphid species using protein profiling and matrix-assisted laser desorption/ionization time-of-flight mass spectrometry, *Entomol. Exp. Applicat.* **117**, 243–247 (2005).

Part V

Targeted Molecules and Analysis of Specific Microorganisms

13

Whole Cell MALDI Mass Spectrometry for the Rapid Characterisation of Bacteria; A Survey of Applications to Major Lineages in the Microbial Kingdom

Ben van Baar

TNO Defence, Security and Safety, Rijswijk, The Netherlands

13.1   Introduction

Mass spectrometry (MS) is now widely becoming established as a toolbox for the investigation of the systems and processes of life. While mass spectrometer operation has long time relied on specialists, the current hardware performance is sufficiently robust to support routine applications with little expert intervention. Much effort is now directed towards information management, as the highly automated analysis processes often produce huge virtual piles of primary data.

Microbiology has become a major application area for MS, in support of studies that range from typing to biological structure and function. The pioneering days of whole cell matrix assisted laser desorption/ionisation (MALDI) MS bacteria typing, as reviewed some years ago [1–3], have passed and more fundamental research as well as routine typing method development have been published (for species covered, see Table 13.1). In addition, methods of matrix-free ionisation have emerged [4–6] that have occasionally been used in a similar way as whole cell MALDI MS. It is noted that two more recent reviews have appeared with a more general scope of bacteria typing than whole cell

Mass Spectrometry for Microbial Proteomics   Edited by Haroun N. Shah and Saheer E. Gharbia
© 2010 John Wiley & Sons, Ltd

MALDI MS [7, 8] and a review that covers the added benefits of MS/MS in conjunction with MALDI [9]. At present, bacteria typing by MS is becoming established as a mature method, along with the well accepted immunochemical and nucleotide based methods. Highly confident typing can be accomplished by tandem MS protein sequencing. Thus, unbiased random peptide sequencing, known as 'bottom-up proteomics', can be employed as a generic typing method that produces protein sequence information [10–12] comparable with the DNA sequence information obtained from multi-locus sequence typing (MLST). Given the current development of technology, DNA and protein sequencing can be accomplished with similar speed and in the timeframe of hours to a day. While DNA sequencing implies the advantage of amplification, bottom-up proteomics has the advantages of being much less biased and equally applicable to bacteria, viruses, toxins, and

Table 13.1   *Overview of bacteria investigated by whole cell MALDI MS*

Genus	Species and strain
Achromobacter	A. denitrificans DSM 30266 [15]; A. insolitus LMG 6003 [15]; A. piechaudii CIP 101223 [167]; DSM 10342 [15]; A. ruhlandii DSM 653 [15]; A. spanios LMG 5911 [15]; A. xylosoxidans CIP 71.32 [167]; CIP 77.15 [167]; DSM 2402 [15]
Acidovorax	A. avenae DSM 7227 [15]; LMG 5376 [15]; A. defluvii DSM 12644 [15]; A. delafieldii DSM 64 [15]; A. facilis DSM 649 [15]; A. conjaci DSM 7481 [15]; A. temperans DSM 7270 [15]
Acinetobacter	A. sp. 10 ATCC 17924 [96]; A. baumannii DSM 30007 [15]; LMC 994 [15]; A. baylyi DSM 14961 [15]; A. bouvetii DSM 14964 [15]; A. calcoaceticus DSM 30006 [15]; A. gerneri DSM 14967 [15]; A. grimontii DSM 14968 [15]; A. haemolyticus DSM 6962 [15]; A. haemolyticus LMG 1033 [15]; A. johnsonii DSM 6963 [15]; LMG 10584 [15], A. junii DSM 6964 [15]; A. lwoffii DSM 2403 [15]; LMG 1138 [15]; LMG 1154 [15], LMG 1300 [15]; A. parvus DSM 16617 [15]; A. radioresistens DSM 6976 [15]; LMG 10614 [15]; A. schindleri DSM 16038 [15]; A. tandoii DSM 14970 [15]; A. tjernbergiae DSM 14971 [15]; A. towneri DSM 14962 [15]; A. ursingii DSM 16037 [15]
Actinomyces	A. meyeri MCCM 01956 [129]; A. odontolyticus DSM 43331 [129]
Aeromonas	A. sp. ATCC 35941 [88, 154]; ATCC 43946 [88, 154]; A. allosaccharophila ATCC 51208 [88, 154]; ATCC 35942 [154]; A. bestiarum ATCC 51108 [88, 154]; A. caviae ATCC 15468 [88, 154]; CIP 76.16 [167]; A. encheleia ATCC 51929 [88, 154]; ATCC 51930 [88, 154]; A. eucrenophila ATCC 23309 [88, 154]; LMG 13060 [154]; LMG 13687 [154]; A. hydrophila ATCC 7966 [88], ATCC 35654 [88]; CIP 76.14 [167]; A. jandaei ATCC 49568 [88, 154]; A. media ATCC 33907 [88, 154]; ATCC 35950 [154]; A. popoffi LMG 17541 [88]; LMG 17542 [154]; LMG 17545 [154]; A. salmonicida ATCC 33658 [88]; A. schubertii ATCC 43700 [88, 154]; A. sobria ATCC 43979 [88]; ATCC 35993 [88, 154]; CIP 74.33 [167]; A. trota ATCC 49657 [88, 154]; ATCC 49659 [154]; A. veronii ATCC 9071 [88, 154]; ATCC 35622 [88, 154]; ATCC 35624 [154]; CIP 103438 [167]
Alcaligenes	A. faecalis CIP 60.80 [167]; DSM 30030 [15]; DSM 13975 [15]
Alishewanella	A. fetalis DSM 16032 [15]
Arsenophonus	A. nasoniae DSM 15247 [15]
Arthrobacter	A. monumenti DSM 16405 [15]
Arthrobacter	Many spp. study [16]

Table 13.1   (continued)

Genus	Species and strain
Bacillus	*B. amyloliquefaciens* FZB42 [115]; *B. anthracis* Sterne [50]; A.16R [95]; *B. atrophaeus* ATCC 49337 [20, 99, 100, 85]; ATCC 9372 (formerly *B. globigii*; [40, 48, 50, 73, 84, 185]); *B. cereus* PEA26 [53]; T [48, 50, 67, 73, 185]; B33 [48, 50]; ATCC 14579 [32, 40, 85, 100]; NCTC 8035 [48, 50]; *B. circulans* ATCC 61 [40, 64]; *B. licheniformis* ATCC 14580 [40, 84]; SAF KL-196 [84]; *B. magaterium* ATCC 14581 [40, 84]; *B. mojavensis* ATCC 51516 [84]; *B. pumilus* A568 [114]; ATCC 7061 [84, 83]; ATCC 27142 [83]; SAF N-029 [83]; SAF R-032 [83]; *B. simplex* ZAN044 [85]; *B. sphaericus* ATCC 7134; [86, 193]; *B. sporothermodurans* IC4 [53]; *B. subtilis* 43 [114]; 49 [114]; 168 [37, 48, 50, 73, 83, 84]; A1/3 [37, 114]; A184 [114]; A190 [114]; A202 [114]; B2 [114]; B3 [114]; b213 [111, 114]; C-1 [113, 114]; JH642 [111]; MC85 [53]; RL45 [53]; ATCC 6051 [40, 84]; ATCC 6633 [37, 111, 112, 114]; ATCC 9943 [111]; ATCC 13952 [111, 114]; ATCC 15841 [23]; ATCC 21332 [111, 114]; ATCC 49760 [85]; ATCC 49822 [85]; DSM 1087 [37]; DSM 1088 [37]; *B. thuringiensis* 4AA-1 [50]; HD-1 (var kurstaki; [48]); ATCC 10792 [31, 50, 84, 85]; ATCC 13366 [50]; ATCC 13367 [50, 85]; ATCC 19265 [85]; ATCC 19267 [31]; ATCC 19269 [50]; ATCC 29730 [50, 85]; ATCC 33679 [50, 85]; ATCC 33680 [85]; ATCC 35646 [50]; ATCC 35646 [85]; ATCC 35866 [85]; ATCC 39152 [85]; ATCC 55172 [50]
Balneatrix	*B. alpica* CIP 103589 [15]
Bergeyella	*B. zoohelcum* LMG 8351 [15]
Blastomonas	*B. natatoria* DSM 3183 [15]; *B. ursincola* DSM 9006 [15]
Bordetella	*B. avium* CIP 103348 [167]; *B. hinzii* CIP 104527 [167]; *B. bronchiseptica* CIP 55.110 [167]
Brevebacillus	*B. laterosporus* NCTC 7579 [86]
Brevundimonas	*B. aurantiaca* DSM 4731 [15]; *B. diminuta* CIP 63.27 [167]; DSM 7234 [15]; *B. intermedia* DSM 4732 [15]; *B. nasdae* DSM 14572 [15]; *B. subvibrioides* DSM 4735 [15]; *B. vesicularis* CIP 101035 [167]; DSM 7226 [15]
Burkholderia	*B. ambifaria* CIP 107266 [167]; LMG 11351 [15, 167]; LMG 17828 [167]; *B. andropogonis* ATCC 23060 [167]; DSM 9511 [15]; *B. anthina* CIP 108228 [167]; LMG 16670 [15, 167]; *B. caledonica* LMG 19076 [15]; *B. caribensis* DSM 13236 [15]; *B. cenocepacia* CIP 108255 [167]; LMG 12614 [15]; LMG 18829 [167]; LMG 18830 [167]; LMG 18863 [167]; LMG 19230 [167]; LMG 19240 [167]; LMG 21440 [167]; LMG 21452 [167]; *B. cepacia* CIP 80.24 [167]; DSM 7288 [15]; LMG 2161 [15, 167]; LMG 6889 [167]; *B. cocovenenans* ATCC 33664 [167]; *B. dolosa* CIP 108406 [167]; DSM 16088 [15]; LMG 18941 [167]; *B. fungorum* LMG 20227 [15]; *B. gladioli* CIP 105410 [167]; DSM 4285 [15]; *B. glathei* CIP 105421 [167]; DSM 50014 [15]; *B. glumae* DSM 9512 [15]; NCPPB 2391 [167]; *B. multivorans* CIP 105495 [167]; LMG 14273 [167]; LMG 14293 [15, 167]; *B. phenazinium* DSM 10684 [15]; *B. phymatum* LMG 21445 [15]; *B. plantarii* ATCC 43733 [167]; DSM 9509 [15]; *B. pyrrocinia* CIP 105874 [167]; LMG 14191 [15]; LMG 21822 [167]; LMG 21823 [167]; *B. sacchari* LMG 19450 [15]; *B. stabilis* CIP 106845 [167]; LMG 6997 [167]; LMG 7000 [167]; LMG 14294 [15]; *B. terricola* LMG 20594 [15]; *B. thailandensis* DSM 13276 [15]; LMG 20219 [167]; *B. tropica* DSM 15359 [15]; *B. tuberum* LMG 21444 [15]; *B. vietnamiensis* CIP 105875 [167]; LMG 6998 [167]; LMG 6999 [167]; LMG 10929 [15]; *B. xenovorans* LMG 21463 [15]
Burkholderia	Many spp. study [104]

(continued overleaf)

Table 13.1   *(continued)*

Genus	Species and strain
Campylobacter	*C. coli* ATCC 33559 [87, 151]; ATCC 43473 [151]; ATCC 43474 [87, 151]; ATCC 43479 [151]; ATCC 43482 [87, 151], ATCC 43485 [151]; *C. fetus* ATCC 19438 [87]; ATCC 25936 [87]; ATCC 27374 [87]; *C. helveticus* ATCC 51209 [151]; *C. jejuni* ATCC 33291 [151]; ATCC 33292 [87]; ATCC 33560 [151]; ATCC 43429 [151]; ATCC 43430 [151]; ATCC 43431 [151]; ATCC 43432 [151]; ATCC 43449 [151]; ATCC 43463 [151]; ATCC 43482 [87]; NCTC 11168 [151]; *C. lari* ATCC 35221 [151]; ATCC 35222 [151]; ATCC 35223 [151]; ATCC 35233 [151]; ATCC 43675 [151]; *C. sputorum* ATCC 33491 [151]; ATCC 33709 [151]; LMG 9104 [151]; LMG 9108 [151]; LMG 9114 [151]; LMG 9125 [151]; LMG 9129 [151]; LMG 9140 [151]; LMG 9226 [151]; LMG 9234 [151]; LMG 9265 [151]; LMG 9269 [151]; *C. upsaliensis* ATCC 49815 [151]
Chryseobacterium	*C. indologenes* CIP 101026 [167]; *C. joostei* LMG18212 [15]; *C. scophthalmum* LMG 13028 [15]
Citrobacter	*C. freundii* NCTC 9750 [32]; *C. koseri* NCTC 10786 [32]; NCTC 10849 [32]
Clostridium	Many spp. study [58]
Comamonas	*C. aquatica* LMG 2370 [15]; *C. kerstersii* DSM 16026 [15]; *C. nitrativorans* DSM 13191 [15]; *C. terrigena* DSM 7099 [15]; *C. testosteroni* DSM 50244 [15]
Coxiella	*C. burnetii* Nine Mile phase I [56, 59, 106]; Australian QD [106], M44 [106], KAV [106], PAV [106], Henzerling [106], Ohio [106]
Cupriavidus	*C. gilardii* CIP 105966 [167]; LMG 3399 [167]; LMG 3400 [167]; *C. pauculus* CIP 105943 [167]; LMG 3245 [167]; LMG 3317 [167]; *C. respiraculi* LMG 21509 [167]; LMG 21510 [167]
Delftia	*D. acidovorans* CIP 103021 [167]; DSM 39 [15]
Elizabethkingia	*E. meningoseptica* CIP 60.57 [167]; DSM 2800 [15]; *E. miricola* DSM 14571 [15]
Empedobacter	*E. brevis* LMG 4011 [15]
Enterobacter	*E. cloacae* NCTC 10005 [32]
Enterococcus	*E. durans* NCTC 8307 [93]; *E. faecalis* ATCC 29212 [81]
Erwinia	Many spp. study [103]
Escherichia	*E. blattae* ATCC 29907 [25]; *E. coli* C1a [17, 52]; HB101 [85]; JM109 [85, 95]; K12 (1090; [98, 191]); RZ1032 [85]; UB5201 [86]; W3110 [85]; ATCC 4157 [25]; ATCC 9637 [31, 146]; ATCC 11175 [21, 31, 32, 96]; ATCC 11775 [25, 65]; ATCC 12014 [34, 107]; ATCC 14948 [34, 107]; ATCC 15223 [34, 107]; ATCC 25404 (K12; [19]); ATCC 25922 [25, 80, 81, 96]; ATCC 25933 [88]; ATCC 27325 [23, 24]; ATCC 33694 [20, 69, 85, 99, 100]; ATCC 35150 (O157:H7; [108]); ATCC 43890 (O157:H7; [85]); ATCC 43894 (O157:H7; [25, 85]); ATCC 43895 (O157:H7; [34, 107]); ATCC 700926 [144]; DSM 5802 [25]; DSM 8695 [25]; DSM 8696 [25]; DSM 8700 [25]; DSM 8711 [25]; DSM 9024 [25]; DSM 9029 [25]; DSM 9030 [25]; DSM 9031 [25]; DSM 9033 [25]; DSM 10722 [25]; DSM 11752 [25]; NCTC 9002 [32]; NCTC 9007 [32]; NCTC 9008 [32]; NCTC 9122 [32]; NCTC 11104 [32]; NCTC 11110 [32]; NCTC 12079 (O157:H7; [32]); NCTC 12080 [32]; NCTC 12900 [32]; *E. fergusonii* ATCC 35469 [25]; *E. hermanni* ATCC 33650 [25]; *E. vulneris* ATCC 33821 [25]
Flavobacterium	*F. flevense* DSM 1076 [15]; *F. gelidilacus* DSM 15343 [15]; *F. hibernum* DSM 12611 [15]; *F. hydatis* DSM 2063 [15]; *F. johnsoniae* DSM 2064 [15]; *F. pectinovorum* DSM 6368 [15]; *F. resinovorum* DSM 7478 [15]; *F. saccharophilum* DSM 1811 [15]

Table 13.1   (continued)

Genus	Species and strain
Fusobacterium	F. nucleatum 25286 [129]
Haemophilus	H. ducreyi 35000 [156]
Helicobacter	H. pylori ATCC 26695 [97]; ATCC 43526 [87]; ATCC 43579 [87]; ATCC 43629 [87]; H. mustelae ATCC 43744 [87]
Inquilinus	I. limosus CIP 8342 [167]; DSM 16000 [15]
Klebsiella	K. aerogenes NCTC 9499 [32]; NCTC 9504 [32]; K. pneumoniae ATCC 23356 [88]
Lactobacillus	L. bulgaricus ATCC BAA-365 [180]; ATCC 11842 [180]; L. plantarum NCIMB 8826 [89]
Listeria	L. grayi ATCC 19120 [137]; ATCC 25401 [137]; L. inoccua sp. [57]; ATCC 33090 [137]; L. ivanovii ATCC 49954 [137]; ATCC 19119 [137]; L. monocytogenes ATCC 9525 [25]; ATCC 15313 [25, 137]; ATCC 19114 [25]; L. seeligeri ATCC 35967 [137]; L. welshimeri ATCC 35897 [137]
Malikia	M. spinosa DSM 15801 [15]
Microbulbifer	M. elongatus DSM 6810 [15]
Micrococcus	M. luteus CIP 103664 [90]
Microcystis	M. sp. PCC 7806 [39]; PCC 7813 [39]; HUB -5-2-4 [39]; HUB 5-3 [39]; HUB 063 [39]; M. aeruginosa isolate [39]; HUB 063 [171]; M. ichthyoblabe isolate [39]; M. wesenbergii isolate [39]
Mycobacterium	M. abscessus ATCC 19977 [135]; ATCC 23016 [135]; ATCC 700869 [135]; M. africanum ATCC 25420 [135]; M. avium ATCC 25291 [135]; ATCC 35717 [135]; ATCC 35718 [135]; ATCC 43544 [135]; ATCC 43545 [135]; ATCC 49884 [135]; ATCC 700735 [102]; ATCC 700736 [54, 102]; ATCC 700737 [102]; ATCC 700898 [102]; M. bovis BCG [54]; M. chelonae ATCC 14472 [135]; ATCC 35751 [135]; ATCC 35752 [135]; M. doricum ATCC BAA-565 [135]; M. fortuitem ATCC 6841 [54, 135]; ATCC 35931 [135]; ATCC 43266 [135]; ATCC 49403 [135]; ATCC 49935 [135]; M. intracellulare ATCC 13950 [54, 102, 135]; ATCC 35761 [102]; ATCC 35762 [102]; ATCC 35764 [102]; ATCC 700662 [135]; ATCC 700664 [135]; M. kansasii ATCC 12478 [54, 102, 135]; ATCC 12479 [102]; ATCC 25100 [135]; ATCC 25101 [135]; ATCC 25414 [135]; ATCC 35775 [102]; ATCC 49913 [102]; M. marinum ATCC 927 [135]; ATCC 11566 [135]; M. microti ATCC 19422 [135]; ATCC 11152 [135]; ATCC 35781 [135]; ATCC 35782 [135]; M. mucogenicum ATCC 49649 [135]; ATCC 49650 [135]; ATCC 49651 [135]; M. tuberculosis H37Ra (ATCC 25177; [54, 102, 135]); ATCC 35815 [102]; 35817 [102]; 35818 [102]; M. tusciae ATCC BAA-564 [135]
Neisseria	N. gonorrhoeae[156]; N. lactamica NCTC 10617 [32]
Oscillatoria	O. sp. PCC 9240 [92]; PCC 6506 [92]; PCC 9029 [92]; PCC 10111 [92]
Pandoraea	P. apista CIP 106627 [167]; LMG 16407 [15]; P. norimbergensis DSM 11628 [15]; LMG 13019 [167]; P. pnomenusa LMG 18087 [167]; LMG 18817 [15]; P. pulmonicola LMG 18106 [15, 167]; P. sputorum LMG 18100 [167]
Pannonibacter	P. phragmitetus LMG 5414 [15]; P. phragmitetus LMG 5430 [15]
Pantoea	P. agglomerans ATCC 33243 [100, 85]
Parvimonas	P. micra ATCC 33270 [129]
Peptostreptococcus	P. anaerobicus ATCC 27337 [129]
Planktothrix	P. agardhii HUB 011 [171]
Plesiomonas	P. shigelloides ATCC 14029 [154]; ATCC 14030 [154]; CIP 63.5 [167]
Porphyromonas	P. gingivalis Boston 381 [129]
Prevotella	P. intermedia DSM 20706 [129]; P. nigrescens DSM 13386 [129]
Proteus	P. mirabilis NCTC 11938 [32]; P. vulgaris NCTC 4175 [32]

(continued overleaf)

Table 13.1   *(continued)*

Genus	Species and strain
Pseudomonas	*P. abietaniphila* CIP 106708 [15]; *P. aeruginosa* B0267 [85]; ATCC 25619 [64]; ATCC 27853 [88]; CIP 76.110 [167]; DSM 50071 [15]; NCTC 10332 [32]; NCTC 11443 [32]; *P. agarici* DSM 11810 [15]; *P. alcaligenes* CIP 101034 [167]; DSM 50342 [15]; *P. amygdali* DSM 7298 [15]; *P. anguilliseptica* DSM 12111 [15]; *P. antarctica* DSM 15318 [15]; *P. asplenii* LMG 2137 [15]; *P. aurantiaca* CIP 106718 [15]; *P. avellanae* DSM 11809 [15]; *P. azotoformans* DSM 106744 [15]; *P. balearica* DSM 6083 [15]; *P. beteli* LMG 978 [15]; *P. boreopolis* LMG 979 [15]; *P. brassicacearum* DSM 13227 [15]; *P. brenneri* DSM 106646 [15]; *P. caricapapayae* LMG 2152 [15]; *P. cedrina* DSM 105541 [15]; *P. chloritidismutans* DSM 13592 [15]; *P. chlororaphis* DSM 50083 [15]; *P. cichorii* DSM 50259 [15]; *P. citronellolis* DSM 50332 [15]; *P. congelans* DSM 14939 [15]; *P. corrugata* DSM 7228 [15]; *P. extremorietalis* DSM 15824 [15]; *P. flavescens* DSM 12071 [15]; *P. fluorescens* isolate [21]; CIP 69.13 [167]; DSM 50090 [15]; *P. fragi* DSM 3456 [15]; *P. frederiksbergensis* DSM 13022 [15]; *P. fulva* LMG 11722 [15]; *P. fuscovaginae* DSM 7231 [15]; *P. geniculata* LMG 2195 [15]; *P. gessardii* CIP 105469 [15]; *P. graminis* DSM 11363 [15]; *P. grimontii* DSM 106645 [15]; *P. hibiscicola* LMG 980 [15]; *P. huttiensis* DSM 10281 [15]; *P. indica* DSM 14015 [15]; *P. jessenii* CIP 105274 [15]; *P. jinjuensis* LMG 21316 [15]; *P. kilonensis* DSM 13647 [15]; *P. koreensis* LMG 21318 [15]; *P. libanensis* CIP 105460 [15]; *P. lundensis* DSM 6252 [15]; *P. lutea* LMG 21974 [15]; *P. luteola* CIP 102995 [167]; DSM 6975 [15]; *P. mandelii* CIP 105273 [15]; *P. marginalis* DSM 13124 [15]; *P. mendocina* CIP 75.21 [167]; DSM 50017 [15]; *P. mephitica* CIP 106720 [15]; *P. migulae* CIP 105470 [15]; *P. monteilii* DSM 14164 [15]; *P. mosselii* CIP 104061 [167]; CIP 105259 [15]; *P. mucidolens* LMG 2223 [15]; *P. multiresinivorans* LMG 20221 [15]; *P. nitroreducens* DSM 14399 [15]; *P. oleovorans* DSM 1045 [15]; *P. orientalis* CIP 105540 [15]; *P. oryzihabitans* CIP 102996 [167]; DSM 6835 [15]; *P. pertucinogena* LMG 1874 [15]; *P. pictorum* LMG 981 [15]; *P. plecoglossicida* DSM 15088 [15]; *P. poae* DSM 14936 [15]; *P. proteolytica* DSM 15321 [15]; *P. pseudoalcaligenes* CIP 66.14 [167]; DSM 50188 [15]; *P. putida* F1 [100]; PP0310 [85]; ATCC 39169 [85]; CIP 52.191 [167]; DSM 291 [15]; DSM 50198 [15]; *P. resinovorans* LMG 2274 [15]; *P. rhizosphaerae* LMG 21640 [15]; *P. rhodesiae* DSM 14020 [15]; *P. savastanoi* LMG 2209 [15]; LMG 5011 [15]; *P. straminea* CIP 106745 [15]; *P. stutzeri* ATCC 39524 [85]; CIP 103022 [167]; DSM 5190 [15]; *P. synxantha* LMG 2190, *P. syringae* DSM 6693 [15]; LMG 1247 [15]; *P. taetrolens* LMG 2336 [15]; *P. thermotolerans* DSM 14292 [15]; *P. thivervalensis* DSM 13194 [15]; *P. tolaasii* LMG 2342 [15]; *P. trivialis* DSM 14937 [15]; *P. umsongensis* LMG 21317 [15]; *P. vancouverensis* CIP 106707 [15]; *P. veronii* DSM 11331 [15]; *P. viridiflava* DSM 11124 [15]
Ralstonia	*R. eutropha* DSM 531 [15]; *R. mannitolilytica* CIP 107281 [167]; LMG 6866 [15]; LMG 18098 [167]; LMG 18102 [167]; *R. pickettii* CIP 73.23 [167]; DSM 6297 [15]; LMG 5942 [167]; LMG 7001 [167]; *R. syzygii* DSM 7385 [15]
Rhizobium	*Rhyzobium* sp. VMA301 [22]; *R. radiobacter* DSM 30147 [15]; *R. rubi* DSM 6772 [15]; *R. tropici* DSM 11418 [15]
Salmonella	*S. choleraesuis* ATCC 14028 [64]; *S. enteritidis* ATCC 13076 [96]; *S. typhimurium* ATCC 6994 [25]; ATCC 13311 [96]
Salmonella	Many spp. study [38]

Table 13.1   *(continued)*

Genus	Species and strain
Serratia	*S. ficaria* NCTC 12148 [32]; *S. grimesii* NCTC 11543 [32]; *S. marcescens* ATCC 13880 [32, 85]
Shewanella	*S. algae* BrY [100]; DSMZ 9167 [15]; *S. baltica* DSM 9439 [15]; *S. fidelis* LMG 20552 [15]; *S. frigidimarina* DSM 12253 [15]; *S. profunda* DSM 15900 [15]; *S. putrefaciens* CIP 80.40 [167]; DSM 6067 [15]
Shigella	*S. boydii* ATCC 9207 [64]; NCTC 9327 [32]; *S. flexneri* PHS-1095 [191]; *S. sonnei* NCTC 9774 [32]; NCTC 8219 [32]; NCTC 8221 [32]
Sphingobacterium	*S. faecium* DSM 11690 [15]; *S. mizutaii* DSM 11724 [15]; *S. multivorum* CIP 100541 [167]; DSM 11691 [15]; *S. spiritivorum* CIP 100541 [167]; DSM 11722 [15]; *S. thalpophilum* DSM 11723 [15]
Sphingobium	*S. chlorophenolicum* DSM 7098 [15]; *S. herbicidovorans* DSM 11019 [15]; *S. xenophagum* DSM 6383 [15]
Sphingomonas	*S. adhaesiva* DSM 7418 [15]; *S. aerolata* DSM 14746 [15]; *S. aquatilis* DSM 15581 [15]; *S. aurantiaca* DSM 14748 [15]; *S. cloacae* DSM 14926 [15]; *S. faeni* DSM 14747 [15]; *S. koreensis* DSM 15582 [15]; *S. melonis* DSM 14444 [15]; *S. parapaucimobilis* DSM 7463 [15]; *S. paucimobilis* CIP 100752 [167]; DSM 1098 [15]; *S. pituitosa* DSM 13101 [15]; *S. trueperi* DSM 7225 [15]; *S. wittichii* DSM 6014 [15]; *S. yabuuchiae* DSM 14562 [15]
Sphingopyxis	*S. macrogoltabida* DSM 8826 [15]; *S. terrae*[15]
Staphylococcus	*S. aureus* ATCC 97-04SV [21]; V8333 [52]; ATCC 6538 [25]; ATCC 8325 [52]; ATCC 12600 [64]; ATCC 13301 [35]; ATCC 14458 [25, 85]; ATCC 14990 [85]; ATCC 19095 [85]; ATCC 25923 [25, 35, 52, 80]; ATCC 27659 [35]; ATCC 29213 [125, 81]; ATCC 43330 [125]; CIP 7625 [90]; NCIMB 6571 [35]; NCIMB 8625 [35]; NCIMB 11852 [35]; NCIMB 11195 [35]; NCTC 6571 [32]; NCTC 8532 [32]; *S. capitis* ATCC 35661 [17, 52]; CIP 8153 [90]; CIP 104191 [90]; *S. caprae* CIP 104519 [90]; *S. cohnii* CIP 8154T [90]; CIP 104023 [90]; NCTC 11041 [35]; *S. epidermidis* CIP 103563 [90]; NCTC 7944 [35]; NCTC 11047 [35]; *S. haemolyticus* ATCC 29970 [52]; CIP 8156 [90]; NCTC 11042 [35]; *S. hominis* CIP 102642 [90]; CIP 105721 [90]; *S. intermedius* CIP 8177 [90]; *S. lentus* CIP 103585 [90]; *S. lugdunensis* CIP 103642 [90]; *S. pasteuri* CIP 103831 [90]; *S. saprophyticus* CIP 104064 [90]; CIP 105260 [90]; NCIMB 8711 [35]; *S. schleiferi* CIP 103643 [90]; CIP 104370 [90]; *S. sciuri* CIP 103824 [90]; *S. simulans* CIP 8164 [90]; *S. warneri* CIP 103960 [90]; *S. xylosus* CIP 8166 [90]
Stenotrophomonas	*S. acidaminiphila* DSM 13117 [15]; *S.africana* CIP 104854 [15]; *S. maltophilia* CIP 60.77 [167]; DSM 50170 [15]; *S. nitritireducens* DSM 12575 [15]; *S. rhizophila* DSM 14405 [15]
Streptococcus	*S. agalactiae* ATCC 13813 [32, 81]; *S. anginosus* DSM 20563 [29]; *S. constellatus* DSM 20575 [29]; *S. gordonii* DSM 6777 [29]; *S. intermedius* DSM 20573 [29]; *S. mitis* DSM 12643 [29]; NCTC 10712 [127]; NCTC 12261 [127]; *S. mutans* DSM 20523 [29]; NCTC 10449 [30]; *S. oralis* ATCC 10557 [127]; ATCC 15914 [127]; ATCC 35037 [127]; DSM 20627 [29]; *S. parasanguinis* DSM 6778 [29]; *S. pneumoniae* NCTC 7465 [32]; *S. pseudopneumoniae* ATCC BAA-960 [127]; *S. pyogenes* ATCC 12112 [52]; ATCC 700294 [81]; NCTC 8198 [32]; *S. salivarius* DSM 20560 [29]; NCTC 8618 [32]; *S. sanguinis* DSM 20567 [29]; *S. thermophilus* ATCC BAA-250 [180]; ATCC BAA-491 [180]
Streptococcus	Many spp. studies [30, 81, 127]
Tannerella	*T. forsythia* ATCC 43037 [128]

(continued overleaf)

Table 13.1   *(continued)*

Genus	Species and strain
Terrimonas	*T. ferruginea*[15]
Vibrio	*V. vulnificus* ATCC 33149 [154]
Weeksella	*W. virosa* LMG 12995 [15]
Wollinella	*W. succinogenes* DSM 1740 [15]
Yersinia	*Y. enterocolitica* ATCC 23716 [108]; ATCC 51871 [23, 24, 85]; NCTC 9499 [25]; NCTC 11503 [25]; *Y. frederiksenii* NCTC 11470 [25]; *Y. intermedia* ATCC 29909 [25]; *Y. kristensenii* NCTC 11471 [25]; *Y. pestis* EV76 [95]

ATCC, American Type Culture Collection (USA; www.lgcstandards-atcc.org); CIP, 'Collection de l'Institut Pasteur' (France; www.pasteur.fr); DSM(Z), 'Deutsche Sammlung von Mikroorganismen und Zellkulturen' (Germany; www.dsmz.de); HUB, Culture Collection of algae at the Humboldt-University, Institute of Biology; LMG, 'Cultuurcollectie van het Laboratorium voor Microbiologie te Gent' (Belgium; bccm.belspo.be/about/lmg.php); MCCM, Medical Culture Collection Marburg (Germany); NCIMB, National Collection of Industrial, Marine and Food Bacteria (UK; www.ncimb.com); NCTC, National Collection of Type Cultures (UK; www.hpacultures.org.uk); PCC, Pasteur Cyanobacteria Collection (France; www.pasteur.fr); SAF, spacecraft assembly facility (Jet Propulsion Laboratory, USA; no www link).

fungi. Somewhat less confident but more rapid and unbiased bacteria typing can be accomplished by a different MS approach: whole cell MALDI MS can be accomplished within a few minutes or even in near real-time [13]. Whole cell MALDI MS is also known as intact cell MS (ICMS), but that name is less appropriate because it alludes to the mass determination of actually intact cells [14]. Rather, whole cell MALDI MS produces a mass spectrum of material that is laser ionised from matrix covered whole cells. This chapter discusses the rapid typing of bacteria by whole cell MALDI MS.

13.2   Scope

The quality of whole cell mass spectra that support the typing of bacteria is determined by the combination of the analysis process and the sample material. As the spectrum varies with sample and process parameters, the first section deals with the ruggedness of the experiments and the data handling required for reproducible whole cell MS of bacteria. With those practical conditions outlined, the reported studies of the actual signatures of several bacteria families and species are discussed. Although the typing is now generally supported by comparison with library spectra, a deeper understanding of the relation of mass spectrum signals to 'biomarker' compounds of the investigated microorganisms is now growing. Therefore, strategies for the identification of such biomarkers in whole cell MALDI MS spectra are discussed. Finally, conclusions on the current position of whole cell MS typing and an outlook on the direction of future endeavours are given.

13.3   Reproducibility

Considering that it is possible to obtain mass spectra of intact bacteria, the utility of whole cell MALDI MS as a typing method is determined by method characteristics: selectivity, sensitivity, and reproducibility. The selectivity is generally considered good, for instance comparable with or better than 16S rRNA typing [15, 16] but it is essentially still being tested in studies on a variety of bacteria. Sensitivity is generally reported as the quantity of bacteria that is deposited in a well of a MALDI target. Reported experience generally indicates that in the order of 10^4–10^5 organisms are minimally required for the acquisition of spectra that display signals of useful intensity [17, 18]. The quantity of organisms actually consumed by laser irradiation, under these 'limit of speciation' conditions, is estimated to range between 10^3 and 10^4 organisms [18]. Despite improvements in MALDI MS technology, this 'limit of speciation' has not changed since the introduction of whole cell MALDI MS, in 1996. The third property, reproducibility, is of key importance: typing implies comparison of the spectrum of an unknown with many spectra of known bacteria. Currently, the processes that lead to a spectrum are not understood to the level that spectra or biomarkers can be predicted from independently available information, for example from bacterial genomes. Therefore, similarity produces a positive type assignment, whereas dissimilarity will generally require follow-up work for actual typing. Of course, spectrum comparison must not be hampered by spectral variation due to the experimental procedure, including sample treatment, or due to data processing. In addition, the prior culture conditions are generally assumed to introduce a degree of uncontrollable variation in whole cell MALDI mass spectra. Here, a division is made into factors concerning the sample and factors concerning the physical chemistry of the MALDI MS process.

13.3.1 Factors Concerning the Sample

Culture conditions may cause differences in several ways, due to the culture media components themselves, to accumulation of excreted material during culture, to biological conversion of cellular and culture medium material, and to different protein expression and metabolite production levels. In principle, any extracellular components can be separated from the cells, prior to MALDI sample application. This separation is desirable, because any material other than the analyte cells can affect the MALDI process. In particular, salt and surfactants are notorious interferents that may cause shifting or suppression of signals.

That uncontrollable experimental variation is caused by bacterial culture conditions is corroborated by many studies. In a first systematic study, in which *Escherichia coli* whole cell mass spectra were monitored over a 48-h growth period, it was demonstrated that intense peaks may appear and vanish while other peaks displayed less dramatic variation [19]. Some other studies early in the development of whole cell MALDI MS corroborated these findings. For example, a strain of *E. coli* and one of *Bacillus atropheus* were employed to obtain replicate spectra at selected points in time during a 10-day period of liquid culture [20]. It was observed that spectrum variation on same culture days was minor, whereas spectrum similarity on different culture days was generally poor. Strains of several Gram-positive and -negative bacteria were used to obtain spectra at selected points in time during a 4-day period of agar plate culture [21]. In these example studies, and others [16, 22], particular biomarker signals were found present under all conditions despite clear variations in the spectra.

Later, more extensive and more systematic studies confirmed these early findings. For example, the PNNL group compared minimal medium M9, tryptic soy broth (TSB), Luria-Bertani (LB) broth, and blood agar plates for automatically extracted peaks of spectra from

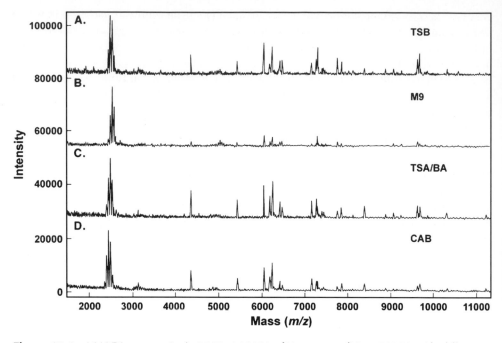

Figure 13.1　*MALDI spectra (m/z 2000–14 000) of* Y. enterocolitica *51871 with different growth media (adapted from Valentine* et al. *[23])*

B. subtilis, *E. coli*, and *Yersinia enterocolitica* [23]; typically observed differences are shown in Figure 13.1. Limited tests with the automated peak extraction and library matching showed that identification of bacteria was correct based on the common biomarker signals and despite the overall spectrum variations [23]. A further study investigated the effects on whole cell MALDI mass spectra of growth rate and culture pH with *E. coli* in a chemostat culture and of temperature with *Y. enterocolitica* [24]. A biofilm sample from the *E. coli* chemostat culture was the only sample to produce a mismatch. Although visual phenotypical changes occurred, the spectra of *E. coli* and *Y. enterocolitica* remained sufficiently similar for correct identification by spectrum comparison and the various spectra clustered correspondingly [24]. Three different media, NB, TSB, and TYB, and three growth times, 18, 24, and 48 h, were tested with *E. coli* ATCC 25922 in a wider study of foodborne pathogens [25]. The different growth times were not causing significant differences, and 15 biomarker signals were found under all conditions tested. However, TSB culture yielded the better spectra in terms of signal-to-noise ratio and number of bacterial peaks [25]. Most recently, a systematic chemometrics approach to the entire sampling and analysis process was reported [26]. A combination of variance analysis (ANOVA) and principal component analysis (PCA) was employed to evaluate contributions from the various experimental factors to variation in (compressed) whole cell MALDI spectra. This study confirmed that growth conditions and culture time were by far the major cause of spectrum variation [26].

　　The growth condition related variation might be due to clutter introduced by the growth medium itself, but also by the compounds differentially expressed or excreted during growth. That culture medium components are an appreciable source of variation in whole cell spectra was challenged by observations made in flow field-flow fractionation (flow FFF; [27]) and its hollow fibre variant [28] in sample preparation for whole cell MALDI MS. These FFF methods achieve cell sorting in the liquid phase, including separation of cells from extracellular components like that of the culture medium. Both studies reported a markedly low variation in whole cell spectra and it was suggested that this was due to separation of culture material from the cells. However, a systematic comparison of growth conditions in combination with FFF and whole cell MALDI MS that might provide more conclusive evidence has not yet been reported.

All evidence concerning the influence of culture conditions on spectrum quality shows that best typing performance of whole cell MALDI MS is obtained with standardised culture conditions. In practice, this consideration is of limited value as not all bacteria can be cultured on a limited set of media. In addition, typing without culture is often desired, for example in the analysis of aerosolised bacteria or in clinical samples. In general, appropriate establishment of robust biomarker signals will provide a sufficient basis for successful typing despite the whole cell MADI mass spectrum variations that spring from culture conditions.

Prior to analysis, sample material is often centrifuged and pelleted from culture, and washed [20, 29, 30]. Alternatively, cells may be reconstituted from previously pelleted and washed lyophilised material [29, 31], they may be taken from plated colonies by careful scraping with an inert metal wire loop [21, 32–36] or a pipette tip [37, 38], they may be taken from liquid culture by careful picking under a microscope [39], or they may even be captured from aerosol prior to or after matrix application [40–46]. While liquid culturing requires additional sample handling, plate culture affords direct transmission of selected colonies to a MALDI MS target plate. For the analysis of spores or highly pathogenic species, additional sample treatment may be required.

Bacterial spore mass spectra may have poor quality, because little or no protein material may have been released from the spore case. The quality of whole cell MALDI mass spectra, in terms of number of peaks and signal-to-noise ratio, generally improves by a treatment that induces proteins release. As the addition of acid to the matrix is common in most MALDI sample preparation, increased acidity is an obvious means to improve protein release [47]. Sonication of spores prior to target application was employed to good effect [48–50]. A corona plasma discharge was successfully applied to spores on the MALDI target plate [50, 51]. That method did also provide improvement of spectra of vegetative bacteria but it is not likely that any whole cells are left in those cases. Improvement of whole cell spectrum quality of Gram-positive bacteria by a brief lysozyme treatment was tested [52]. An optimum lysozyme concentration had to be established, and lysozyme signals started to appear in the spectra. With an established optimum sample treatment protocol in place, the authors concluded that lysozyme treatment did increase the number of peaks and the spectral range for Gram-positive bacteria [52]. Alternatively, a wet heat treatment can be applied prior to sample application to afford protein release on target [53]. This wet heat treatment was shown to be particularly useful for the analysis of washed spores recovered from complex sample matrices, for example foodstuff.

Four methods for improved protein release from Gram-positive bacteria, i.c. nonsporulating *Arthrobacter* spp., were compared [16]. Heat and ethanol treatment yielded similar spectra but with ~40-fold less intense signals than in spectra of untreated cells. Lysozyme application yielded more signals in the 8–10kDa range, but overall signal intensity was

~20 times lower than for untreated cells. Mechanical cell disruption did not produce useful spectra [16]. This study shows that the possible benefit of extra treatment for Gram-positive bacteria must first be explored before it is generally applied.

In principle, whole cell MALDI MS can be equally applied to nonpathogenic and highly pathogenic bacteria. However, MALDI mass spectrometers are rarely found inside a biological containment and whole cell samples of BSL-II to –IV organisms require inactivation prior to analysis. Common autoclaving destroys the cell structure and it is only suitable for obtaining cell lysates. Of course, the observation that crude cell lysates can also be subjected to MALDI MS may provide an escape from the biosafety issue [54]. Gamma-irradiation is a validated and widely accepted option [55, 56] but the strong sources required to confidently kill pathogens within ~1 h are not widely available. More recently, ways of killing pathogens by chemicals that do not interfere with the MALDI analysis have been reported.

The effects of some common sterilisation procedures on whole cell mass spectra were investigated for the Gram-positive *Listeria innocua* and the Gram-negative *E. coli* O157:H7 [57]. Formaldehyde, phenol, and lysozyme treatment resulted in poor spectra (few signals, high signal-to-noise), while boiling in distilled water or suspension in 70% ethanol resulted in good spectra. The observation that ethanol yielded better spectra was corroborated by an earlier study, where 40% ethanol was employed [21]. Cell disruption by ethanol was excluded for the investigated *E. coli* and *Listeria innocua*, as ethanol supernatant did not display any protein signals [57]. The earlier study employed microscopy to reveal that cells were desiccated and more evenly distributed over the matrix crystals [21]. It was concluded that sterilisation in 70% ethanol worked well but that possible persistence of certain bacteria might require closer investigation for highly pathogenic species [57]. Other means of rendering pathogens harmless and compatible with whole cell MALDI MS were also investigated.

Treatment with trifluoroacetic acid (TFA) was employed in an investigation of potentially extremely toxic and pathogenic *Clostridium* spp. [58]. Trichloroacetic acid (TCA) and acetonitrile were employed in a procedure for obtaining safe extracts from pathogenic bacteria for MALDI MS [59]. A generally applicable and whole cell MALDI MS compatible inactivation method for sporulated and vegetative bacteria was developed and validated in a systematic study involving *Yersinia*, *Bacillus* and *Burkholderia* spp. [60]. Treatment with 80 vol% TFA proved sufficient for killing vegetative cells but spores required additional centrifugation and filtering to ascertain adequate killing. *E. coli* and *Bacillus subtilis* were employed as surrogates in a comparison of treated and untreated material; while spectra changed significantly by the treatment, key biomarker signals were still present in good abundance [60]. Thus, radiation and acid treatment provide ways of rendering pathogenic bacteria harmless prior to whole cell MALDI MS outside a BSL containment laboratory.

Whole cell MALDI MS application may benefit from fairly simple sample treatment for the isolation of organisms from complex sample matrices, for instance foodstuff or urine. Successful bioaffinity clean-up for whole cell MALDI MS has been reported by several research groups. Effectively, the biomolecular interaction analysis (BIA)/MS [61], surface enhanced laser desorption/ionisation (SELDI) [4, 62] and iMALDI [63] variants of MALDI are based on a similar bioaffinity-extraction, not of the intact cells but of preselected proteins or digest peptides, respectively. While BIA/MS, SELDI and iMALDI do target a limited set of predetermined antigen molecules, bioaffinity can be employed as a less biased clean-up method.

Magnetic beads coated with polyclonal anti-*Salmonella* antibodies were used for the extraction of *Salmonella choleraesuis* from a mixture of bacteria; after incubation, the beads were washed, applied to a MALDI target plate, and subjected to MALDI MS [64]. It was shown that the spectra of *S. choleraesuis* attached to beads were very similar to those of the directly suspended bacteria. Relevant signals were found up to ~52 kDa, although the higher mass signals (above ~40 kDa) were absent from bacteria cultured less than 24 h. Extraction of *S. choleraesuis* from river water and urine yielded spectra with little or no interference, and distinct signals of the bacteria were present along interfering signals from an extract of chicken blood [64]. Antibody coated magnetic beads were also used for the selective extraction of enterohaemorrhagic *E. coli* O157:H7, where subsequent application of the beads allowed rapid and successful whole cell MALDI MS typing [34].

Although the above antibody based clean-up can be combined with whole cell MALDI MS, the bias introduced by antibody selectivity is generally at variance with the unbiased MS analysis. Bundy and Fenselau introduced the use of much less selective lectin and carbohydrate affinity for the isolation of bacteria prior to whole cell MALDI MS [65, 66]. The carbohydrate structures, for example O-antigens, at the cell surface of many bacteria do have an affinity for chemically similar structures in lectins or carbohydrates. Immobilisation of these affinity molecules on the MALDI target affords extraction capability and subsequent application of matrix then allows whole cell MALDI MS. Although the immobilisation was provisional, employing a standard target, this kind of processing was shown successful for the extraction and analysis of *E. coli* from urine by a lectin [65] and of *Salmonella typhimurium* from chicken package exudate and milk by a human blood group antigen [66]. In a later study, a less provisional way for lectin affinity capture was demonstrated with commercially available materials, for lectins bound to glass slides and capture of *Bacillus* spores [67].

Although these binding approaches were designed to serve clean-up and sample enrichment, they may be modified to support a research application. Interestingly, a recent whole cell MALDI MS study investigated the binding of *Bifidobacterium* spp. to human plasminogen and it was shown that plasminogen produced a signal in the *Bifidobacterium* spectra [68]. This suggests that immobilisation of many human cell wall structures, for example receptors, may provide new insights in pathogen binding.

13.3.2   Factors Concerning the MALDI MS Process

An important test of the influence of the MALDI MS process on spectrum reproducibility was delivered by an inter-laboratory comparison in which the sample parameters were well controlled and data processing was done in the same way [69]. The spectra obtained from three laboratories were similar but not identical, even when signal intensities were not considered. Nevertheless, a set of signals was observed in all cases and *E. coli* was correctly identified against a previously accumulated library of spectra. Across the laboratories, the lowest common denominator of matching still produced 82% similarity in the observed mass signals [69].

For a consideration of the MALDI MS analysis process, a schematic breakdown is employed here (see Figure 13.2). Ionisation is a prime prerequisite for the physics of mass

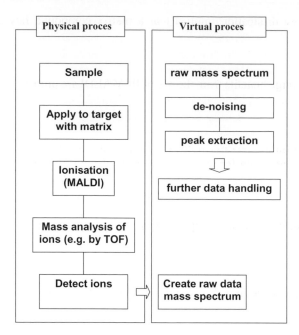

Figure 13.2   *Schematic breakdown of the MALDI MS analysis process, including general data processing; the 'physical process' encompasses the actual experiment, while the 'virtual process' concerns electronic data handling*

analysis and detection to work. Although mass analysis and detection can be accomplished in several ways, the actual hardware technology is usually not a factor for immediate experimental control. Rather, the technology imposes certain limitations on the experiment, for example with respect to sensitivity or accessible mass range. For whole cell MS, the time-of-flight (TOF) principle is widely employed for mass analysis. TOF MS imposes no mass range limitation when combined with the present ionisation method performance. As TOF constitutes a nonscanning method of mass analysis, highly efficient transmission can be designed to provide below attogram (10^{-18} g) sensitivity. In addition, TOF mass analysis requires specific detection technology and electronics, to match time separation of the arrival of ions of different mass. A discussion of TOF technology is beyond the scope of this chapter, and a detailed discussion can be found in an appropriate textbook [70]. The other steps of the process, sample preparation, ionisation and data analysis are under immediate experimental control and they deserve further consideration here.

13.3.2.1   *Sample Application and Ionisation*

Sample application and ionisation are closely linked for MALDI MS. Plain laser ionisation of biological material leads to atomisation, where much of the information on chemical structure is lost. In a Nobel Prize winning discovery it was shown that the addition of a matrix compound to a sample affords dissipation of incident laser energy with concomitant formation of ions from the larger molecules that still carry relevant information, like proteins and sugars [71, 72]. Much of the MALDI work is conducted with UV lasers and the required energy dissipation implies that a matrix compound has a significant UV absorbance. Although IR lasers have a different matrix compatibility, the still limited hardware ruggedness hampers widespread use. A single study compared 355 nm UV MALDI and 2.94 μm IR MALDI for bacteria, albeit with the same matrix; the comparison demonstrated that large benefits from the use of IR MALDI are probably not to be expected in this area of application [73]. Except for dissipation of laser energy, the matrix also has to provide or accept protons, as positive or negative ionisation of biochemicals usually involves respective protonation or deprotonation. However, the physical chemistry involved in the ionisation process is not yet understood to the extent that reliable predictions can be made, not even for well defined cases of ionisation of pure biochemicals in a given matrix. Despite thorough fundamental studies of MALDI [74–79], matrix choice and sample application are still based on experience rather than on science [53].

Matrices commonly found to perform well for whole cell MALDI MS of bacteria are α-cyano-4-hydroxycinnamic acid (CHCA; [29–31, 33, 54, 57, 80–82]), ferulic acid (FA; [20, 21, 83–85]), sinapinic acid (SA; [17, 27, 48, 54, 73, 81, 86–89]), 2,5-dihydroxybenzoic acid (DHB; [81, 90–92]), a mixture of DHB and 2-hydroxy-5-methoxybenzoic acid [37], 2-mercaptobenzothiazole (MBT; [80]), 5-ethyl-2-mercaptothiazole (MT; [80]), and 5-chloro-2-mercaptobenzothiazole (CMBT; [32, 35, 36, 82]). This is not an exhaustive list because other compounds have been successfully employed as a matrix. Some studies have shown that use of different matrices for Gram-positive and -negative bacteria may benefit spectrum quality in terms of number of signals and signal intensity [17, 82]. However, careful experimental optimisation has shown that a single matrix may be used for Gram-positive and -negative bacteria [57]. Matrix compounds are generally dissolved to saturation, with solvent composition ranging from deionised water to intricate mixtures of water and additives, for example 18-crown-6 [93] or fructose [54]. That the choice of matrix may have dramatic consequences was well illustrated in a study of *Bacillus* spp. spore extracts, where SA failed to produce signals of distinctive *B. anthracis* biomarkers observed with CHCA [47]. Another study of *Bacillus* spp. showed that FA yielded substantially more signals above 10 kDa than other matrices [84]. Yet another study of *Bacillus* spp. reported comparable performance for CHCA and SA and relatively better performance of DHB in the high mass range [53]. It was noted, however, that seemingly minor differences in sample composition, for example due to washing or sporulation, might hamper comparison of results across the various studies [53]. In conclusion, there is no generic choice of a matrix for whole cell MALDI MS of bacteria and that choice will often be determined by secondary considerations, for example the availability of a spectrum library supported by a particular matrix.

Although some investigators did straightforwardly and successfully employ resuspension of cells in a matrix solution and application to the MALDI target [17, 30, 36, 83], several studies have made a comparison of typical methods of sample application. Cell concentration is a factor rarely considered. One recent study, which employed careful cell counting, showed that there is a concentration range in which signal-to-noise ratio improves with cell number; that range spanned three or more orders of magnitude (typically ~10^6–~10^8 cfu ml^{-1}) and depended on the applied matrix [16]. Given the early observations of dramatic effects of sample application on MALDI mass spectra of proteins, sample application in whole cell MALDI MS requires some consideration. Empirically, homogeneous crystallisation of the matrix material was found a key issue [94] but the physical and/or chemical reasons for these observations are not yet clear.

Three methods of sample application to the MALDI target were compared for whole cell analysis, employing different matrix solvents and a variety of Gram-positive and -negative bacteria [86]. An evaluation of spectrum quality, in terms of signal intensity and signal-to-noise ratio in the high mass range (over 20 kDa), showed that (1) a matrix solution with relatively low water content (3 : 1 acetonitrile : 0.1% aqueous TFA) and (2) mixing and spotting of a bacteria suspension with matrix solution (1 : 9 v : v) provided the best results. A closer inspection by electron microscopy showed that the more homogeneous distribution of bacteria among matrix crystals was obtained with the above application conditions and that most cells remained intact [86].

A range of Gram-positive and -negative bacteria was employed to establish a general protocol for whole cell MALDI MS analysis of bacteria [95] (Figure 13.3). The optimum treatment employed several steps of washing, among others with chloroform/methanol mixture, and use of the dried droplet method with CHCA matrix. Employing these condi-

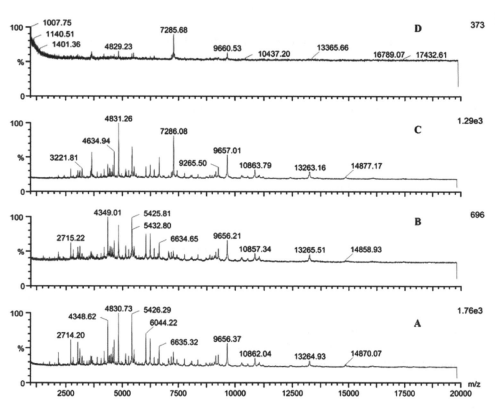

Figure 13.3 *MALDI mass spectra of* Yersinia pestis *analysed with CHCA in different matrix solvents and employing bacterial samples treatment with TFA (0.1%) and chloroform–methanol (1 : 1). Solvents: (A) acetonitrile–methanol–water (1 : 1 : 1) with 0.1% formic acid and 0.01 M 18-crown-6; (B) acetonitrile–ethanol–water (1 : 1 : 1) with 0.1% formic acid and 0.01 M 18-crown-6; (C) 2-propanol–water (1 : 1); (D) acetonitrile–water (1 : 2) containing 0.1% TFA (adapted from Liu et al. [95])*

tions, spectra of several bacteria were found to be highly reproducible and strain level typing of 10 *E. coli* and nine *S. aureus* strains was accomplished fairly easy [95].

Several matrix compounds, matrix solvents, and sample application methods were tested in method development for whole cell MALDI MS of bacteria isolated from wastewater sludge by a sonication based microextraction [96]. Optimum performance was established for CHCA in a mixture of acetonitrile, isopropanol, and 0.1% trifluoroacetic acid (49:49:2 v:v:v), combined with ethanol treatment and sample application in a three-layer fashion. Genus and species level biomarker signals for *E. coli* were confirmed, while biomarker signals for *Salmonella enteritidis*, *S. typhimurium*, and *Acinetobacter* spp. were established. Two strains each of *E. coli* and *Bacillus thuringiensis* were employed in a cross-laboratory study of the reproducibility of whole cell and cell extract spectra employing several compositions of HCCA matrix solutions and applying subsequent layers of matrix and a matrix/analyte mixture [31]. Cells were applied to the MALDI target after suspension in aqueous 0.1% TFA, which was thought to act as an extractant. It was found that spectrum reproducibility in terms of intensity and presence of individual signals was poor but that particular and specific signals –'biomarkers' – were always present in the spectra of the individual organisms studied. It was shown that on-target treatment of a variety of deposited bacteria with ethanol, prior to FA matrix application, produced significantly more intense signals at higher mass (above 20 kDa) and general improvement of spectrum quality [21]. Subsequent electron microscopy of the preparations showed that most cells remained intact on ethanol exposure. It was concluded that ethanol application resulted in a more homogeneous distribution of the bacteria over the crystallised matrix.

Most studies exclusively employ positive ion whole cell MALDI MS. Corresponding negative ion spectra are generally reported to have lower numbers of signals [93, 97]. It was noted, however, that negative ion spectra were useful in finding doubly charged ions in positive ion spectra and in distinguishing metal cation adducts from protonated molecules [97].

13.3.2.2   Data Analysis

Data analysis was initially done by human intervention and employing common spreadsheet software. Various methods for automated information extraction and matching have been described and a few commercial packages for data extraction, library accumulation, and spectrum comparison have become available. Due to poor reproducibility of individual peak intensities and to measurement error in mass assignment, unsupervised spectrum matching to a library of spectra from known organisms is not trivial. General improvement of the matching process and of the probability of correct matching was, and still is pursued by adjustments to the measurement process and by bringing bioinformatics into the data handling process.

In early studies, peak extraction from the spectra was simply omitted in a qualitative comparison or it was done by visual inspection and compilation of peak lists for 'biomarker' signals. In a first attempt at more objective unsupervised data handling, spectra were each divided into a number of 500 Da intervals, for which a cross-correlation was calculated and corrected against the auto-correlation [98]. Variations in mass resolution and in accuracy of peak position were dealt with by prior baseline correction and smoothing. Although the combined cross-correlations were then used as a measure of spectrum similarity, the authors noted that comparison of the separate intervals or of differently divided intervals might provide a better similarity [98]. The authors also employed this cross-correlation method successfully in a monitoring study of *E. coli* growth [19]. Completely independent spectrum handling was first accomplished by a 'fingerprint extraction' algorithm that condensed the information from replicate spectra in a derived spectrum with only distinct peaks [99]. The spectra from replicate samples give a frequency of occurrence, to account for variability in expression, and average peak positions and intensities, to account for experimental variations in MALDI MS. The approach was tested to good effect with *B. atropheus*, *E. coli*, and a mixture of those two organisms [99]. In a follow-up study, the same group developed a probabilistic matching method for the extracted fingerprints [100]. Both fingerprint extraction and probabilistic matching were then successfully tested with a set of six organisms, *Bacillus atropheus*, *Bacillus cereus*, *E. coli*, *Pantoea agglomerans*, *Shewenella alga*, and *Pseudomonas putida*, and blind mixture analysis. No false negatives occurred in the mixture analyses. The only false positives occurred with *B. cereus* because this species displayed very few biomarker signals in its extracted fingerprint [100]. In a later study the ruggedness of the approach was further tested with improved probabilistic fingerprint matching and using a library of 32 strains across 14 species and blind data analysis of virtual mixtures of the library organisms [85]. All positive identifications were correct at the genus and species levels, while mistyping occasionally occurred at the strain level and while most bacteria were also correctly eliminated when not present [85]. Primary data processing, in particular de-noising and standardisation, was investigated [101]. A standardisation algorithm was devised to eliminate a nonuniform spectrum baseline and cross-spectrum variability in maximum intensity. A subsequently applied de-noising algorithm produced spectra for Random Forest (RF) classification. A training set of 60 whole cell MALDI mass spectra of four selected bacteria species, among others *Streptococcus pneumoniae*, was employed to derive a limited number of relevant signals. These signals were then used to match 240 standardised and de-noised test set spectra, in order to evaluate the data processing chain. Classification of test spectra to the training set was correct in all cases and some additional evaluation was done to test the ruggedness of the approach [101]. Another study explored several methods of discriminant analysis and RF classification, to establish a method for adequate distinction of *Mycobacterium* spp. mass spectra [102] (see Figure 13.4 for a visual impression of similarity). RF classification was superior to other methods of spectrum comparison and allowed accurate strain level typing [102]. Hierarchical clustering with a maximum of 100 determinant signals in the 3–15 kDa mass range, condensed from replicate whole cell MALDI mass spectra, was successfully employed in the classification of *Erwinia* spp. [103]. Careful reproducibility checking and cross-referencing to SNP genotyping information were used to accumulate a high quality mass spectrum library. The constructed software was integrated into a single commercial data handling package for whole cell MALDI MS ('BioTyper'; [103]). The demonstrated spectrum reduction, matching, and clustering [85, 100, 103] provide a completely transparent process, that allows tracking of the causative biomarker signals for a particular biomarker distinction or overlap.

A recent study compared the performance of two data processing methods by commercial bioinformatics packages, the general package BioNumerics and the SARAMIS package tailored to whole cell MALDI MS [104]. The first represents mass spectra as densitograms of one-dimensional (1D) gels, while the other reduces those spectra to a list

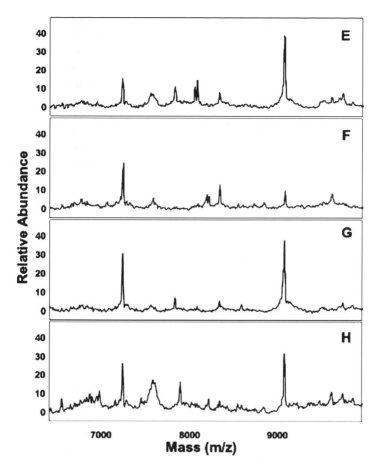

Figure 13.4 *Expanded view of the m/z range 6000–10 000 Da from MALDI TOF mass spectra of four M. tuberculosis strains (adapted from Hettick et al. [102])*

of extracted masses. Spectrum distinction and clustering was similar for both packages, but the discrimination of reduced spectra was greatly enhanced by fast and easy exclusion of overlapping signals from the comparison [104].

Some other methods have emerged, where the causative link of assigned identity and biomarker signals is generally no longer easily traced. These methods employ various ways of pattern matching but in a more complicated way than the above-mentioned correlation method [98]. In one study, multivariate statistical analysis by unsupervised clustering was applied to determine the similarity of transformed spectra from selected standards and from unknowns against a library of these standards [30]. The initial spectrum transformation also involved combination of the information from replicate spectra, like in the above-mentioned study [99], for which the authors noted that it was decisive for successful use of multivariate analysis. Whole cell MALDI MS with this multivariate analysis was used successfully in the typing of *Streptococcus* spp. to the isolate level [30]. Multivariate analysis was also employed in the classification of whole cell MALDI mass spectra of marine organisms isolated from sponges [105]. The established classification was then verified by 16S rRNA typing and it was noted that MALDI MS offered the better resolution of these two methods. The authors noted that the apparent phylogenetic classification provided by whole cell MALDI MS always requires genetic verification of a few of the typed strains to allow true phylogenetic ordering [105]. Another study successfully employed multivariate analysis by supervised pattern recognition, through partial least-squares discriminant analysis, for the distinction of *C. burnetii* isolates [106]. For the dataset studied, decomposition of the de-noised and binarised mass spectra into latent variables generated adequate models for further use in the typing of unknowns without reanalysis or spectral correction [106]. Discrete wavelet transformation (DWT) was employed as a method for standardisation and de-noising of spectra, to afford size compression, and PCA clustering were evaluated for whole cell MALDI mass spectrum classification [107]. The coefficients for optimum wavelet transformation were determined by either of the two methods. The authors concluded that nonlinear DWT and coefficient selection by Fisher's criterion method provided the best performance in spectrum reduction and organism classification [107]. Another group evaluated the use of discriminant analysis and SIMCA classification, using one *Y. enterocolitica* and two *E. coli* strains [108]. The data processing chain provided correct species and subspecies classification with a test set of spectra. Subsequent validation of performance was done with whole cell MALDI mass spectra of *E. coli* and *Y. enterocolitica* from spiked meat samples and of wild-type *E. coli* isolates. Independent analysis by a reference method was employed to verify MS classification and the assignments were correct in all cases. The authors pointed out that multiple protein signals were more reliable than a single epitope recognition by immunological methods; in addition, the method was found tolerant to the positive identification of wild-type strains not necessarily accessible by immunological methods [108]. Mass value alignment and two steps of hierarchical clustering analysis (HCA) with an intermediate data reduction by statistical significance evaluation (ANOVA) were employed in a study of whole cell MALDI MS application to six species, 57 well characterised isolates, of human pathogens [33] (*Streptococcus* group B, *K. pneumoniae*, *Salmonella* spp., *P. aeruginosa*, *S. aureus* and *E. coli*). The first HCA classification step already separated the six species and the ANOVA analysis reduced the spectra to 35 determinant signals for the second HCA step. Further statistical analysis reduced the classifier signals to six. In competitive comparison with this HCA approach, a genetic algorithm was employed in the construction of a model from a 'known' spectra training set for later use in identifying unknowns. Both lines of work supported the authors' conclusion that bacterial species in mixed flora can be speciated at detection levels below those regarded clinically significant for infection [33]. These data processing and classification methods may require some human intervention, for example in defining a starting library or in a decision to add a newly recorded and classified spectrum to a class. Other advance methods, based on neural or learning methods can be operated in a self-consistent way and in support of whole cell MALDI MS as a typing method in its own right.

Only a few of these advanced methods have been tested for whole cell MALDI MS. The performance of a hybrid neural network program was tested with a database of whole cell MALDI mass spectra of 35 strains, from 20 species and mainly enterobacteria, for the typing of over 200 isolates [32]. The hybrid neural network expands the library and sample spectra in an *n*-dimensional vector space and employs the spatial distances as a measure of similarity. The algorithm performed well, even in the distinction of species for which biochemical typing notoriously failed, for example *E. coli* O 122 and *Citrobacter freundii* [32]. Mathematically complicated learning vector quantisation methods were evaluated for their utility in classifying whole cell MALDI mass spectra, employing spectra of a variety of *Listeria* spp. as test objects [109]. It was shown that some of these methods perform better than 'classical' methods, for example linear discriminant analysis. The authors concluded that the use of learning vector quantisation methods for MALDI mass spectrum classification was promising and should be explored further [109]. Given the ongoing development of MALDI MS instrumentation and of bioinformatics technology, more advanced data processing and spectrum classification software will emerge in the near future.

13.3.2.3   *Spectrum Libraries*

Spectrum libraries are generally required for whole cell MALDI MS typing of unknown bacteria in real-world samples. Many of the exploratory and method development studies typically reported construction of a library of spectra from 20 to ~100 selected and well characterised strains, where the diversity of the selected bacteria and the size of the library was related to the study target. A library commonly includes data from multiple spectra of the same strain, to represent different culture conditions, variation in experimental MALDI MS conditions, and possible other sources of spectrum variability. This initial core library was then commonly expanded to include newly typed strains and isolates. While this approach may work well with the same equipment and in the same laboratory, general utility represents a higher scientific value and supports establishment among other typing methods. The general utility of whole cell MALDI mass spectrum libraries was evaluated in a few studies on mixture analysis, in some multi-laboratory studies, and by comparison of several typing methods.

   While commercial data analysis packages for whole cell MALDI MS of bacteria provide a readily usable experimental toolbox, their added value is also determined by the accompanying spectrum libraries. Three packages are currently widely employed: MicrobeLynx (Waters Corp., formerly Micromass Ltd, Manchester, UK), BioTyper (Bruker Daltonik, Bremen, Germany) and Saramis (Anagnostec, Berlin, Germany). The corresponding libraries have been accumulated by application of different experimental constraints, typically concerning mass range, matrix composition, and specific culturing and sample treatment protocols.

   A large, commercially available spectrum library was compiled, including carefully obtained whole cell MALDI MS spectra of hundreds of bacteria [82]. Culture conditions and sample handling were well controlled and distinct matrices were used for Gram-positive and -negative species (CMBT and CHCA, respectively). Replicate culturing was done and replicate spectra were taken, in order to eliminate spectrum variation. Data processing was done with the first commercial package tailored to whole cell MALDI MS, MicrobeLynx. The established library was then employed in typing 25 clinical isolates of *Klebsiella pneumoniae* and *Proteus mirabilis* and over 250 field isolates [82]. This major study illustrates the careful efforts required for clinical application of whole cell MALDI MS and it demonstrates the utility and speed of typing. At the time of publication, July 2004, the library was reported to include hundreds of spectra [82].

Compilation of another large, commercially available spectrum library was reported with the development of whole cell MALDI MS classification software [103]. Uniform sample inactivation, extraction, and application were employed to get highly consistent performance, with signals monitored in the 3–15 kDa mass range. At the time of publication, July 2008, the BioTyper database was reported to comprise more than 2800 spectra [103].

Two studies that focused on the typing of *Salmonella* spp. [38] and *Burkholderia* spp. [104] included reports on library compilation of spectra using the Saramis software package. Although no size indication of the library was given, these studies demonstrated that library compilation requires a major effort in warranting spectrum quality and in cross-referencing to other typing data.

Although these commercially available libraries may help the implementation of whole cell MALDI MS in the clinical and/or microbiological laboratory, daily practice will require further additions by the user's laboratory. When extended carefully and with due attention for the spectrum quality, such spectrum collections may come to represent considerable commodity value.

13.4 Whole Cell MALDI MS of Particular Bacteria Genera and Species

Typing of bacteria is still an expanding area of application of whole cell MALDI MS. Although the use of spectrum libraries is common in routine applications, much of the published work concerns a single selected genus or species of bacteria. Systematic investigations of large numbers of strains and isolates have shown that particular applications can straightforwardly be implemented, for example in support of food quality control or clinical diagnosis of infectious diseases. In addition, systematic in-depth investigations of the biomarker signals in spectra of selected species and strains have provided some insight to the link between whole cell MALDI MS 'chemotypes' and more generally employed genotypes. This section provides an overview of findings for particular groups of bacteria, ranging from Gram-positive to -negative and further subdivided quite arbitrarily.

13.4.1 *Bacillus* spp.

Whole cell MALDI MS of *Bacillus* spp., both of the spores and of vegetative bacteria, has been the focus of many studies. The threat agent status of *B. anthracis* posed a clear goal for method development [55, 110], with other members of the *Bacillus* genus employed as surrogates. A few studies employed *Bacillus* spp. for whole cell MS method development but some studies also addressed spectrum biomarker identities.

Vegetative *Bacillus* spp. were included in studies of mixture analysis by whole cell MALDI MS and using statistical spectrum evaluation [85, 100]. Some of the *Bacillus* spp. spectra initially provided a problem because the spectra did not carry many biomarkers [100] but correct typing was accomplished after algorithm improvement [85]. Whole cell MALDI MS typing of putative *B. pumilus* spore isolates from spacecraft was compared with metabolic profiling, 16S rRNA sequencing, *gyrB* DNA sequencing, and DNA-hybridisation [83]. Characteristic signals were observed in the 3–35 kDa mass range, with high mass signals of considerably lower intensity than low mass signals (Figure 13.5). The isolate spectra showed a poor linear correlation with the *B. pumilus* type strain spectrum

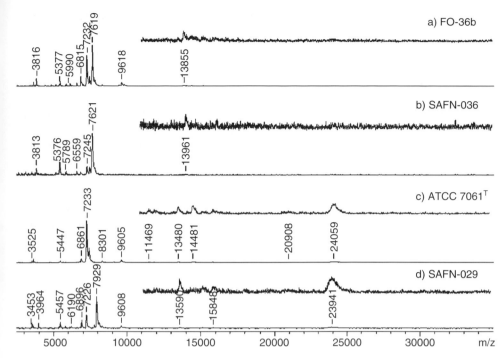

Figure 13.5 *MALDI TOF MS protein profiles comparing four B. pumilus strains in the mass range m/z 3000–35 000, with inset in green m/z 9500–25 000 amplified 10×. Adapted with permission from Dickinson, D.N., et al., MALDI–TOFMS Compared with other Polyphasic Taxonomy Approaches for the Identification and Classification of Bacillus pumilus spores, J. Microbial. Methods 58(2004), 1–12.*

and a yet poorer correlation with spectra from other *Bacillus* spp. but most of the isolate spectra correlated well within the group. For the *Bacillus* spp. studied, the authors assessed that whole cell MALDI TOF MS is far more accurate than metabolic profiling, more discriminating than 16S rRNA sequence analysis, and faster than but complementary to the applied *gyrB* sequencing and DNA–DNA hybridisation [83].

Whole cell mass spectra of vegetative *Bacillus* spp. may have abundant and characteristic signals in the 0.5–5 kDa mass range. Some investigators noted the lipophilic nature of the corresponding compounds, without further identification [48]. A first confirmed structure attribution for these low mass signals in mass spectra of six *B. subtilis* strains was achieved by MALDI post source decay (PSD) type MS/MS, for many of the then known secondary *Bacillus* metabolites [111]. These metabolites were identified as cyclic lipopeptides, typically of the surfactin, iturin, and fengycin classes, and their localisation in cells was established. It was shown that these compounds can afford strain level discrimination, using the variation in amino acid and fatty acid composition [111]. Further extensive investigation of *B. subtilis* ATCC 6633 secondary metabolism also identified mycosubtilins in whole cell mass spectra [112]. All these lipopeptides were investigated in great detail, in a study that corroborated the earlier findings on a few *B. subtilis* strains in the single C-1 strain that was isolated from petroleum sludge [113]. A subsequent investigation reported similar findings on more than 10 additional *B. subtilis* strains and reviewed the lipopeptide secondary metabolites observed in the whole cell MALDI mass spectra of *B. subtilis* and of *Microcystis* spp. [114]. A later study, by another research group and conducted with slightly different atmospheric pressure MALDI, extended these findings on lipopeptides to six other *Bacillus* species and demonstrated that these low mass signals can also be found in the spectra of unwashed spores [40]. A combined genetic and whole cell MALDI MS study on *B. amyloliquefaciens* showed the presence of fengycin, surfactin, and bacillomycin lipopeptides and the corresponding operons required for their expression [115]. A recent study reconfirmed these findings and focused on yet other *Bacillus* secondary metabolites that appear in whole cell mass spectra, in the 3–5 kDa range: lantibiotics and bacteriocins [37]. These two types of compounds may contain D-stereoisomer amino acids and nonpeptide moieties, in particular fatty acids. The findings were independently confirmed by comparison of spectra with those from deletion mutants of the corresponding *B. subtilis* strains [37]. These secondary metabolites and their corresponding biomarker signals can generally aid in the typing of vegetative *Bacillus* spp.

Spores of *B. cereus*, *B. atropheus* (then *B. globigii*), *B. thuringiensis*, and *B. cereus* were investigated by whole cell MALDI MS [48]. Low mass lipophilic compounds were found responsible for many of the low mass signals (below ~5 kDa). Although these signals afforded species discrimination, results improved markedly by an on-target corona discharge treatment that released more high mass compounds [48]. Correlation of the accurate mass of a few signals with protein and gene databases afforded some putative protein identities for *B. subtilis* (see Table 13.2). Similar results were obtained in a later study which focused on spores of the *B. cereus* group of bacteria, which group encompasses some species with a high homology in their genomic DNA (*B. anthracis*, *B. cereus*, *B. thuringiensis*, and *B. mycoides*); accurate masses were again employed to putatively assign a few protein identities [50] (included in Table 13.2). This work indicated the general possibility of the presence of small acid soluble proteins (SASPs) in the whole cell MALDI mass spectra of *Bacillus* spp., which was later confirmed by many successive studies that also employed other methods of analysis [116, 117] and, for example, by distinction of *B. anthracis* by MALDI MS/MS [118–120] or LC-ESI MS(/MS) [119–121] analyses that specifically targeted SASPs.

Protein sequences of SASPs were obtained from whole cell MALDI MS/MS experiments [185]. In addition, it was shown that SASP signals afforded distinction of *B. atropheus* (BG) and *B. cereus* in whole cell MALDI MS spectra of a 1:1 mixture. The performance of whole cell MALDI MS typing was investigated with a panel of spores from various *Bacilli* and applying a matrix that provided high mass signals (>10 kDa) [84]. Linear correlation of the spectra afforded adequate strain and isolate specific typing. In addition, a subsequent more detailed investigation of *B. subtilis* 168 spore extracts by LC-ESI MS/MS yielded 10 protein identities from which six peaks in the whole cell MALDI mass spectrum were attributed (see Table 13.2; [84]). Homologues of these 10 spore proteins, and others, were also reported present in a comparative proteomics study of *B. anthracis* Sterne and *B. subtilis* PY79 [122] but no confirmation by whole cell MALDI MS was reported for those two species. These detailed studies on SASPs demonstrate that there is a solid link of whole cell MALDI mass spectra to bacterial genomes.

Table 13.2 *Whole cell MALDI mass spectrum biomarker signals assigned to chemical compounds in various studies*

[M + H]⁺ mass (Da)[a,b]	Corresponding protein	NCBI reference[c,d]	Reference(s)[e]	Strain(s)[f]	Evidence
B. amyloliquefaciens					
950–1550 (ΔCH₂)	Fengycin, surfactin, bacillomycin	Lipopeptides	[115]	FZB42	MS/MS
B. atrophaeus[g]					
950–1550 (ΔCH₂)	Fengycin, surfactin	Lipopeptides	[40]	ATCC 9372	MS/MS
7068.8	Small, acid-soluble spore protein (SASP)	P84583, 2..70	[185]	ATCC 9372	MS/MS
7333.0	Small, acid-soluble spore protein (SASP)	P84584, 2..71	[185]	ATCC 9372	MS/MS
B. cereus					
6711.5	Small, acid-soluble spore protein 2 (SasP-2)	P0A4F4, 2..65	[185]	T	MS/MS
6835.6	Small, acid-soluble spore protein A (SspA)	Q63FF2, 2..67	[185]	T	MS/MS
7081.8	Small, acid-soluble spore protein alpha/beta family	Q739S3, 2..68	[185]	T	MS/MS
7336.1	Small, acid-soluble spore protein (SasP-1)	P0A4F3, 2..70	[50]	B33	MS
7351.9	Spore germination protein F	gi|2984723, 1..71	[50]	NCTC 8035	MS
7451	Small, acid-soluble spore protein	–	[50]	B33	MS
B. circulans					
950–1550 (ΔCH₂)	Fengycin, surfactin	Lipopeptides	[40]	ATCC 61	MS, analogy
B. licheniformis					
950–1550 (ΔCH₂)	Lichenysin	Lipopeptides	[40]	ATCC 14580	MS, analogy
B. megaterium					
950–1550 (ΔCH₂)	Fengycin, surfactin	Lipopeptides	[40]	ATCC 14581	MS, analogy
B. sphaericus					
950–1550 (ΔCH₂)	Fengycin, surfactin	Lipopeptides	[40]	Unspecified	MS, analogy
32657	S-layer protein		[193]	DSM 28	MS, ESI-MS

(continued overleaf)

Table 13.2 (continued)

[M + H]+ mass (Da)[a,b]	Corresponding protein	NCBI reference[c,d]	Reference(s)[e]	Strain(s)[f]	Evidence
B. subtilis					
950–1550 (ΔCH$_2$)	Iturin, fengycin, surfactin, mycosubtilin, bacillomycin	Lipopeptides	[37, 40, 111–114]	>5	MS/MS
2986.4	Ericin A	Extensive PTM	[37]	ATCC 6633	MS
3342.5	Ericin S	Extensive PTM	[37]	ATCC 6633	MS
3319.6	Subtilin	Extensive PTM	[37]	ATCC 6633	MS
3400.6	Subtilosin	Extensive PTM	[37]	ATCC 6633	MS
3419.6	Succ-subtilin	Extensive PTM	[37]	ATCC 6633	MS
5564.3	Small, acid-soluble spore protein A (SspA)	gi|16080009, 12..65	[84]	168	MS/MS
5685.5	Small, acid-soluble spore protein C (SspC)	P02958, 15..68	[84]	168	MS/MS
5971.7	Small, acid-soluble spore protein D (SspD)	gi|16078411, 5..60	[84]	168	MS/MS
6846.2	YvrF	gi|2832809, 1..61	[48]	168	MS
6950.3	YdiQ	gi|2632921, 1..62	[48]	168	MS
7774.6	Spore coat protein CotT	P11863, 20..82	[84]	168	MS/MS
9137.5	YukE	gi|2635686, 1..80	[48]	168	MS
9398.6	Nonspecific DNA binding protein HBsu	gi|16079336, 2..88	[84]	168	MS/MS
11622.0	Spore coat protein CotJB	gi|16077757, 2..100	[84]	168	MS/MS
B. thuringiensis					
6828.9	Small spore protein (SspF)	gi|75759606, 1..59	[50]	ATCC 55172	MS
Brevibacillus sp.					
1200–1400 range	Streptocidins	Cyclic oligopeptides	[141]	Isolate	MS/MS
C. coli					
10030.6	DNA-binding protein HU (HupB)	gi|58979413, 1..96	[151, 152]	>5	MS
12854.7	50S ribosomal protein L7/L12	gi|57504728, 2..125	[151]	>5	MS

Species / ID	Protein	Accession	Reference	Strain / notes	Method
C. jejuni					
10274.9	DNA-binding protein HU (HupB)	gi|58979485, 1..98	[151, 152]	>5	MS
13736.1	30S ribosomal protein S13 subunit (RpsM)	gi|32363423, 1..121	[151]	>5	MS
C. lari					
9618.2	Chaperonin GroES	gi|57240445, 1..86	[151]	>5	MS
12968.8	50S ribosomal protein L7/L12	gi|57240976, 2..125	[151]	>5	MS
C. upsaliensis					
10271.9	DNA-binding protein HU (HupB)	gi|58979437, 1..98	[151, 152]	>5	MS
C. perfringens					
4300–10000 range	Several ribosomal proteins		[58]	>5	MS
C. burnetii					
3615.0	Small-cell-variant protein A (ScvA)	Q45966	[56]	Nine Mile, phse 1	MS
10504.0	Chaperonin GroES	P19422	[56]	Nine Mile, phse 1	MS
11132.0	Hypothetical protein (located on plasmid QpH1)	gi|10956008, 1..109	[56]	Nine Mile, phse 1	MS
Oscillatoria spp.					
166.1226	Anatoxin A		[92]	PCC 9240	MS, NMR
180.1383	Homoanatoxin A		[92]	>2	MS, NMR
E. coli					
4759.9	Hypothetical protein Z4042	gi|15803251, 2..44	[34]	ATCC 43895	MS
7272.1	Cold shock protein C (CspC)	gi|218369594, 2..69	[20, 21, 34, 146, 150]	>3	MS, digest map
7333.2	Cold shock-like protein E (CspE)	gi|15829916, 2..69	[19, 21, 34, 146]	>3	MS, digest map
8326.2	8.3 kDa protein in DINF-QOR intergenic region	P32691, 1..69	[19]	ATCC 25404	MS
9066.3	*hns* deletion-induced protein B (HdeB)	P0AET2, 30..108	[20, 21, 96, 191]	>5	MS
9226.5	DNA-binding protein HU, beta unit	gi|215263593, 1..90	[96]		MS, digest map
9535.9	DNA-binding protein HU-alpha (NS2)	P0ACF2, 1..90	[20, 96, 146]	ATCC 9637	MS, digest map

(continued overleaf)

305

Table 13.2 (continued)

[M + H]$^+$ mass (Da)^{a,b}	Corresponding protein	NCBI reference^{c,d}	Reference(s)^e	Strain(s)^f	Evidence
9741.9	*hns* deletion-induced protein A (HdeA)	P0AET0, 22..110	[20, 96, 191]	>3	
9743	F plasmid related protein	–	[147]	>5	MS
5000–10000 range	Several ribosomal proteins		[19, 34, 142]	>5	
15153.0	DNA binding protein H-NS	gi|43087, 2..135	[34]	ATCC 43895 C1a, pRHU-845 (pap+)	MS
16555.4	Fimbrial major pilin protein PapA	P04127, 23..184	[17]		MS
29168.3	Plasmid located beta-lactamase Bla	gi|125743122, 22..286^i	[144]	ATCC 700926, pUC19	MS/MS
H. pylori					
5336.1	Hypothetical protein	O25198, 1..44	[97]	ATCC 26995	MS
5426.4	Hypothetical protein	O25662, 1..48	[97]	ATCC 26995	MS
5675.5	Hypothetical protein	Q48270, 1..48	[97]	ATCC 26995	MS
6058.9	Hypothetical protein	O25451, 1..48	[97]	ATCC 26995	MS
7512.6	Oxalocrotonate tautomerase	O25581, 1..68	[97]	ATCC 26995	MS
8217.6	Translation initiation factor IF1	P55974, 2..72	[97]	ATCC 26995	MS
8319.1	Hypothetical protein	P94821, 1..74	[97]	ATCC 26995	MS
8590.7	Acyl carrier protein	P56464, 1..78	[97]	ATCC 26995	MS
8976.0	Cytotoxin associated protein A	Q9Z5L4, 1..81	[97]	ATCC 26995	MS
9113.8	Hypothetical protein	O25449, 1..79	[97]	ATCC 26995	MS
10191.5	Helicase HelA	Q9X5H7, 1..88	[97]	ATCC 26995	MS
10382.5	Hypothetical protein	O25689, 1..89	[97]	ATCC 26995	MS
10517.6	Hypothetical protein	O24902, 1..86	[97]	ATCC 26995	MS
11866.0	CagC	P94838, 1..115	[97]	ATCC 26995	MS
13287.6	Hypothetical protein	O26052, 1..115	[97]	ATCC 26995	MS
13597.4	Cag pathogenicity island protein	O25269, 1..114	[97]	ATCC 26995	MS
14543.5	Flagellar protein FliS	O25448, 1..126	[97]	ATCC 26995	MS
4000–14000 range	Several ribosomal proteins		[97]	ATCC 26995	MS
26492.5	Urease alpha-subunit	–	[97]	ATCC 26995	MS
K. pneumoniae					
7272.1	Cold shock protein C (CspC)	gi|206580814, 2..69	[64, 150]	Unspecified	MS

Organism / Mass	Protein / Compound	Type / Accession	Ref.	Strain / Detail	Method
L. lactis					
2902	Lacticin 481	Oligopeptide	[140]	>5	MS
3354	Nisin A	Oligopeptide	[140]	>5	MS
L. plantarum					
2500–20000 range	Several ribosomal proteins		[89]	NCIMB 8826	MS
Microcystis spp.					
593.3	Pheophorbide	Nonpeptide	[173]	>5	MS
871.6	Pheophytine	Nonpeptide	[173]	>5	MS
500–1800 range	Microcystins, anabaenopeptins, microginins, aeruginosins, and cyanopeptolin	Oligopeptides	[39, 91, 172–174]	>5	MS/MS
P. aeruginosa					
7606.5 (observed 7643)	Cold shock protein A (CspA)	gi|1778825, 1..69	[191]	Not specified	MS
P. putida					
7669.6	Cold acclimation protein B (CapB)	gi|167035306, 1..69	[191]	Not specified	MS
S. enterica subsp. *arizonae*					
4300–30000 range	Species and subspecies specific ribosomal proteins		[38]	>5	MS
S. enterica subsp. *diarizonae*					
4300–30000 range	Species and subspecies specific ribosomal proteins		[38]	>5	MS
S. e. sbsp *e.* sv Choleraesuis[h]					
7272.1	Cold shock protein C (CspC)	gi|62128034, 2..69	[64]	ATCC 14028	MS
S. e. sbsp *e.* sv Dublin[h]					
7272.1	Cold shock protein C (CspC)	gi|16420370, 2..69	[64, 150]	Unspecified	MS
S. e. sbsp *e.* sv Enteritidis[h]					
4300–30000 range	Species and subspecies specific ribosomal proteins		[38]	>5	MS
S. e. sbsp *e.* sv Hadar[h]					
4300–30000 range	Species and subspecies specific ribosomal proteins		[38]	>5	MS

(continued overleaf)

Table 13.2 *(continued)*

[M + H]+ mass (Da)[a,b]	Corresponding protein	NCBI reference[c,d]	Reference(s)[e]	Strain(s)[f]	Evidence
S. e. sbsp e. sv Infantis[h]					
4300–30000 range	Species and subspecies specific ribosomal proteins		[38]	>5	MS
S. e. sbsp e. sv Typhimurium[h]					
6483.5	Protein Gns	P64268, 1..57	[38]	>5	MS
6572.5	Ribosome modulation factor Rmf	gi|16764425, 1..55	[38]	>5	MS
6856.9	Carbon storage regulator CsrA	gi|11415011, 1..61	[38]	>5	MS
7272.1	Cold shock protein C (CspC)	gi|16420370, 2..69	[38, 64, 150]	>5	MS
7321.2	Cold shock protein E (CspE)	gi|16764006, 2..69	[38]	>5	MS
7660.9	Major cold shock protein H (CspH)	gi|16760655, 1..70	[38]	>5	MS
7885.8	Cold shock protein D (CspD)	gi|16759816, 1..73	[38]	>5	MS
7993.0	Uncharacterised protein YibT	Q8ZL66, 2..69	[38]	>5	MS
8119.4	Translation initiation factor IF-1	P69226, 2..72	[38]	>5	MS
9120.3	Phosphocarrierprotein HPr	P0AA07, 1..85	[38]	>5	MS
9192.2	Cell division protein ZapB	Q8ZKP1, 2..79	[38]	>5	MS
9240.6	DNA-binding protein HU-beta (HupB)	P0A1R8, 1..90	[38]	>5	MS
9521.9	DNA-binding protein HU-alpha (HupA)	P0A1R6, 1..90	[38]	>5	MS
9925.3	Glutathioredoxin-1 (Grx1)	P0A1P8, 1..87	[38]	>5	MS
10187.7	10kDa chaperonin GroES	gi|16505459, 2..97	[38]	>5	MS
10622.1	Integration host factor subunit beta (HimB)	P64394, 1..94	[38]	>5	MS
10858.4	Probable sigma(54) modulation protein (YhbH)	P26983, 1..95	[38]	>5	MS
11237.7	Integration host factor subunit alpha (HimA)	P0A1S0, 1..99	[38]	>5	MS
12522.2	Ribosome associated inhibitor A (RaiA)	gi|16765980, 2..112	[38]	>5	MS
13445.3	Uncharacterised protein YjgF	gi|16767703, 2..128	[38]	>5	MS

Mass[a]	Description	Database entry[c] [d]	Reference[e]	Strains[f]	Method
15412.4	DNA-binding protein H-NS (HnsA)	P0A1S2, 2..137	[38]	>5	MS
18973.6	Putative cytoplasmic protein YciE	gi|16765074, 1..168	[38]	>5	MS
19546.3	Inorganic pyrophosphatase (PpA)	P65748, 2..176	[38]	>5	MS
4300–30000 range	Species and subspecies specific ribosomal proteins		[38]	>5	MS
S. e. sbsp e. sv Virchow[h]					
4300–30000 range	Species and subspecies specific ribosomal proteins		[38]	>5	MS
S. enterica subsp. *houtenae*					
4300–30000 range	Species and subspecies specific ribosomal proteins		[38]	>5	MS
S. enterica subsp. *indica*					
4300–30000 range	Species and subspecies specific ribosomal proteins		[38]	3	MS
S. enterica subsp. *salamae*					
4300–30000 range	Species and subspecies specific ribosomal proteins		[38]	>5	MS
S. flexneri					
9066.2	*hns* deletion-induced protein B (HdeB)	gi|84028830, 30..108	[191]	PHS-1059	MS
9741.9	*hns* deletion-induced protein A (HdeA)	gi|84028827, 22..110	[191]	PHS-1059	MS
S. pyogenes					
4000–14000 range	Strain specific ribosomal proteins		[81]	>5	MS
S. pneumoniae					
5400–11000 range	Strain specific ribosomal proteins		[127]	>5	MS

[a] Calculated average mass.

[b],[c] '(ΔCH_2)' indicates that this entry concerns one member of a series of homologues.

[c] If not available from the original reference, an NCBI database entry was matched with the provided description.

[d] *n..m*, sequence indication accounting for PTM.

[e] Reference that reported the sequence match.

[f] >number>, for more than the number of different strains and/or isolates.

[g] Currently *B. atropheus* ATCC 9372.

[h] *S. e.* sbsp. *e.* sv, *Salmonella enterica* subspecies *enterica* serovar.

[i] Accounting for a 21 amino acid signal peptide (in analogy to entry P0AD63).

13.4.2   *Staphylococcus* spp.

Rapid typing of *Staphylococci* is of particular interest, because antibiotic resistant *S. aureus* is an opportunistic pathogen that also appears as a commensal bacterium. An initial whole cell MALDI MS study with 20 isolates explored the differentiation of independently typed *S. aureus* strains and other *Staphylococcus* spp. [37]. Characteristic signals were observed in the 500–10 000 Da mass range, with specificity ranging from the genus level to strain level and extending to *S. aureus* methicillin susceptibility. As methicillin-resistant *Staphylococcus aureus* (MRSA) and methicillin-sensitive *Staphylococcus aureus* (MSSA) were readily distinguished by their spectra, a more extensive study was conducted by the same research group, employing another 26 independently characterised *S. aureus* isolates and investigating possible culture medium effects and inter-laboratory reproducibility [123]. The choice of culture medium seriously affected the number of significant peaks, ranging from 50 to 120, with the smaller number of peaks corresponding to better spectrum reproducibility. It was concluded (1) that standardisation of culturing and analysis was required to warrant intra- and inter-laboratory reproducibility in Staphylococcus whole cell MALDI MS and (2) that MRSA was readily distinguished from other *Staphylococci* including MSSA [123]. These results were corroborated by a study from a different research group who employed similar experimental conditions and a different set of *S. aureus* isolates [124].

Other studies provided more support for the observations on *S. aureus* whole cell MALDI MS and showed that even more detail was accessible. The 1–10 kDa range of whole cell lysate mass spectra obtained from a MRSA and a MSSA was compared, after subculturing during a 3-month period and across two employed culture media [125]. Further comparison with spectra from lysates of clinical isolates did not reveal a uniform signature or specific biomarkers for *S. aureus* in the selected mass range or for its resistant or sensitive species. However, whole cell lysate spectrum consistency between strains or isolates was found sufficient for epidemiological application of whole cell MALDI MS [125]. Differences in teicoplanin antibiotic resistance among four isogenic *S. aureus* strains were investigated by whole cell MALDI MS, pulsed field gel electrophoresis (PFGE) of genomic DNA, and detailed peptidoglycan analysis [36]. The mass spectra correctly reflected the teicoplanin sensitivity or resistance, while PFGE and the peptidoglycan muropeptide patterns did not show significant differences. This differentiation capability was attributed to yet unidentified cell surface compounds [36]. Whole cell MALDI MS typing of *Micrococcaceae* family bacteria, in particular coagulase-negative *Staphylococcus* spp., was investigated by spectrum comparison for 51 qualified strains and 100 clinical isolates [90]. The qualified strains were cultured on four different media and spectrum comparison was employed to find recurrent biomarker signals for each different strain and for subsequent typing of the isolate strains. The authors proposed a two tiered typing process for the clinical microbiology laboratory, encompassing family assignment from culturing characteristics and subsequent species or strain typing by whole cell MALDI MS [90].

None of these whole cell MALDI MS studies on *Staphylococcus* spp. identified single biomarker proteins, despite the generally good species and subspecies typing performance. Even the distinction between MRSA and less pathogenic *Staphylococcus* species has not yet prompted investigation of a causal relationship between particular biomarker signals and properties. A recent review on MS investigation of *Staphylococcus* species [126] indicates that the search for such a causal link may turn out to be fruitful.

13.4.3 *Streptococcus* spp.

Species typing of *Streptococcus* spp. by molecular biology and other conventional methods is fairly difficult [80, 81]. The degree to which blood is lysed by *Streptococcus* spp. is still widely applied as an initial classification in their clinical typing as alpha-, beta-, or non-haemolytic. Further differentiation in typing by classical methods is much more difficult, if not impossible. As a consequence, several studies have been directed at the differentiation attainable by whole cell MALDI MS.

Seventy qualified beta-haemolytic *Streptococcus* isolates, of the Lancefield antigenic groups A, B, C, and G, were studied with the help of three MALDI matrices, MT, MBT, and CHCA [80]. CHCA produced the better spectra and specific signals for the studied Lancefield groups were typically found in the 350–1600 Da mass range by visual inspection of the spectra [80]. Whole cell mass spectra from selected *Streptococcus* spp., nine reference strains and 177 isolates, were investigated for their distinctive typing power relative to 16S rRNA sequencing and biochemical phenotyping [30]. With the CHCA matrix, significant signals were observed between 1.5 kDa and 10 kDa. Multivariate statistical analysis of the spectra allowed subspecies level typing to the extent that three reference strains typed by DNA methods required reclassification [30]. In a larger study, 30 qualified strains including 10 reference strains, and 99 clinical isolates were employed to compare differentiation by gene analysis methods with that of whole cell MALDI MS [29]. The failure of common spectrum matching to accurately distinguish two species, *S. oralis* and *S. mitis*, was attributed to the observed intraspecies variation being larger than the interspecies variation. Data analysis by a statistical learning algorithm was required to achieve correct strain typing [29]. From these studies with a variety of streptococci it is clear that control of experimental conditions, a spectrum library, and intricate data analysis are required to achieve typing.

In addition, some studies addressed subspecies typing for highly pathogenic beta-haemolytic *Streptococcus* species, for example *S. pyogenes* [81] and *S. pneumoniae* [127]. Whole cell MALDI MS of *S. pyogenes* isolates and genome strain were investigated as a method for pathogenic strain differentiation, with comparison with an *emm* gene sequencing method [81]. A variety of experimental conditions were employed to establish consistent and specific biomarkers of *S. pyogenes* and to achieve adequate strain typing. Many of these biomarker signals in the 4–14 kDa mass range (see Figure 13.6) were found to match ribosomal proteins of *S. pyogenes* [81]. A similar study was reported for *S. pneumoniae*, but additional attention was given to strain distinction for epidemiological purposes [127]. Biomarker signals in the 5–11 kDa mass range were found to match ribosomal proteins of *S. pneumoniae*, whereas characteristic signals in the 2.4–5 kDa range remained unidentified. Spectrum clustering did reflect independently established epidemiological relationships between isolates from corresponding outbreaks of *S. pneumoniae* conjunctivitis [127].

Whole cell MALDI MS was employed for *Streptococcus* spp. typing in a study that compared cultures of subgingival biofilm isolates from periodontitis patients with those of healthy controls [128]. Only for *S. sanguinis* was the prevalence among healthy subjects

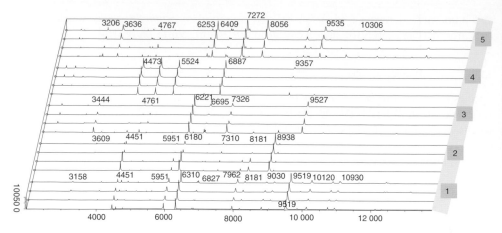

Figure 13.6   *Multi-spectrum view, with quadruplicate analyses for each organism, of MALDI TOF de-noised mass spectra (m/z 2000–14 000 Da) for whole cells of (1) S. pyogenes, (2) S. agalactiae, (3) E. faecalis, (4) S. aureus and (5) E. coli, revealing differences among genus and species (reproduced from Moura et al. [81], Copyright 2008 Blackwell Publishing Ltd)*

significantly higher than for patients. The authors concluded that loss of *S. sanguinis* colonisation seems to be associated with aggressive periodontitis [128]. A subsequent study by the same group did not look at Streptococcus spp., but at the Gram-positive and -negative anaerobe species often positively associated with aggressive periodontitis [129]. Employing nine reference strains for these species, whole cell MALDI MS typing was found to be in general good agreement with 16S rRNA analysis and biochemical typing. It was concluded that whole cell MALDI MS was particularly useful for the typing of anaerobes that resisted biochemical typing [129].

13.4.4   *Mycobacterium* spp.

Well before the advent of MALDI, MS analysis was shown to have potential for the differentiation of *M. tuberculosis* and *M. leprae* from other species, initially by pyrolysis MS (pyMS; [130–132]) and more recently by single particle laser desorption/ionisation (LDI) [133]. These methods typically employed known biomarkers, like mycolic acid profiles in pyMS [132] and an *M. tuberculosis* specific sulfolipid biomarker in negative ion LDI [133]. Quite complementary, whole cell MALDI MS was shown to produce protein information.

Whole cells and extracts of six *Mycobacterium* spp. were investigated by whole cell MALDI MS and exploring the 1–20 kDa range [54]. Genus specific signals were found in the 1300–1800 Da mass range, whereas a few species specific signals for each of the six species studied were found above 5 kDa [54]. At least for the *M. tuberculosis* H37Rv studied, it is quite remarkable that so few proteins below 20 kDa were observed, because a quarter of the ~4000 proteins predicted from the genome falls in that MW range [134]. A follow-up study with 16 strains across four species confirmed these earlier findings and it also demonstrated that there were no distinct strain markers [102]. However, strain classification was possible within a species by employing linear discriminant analysis. In

(a)   (b)

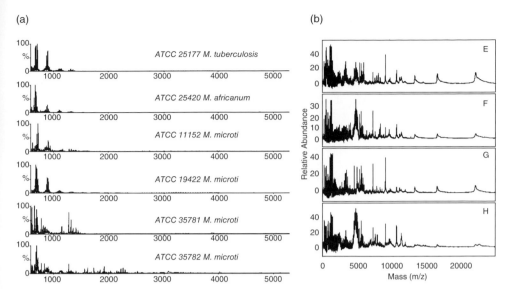

Figure 13.7 *Spectra of* M. tuberculosis *complex species: (a) adapted from Pignone et al. [135] and (b) adapted from Hettick, J.M., et al., Discrimination of Intact Mycobacteria at the Strain Level: A Combined MALDI–TOF MS and Biostatistical Analysis, Proteomics 6, 6416–6425. Copyright (2006) John Wiley & Sons.*

addition, it was shown that a RF data analysis distinguished *M. tuberculosis* mass spectra correctly from nontuberculous *Mycobacteria* in all cases [102]. A study by a different research group covered whole cell MALDI MS of 37 *Mycobacterium* spp. that were independently characterised by 16S rRNA and *hsp65* sequencing [135]. Although the same matrix and similar sample treatment was used as in other studies [54, 102], signals were only found in the 0.5–4 kDa mass range with most signals below 1 kDa (Figure 13.7). In addition, all *Mycobacterium* spp. mass spectrum matching afforded distinction at the species level, except for *M. tusciae* that was only classified correctly at the genus level [135]. Although none of the biomarker signals were actually identified in these reported investigations of *Mycobacteria*, a contingency structure assignment for many of the 1–2 kDa range signals was made as 'mycolic acids' [54, 102, 135].

13.4.5 Other Gram-Positive Bacteria

Typing of *Clostridium* spp. is difficult, because the genus spans over 150 species that are only defined by four common properties: Gram-positive stain, spore-forming nature, anaerobic growth, and absence of the dissimilatory reduction of sulfate [58]. Despite the fact that the genus *Clostridium* contains pathogens and threat agents, and despite the fact that some *Clostridium* spp. are ubiquitous in the environment, only one MALDI MS study has appeared so far, concerning whole cell extracts for safety. The extracts were checked for lack of residual growth, diluted and applied to the MALDI target plate. A reference database of 64 *Clostridium* strains, across 31 species and including various *C. perfringens*, *C. botulinum* and *C. tetani*, was established. Reference spectra were acquired from mixtures of vegetative and sporulated bacteria, to avoid mistyping. Protein signals in the range of 2–20 kDa were observed, and all data handling was done by a commercial package

(BioTyper). The 64 reference strains were well distinguished by the whole cell MALDI MS spectra obtained. Protein assignments were only made for *C. perfringens*, for which organisms several ribosomal proteins were observed. Typing results of field isolates corresponded well with results from 16S rRNA typing and, for example, an undefined *C. botulinum* strain was correctly positioned as a BoNT/A, B, or F producer. The authors concluded that MALDI MS of whole cell extracts was accurate, reliable, and fast in typing *Clostridium* spp. [58].

Five *Listeria* spp. from a single subgroup, *Lactococcus lactis*, *Leuconostoc mesenteroides*, and a *Micrococcus* were included in a larger study on foodborne pathogens that applied a unified protocol for preparation of Gram-positive and -negative bacteria [25]. The spectra allowed distinction at the genus level and the species level, with six signals specific for the genus *Listeria*. *L. monocytogenes* and *L. innocua* were well speciated, despite their similarity in other methods [25]. A database of condensed whole cell mass spectra was accumulated and made available via the internet [136].

Whole cell extracts of the six *Listeria* type strains and an additional *L. grayi* strain were employed to establish a basis for the MALDI MS typing of over 100 *Listeria* isolates [137]. It was observed that *L. monocytogenes* was typable at the level of clonal lineage and, hence, useful in epidemiological outbreak investigations. MALDI MS derived lineage agreed with those from PFGE and MALDI MS performance was found nearly identical across three mass spectrometers [137].

Enterococcus durans was employed in a study of MALDI MS experimental conditions, but without any comparison with other organisms [93]. Source tracking of environmental *Enterococcus* spp. isolates by whole cell MALDI MS with SA matrix was investigated with samples from six animal sources and from humans [138]. Significant signals were only found below ~2 kDa, unless samples were treated with lysozyme. Prolonged (20 h) lysozyme exposure was required to obtain signals up to 10 kDa; this observation was related to the Gram-positive nature of *Enterococcus* spp. When using this prolonged sample treatment, spectrum quality and repeatability was sufficient to obtain adequate clustering of the isolate spectra by source. It was concluded that the whole cell MALDI MS was useful for source tracking by *Enterococcus* spp. but that the prolonged sample treatment time was undesirable for the application [138]. In a later study of *Enterococcus faecalis* oral isolates [139], other investigators applied an instant sample preparation earlier designed for Gram-positive species and employing CBMT as the matrix [82]. This afforded spectra with significant *E. faecalis* signals between 2 kDa and 20 kDa in whole cell MALDI MS [139]. The performance of whole cell MALDI MS for *E. faecalis* virulence determination was assessed for endodontic isolates that had also been characterised by PCR for four virulence genes (*efaA*, *gelE*, *esp* and *ace*) and cytolysin assay. This virulence seems to be associated with development of endocarditis after medical treatment. Correlation of spectra with presence of virulence was observed and the mass spectra were also found to reflect geographic clustering. The authors suggested that whole cell MALDI MS might be used in directing antibiotic prophylaxis in root canal treatment [139].

Lactococcus spp. are important in food production, both because of their probiotic activity and because of their antimicrobial peptide production. This bacteriocin production was investigated by whole cell MALDI MS of lacticin 481, nisin, and coagulin producing *Lactococcus* strains and comparison with *Bacillus coagulans* and genetically modified bacteriocin producing *E. coli* [140]. The bacteria were taken from liquid or solid culture and the target bacteriocins were readily detected in the whole cell mass spectra. In addition, it was shown that designed mutations in these bacteriocins were also readily detectable [140].

A streptocidin producing *Brevibacillus* species was isolated from a bioreactor *Streptomyces* culture, after detection of streptocidin in the culture by metabolic profiling through whole cell MALDI MS [141]. MALDI PSD was employed to elucidate the amino acid sequence of the cyclic decapeptides [141]. This study demonstrated that secondary metabolites from Brevibacillus were accessible in an experimental fashion similar to that for *Bacillus* spp. (see above) and *Cyanobacteria* (see below).

Optimum experimental conditions for whole cell MALDI MS typing of 16S rRNA gene sequenced *Arthrobacter* spp. were determined and applied to the speciation of *Arthrobacter* isolates [16]. The whole cell spectra obtained provided subspecies resolution when compared with the strain distinction attained 16S rRNA typing [16].

13.4.6   *Escherichia coli*

Escherichia coli is probably the most widely studied species when it comes to whole cell MALDI MS. It was often used in the verification of experimental procedures when investigating other species [25, 88, 140] or in method development [21, 34, 65, 98, 142, 143]. For example, whole cell MALDI MS was shown to reveal ampicillin resistance in a comparison of wild type *E. coli* strains and a plasmid transformed strain: the beta-lactamase signal, ~29 kDa, was clearly observed in the spectrum of the resistant strain, along with two other plasmid specific signals [144]. Apart from its status as a common method development object, quite a number of investigations focused specifically on the whole cell mass spectra of *E. coli*.

Extraction of enterohaemorrhagic *E. coli* O157:H7 from meat and meat extracts by immunomagnetic beads and application of these beads to a MALDI target was successfully employed for whole cell MS analysis [34]. By comparison with other *E. coli* strains, this method was shown to be highly selective for the particular enterohaemorrhagic strain. Method detection levels of $2 \times 10^6 \, cfu \, ml^{-1}$ were reported. In addition, observed mass spectral peaks were mass-matched to the *E. coli* O157:H7 genome, applying possible post-translational modifications (PTMs), and corresponding proteins were tentatively identified against the genome (also see Table 13.2) [34]. *E. coli* O157:H7 was also included in a later and more general study of foodborne pathogens (see below; [25]). Here, spectra from nine O157:H7 and 16 other *E. coli* strains were compared to show that O157:H7 strains were readily distinguished by the presence of an exclusive and prominent signal at *m/z* 9740 and the absence of *m/z* 9060 common to all other *E. coli* studied [25]. In hindsight, this signal may be due to the verotoxin B chain (gi|5738240 2..89; $M_{av}[M + H] = 9744.0 \, Da$), while it does also correspond with the mass of a reported but yet unidentified F-plasmid related protein (also see Table 13.2).

E. coli was used in a study of the source tracking potential of whole cell MALDI MS and a PCR based method for assay of possible faecal contamination of freshwater [145]. Animal and human faecal samples and wastewater treatment plant samples were used for isolation and speciation of *E. coli*. Repeatability, tested by repeated analysis of a single culture, was better for the PCR, but the mass spectra afforded better source specific grouping. Potential source tracking biomarker signals were established from the mass spectra, but the corresponding compounds were not identified. Avian, bovine, and canine sources were correctly grouped by MALDI, while human sources were not fully grouped [145].

Culturing of *E. coli* K12 was used in a study directed at following changes in the bacteria over time during 2 days [19]. Three culture conditions, all including Luria Bertani medium, were employed and further sample preparation was such that the number of cells per MALDI spot was the same. The intensities of many peaks were clearly dependent on growth time, with a marked shift from low mass (below ~8 kDa) to higher mass over time. Several *E. coli* ribosomal proteins were identified in the whole cell spectra, by comparison with earlier studies of *E. coli* K12 ribosomal extracts [19].

E. coli strains derived from K12 were employed as the test set in a data handling study, but no particular biomarker identities were elucidated [98]. Cell extract MALDI mass spectra of *E. coli* were studied more closely by HPLC fractionation and MALDI MS of the fractions and their trypsin digests, in order to elucidate protein biomarkers [146]. Although the identification of three biomarkers (Table 13.2) was accomplished with cell extracts rather than with whole cells, $[M + H]^+$ signals of the identified proteins have also been reported in whole cell MALDI mass spectra of *E. coli* in other studies [20, 21, 96]. One study identified a signal in whole cell MALDI mass spectra of *E. coli* that was representative for the presence of the F plasmid [147].

In a study on possible clinical applications, 14 genetically characterised *E. coli* isolates were speciated by whole cell MALDI MS and compared with strains of 10 other *Enterobacteriaceae* and a strain of *Staphylococcus capitis* [17]. It was observed that a particular biomarker signal indicated the presence or absence of a virulence plasmid (pRHU-845) in *E. coli* C1a and the signal was attributed to a particular pilin protein. The authors judged that species level typing of *Enterobacteriaceae* by whole cell MALDI MS was well accomplished, but that strain level typing required more research effort [17].

E. coli and *Pichia pastoris* were studied by whole cell MALDI MS to establish whether the altered protein expression from genetic alteration was rapidly verifiable [148]. Such proteins were readily analysed in whole cell preparations, although the bacterial 50S ribosomal protein and GroEL were occasionally also observed in the spectra. The method was found suitable for high-throughput screening of genetically modified bacteria [148]. This finding was confirmed by a later study that particularly addressed recombinant *E. coli* JM101 [149]. The over expression of recombinant protein production was arabinose inducible, and protein production was followed in near real-time by whole cell MALDI MS (Figure 13.8). The MALDI MS findings were conveniently verified by the use of GFP as the recombinant protein. Although the GFP response in MALDI did not allow as sensitive a detection as fluorescence, agreement between expression level assessment was good. The authors concluded that whole cell MALDI MS was suitable for recombinant protein expression in near real-time [149].

13.4.7   Gram-Negative Food- and Waterborne Pathogen Proteobacteria Other Than *E. coli*

An early study was directed at establishment of *Enterobacteriaceae* biomarkers in the 2–20 kDa range and using CHCA matrix, considering that the phenotypical differences were small within this family [150]. *E. coli* (TG1 and O157:H7), *K. pneumoniae*, *S. typhimurium*, *S. dublin*, and *Providencia rettgeri* shared six signals, of which the *m/z* 7274 signal was elsewhere identified for *E. coli* as CspC ($[M + H]^+$ av. 7272.1 Da; Table 13.2);

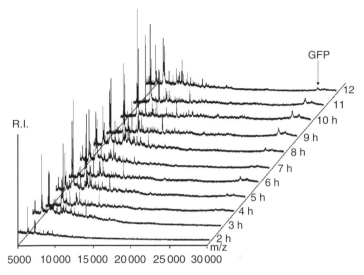

Figure 13.8 *Time-incremented whole cell MALDI MS spectra of recombinant E. coli
JM101 taken from 2 h after incubation with arabinose through to the 12th hour; each
spectrum normalised to the most intense, at the 9th hour (in aquamarin; adapted from
Jones et al. [149])*

the same CspC seems to be shared by some of the *Enterobacteriaceae*. This finding was
confirmed in a later study, by a different group, on antibody based extraction of *S. choleraesuis* from several matrices [64] (Table 13.2).

Three strains of *Salmonella bongori* and 123 strains of the six subspecies of *Salmonella
enterica* were investigated by whole cell MALDI MS, using the 20–40 kDa mass range
observed with SA as the matrix [38]. A database of the calculated masses of 'housekeeping
proteins' was constructed, including ribosomal and DNA binding proteins and accounting
for PTM, and spectra of known strains were checked against this database. Spectra of
known strains for which the genome was not available were evaluated by cross-compar-
ison and considering likely single nucleotide polymorphisms relative to *S. enterica* subsp.
enterica serovar Typhimurium. Although this way of working required a major evaluation
effort, it allowed a highly efficient reduction of the information in *Salmonella* spectra to
a mapping by presence or absence of the housekeeping protein signals. Genus, species,
and subspecies specific biomarker ions of *Salmonellae* were established and verified ([38];
also see Table 13.2).

Gram-negative and -positive food pathogens, including species of the genera *Escherichia*,
Yersinia, *Salmonella* and *Listeria*, were analysed by whole cell MALDI MS [25]. Spectra
were condensed to biomarker lists, of intensity and *m/z*, by mass spectrometer manufacturer software (Date Explorer, Applied BioSystems) before inclusion in a spectrum library.
It was found that the four *Yersinia* species studied did not share any genus level biomarker
signal, whereas the five *Escherichia* genus bacteria only shared two biomarkers. Through
the distinction of the verotoxin producing *E. coli* O157:H7 from other foodborne pathogens, it was shown that the developed method worked well [25].

Salmonella typhimurium was extracted from chicken package exudate and milk, by an on-target carbohydrate-affinity method, and whole cell MALDI mass spectra were obtained [66].

At a first investigation of *Campylobacter* spp. by whole cell MALDI MS, with SA as the matrix, strains of *C. coli*, *C. fetus*, and *C. jejuni* were found to display biomarker signals in the range of 0.5–30 kDa [87]. High mass biomarkers vanished within an hour of sample application to the target, while biomarkers in the 10 kDa range remained. Species differentiation was readily achieved for the *Campylobacter* strains studied [87].

A subsequent study employed whole cell extracts of well over a hundred strains of *Campylobacter*, across six species, and a number of isolates from animals, humans, and from foodstuff [151]. The extracts were prepared by a bead beating procedure, from which the clear supernatants were taken for analysis. Strains of known identity were employed to establish a set of species specific biomarker signals and this set was then employed in the typing of isolates and of unspeciated *Campylobacter*. Several biomarker signals were tentatively attributed to known proteins, notably ribosomal and DNA-binding proteins (Table 13.2). Two separate studies, by the same group, focused on DNA- and MS/MS-protein sequencing of the observed DNA-binding protein HU in the same *Campylobacter* species [152] and on subspeciation of *C. jejuni* by proteomics methods [153]. Some of the unspeciated samples in the whole extract study were found to be mixtures of two of the reference species; these findings were independently confirmed, and speciation from a binary mixture was established possible at levels of 5–10% [151].

Aeromonas spp. are waterborne pathogens that may be found in water reservoirs, including aquaria. An exploratory whole cell MALDI MS study involved 15 *Aeromonas* spp. type strains and 17 well characterised isolates for the optimisation of experimental conditions and data handling [88]. The spectra each had between 17 and 25 biomarker signals, of which four were found potential biomarkers for the genus [88]. In a subsequent study, by the same group, other *Aeromonas* strains and a few known interfering species were added to the compiled biomarker library; this compilation was then used for the blind speciation of over fifty EPA 1605 method *Aeromonas* isolates from drinking water distribution systems [154]. The whole cell MS results were compared with results from biochemical typing and it was concluded that mistyping by MS matched ambiguities in the biochemical profile [154]. Whole cell MALDI MS was included as a method in the polyphasic approach, along with 16S rRNA typing *gyrB* and *rpoD* sequencing, and biochemical testing, used to speciate a new *Aeromonas* species, *A. aquariorum*, that was prevalent in ornamental fish [155]. This study confirmed the speciation capability for *Aeromonas*, established by the above-mentioned work, and it also revealed the outlier status of the new species relative to known *Aeromonas* spp. [155]. This *de novo* speciation demonstrates the maturity of whole cell MALDI MS for *Aeromonas* typing.

13.4.8   Typical Sexually Transmitted Pathogens: *Neisseria* spp. and *Haemophilus* spp.

The genus *Neisseria* harbours several pathogens, of which *N. gonorrhoeae* and *N. meningitidis* are best known. The first whole cell mass spectrum of *N. gonorrhoeae* was included in a pioneering typing study on equally sexually transmissible *Haemophilus ducreyi*, in which work it was shown that these bacteria were easily distinguished [156].

Much later, SELDI MS with RP H50 ProteinChip was employed for the exploration of *N. gonorrhoeae* typing [4]. This study encompassed a large number of isolates of the target species and closely related organisms. A neural network data analysis was employed to establish potential biomarkers in the 5.3–8.1 kDa range [4]. More recently, on-target digestion of heat-killed *Neisseria* spp., including *N. gonorrhoeae* and *N. meningitidis*, was employed to identify the species by atmospheric pressure MALDI MS and MS/MS [157]. Although the observed peptide sequences support highly confident typing, it is remarkable that no correlation with much faster whole cell MALDI mass spectra was established in the study [157]. Although whole cell MALDI MS of *Neisseria* was not extensively studied, the early whole cell mass spectrum, the SELDI study and the whole cell digest MALDI study indicate that whole cell MALDI MS should work well for this genus.

Speciation within the *Haemophilus* genus by MALDI MS was first proved possible in a study using acetonitrile/water extracts and whole cells of the infectious disease agents *H. influenzae*, *H. parainfluenzae*, *H. aphrophilus*, and *H. ducreyi* [156]. Distinctive signals were found in the 5–20 kDa mass range. Additional investigation of 13 *H. ducreyi* isolates showed that subsequent categorisation correctly reflected chancroid outbreak clusters. It was concluded that whole cell MALDI MS allowed rapid typing by elimination of the then commonly required subculturing [156]. In a more recent study, direct application of colonies to the MALDI target and a genetic algorithm based data analysis afforded correct typing of *H. influenzae* among many other pathogens [33]. A more specific whole cell MALDI MS study of *Haemophilus* spp. would be required to support assignment of spectral biomarkers.

13.4.9   Gram-Negative Biothreat Agent Bacteria

Already in the early days of whole cell MALDI MS, the method was explored for the specific purpose of rapid identification of biothreat agent bacteria. Whole cell MALDI MS spectra of selected *Yersinia pestis*, *Brucella* spp., *Bacillus anthracis* and *Francisella tularensis* strains were included in early studies [55, 110]. Although a few later studies included some of these threat agents, for example *Y. pestis* and *B. anthracis* in optimisation experiments [95], no systematic study was published [8]. The term 'biothreat' applies to agents traditionally seen as biological weapons agents. However, the distinction of biothreat agents from more common infectious disease bacteria is not very strict, because some of the biothreat agents are endemic disease agents in particular areas of the world. An open definition of biothreat agents, provided by the USA CDC and encompassing a weighted division into three categories [158], has become widely accepted. A recent review covers more general use of MS for the identification of such agents [8], but the particular use of whole cell MALDI MS for the rapid typing of specific biothreat bacteria other than *B. anthracis* (see above) will also be discussed here.

MALDI MS typing of *Coxiella burnetii*, the causative agent of Q-fever, was the subject of a few studies. Purified and gamma-irradiated *C. burnetii* were subjected to whole cell MALDI MS [56]. The 1–10 kDa and 1–25 kDa mass ranges were separately explored by using two matrices. Over the entire mass range, 16 biomarker signals were observed of which three were tentatively identified against the genome (Table 13.2). This investigation was later extended to include multiple strains and *C. burnetii* phase [106]. All strains investigated had three signals in common and strain distinction was well achieved by

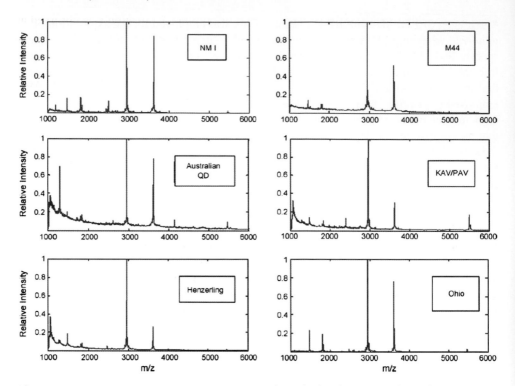

Figure 13.9 *Average MALDI TOF MS spectra of purified* C. burnetii *cultures by strain for day 1 of growth (intensity normalized to m/z = 2951 base peak; adapted from Pierce* et al. *[106])*

multivariate pattern analysis ([106]; Figure 13.9). A second research group employed cell extracts of the highly pathogenic *C. burnetii* for MALDI MS typing [59, 159], following a previous exploratory proteomics study of the organism [160]. After MALDI MS of the extracts, the proteins were digested and partly sequenced by LC-ESI MS/MS. Although the observation of some ribosomal and DNA-binding proteins provided some analogy with whole cell MALDI MS, some highly distinctive membrane located proteins were found as well [40].

Yet other biothreat agent bacteria were investigated by methods closely related to whole cell MALDI MS. For example, for *F. tularensis* the typing by SELDI MS of cell lysates prepared on ProteinChip arrays was studied [161, 162], proteomics MS studies were conducted [163–165], and bio-affinity MALDI MS/MS was investigated [63]. However, it is presently not known how results from such studies might translate to whole cell MALDI MS spectra. Therefore, further discussion of these agents is not included here.

13.4.10 Other Gram-Negative Bacteria

Unspecified strains of *Pseudomonas putida, P. aeruginosa* and *P. mendocina* were employed in an early whole cell MALDI MS study, to successfully demonstrate species specificity of the spectra [166]. *Pseudomonas* spp. were included in a study of mixture analysis by whole cell MALDI MS and using statistical spectrum evaluation [85]. In a blind study, with sample spectra in the 4–14 kDa mass range, distinction was observed at the genus and species level, whereas strain level typing was only partly successful [85]. Various other studies included *Pseudomonas* spp. as selected vehicles for method development [27].

Nonfermenting Gram-negative bacteria isolated from cystic fibrosis (CF) patients were investigated by whole cell MALDI MS, after compilation of a library of spectra of 58 reference strains [167]. Of 559 isolates tested, 10 were misidentified against the compiled library. The authors concluded that the number of misidentifications might be reduced by employing a yet larger library but that clinical management of CF patients would already improve with the established performance of whole cell MALDI MS [167]. An almost simultaneous study covered whole cell MALDI MS of nonfermenting bacteria, starting from a set of 248 reference strain spectra and comparing performance with that of 16S rRNA typing [15]. Reproducibility of the spectra was tested by variation of sample storage time and use of three different mass spectrometers; results of typing were identical under these variations. Subsequent blind analysis of 80 isolates was done by whole cell MALDI MS and by partial 16S rRNA gene sequencing, for reference. Agreement was found for 47 out of 57 16S rRNA types at species level, and 20 out of 21 at genus level, while two isolates remained untyped by the two methods. A more detailed test was done for characterised species from the *Burkholderia cepacia* complex (BCC), in which case 16S rRNA sequencing did not differentiate all strains while whole cell MALDI MS did [15].

Species belonging to the BCC, and some closely resembling bacteria, were investigated by whole cell MALDI MS and two ways of data processing [104]. In a cluster analysis, all spectra of BCC strains grouped together separate from the non-BCC strains. Within the BCC group, spectra clustered in a species specific way, although the *B. anthina* and *B. pyrrocinia* strains constituted one single cluster. The authors noted that no other single method allowed distinction for this taxonomically difficult and clinically important BCC with the speed and simplicity of whole cell MALDI MS [104].

The plant pathogen genus *Erwinia* was covered in a study on classification of whole cell MALDI mass spectra [103]. Typing of these phytopathogens was achieved by whole cell MALDI MS for isolates from various sources [103]. A *Rhyzobium* strain from plant root nodules was employed in a study that optimised culture conditions and other experimental parameters of whole cell MALDI MS [22]. It was concluded that spectrum matching performed best for *Rhizobium* sampled from culture at the late log or early stationary phase [22]. Plant pathogens have not been as widely investigated by whole cell MALDI MS as human and animal pathogenic bacteria.

Helicobacter pylori is a common cause of gastritis. In a first whole cell MALDI MS investigation that included four *Helicobacter* strains, *H. pylori* and *H. mustellae* were found to display biomarker signals in the range of 0.5–65 kDa, when employing SA as the matrix [87]. In two different studies, one using MALDI MS of cell extracts and several matrices [168] and the other employing SA as the matrix [97], *H. pylori* biomarker masses only ranged up to ~38 and ~27 kDa, respectively. It was noted that the high mass biomarkers (above 40 kDa) vanished within an hour of whole cell sample application to the target plate. Strain differentiation was not achieved for the three strains of *H. pylori* studied [87]. A single strain of *H. pylori* was investigated in a study that focused on correspondence of whole cell MALDI mass spectra and a genome, when accounting for PTMs [97]. With SA as the matrix, *H. pylori* peaks were found in the 0.5–~27 kDa mass range. Many peaks in the spectrum were tentatively identified, among these a few from the CAG pathogenicity island of *H. pylori*. Notably, some of the identified proteins were explicitly not found in a MALDI MS investigation of *H. pylori* extracts [168] and only one of the proteins was reported in two proteomics studies on immunogenic proteins from *H. pylori* [168, 169]. With all matching proteins from the whole cell mass spectra, it was shown that the significance of spectrum-to-genome matching increased markedly when PTMs were properly accounted for [97]. These three whole cell MALDI MS studies, in particular the genome mapping study [97], show that the method was thoroughly explored for typing of *H. pylori*.

13.4.11 Pathogenic *Cyanobacteria*

Cyanobacteria are notorious for their seafood contamination and for toxic blooms. Most of their pathogenic nature relates to excreted secondary metabolites, many of which are toxins. Although there is a wide variety of those toxins, commonly referred to by their action as paralytic, diarrhoeic, or amnesic shellfish poisons (PSP, DSP and ASP, respectively), single *Cyanobacteria* species are generally associated with particular classes of toxins. Although analytical methods have been developed for target analysis of the toxins, some whole cell MALDI MS studies have been reported.

The first study on whole cell MALDI MS of *Cyanobacteria* demonstrated that the spectra predominantly carry signals from the secondary metabolites [170]. Structure information of these compounds was accessible by PSD, and a new compound was reported. The authors concluded that species typing and toxicity were easily established by whole cell MALDI MS. The authors concluded that it was possible to directly identify the toxic bacteria species in natural freshwater populations by whole cell MALDI MS experiments [170]. This finding was tested in more detail, using previously lyophilised and reconstituted *Planktothrix agardhii* and *Microcystis aeruginosa* [171]. Many of the cyclic oligopeptide secondary metabolites, microcystins, anabaenopeptins, microginins, aeruginosins, and cyanopeptolins, were identified and their occurrence was assessed in several freshwater isolates [171]. In a wider study, standards and over 250 freshwater isolates of *Microcystis* spp. were investigated by application of material from single aqueous colonies to the MALDI target plate [39]. Secondary metabolite classification was achieved by using class-specific ions in PSD spectra. Single colonies were found to have a less diverse collection of oligopeptides than the whole population in a sample. For three independently classified *Microcystis* spp., the oligopeptide profiles were fairly homogeneous, with microcystin-RR, -YR and -LR found dominant in most colonies [39]. This study was followed up with the development of an rRNA internal transcribed spacer typing method for *Microcystis* spp. with complementary use of whole cell MALDI MS [172]. The developed whole cell MALDI MS method was extended and validated by a study spanning 190 isolates and using PCA for chemotype clustering [173]. A total of 46 different peptides in the 500–1800 Da mass range were found, of which 21 yet unknown, and these were classified by PSD. No clear correlation was found between chemotypes and geographical location [173]. A later study addressed single colonies of *Microcystis* spp. obtained directly from water reservoir samples [91]. Here, 90 different peptides were tracked in the 500–1500 Da mass range, of which 61 were known *Microcystis* products or homologues among which were several microcystin toxins [91]. Recently, whole cell MALDI MS was employed for the typing of *Microcystis aeruginosa* isolates from East-African freshwaters [174]. The oligopeptide patterns from 24 isolates were grouped into 10 whole cell MALDI MS chemotypes. For the microcystin producing strains, results were in full agreement with independent ELISA for these toxins. A comparison of peptide patterns with genetic sub-typing was not given [174].

Microfilaments from *Oscillatoria* spp. were analysed to verify the presence of low mass anatoxin A and homoanatoxin A without the more elaborate sample preparation required for compared methods of analysis [92]. Use of high mass resolution and matrix suppression was required to achieve successful confirmation of the presence of either of the two neurotoxins but the speed of the overall analysis process provided a clear advantage over other methods in screening large amounts of samples [92].

13.5 Strategies for the Identification of Biomarkers in Whole Cell MALDI MS Spectra

Whereas whole cell mass spectra provide a signature for bacteria typing by matching the spectrum of an unknown against a library of known organisms, the origin of the individual signals within the signature represents a deeper layer of information. In the mid 1990s, the initial but then unproved conviction was that these signals result from proteins and peptides. This conviction has generally been proved right by studies published since, although some exceptions have also been reported. The deeper layer of information gives access to some generalisation of whole cell MALDI MS spectra of bacteria that may allow limited interpretation of spectra from independently available genome data in addition to, or even instead of, spectrum library compilation.

Elucidation of potential biomarker structure is not straightforward and can be pursued via theoretical and experimental routes. As a theoretical approach, considerations on the protein expression and on the physical chemistry of the MALDI MS process have been employed to good effect. Several experimental approaches have proved useful. Whereas extension of whole cell MALDI MS to MS/MS provides the most reliable link between biomarker identity and spectrum signal, off-line experimental approaches that employ separation and other types of MS will often be required for higher mass compounds (over ~5 kDa). Theoretical and experimental strategies are discussed in more detail in the following subsections.

13.5.1 Protein Database Consideration

From their earlier detailed work [97, 175–177], Pineda *et al.* converged on a general theoretical approach to whole cell MALDI mass spectra of bacteria [178]. They considered that positive ion MALDI MS requires protonation of any material present and that the more abundant compounds are more likely to produce higher intensity signals. Thus, basic peptides and proteins with relatively high expression levels must be highly likely to result in whole cell MALDI MS signals. Subsequent evaluation of genome information showed that many ribosomal and DNA binding proteins fulfil these conditions [175, 178]. Subsequent compilation of a genome derived ribosomal protein sequence database [179] that also accounts for PTM, afforded experimental validation of these findings by the authors [178] and by other research groups [81, 127, 142]. Ribosomal proteins as biomarkers in whole cell MALDI mass spectra were studied in detail for genome sequenced

Lactobacillus plantarum, to verify the predictive value [89]. Whole cells and cell lysates were subjected to MALDI MS, whereas 2D PAGE separation and subsequent MALDI MS was applied to isolated ribosomal protein fractions. The ribosomal proteins thus confirmed as biomarkers were employed to successfully characterise *L. plantarum* in an industrial yogurt culture and to reveal sequence differences in the particular ribosomal proteins [89]. In a follow-up study with genome sequenced *Lactobacillus bulgaricus* and *Streptococcus thermophilus*, MALDI MS of cell lysates was employed to verify ribosomal sequences [180]. Subsequently, LC-ESI product ion MS/MS was employed to elucidate deviant ribosomal protein sequences from closely related strains and to correct some protein database entries [180]. This work was yet further extended with a similar investigation of *Pseudomonas putida* KT2440 and other strains [181]. A different research group applied the interpretative approach to subspeciation of *Salmonellae*, to the extent that many observed subspecies specific biomarkers were tied to defined proteins [38]. This work demonstrated the predictive value because signals were also successfully assigned for subspecies of which the genome was not yet available [38]. In cases where putative ribosomal proteins may coincide with other proteins predicted from a genome, stable isotope labelling and high mass resolution may be employed for verification [182]. With the experimental evidence accumulated over the last 5 years, it has generally been shown that ribosomal proteins are often represented in whole cell MALDI mass spectra of bacteria. Although the experiments do not absolutely follow *ab initio* predictions on the basis of genomic information, the approach provides a good and general working hypothesis. It is noted that ribosomal proteins can be involved in particular forms of antibiotics resistance (see, e.g. [183] for *E. coli*), which may then also be reflected in the whole cell MALDI mass spectrum. Despite the success of the ribosomal proteins hypothesis, no other theoretical approaches for the interpretation of whole cell MALDI mass spectra have yet emerged.

13.5.2   On-Target Treatment and Analysis

As an experimental approach, whole cell MALDI MS/MS would seem the most appropriate choice, because it does provide a direct link of product ion MS/MS spectra derived peptide sequence to observed MALDI MS signals. This was shown with many of the now known biomarkers in whole cell MALDI mass spectra, for example those of *Bacillus* spp. (see above). In general, direct protein identification in combination with whole cell MALDI MS can be greatly facilitated by employing increased mass resolution in MS and MS/MS analysis [142, 184]. However, the physical chemistry of most commonly employed product ion MS/MS processes does not afford sufficient fragmentation for the singly charged precursor ions of higher mass (over ~5 kDa) that are typically found in MALDI mass spectra. The precursor ion mass range amenable to MS/MS can be extended by the application of presently still unconventional laser induced dissociation MS/MS, as it was shown for ions generated from intact *Bacillus* spores [185]. The limited availability of high mass and/or high mass resolution MS/MS options may require the application of indirect approaches with alternative sample treatment and analysis.

Another fairly simple and MALDI MS based approach employs the fact that some sample material remains intact after analysis. As samples have commonly been applied to a target plate, the remaining material can be subjected to chemical treatment and re-measurement. On-target enzymatic digestion can be employed to demonstrate that particular signals in whole cell mass spectra are related to proteins because these signals will be lost [147]. Of course, on-target enzymatic digestion can also be employed to obtain peptides from high mass proteins. These peptides may by themselves provide access to protein identity because they constitute a peptide map. This was demonstrated, for example, for proteins from *Bacillus* spp. [186, 187] and from *E. coli* [45]. While deconvolution of peptide maps of complicated mixtures may require the very high mass resolution typical for Fourier transform mass spectrometry (FTMS), the individual peptides will also be amenable to MALDI MS/MS sequencing. On-target digestion and (re-)analysis by MALDI MS/MS was, for example, successfully employed in the identification of proteins from *Bacillus* spores [118, 187], spore mixtures [188, 189], and vegetative cells by their digest peptide amino acid sequences [186, 190].

13.5.3 'Off-Target' Analysis and Correlation with Proteomics Studies

The analysis of proteins from culture filtrates, cell washing liquids or other sample extracts is the most widely applied way of identifying spectrum biomarkers. These liquid preparations can be subjected to proteomics sequencing methods that involve chromatographic separation and MS/MS sequencing of the separated material, like in LC-ESI MS/MS [153]. One of the first studies to employ such an approach encompassed LC-ESI MS and Edman sequencing of purified protein fractions, for some biomarker signals in the whole cell MALDI mass spectra of *E. coli*, *S. flexneri*, *P. aeruginosa* and *P. putida* [191] (also see Table 13.2). Mild acid extraction of *E. coli* cells, with subsequent fractionation and MALDI MS of fractions and their trypsin digests was applied to good effect in biomarker identification [146]. In a rare and more elaborate approach it was shown that extracts prepared from material that had previously been spotted on the MALDI target plate and analysed could still afford sequence information in subsequent LC-ESI MS/MS analysis [84]. Separate LC-ESI MS analysis of *Bacillus* spore extracts was employed by several research groups [116, 192], after it turned out that highly distinctive mass signals of these spores corresponded to readily extracted SASPs. Conversely, the finding that an S-layer protein was found in whole cell ESI MS/MS analysis of *B. sphaericus* was confirmed by subsequent whole cell MALDI MS [193]. This emphasises that a certain degree of correspondence between the different methods can be employed in biomarker structure elucidation. Although this kind of whole cell derivative sample analysis generally does not cover all observed biomarker signals, and indeed does not even warrant identification of a single biomarker signal, it has widely been applied to access whole cell MALDI MS biomarker identity.

As an alternative to the above peptide sequencing of proteins in whole cell derived liquids, more elaborate proteomics methods may be employed to track spectrum biomarker identity. These proteomics methods can simply be applied to any fractions obtained from washing, culture filtration, or more invasive extraction. Some investigators have followed this route to protein biomarker identification. For example, two-dimensional gel electrophoresis was employed to find the gel analogue of an observed high mass biomarker, ~126 kDa, of *B. sphaericus*. The gel separated protein was then subjected to in-gel digestion and MALDI MS peptide mapping, to obtain its identity: *B. sphaericus* surface layer protein [86]. Although this kind of approach has been used more frequently, a discussion of these proteomics studies is outside the scope of this chapter.

13.5.4   General Consideration of Biomarker Identification Strategies

None of the above approaches, neither theoretical nor empirical, has yet yielded a handle to the confident and systematic attribution of all spectrum biomarkers for even a single species. The theoretical approach is hampered, among others, by the limited number of available bacterial genomes, by a fairly poor understanding of protein expression levels, by PTM of the genetically encoded proteins, by a yet limited capability for the prediction of protein properties from chemical structure, and by a limited understanding of protein and peptide ionisation in MALDI. Where MALDI MS/MS is not directly applicable, the experimental approach is flawed by the required correspondence between MALDI mass spectra signals and otherwise obtained protein identity. Only thorough experimentation and cross-verification of all information provides reliable spectrum biomarker identification for whole cell MALDI MS.

13.6   Conclusions and Outlook

While the scope of whole cell MALDI MS typing was initially limited to potential hospital applications and biodefence threats, the method has now successfully been applied to samples from widely different areas, for example spacecraft and related clean rooms [83], foodstuff [25, 151], freshwater [64, 145], and wastewater [96]. Although it is outside the scope of this chapter, it is noted that whole cell MALDI MS application has also expanded in the area of typing of fungi and yeast [194–198], archaea [199] and mammalian cell lines [200]. Given the performance of whole cell MALDI MS in these areas, further expansion and wider use are to be expected.

Scientific understanding of the method has advanced over the last 5 years. The fairly independent developments in the field of proteomics MS, and the increasing availability of bacterial genome information afforded a basis for the attribution of chemical structures to some of the biomarker signals observed in whole cell MALDI mass spectra. This development strengthens whole cell MALDI MS as a method, because it supports the link to a layer of information beyond the 'fingerprint'. In addition, more extensive use of MS/ MS with whole cell MALDI will improve the confidence in typing. As a research trend, a growing number of bacteria spectra will be related to genome information.

Over the last 5 years, MALDI MS equipment and the corresponding data handling software for whole cell MALDI MS have become more mature and better tailored to applications in microbiology and to routine assays. The method is generally faster than many other typing methods and it can be set up for high throughput screening. The first exploratory studies on large scale clinical applications have appeared and MALDI MS is likely to become an indispensable method in clinical microbiology.

Despite the increased maturity of whole cell MALDI MS, several desirable developments may serve to improve the method and to make it even more popular. A gain in sensitivity, perhaps attainable by improved ionisation yield, better ion transmission or more sensitive ion detectors would lower detection limits. Miniaturisation of equipment would open up the way to point-of-care diagnostics and field detection applications. Better and on-line interfacing with separation and clean-up methods would support ways of analysing complicated samples. While these development outlooks are not exhaustive, it is obvious that sufficient challenges remain for further method exploration and development.

References

1) van Baar, B.L.M., Characterisation of Bacteria by Matrix-Assisted Laser Desorption/Ionisation and Electrospray Mass Spectrometry, FEMS Microbiol. Rev. 24 (2000), 193–219.
2) Fenselau, C. and Demirev, P.A., Characterization of Intact Microorganisms by MALDI Mass Spectrometry, Mass Spectrom. Rev. 20 (2001), 157–171.
3) Lay Jr, J.O., MALDI-TOF Mass Spectrometry of Bacteria, Mass Spectrom. Rev. 20 (2001), 172–194.
4) Schmid, O., Ball, G., Lancashire, L., Culak, R. and Shah, H., New Approaches to Identification of Bacterial Pathogens by Surface Enhanced Laser Desorption/Ionization Time of Flight Mass Spectrometry in Concert with Artificial Neural Networks, with Special Reference to *Neisseria gonorrhoeae*, J. Med. Microbiol. 54 (2005), 1205–1211.
5) Dattelbaum, A.M. and Iyer, S., Surface-Assisted Laser Desorption/Ionization Mass Spectrometry, Expert Rev. Proteomics 3 (2006), 153–161.
6) Peterson, D.S., Matrix-Free Methods for Laser Desorption/Ionization Mass Spectrometry, Mass Spectrom. Rev. 26 (2007), 19–34.
7) Fox, A., Mass Spectrometry for Species or Strain Identification After Culture or Without Culture: Past, Present, and Future, J. Clin. Microbiol. 44 (2006), 2677–2680.
8) Demirev, P.A. and Fenselau, C., Mass Spectrometry in Biodefense, J. Mass Spectrom. 43 (2008) 1441–1457.
9) Hardouin, J., Protein Sequence Information by Matrix-Assisted Laser Desorption/Ionization In-Source Decay Mass Spectrometry, Mass Spectrom. Rev. 26 (2007), 672–682.
10) Dworzanski, J.P. and Snyder, A.P., Classification and Identification of Bacteria Using Mass Spectrometry-Based Proteomics, Expert Rev. Proteomics 2 (2005), 863–878.
11) Norbeck, A.D., Callister, S.J., Monroe, M.E., Jaitly, N., Elias, D.A., Lipton, M. S. and Smith, R.D., Proteomic Approaches to Bacterial Differentiation, J. Microbiol. Meth. 67 (2006), 473–486.
12) Dworzanski, J.P., Deshpande, S.V., Chen, R., Jabbour, R.E., Snyder, A.P., Wick, C.H. and Li, L., Mass Spectrometry-Based Proteomics Combined with Bioinformatic Tools for Bacterial Classification, J. Proteome Res. 5 (2006), 76–87.
13) van Wuijckhuijse, A.L. and van Baar, B.L.M., Recent Advances in Real-time Mass Spectrometry Detection of Bacteria, in Principles of Bacterial Detection: Biosensors, Recognition Receptors and Microsystems, Zourob, M., Elwary, S. and Turner, A. (Eds), Springer, New York, 2008, 929–954.
14) Peng, W.-P., Yang, Y.-C., Kang, M.-W., Lee, Y.T. and Chang, H.-C., Measuring Masses of Single Bacterial Whole Cells with a Quadrupole Ion Trap, J. Am. Chem. Soc. 126 (2004), 11 766–767.
15) Mellmann, A., Cloud, J., Maier, T., Keckevoet, U., Ramminger, I., Iwen, P., Dunn, J., Hall, G., Wilson, D., Lasala, P., Kostrzewa, M. and Harmsen, D., Evaluation of Matrix-Assisted laser Desorption/Ionization Time-of-Flight Mass Spectrometry (MALDI-TOF MS) in Comparison to 16S rRNA Gene Sequencing for Species Identification of Nonfermenting Bacteria, J. Clin. Microbiol. 46 (2008), 1946–1954.
16) Vargha, M., Takáts, Z., Konopka, A. and Nakatsu, C.H., Optimization of MALDI-TOF MS for Strain Level Differentiation of *Arthrobacter* Isolates, J. Microbiol. Meth. 66 (2006), 399–409.
17) Evason, D.J., Claydon, M.A. and Gordon, D.B., Exploring the Limits of Bacterial Identification by Intact Cell-Mass Spectrometry, J. Am. Soc. Mass Spectrom. 12 (2001), 49–54.
18) Conway, G.C., Smole, S.C., Sarracino, D.A., Arbeit, R.D. and Leopold, P.E., Phyloproteomics: Species Identification of *Enterobacteriaceae* Using Matrix-Assisted Laser Desorption/Ionization Time-of-Flight Mass Spectrometry, J. Mol. Microbiol. Biotechnol. 3 (2001), 103–112.
19) Arnold, R.J., Karty, J.A., Ellington, A.D. and Reilly, J.P., Monitoring the Growth of a Bacteria Culture by MALDI-MS of Whole Cells, Anal. Chem. 71 (1999), 1990–1996.
20) Saenz, A.J., Petersen, C.E., Valentine, N.B., Gantt, S.L., Jarman, K.H., Kingsley, M.T. and Wahl, K.L., Reproducibility of Matrix-Assisted Laser Desorption/Ionization Time-of-Flight Mass Spectrometry for Replicate Bacterial Culture Analysis, Rapid Commun. Mass Spectrom. 13 (1999), 1580–1585.

21) Madonna, A.J., Basile, F., Ferrer, I., Meetani, M.A., Rees, J.C. and Voorhees, K.J., On-probe Sample Pretreatment for the Detection of Proteins above 15 kDa from Whole Cell Bacteria by Matrix-Assisted Laser Desorption/Ionization Time-of-Flight Mass Spectrometry, Rapid Commun. Mass Spectrom. 14 (2000), 2220–2229.

22) Mandal, S.M., Pati, B.R., Ghosh, A.K. and Das, A.K., Influence of Experimental Parameters on Identification of Whole Cell *Rhizobium* by Matrix-Assisted Laser Desorption/Ionization Time-of-Flight Mass Spectrometry, Eur. J. Mass Spectrom. 13 (2007), 165–171.

23) Valentine, N., Wunschel, S., Wunschel, D., Petersen, C. and Wahl, K., Effect of Culture Conditions on Microorganism Identification by Matrix-Assisted Laser Desorption Ionization Mass Spectrometry, Appl. Environ. Microbiol. 71 (2005), 58–64.

24) Wunschel, D.S., Hill, E.A., McLean, J.S., Jarman, K., Gorby, Y.A., Valentine, N. and Wahl, K., Effects of Varied pH, Growth Rate and Temperature Using Controlled Fermentation and Batch Culture on Matrix Assisted Laser Desorption/Ionization Whole Cell Protein Finger-prints, J. Microbiol. Methods 62 (2005), 259–271.

25) Mazzeo, M.F., Sorrentino, A., Gaita, M., Cacace, G., di Stasio, M., Facchiano, A., Comi, G., Malorni, A. and Siciliano, R.A., Matrix-Assisted Laser Desorption Ionization – Time of Flight Mass Spectrometry for the Discrimination of Food-Borne Microorganisms, Appl. Environ. Microbiol. 72 (2006), 1180–1189.

26) Chen, P., Lu, Y. and Harrington, P.B., Biomarker Profiling and Reproducibility Study of MALDI-MS Measurements of *Escherichia coli* by Analysis of Variance-Principal Component Analysis, Anal. Chem. 80 (2008), 1474–1481.

27) Lee, H., Williams, S.K., Wahl, K.L. and Valentine, N.B., Analysis of Whole Bacterial Cells by Flow Field-Flow Fractionation and Matrix-Assisted Laser Desorption/Ionization Time-of-Flight Mass Spectrometry, Anal. Chem. 75 (2003), 2746–2752.

28) Reschiglian, P., Zattoni, A., Cinque, L., Roda, B., dal Piaz, F., Roda, A., Moon, M.H. and Min, B.R., Hollow-Fiber Flow Field-Flow Fractionation for Whole Bacteria Analysis by Matrix-Assisted Laser Desorption/Ionization Time-of-Flight Mass Spectrometry, Anal. Chem. 76 (2004), 2103–2111.

29) Friedrichs, C., Rodloff, A.C., Chhatwal, G.S., Schellenberger, W. and Eschrich, K., Rapid Identification of Viridans *Streptococci* by Mass Spectrometric Discrimination, J. Clin. Micro-biol. 45 (2007), 2392–2397.

30) Rupf, S., Breitung, K., Schellenberger, W., Merte, K., Kneist, S. and Eschrich, K., Differentia-tion of Mutans *Streptococci* by Intact Cell Matrix-Assisted Laser Desorption/Ionization Time-of-Flight Mass Spectrometry, Oral Microbiol. Immunol. 20 (2005), 267–273.

31) Wang, Z., Russon, L., Li, L., Roser, D.C. and Long, S.R., Investigation of Spectral Repro-ducibility in Direct Analysis of Bacteria Proteins by Matrix-Assisted Laser Desorption/Ionization Time-of-Flight Mass Spectrometry, Rapid Commun. Mass Spectrom. 12 (1998), 456–464.

32) Bright, J.J., Claydon, M.A., Soufian, M. and Gordon, D.B., Rapid Typing of Bacteria Using Matrix-Assisted Laser Desorption Ionisation Time-of-Flight Mass Spectrometry and Pattern Recognition Software, J. Microbiol. Meth. 48 (2002), 127–138.

33) Hsieh, S.Y., Tseng, C.L., Lee, Y.S., Kuo, A.J., Sun, C.F., Lin, Y.H. and Chen, J.K., Highly Efficient Classification and Identification of Human Pathogenic Bacteria by MALDI-TOF MS, Mol. Cell. Proteomics 7 (2008), 448–456.

34) Ochoa, M.L. and Harrington, P.B., Immunomagnetic Isolation of Enterohemorrhagic *Escherichia coli* O157:H7 from Ground Beef and Identification by Matrix-Assisted Laser Desorption/Ionization Time-of-Flight Mass Spectrometry and Database Searches, Anal. Chem. 77 (2005), 5258–5267.

35) Edwards-Jones, V., Claydon, M.A., Evason, D.J., Walker, J., Fox, A.J. and Gordon, D.B., Rapid Discrimination between Methicillin-Sensitive and Methicillin-Resistant *Staphylococcus aureus* by Intact Cell Mass Spectrometry, J. Med. Microbiol. 49 (2000), 295–300.

36) Majcherczyk, P.A., McKenna, T., Moreillon, P. and Vaudaux, P., The Discriminatory Power of MALDI-TOF Mass Spectrometry to Differentiate between Isogenic Teicoplanin-Suscepti-ble and Teicoplanin-Resistant Strains of Methicillin-Resistant *Staphylococcus aureus*, FEMS Microbiol. Lett. 255 (2006), 233–239.

37) Stein, T., Whole-Cell Matrix-Assisted Laser Desorption/Ionization Mass Spectrometry for Rapid Identification of Bacteriocin/Lantibiotic-Producing Bacteria, Rapid. Commun. Mass Spectrom. 22 (2008), 1146–1152.

38) Dieckmann, R., Helmuth, R., Erhard, M. and Malorny, B., Rapid Classification and Identification of *Salmonellae* at the Species and Subspecies Levels by Whole-Cell Matrix-Assisted Laser Desorption Ionization–Time of Flight Mass Spectrometry, Appl. Environ. Microbiol. 74 (2008), 7767–7778.

39) Fastner, J., Erhard, M. and von Döhren, H., Determination of Oligopeptide Diversity Within a Natural Population of *Microcystis* spp. (*Cyanobacteria*) by Typing Single Colonies by Matrix-Assisted Laser Desorption Ionization-Time of Flight Mass Spectrometry, Appl. Environ. Microbiol. 67 (2001), 5069–5076.

40) Stowers, M.A., van Wuijckhuijse, A.L., Marijnissen, J.C.M., Scarlett, B., van Baar, B.L.M. and Kientz, C.E., Application of Matrix-assisted Laser Desorption/Ionization to On-line Aerosol Time-of-flight Mass Spectrometry, Rapid Commun. Mass Spectrom. 14 (2000), 829–833.

41) Madonna, A.J., Voorhees, K.J., Taranenko, N.I., Laiko, V.V. and Doroshenko, V.M., Detection of Cyclic Lipopeptide Biomarkers from *Bacillus* Species Using Atmospheric Pressure Matrix-Assisted Laser Desorption/Ionization Mass Spectrometry, Anal. Chem. 75 (2003), 1628–1637.

42) Antoine, M.D., Carlson, M.A., Drummond, W.R., Doss III, O.W., Hayek, C.S., Saksena, A. and Lin, J.S., Mass Spectral Analysis of Biological Agents Using the BioTOF Mass Spectrometer, Johns Hopkins APL Tech. Digest 25 (2004), 20–26.

43) Jackson, S.N., Mishra, S. and Murray, K.K., On-line Laser Desorption/Ionization Mass Spectrometry of Matrix-Coated Aerosols, Rapid Commun. Mass Spectrom. 18 (2004), 2041–2045.

44) van Wuijckhuijse, A.L., Stowers, M.A., Kleefsman, W.A., van Baar, B.L.M., Kientz, C.E. and Marijnissen, J.C.M., Matrix-Assisted Laser Desorption/Ionisation Aerosol Time-of-Flight Mass Spectrometry for the Analysis of Bioaerosols: Development of a Fast Detector for Airborne Biological Pathogens, J. Aerosol Sci. 36 (2005), 677–687.

45) Kim, J.-K., Jackson, S.N. and Murray, K.K., Matrix-Assisted Laser Desorption/Ionization Mass Spectrometry of Collected Bioaerosol Particles, Rapid Commun. Mass Spectrom. 19 (2005), 1725–1729.

46) Dugas Jr, A.J. and Murray, K.K., On-Target Digestion of Collected Bacteria for MALDI Mass Spectrometry, Anal. Chim. Acta 627 (2008), 154–161.

47) Elhanany, E., Barak, R., Fisher, M., Kobiler, D. and Altboum, Z., Detection of Specific *Bacillus anthracis* Spore Biomarkers by Matrix-Assisted Laser Desorption/Ionization Time-of-Flight Mass Spectrometry, Rapid Commun. Mass Spectrom. 15 (2001), 2110–2116.

48) Cain, T.C., Lubman, D.M. and Weber Jr, W.J., Differentiation of Bacteria Using Protein Profiles from Matrix-assisted Laser Desorption/Ionization Time-of-flight Mass Spectrometry, Rappid Commun. Mass Spectrom. 8 (1994), 1026–1030.

49) Hathout, Y., Demirev, P.A., Ho, Y.-P., Bundy, J.L., Ryzhov, V., Sapp, L., Stutler, J., Jackman, J. and Fenselau, C., Identification of *Bacillus* Spores by Matrix-Assisted Laser Desorption Ionization-Mass Spectrometry, Appl. Environ. Microbiol. 65 (1999), 4313–4319.

50) Ryzhov, V., Hathout, Y. and Fenselau, C., Rapid Characterization of Spores of *Bacillus cereus* Group Bacteria by Matrix-Assisted Laser Desorption-Ionization Time-of-Flight Mass Spectrometry, Appl. Environ. Microbiol. 66 (2000), 3828–3834.

51) Birmingham, J., Demirev, P., Ho, Y.-P., Thomas, J., Bryden, W. and Fenselau, C., Corona Plasma Discharge for Rapid Analysis of Microorganisms by Mass Spectrometry, Rapid Commun. Mass Spectrom. 13 (1999), 604–606.

52) Smole, S.C., King, L.A., Leopold, P.E. and Arbeit, R.D., Sample Preparation of Gram-Positive Bacteria for Identification by Matrix Assisted Laser Desorption/Ionization Time-of-Flight Mass Spectrometry, J. Microbiol. Meth. 48 (2002), 107–115.

53) Horneffer, V., Haverkamp, J., Janssen, H.G., ter Steeg, P.F. and Notz, R., MALDI-TOF-MS Analysis of Bacterial Spores: Wet Heat-Treatment as a New Releasing Technique for Biomarkers and the Influence of Different Experimental Parameters and Microbiological Handling, J. Am. Soc. Mass Spectrom. 15 (2004), 1444–1454.

54) Hettick, J.M., Kashon, M.L., Simpson, J.P., Siegel, P.D., Mazurek, G.H. and Weissman, D.N., Proteomic Profiling of Intact Mycobacteria by Matrix-Assisted Laser Desorption/Ionization Time-of-Flight Mass Spectrometry, Anal. Chem. 76 (2004), 5769–5776.

55) Krishnamurthy, T., Ross, P.L. and Rajamani, U., Detection of Pathogenic and Non-Pathogenic Bacteria by Matrix-Assisted Laser Desorption/Ionization Time-of-Flight Mass Spectrometry, Rapid Commun. Mass Spectrom. 10 (1996), 883–888.

56) Shaw, E.I., Moura, H., Woolfitt, A.R., Ospina, M., Thompson, H.A. and Barr, J.R., Identification of Biomarkers of Whole *Coxiella burnetii* Phase I by MALDI-TOF Mass Spectrometry, Anal. Chem. 76 (2004), 4017–4022.

57) Williams, T.L., Andrzejewski, D., Lay Jr, J.O. and Musser, S.M., Experimental Factors Affecting the Quality and Reproducibility of MALDI TOF Mass Spectra Obtained from Whole Bacteria Cells, J. Am. Soc. Mass Spectrom. 14 (2003), 342–351.

58) Grosse-Herrenthey, A., Maier, T., Gessler, F., Schaumann, R., Bohnel, H., Kostrzewa, M. and Kruger, M., Challenging the Problem of Clostridial Identification with Matrix-Assisted Laser Desorption/Ionization – Time-of-Flight Mass Spectrometry (MALDI-TOF MS), Anaerobe 14 (2008), 242–249.

59) Hernychova, L., Toman, R., Ciampor, F., Hubalek, M., Vackova, J., Macela, A. and Skultety, L., Detection and Identification of *Coxiella burnetii* Based on the Mass Spectrometric Analyses of the Extracted Proteins, Anal. Chem. 80 (2008), 7097–7104.

60) Lasch, P., Nattermann, H., Erhard, M., Stammler, M., Grunow, R., Bannert, N., Appel, B. and Naumann, D., MALDI-TOF Mass Spectrometry Compatible Inactivation Method for Highly Pathogenic Microbial Cells and Spores, Anal. Chem. 80 (2008), 2026–2034.

61) Nedelkov, D. and Nelson, R.W., Practical Considerations in BIA/MS: Optimizing the Biosensor – Mass Spectrometry Interface, J. Mol. Recogn. 13 (2000), 140–145.

62) Kiehntopf, M., Siegmund, R. and Deufel, T., Use of SELDI-TOF Mass Spectrometry for Identification of New Biomarkers: Potential and Limitations, Clin. Chem. Lab. Med. 45 (2007), 1435–1449.

63) Jiang, J., Parker, C.E., Fuller, J.R., Kawula, T.H. and Borchers, C.H., An Immunoaffinity Tandem Mass Spectrometry (iMALDI) Assay for Detection of *Francisella tularensis*, Anal. Chim. Acta 605 (2007), 70–79.

64) Madonna, A.J., Basile, F., Furlong, E., Voorhees, K.J., Detection of Bacteria from Biological Mixtures Using Immunomagnetic Separation Combined with Matrix-Assisted Laser Desorption/Ionization Time-of-Flight Mass Spectrometry, Rapid Commun. Mass Spectrom. 15 (2001), 1068–1074.

65) Bundy, J.L. and Fenselau, C., Lectin-Based Affinity Capture for MALDI-MS Analysis of Bacteria, Anal. Chem. 71 (1999), 1460–1463.

66) Bundy, J.L. and Fenselau, C., Lectin and Carbohydrate Affinity Capture Surfaces for Mass Spectrometric Analysis of Microorganisms, Anal. Chem. 73 (2001), 751–757.

67) Afonso, C. and Fenselau, C., Use of Bioactive Glass Slides for Matrix-Assisted Laser Desorption/Ionization Analysis: Application to Microorganisms, Anal. Chem. 75 (2003), 694–697.

68) Candela, M., Fiori, J., Dipalo, S., Naldi, M., Gotti, R. and Brigidi, P., Rapid MALDI-TOF-MS Analysis in the Study of Interaction between Whole Bacterial Cells and Human Target Molecules: Binding of *Bifidobacterium* to Human Plasminogen, J. Microbiol. Methods 73 (2008), 276–278.

69) Wunschel, S.C., Jarman, K.H., Petersen, C.E., Valentine, N.B., Wahl, K.L., Schauki, D., Jackman, J., Nelson, C.P. and White 5th, E., Bacterial Analysis by MALDI-TOF Mass Spectrometry: An Inter-Laboratory Comparison, J. Am. Soc. Mass Spectrom. 16 (2005), 456–462.

70) Cotter, R.C., Time-of-Flight Mass Spectrometry: Instrumentation and Applications in Biological Research, American Chemical Society, Washington, DC, 1997.

71) Tanaka, K., Waki, H., Ido, Y., Akita, S., Yoshida, Y. and Yoshida, T., Protein and Polymer Analyses up to m/z 100000 by Laser Ionization Time-of flight Mass Spectrometry, Rapid Commun. Mass Spectrom. 2 (1988), 151–153.

72) Karas, M. and Hillenkamp, F., Laser Desorption Ionization of Proteins with Molecular Masses Exceeding 10000 Daltons, Anal. Chem. 60 (1988), 2299–2301.

73) Ryzhov, V., Bundy, J.L., Fenselau, C., Taranenko, N., Doroshenko, V. and Prasad, C.R., Matrix-Assisted Laser Desorption/Ionization Time-of-Flight Analysis of *Bacillus* Spores Using a 2.94 Micron Infrared Laser, Rapid Commun. Mass Spectrom. 14 (2000), 1701–1706.

74) Niu, S., Zhang, W. and Chait, B.T., Direct Comparison of Infrared an Ultraviolet Wavelength Matrix-Assisted Laser Desorption/Ionisation Mass Spectrometry of Proteins, J. Am. Soc. Mass Spectrom. 9 (1998), 1–7.

75) Puretzky, A.A., Geohegan, D.B., Hurst, G.B., Buchanan, M.V. and Luk'yanchuk, B.S., Imaging of Vapor Plumes Produced by Matrix Assisted Laser Desorption: A Plume Sharpening Effect, Phys. Rev. Lett. 83 (1999), 444–447.

76) Luo, G., Marginean, I. and Vertes, A., Internal Energy of Ions Generated by Matrix-Assisted Laser Desorption/Ionization, Anal. Chem. 74 (2002), 6185–6190.

77) Dreisewerd, K., The Desorption Process in MALDI, Chem. Rev. 103 (2003), 395–425.

78) Knochenmuss, R. and Zenobi, R., MALDI Ionization: the Role of In-Plume Processes, Chem. Rev. 103 (2003), 441–452.

79) Zhigilei, L.V., Yingling, Y.G., Itina, T.E., Schoolcraft, T.A. and Garrison, B.J., Molecular Dynamics Simulations of Matrix-Assisted Laser Desorption – Connections to Experiment, Int. J. Mass Spectrom. 226 (2003), 85–106.

80) Kumar, M.P., Vairamani, M., Raju, R.P., Lobo, C., Anbumani, N., Kumar, C.P., Menon, T. and Shanmugasundaram, S., Rapid Discrimination between Strains of Beta Haemolytic Streptococci by Intact Cell Mass Spectrometry, Indian J. Med. Res. 119 (2004), 283–288.

81) Moura, H., Woolfitt, A.R., Carvalho, M.G., Pavlopoulos, A., Teixeira, L.M., Satten, G.A. and Barr, J.R., MALDI-TOF Mass Spectrometry as a Tool for Differentiation of Invasive and Noninvasive *Streptococcus pyogenes* Isolates, FEMS Immunol. Med. Microbiol. 53 (2008), 333–342.

82) Keys, C.J., Dare, D.J., Sutton, H., Wells, G., Lunt, M., McKenna, T., McDowall, M. and Shah, H.N., Compilation of a MALDI-TOF Mass Spectral Database for the Rapid Screening and Characterisation of Bacteria Implicated in Human Infectious Diseases, Inf. Gen. Evol. 4 (2004), 221–242.

83) Dickinson, D.N., la Duc, M.T., Satomi, M., Winefordner, J.D., Powell, D.H. and Venkateswaran, K., MALDI-TOFMS Compared with other Polyphasic Taxonomy Approaches for the Identification and Classification of *Bacillus pumilus* Spores, J. Microbiol. Methods 58 (2004), 1–12.

84) Dickinson, D.N., la Duc, M.T., Haskins, W.E., Gornushkin, I., Winefordner, J.D., Powell, D.H. and Venkateswaran, K., Species Differentiation of a Diverse Suite of *Bacillus* Spores by Mass Spectrometry-Based Protein Profiling, Appl. Environ. Microbiol. 70 (2004), 475–482.

85) Wahl, K.L., Wunschel, S.C., Jarman, K.H., Valentine, N.B., Petersen, C.E., Kingsley, M.T., Zartolas, K.A. and Saenz, A.J., Analysis of Microbial Mixtures by Matrix-Assisted Laser Desorption/Ionization Time-of-Flight Mass Spectrometry, Anal. Chem. 74 (2002), 6191–6199.

86) Vaidyanathan, S., Winder, C.L., Wade, S.C., Kell, D.B. and Goodacre, R., Sample Preparation in Matrix-Assisted Laser Desorption/Ionization Mass Spectrometry of Whole Bacterial Cells and the Detection of High Mass (>20 kDa) Proteins, Rapid Commun. Mass Spectrom. 16 (2002), 1276–1286.

87) Winkler, M.A,; Uher, J. and Cepa, S., Direct Analysis and Identification of *Helicobacter* and *Campylobacter* Species by MALDI-TOF Mass Spectrometry, Anal. Chem. 71 (1999), 3416–3419.

88) Donohue, M.J., Smallwood, A.W., Pfaller, S., Rodgers, M. and Shoemaker, J.A., The Development of a Matrix-Assisted Laser Desorption/Ionization Mass Spectrometry-Based Method for the Protein Fingerprinting and Identification of *Aeromonas* Species Using Whole Cells, J. Microbiol. Meth. 65 (2006), 380–389.

89) Sun, L., Teramoto, K., Sato, H., Torimura, M., Tao, H. and Shintani, T., Characterization of Ribosomal Proteins as Biomarkers for Matrix-Assisted Laser Desorption/Ionization Mass Spectral Identification of *Lactobacillus plantarum*, Rapid Commun. Mass Spectrom. 20 (2006), 3789–3798.

90) Carbonnelle, E., Beretti, J.L., Cottyn, S., Quesne, G., Berche, P., Nassif, X. and Ferroni, A., Rapid Identification of *Staphylococci* Isolated in Clinical Microbiology Laboratories by Matrix-Assisted Laser Desorption Ionization – Time of Flight Mass Spectrometry, J. Clin. Microbiol. 45 (2007), 2156–2161.

91) Welker, M., Marsálek, B., Sejnohová, L. and von Döhren, H., Detection and Identification of Oligopeptides in *Microcystis* (Cyanobacteria) Colonies: Toward an Understanding of Metabolic Diversity, Peptides 27 (2006), 2090–2103.

92) Araoz, R., Guerineau, V., Rippka, R., Palibroda, N., Herdman, M., Laprevote, O., von Döhren, H., de Marsac, N.T. and Erhard, M., MALDI-TOF-MS Detection of the Low Molecular Weight Neurotoxins Anatoxin-A and Homoanatoxin-A on Lyophilized and Fresh Filaments of Axenic *Oscillatoria* Strains, Toxicon 51 (2008), 1308–1315.

93) Evason, D.J., Claydon, M.A. and Gordon, D.B., Effects of Ion Mode and Matrix Additives in the Identification of Bacteria by Intact Cell Mass Spectrometry, Rapid Commun. Mass Spectrom. 14 (2000), 669–672.

94) Amado, F.M.L., Domingues, P., Santana-Marques, M.G., Ferrer-Correia, A.J. and Tomer, K.B., Discrimination Effects and Sensitivity Variations in Matrix-Assisted Laser Desorption/Ionization, Rapid. Commun. Mass Spectrom. 11 (1997), 1347–1352.

95) Liu, H., Du, Z., Wang, J. and Yang, R., Universal Sample Preparation Method for Characterization of Bacteria by Matrix-Assisted Laser Desorption Ionization – Time of Flight Mass Spectrometry, Appl. Environ. Microbiol. 73 (2007), 1899–1907.

96) Ruelle, V., El Moualij, B., Zorzi, W., Ledent, P. and de Pauw, E., Rapid Identification of Environmental Bacterial Strains by Matrix-Assisted Laser Desorption/Ionization Time-of-Flight Mass Spectrometry, Rapid Commun. Mass Spectrom. 18 (2004), 2013–2019.

97) Demirev, P.A., Lin, J.S., Pineda, F.J. and Fenselau, C., Bioinformatics and Mass Spectrometry for Microorganism Identification: Proteome-Wide Post-Translational Modifications and Database search Algorithms for Characterization of Intact *H. pylori*, Anal. Chem. 73 (2001), 4566–4571.

98) Arnold, R.J. and Reilly, J.P., Fingerprint Matching of *E. coli* Strains with Matrix-Assisted Laser Desorption/Ionization Time-of-Flight Mass Spectrometry of Whole Cells Using a Modified Correlation Approach, Rapid Commun. Mass Spectrom. 12 (1998), 630–636.

99) Jarman, K.H., Daly, D.S., Petersen, C.E., Saenz, A.J., Valentine, N.B. and Wahl, K.L., Extracting and Visualising Matrix-Assisted Laser Desorption/Ionization Time-of-Flight Mass Spectral Fingerprints, Rapid Commun. Mass Spectrom. 13 (1999), 1586–1594.

100) Jarman, K.H., Cebula, S.T., Saenz, A.J., Petersen, C.E., Valentine, N.B., Kingsley, M.T. and Wahl, K.L., An Algorithm for Automated Bacterial Identification Using Matrix-Assisted Laser Desorption/Ionization Mass Spectrometry, Anal. Chem. 72 (2000), 1217–1223.

101) Satten, G.A., Datta, S., Moura, H., Woolfitt, A.R., Carvalho, M., da, G., Carlone, G.M., De, B.K., Pavlopoulos, A. and Barr, J.R., Standardization and Denoising Algorithms for Mass Spectra to Classify Whole-Organism Bacterial Specimens, Bioinformatics 20 (2004), 3128–3136.

102) Hettick, J.M., Kashon, M.L., Slaven, J.E., Ma, Y., Simpson, J.P., Siegel, P.D., Mazurek, G.N. and Weissman, D.N., Discrimination of Intact Mycobacteria at the Strain Level: A Combined MALDI-TOF MS and Biostatistical Analysis, Proteomics 6 (2006), 6416–6425.

103) Sauer, S., Freiwald, A., Maier, T., Kube, M., Reinhardt, R., Kostrzewa, M. and Geider, K., Classification and Identification of Bacteria by Mass Spectrometry and Computational Analysis, PLoS ONE 3 (2008), e2843.

104) Vanlaere, E., Sergeant, K., Dawyndt, P., Kallow, W., Erhard, M., Sutton, H., Dare, D., Devreese, B., Samyn, B. and Vandamme, P., Matrix-Assisted Laser Desorption Ionisation – Time-of-Flight Mass Spectrometry of Intact Cells Allows Rapid Identification of *Burkholderia cepacia* Complex, J. Microbiol. Meth. 75 (2008) 279–286.

105) Dieckmann, R., Graeber, I., Kaesler, I., Szewzyk, U. and von Döhren, H., Rapid Screening and Dereplication of Bacterial Isolates from Marine Sponges of the Sula Ridge by Intact-Cell – MALDI-TOF Mass Spectrometry (ICM-MS), Appl. Microbiol. Biotechnol. 67 (2005), 539–548.

106) Pierce, C.Y., Barr, J.R., Woolfitt, A.R., Moura, H., Shaw, E.I., Thompson, H.A., Massung, R.F. and Fernandez, F.M., Strain and Phase Identification of the U.S. Category B Agent *Coxiella burnetii* by Matrix Assisted Laser Desorption/Ionization Time-of-Flight Mass Spectrometry and Multivariate Pattern Recognition, Anal. Chim. Acta 583 (2007), 23–31.

107) Chen, P., Lu, Y. and Harrington, P.B., Application of Linear and Nonlinear Discrete Wavelet Transforms to MALDI-MS Measurements of Bacteria for Classification, Anal. Chem. 80 (2008), 7218–7225.

108) Parisi, D., Magliulo, M., Nanni, P., Casale, M., Forina, M. and Roda, A., Analysis and Classification of Bacteria by Matrix-Assisted Laser Desorption/Ionization Time-of-Flight Mass Spectrometry and a Chemometric Approach, Anal. Bioanal. Chem. 391 (2008), 2127–2134.

109) Villmann, T., Schleif, F.M., Kostrzewa, M., Walch, A. and Hammer, B., Classification of Mass-Spectrometric Data in Clinical Proteomics Using Learning Vector Quantization Methods, Brief Bioinform. 9 (2008), 129–143.

110) Krishnamurthy, T. and Ross, P.L., Rapid Identification of Bacteria by Direct Matrix-Assisted Laser Desorption/Ionization Mass Spectrometric Analysis of Whole Cells, Rapid Commun. Mass Spectrom. 10 (1996), 1992–1996.

111) Leenders, F., Stein, T.H., Kablitz, B., Franke, P. and Vater, J., Rapid Typing of *Bacillus subtilis* Strains by Their Secondary Metabolites Using Matrix-Assisted Laser Desorption/Ionization Mass Spectrometry of Intact Cells, Rapid Commun. Mass Spectrom. 13 (1999), 943–949.

112) Duitman, E.H., Hamoen, L.W., Rembold, M., Venema, G., Seitz, H., Saenger, W., Bernhard, F., Reinhard, R., Schmidt, M., Ullrich, C., Stein, T., Leenders, F. and Vater, J., The Mycosubtilin Synthetase of *Bacillus subtilis* ATCC6633: A Multifunctional Hybrid between a Peptide Synthetase, an Amino Transferase, and a Fatty Acid Synthase, Proc. Natl Acad. Sci. USA 96 (1999), 13 294–299.

113) Vater, J., Kablitz, B., Wilde, C., Franke, P., Mehta, N. and Cameotra, S.S., Matrix-Assisted Laser Desorption Ionization–Time of Flight Mass Spectrometry of Lipopeptide Biosurfactants in Whole Cells and Culture Filtrates of *Bacillus subtilis* C-1 Isolated from Petroleum Sludge, Appl. Environ. Microbiol. 68 (2002), 6210–6219.

114) Vater, J., Gao, X., Hitzeroth, G., Wilde, C. and Franke, P., 'Whole cell' – Matrix-Assisted Laser Desorption Ionization – Time of Flight – Mass Spectrometry, An Emerging Technique for Efficient Screening of Biocombinatorial Libraries of Natural Compounds – Present State of Research, Comb. Chem. High Throughput Screen. 6 (2003), 557–567.

115) Koumoutsi, A., Chen, X.-H., Henne, A., Liesegang, H., Hitzeroth, G., Franke, P., Vater, J. and Borriss, R., Structural and Functional Characterization of Gene Clusters Directing Nonribosomal Synthesis of Bioactive Cyclic Lipopeptides in Bacillus amyloliquefaciens Strain FZB42, J. Bacteriol. 186 (2004), 1084–1096.

116) Hathout, Y., Setlow, B., Cabrera-Martinez, R.M., Fenselau, C. and Setlow, P., Small, Acid-Soluble Proteins as Biomarkers in Mass Spectrometry Analysis of *Bacillus* Spores, Appl. Environ. Microbiol. 69 (2003), 1100–1107.

117) Whiteaker, J.R., Warscheid, B., Pribil, P., Hathout, Y. and Fenselau, C., Complete Sequences of Small Acid-Soluble Proteins from *Bacillus globigii*, J. Mass Spectrom. 39 (2004), 1113–1121.

118) Pribil, P.A., Patton, E., Black, G., Doroshenko, V. and Fenselau, C., Rapid Characterization of *Bacillus* Spores Targeting Species-Unique Peptides Produced with an Atmospheric Pressure Matrix-Assisted Laser Desorption/Ionization Source, J. Mass Spectrom. 40 (2005), 464–474.

119) Callahan, C., Castanha, E.R., Fox, K.F. and Fox, A., The *Bacillus cereus* Containing Sub-Branch Most Closely Related to *Bacillus anthracis*, Have Single Amino Acid Substitutions in Small Acid-Soluble Proteins, While Remaining Sub-Branches Are More Variable, Mol. Cell. Probes 22 (2008), 207–211.

120) Castanha, E.R., Vestal, M., Hattan, S., Fox, A., Fox, K.F. and Dickinson, D., *Bacillus cereus* Strains Fall into Two Clusters (One Closely and One More Distantly) Related to *Bacillus anthracis* According to Amino Acid Substitutions in Small Acid-Soluble Proteins as Determined by Tandem Mass Spectrometry, Mol. Cell. Probes 21 (2007), 190–201.

121) Castanha, E.R., Fox, A. and Fox, K.F., Rapid Discrimination of *Bacillus anthracis* from Other Members of the *B. cereus* Group by Mass and Sequence of 'Intact' Small Acid Soluble Proteins (SASPs) Using Mass Spectrometry, J. Microbiol. Methods 67 (2006), 230–240.

122) Lai, E.-M., Phadke, N.D., Kachman, M.T., Giorno, R., Vazquez, S., Vazquez, J.A., Maddock, J.R. and Driks, A., Proteomic Analysis of the Spore Coats of *Bacillus subtilis* and *Bacillus anthracis*, J. Bacteriol. 185 (2003), 1443–1454.

123) Walker, J., Fox, A.J., Edwards-Jones, V. and Gordon, D.B., Intact Cell Mass Spectrometry (ICMS) Used to Type Methicillin-Resistant *Staphylococcus aureus*: Media Effects and Interlaboratory Reproducibility, J. Microbiol. Meth. 48 (2002), 117–126.

124) Du, Z., Yang, R., Guo, Z., Song, Y. and Wang, J., Identification of *Staphylococcus aureus* and Determination of Its Methicillin Resistance by Matrix-Assisted Laser Desorption/Ionization Time-of-Flight Mass Spectrometry, Anal. Chem. 74 (2002), 5487–5491.

125) Bernardo, K., Pakulat, N., Macht, M., Krut, O., Seifert, H., Fleer, S., Hünger, F. and Krönke, M., Identification and Discrimination of *Staphylococcus aureus* Strains Using Matrix-Assisted Laser Desorption/Ionization – Time of Flight Mass Spectrometry, Proteomics 2 (2002), 747–753.

126) Ho, Y.P. and Reddy, P.M., Mass Spectrometry-Based Approaches for the Detection of Proteins of *Staphylococcus* Species, Infect. Disord. Drug Targets 8 (2008), 166–182.

127) Williamson, Y.M., Moura, H., Woolfitt, A.R., Pirkle, J.L., Barr, J.R., da Gloria Carvalho, M., Ades, E.P., Carlone, G.M. and Sampson, J.S., Differentiation of *Streptococcus pneumoniae* Conjunctivitis Outbreak Isolates by Matrix-Assisted Laser Desorption Ionization – Time of Flight Mass Spectrometry, Appl. Environ. Microbiol. 74 (2008), 5891–5897.

128) Stingu, C.S., Eschrich, K., Rodloff, A.C., Schaumann, R. and Jentsch, H., Periodontitis is Associated with a Loss of Colonization by *Streptococcus sanguinis*, J. Med. Microbiol. 57 (2008), 495–499.

129) Stingu, C.S., Rodloff, A.C., Jentsch, H., Schaumann, R. and Eschrich, K., Rapid Identification of Oral Anaerobic Bacteria Cultivated from Subgingival Biofilm by MALDI-TOF-MS, Oral Microbiol. Immunol. 23 (2008), 372–376.

130) Wieten, G., Haverkamp, J., Meuzelaar, H.L.C., Bondwijn, H.W. and Berwald, L.G., Pyrolysis Mass Spectrometry: A New Method to Differentiate Between the *Mycobacteria* of the 'Tuberculosis Complex' and Other *Mycobacteria*, J. Gen. Microbiol. 122 (1981), 109–118.

131) Wieten, G., Haverkamp, J., Berwald, L.G., Groothuis, D.G. and Draper, P., Pyrolysis Mass Spectrometry: Its Applicability to Mycobacteriology, Including *Mycobacterium leprae*, Ann. Microbiol. (Paris) 133 (1982), 15–27.

132) Kusaka, T. and Mori, T., Pyrolysis Gas Chromatography-Mass Spectrometry of Mycobacterial Mycolic Acid Methyl Esters and Its Application to the Identification of *Mycobacterium leprae*, J. Gen. Microbiol. 132 (1986), 3403–3406.

133) Tobias, H.J., Schafer, M.P., Pitesky, M., Fergenson, D.P., Horn, J., Frank, M. and Gard, E.E., Bioaerosol Mass Spectrometry for Rapid Detection of Individual Airborne *Mycobacterium tuberculosis* H37Ra Particles. Appl. Environ. Microbiol. 71 (2005), 6086–6095.

134) Mattow, J., Jungblut, P.R., Müller, E.C. and Kaufmann, S.H., Identification of Acidic, Low Molecular Mass Proteins of *Mycobacterium tuberculosis* strain H37Rv by Matrix-Assisted Laser Desorption/Ionization and Electrospray Ionization Mass Spectrometry, Proteomics 1 (2001), 494–507.

135) Pignone, M., Greth, K.M., Cooper, J., Emerson, D. and Tang, J., Identification of Mycobacteria by Matrix-Assisted Laser Desorption Ionization–Time-of-Flight Mass Spectrometry, J. Clin. Microbiol. 44 (2006), 1963–1970.

136) bioinformatica.isa.cnr.it/Bact_Dbase.htm

137) Barbuddhe, S.B., Maier, T., Schwarz, G., Kostrzewa, M., Hof, H., Domann, E., Chakraborty, T. and Hain, T., Rapid Identification and Typing of *Listeria* Species by Matrix-Assisted Laser Desorption Ionization – Time of Flight Mass Spectrometry, Appl. Environ. Microbiol. 74 (2008), 5402–5407.

138) Giebel, R.A., Fredenberg, W. and Sandrin, T.R., Characterization of Environmental Isolates of *Enterococcus* spp. by Matrix-Assisted Laser Desorption/Ionization Time-of-Flight Mass Spectrometry, Water Res. 42 (2008), 931–940.

139) Reynaud af Geijersstam, A., Culak, R., Molenaar, L., Chattaway, M., Røslie, E., Peciuliene, V., Haapasalo, M. and Shah, H.N., Comparative Analysis of Virulence Determinants and Mass Spectral Profiles of Finnish and Lithuanian Endodontic *Enterococcus faecalis* Isolates, Oral Microbiol. Immunol. 22 (2007), 87–94.

140) Hindré, T., Didelot, S., le Pennec, J.P., Haras, D., Dufour, A. and Vallée-Réhel, K., Bacteriocin Detection from Whole Bacteria by Matrix-Assisted Laser Desorption Ionization – Time of Flight Mass Spectrometry, Appl. Environ. Microbiol. 69 (2003), 1051–1058.

141) Hitzeroth, G., Vater, J., Franke, P., Gebhardt, K. and Fiedler, H.P., Whole Cell Matrix-Assisted Laser Desorption/Ionization Time-of-Flight Mass Spectrometry and In Situ Structure Analysis

of Streptocidins, a Family of Tyrocidine-Like Cyclic Peptides, Rapid Commun. Mass Spectrom. 19 (2005), 2935–2942.

142) Chong, B.E., Wall, D.B., Lubman, D.M. and Flynn, S.J., Rapid Profiling of *E. coli* Proteins Up to 500 kDa from Whole Cell Lysates Using Matrix-assisted Laser Desorption/Ionization Time-of-flight Mass Spectrometry, Rapid Commun. Mass Spectrom. 11 (1997), 1900–1908.

143) Jones, J.J., Stump, M.J., Fleming, R.C., Lay Jr., J.O. and Wilkins, C.L., Investigation of MALDI-TOF and FT-MS Techniques for Analysis of *Escherichia coli* Whole Cells, Anal. Chem. 75 (2003), 1340–1347.

144) Camara, J.E. and Hays, F.A., Discrimination Between Wild-Type and Ampicillin-Resistant *Escherichia coli* by Matrix-Assisted Laser Desorption/Ionization Time-of-Flight Mass Spectrometry, Anal. Bioanal. Chem. 389 (2007), 1633–1638.

145) Siegrist, T.J., Anderson, P.D., Huen, W.H., Kleinheinz, G.T., McDermott, C.M. and Sandrin, T.R., Discrimination and Characterization of Environmental Strains of *Escherichia coli* by Matrix-Assisted Laser Desorption/Ionization Time-of-Flight Mass Spectrometry (MALDI-TOF-MS), J. Microbiol. Meth. 68 (2007), 554–562.

146) Dai, Y., Li, L., Roser, D.C. and Long, S.R., Detection and Identification of Low-Mass Peptides and Proteins from Solvent Suspensions of *Escherichia coli* by High Performance Liquid Chromatography Fractionation and Matrix-Assisted Laser Desorption/Ionization Mass Spectrometry, Rapid Commun. Mass Spectrom. 13 (1999), 73–78.

147) Karty, J.A., Lato, S. and Reilly, J.P., Detection of the Bacteriological Sex Factor in *E. coli* by Matrix-Assisted Laser Desorption/Ionization Time-of-Flight Mass Spectrometry, Rapid Commun. Mass Spectrom. 12 (1998), 625–629.

148) Jebanathirajah, J.A., Andersen, S., Blagoev, B. and Roepstorff, P., A Rapid Screening Method to Monitor Expression of Recombinant Proteins from Various Prokaryotic and Eukaryotic Expression Systems Using Matrix-Assisted Laser Desorption Ionization – Time-of-Flight Mass Spectrometry, Anal. Biochem. 305 (2002), 242–250.

149) Jones, J.J., Wilkins, C.L., Cai, Y., Beitle, R.R., Liyanage, R. and Lay Jr, J.O., Real-Time Monitoring of Recombinant Bacterial Proteins by Mass Spectrometry, Biotechnol. Prog. 21 (2005), 1754–1758.

150) Lynn, E.C., Chung, M.-C., Tsai, W.-C. and Han, C.-C., Identification of Enterobacteriaceae Bacteria by Direct Matrix-assisted Laser Desorption/Ionization Mass Spectrometric Analysis of Whole Cells, Rapid Commun. Mass Spectrom. 13 (1999), 2022–2027.

151) Mandrell, R.E., Harden, L.A., Bates, A., Miller, W.G., Haddon, W.F. and Fagerquist, C.K., Speciation of *Campylobacter coli*, *C. jejuni*, *C. helveticus*, *C. lari*, *C. sputorum*, and *C. upsaliensis* by Matrix-Assisted Laser Desorption Ionization – Time of Flight Mass Spectrometry, Appl. Environ. Microbiol. 71 (2005), 6292–6307.

152) Fagerquist, C.K., Miller, W.G., Harden, L.A., Bates, A.H., Vensel, W.H., Wang, G. and Mandrell, R.E., Genomic and Proteomic Identification of a DNA-Binding Protein Used in the 'Fingerprinting' of *Campylobacter* Species and Strains by MALDI-TOF-MS Protein Biomarker Analysis, Anal. Chem. 77 (2005), 4897–4907.

153) Fagerquist, C.K., Bates, A.H., Heath, S., King, B.C., Garbus, B.R., Harden, L.A. and Miller, W.G., Sub-speciating *Campylobacter jejuni* by Proteomic Analysis of Its Protein Biomarkers and Their Post-Translational Modifications, J. Proteome Res. 5 (2006), 2527–2538.

154) Donohue, M.J., Best, J.M., Smallwood, A.W., Kostich, M., Rodgers, M. and Shoemaker, J.A., Differentiation of *Aeromonas* Isolated from Drinking Water Distribution Systems Using Matrix-Assisted Laser Desorption/Ionization-Mass Spectrometry, Anal. Chem. 79 (2007), 1939–1946.

155) Martinez-Murcia, A.J., Saavedra, M.J., Mota, V.R., Maier, T., Stackebrandt, E. and Cousin, S., *Aeromonas aquariorum* sp. nov., Isolated from Aquaria of Ornamental Fish, Int. J. Syst. Evol. Microbiol. 58 (2008), 1169–1175.

156) Haag, A.M., Taylor, S.N., Johnston, K.H. and Cole, R.B., Rapid Identification and Speciation of *Haemophilus* Bacteria by Matrix-Assisted Laser Desorption/Ionization Time-of-Flight Mass Spectrometry, J. Mass Spectrom. 33 (1998), 750–756.

157) Gudlavalleti, S.K., Sundaram, A.K., Razumovski, J. and Doroshenko, V., Application of Atmospheric Pressure Matrix-Assisted Laser Desorption/Ionization Mass Spectrometry for Rapid Identification of Neisseria Species, J. Biomol. Techn. 19 (2008), 200–204.

158) See the CDC website at: emergency.cdc.gov/agent/agentlist-category.asp.

159) Skultéty, L., Hernychová, L., Beregházyová, E., Slabá, K. and Toman, R., Detection of Specific Spectral Markers of *Coxiella burnetii* Isolates by MALDI-TOF Mass Spectrometry, Acta Virol. 51 (2007), 55–58.

160) Skultety, L., Hernychova, L., Toman, R., Hubalek, M., Slaba, K., Zechovska, J., Stofanikova, V., Lenco, J., Stulik, J. and Macela, A., *Coxiella burnetii* Whole Cell Lysate Protein Identification by Mass Spectrometry and Tandem Mass Spectrometry, Ann. NY Acad. Sci. 1063 (2005), 115–122.

161) Lundquist, M., Caspersen, M.B., Wikström, P. and Forsman, M., Discrimination of *Francisella tularensis* Subspecies Using Surface Enhanced Laser Desorption Ionization Mass Spectrometry and Multivariate Data Analysis, FEMS Microbiol Lett. 243 (2005), 303–310.

162) Seibold, E., Bogumil, R., Vorderwülbecke, S., al Dahouk, S., Buckendahl, A., Tomaso, H. and Splettstoesser, W., Optimized Application of Surface-Enhanced Laser Desorption/Ionization Time-of-Flight MS to Differentiate *Francisella tularensis* at the Level of Subspecies and Individual Strains, FEMS Immunol. Med. Microbiol. 49 (2007), 364–373.

163) Hernychova, L., Stulik, J., Halada, P., Macela, A., Kroca, M., Johansson, T. and Malina, M., Construction of a *Francisella tularensis* Two-Dimensional Electrophoresis Protein Database, Proteomics 1 (2001), 508–515.

164) Hubálek, M., Hernychová, L., Brychta, M., Lenco, J., Zechovská, J. and Stulík, J., Comparative Proteome Analysis of Cellular Proteins Extracted from Highly Virulent *Francisella tularensis* ssp. tularensis and Less Virulent *F. tularensis* ssp. holarctica and *F. tularensis* ssp. mediaasiatica, Proteomics 4 (2004), 3048–3060.

165) Janovská, S., Pávková, I., Hubálek, M., Lenco, J., Macela, A. and Stulík, J., Identification of Immunoreactive Antigens in Membrane Proteins Enriched Fraction from *Francisella tularensis* LVS, Immunol. Lett. 108 (2007), 151–159.

166) Holland, R.D., Wilkes, J.G., Rafii, F., Sutherland, J.B., Persons, C.C., Voorhees, K.J. and Lay Jr., J.O., Rapid Identification of Intact Whole Bacteria Based on Spectral Patterns using Matrix-Assisted Laser Desorption/Ionization with Time-of-flight Mass Spectrometry, Rapid Commun. Mass Spectrom. 10 (1996), 1227–1232.

167) Degand, N., Carbonnelle, E., Dauphin, B., Beretti, J.L., le Bourgeois, M., Sermet-Gaudelus, I., Segonds, C., Berche, P., Nassif, X. and Ferroni, A., Matrix-Assisted Laser Desorption Ionization-Time of Flight Mass Spectrometry for Identification of Nonfermenting Gram-Negative *Bacilli* Isolated from Cystic Fibrosis Patients, J. Clin. Microbiol. 46 (2008), 3361–3367.

168) Nilsson, C.L., Fingerprinting of *Helicobacter pylori* Strains by Matrix-assisted Laser Desorption/Ionization Mass Spectrometric Analysis, Rapid Commun. Mass Spectrom. 13 (1999), 1067–1071.

169) Nilsson, C.L., Larsson, T., Gustafsson, E., Karlsson, K.A. and Davidsson, P., Identification of Protein Vaccine Candidates from *Helicobacter pylori* Using a Preparative Two-Dimensional Electrophoretic Procedure and Mass Spectrometry, Anal. Chem. 72 (2000), 2148–2153.

170) Erhard, M., von Döhren, H. and Jungblut, P., Rapid Typing and Elucidation of New Secondary Metabolites of Intact *Cyanobacteria* Using MALDI-TOF Mass Spectrometry, Nat. Biotechnol. 15 (1997), 906–909.

171) Erhard, M., von Döhren, H. and Jungblut, P.R., Rapid Identification of the New Anabaenopeptin G from *Planktothrix agardhii* HUB 011 using Matrix-assisted Laser Desorption/Ionization Time-of-flight Mass Spectrometry, Rapid Commun. Mass Spectrom. 13 (1999), 337–343.

172) Janse, I., Kardinaal, W.E.A., Meima, M., Fastner, J., Visser, P.M. and Zwart, G., Toxic and Nontoxic *Microcystis* Colonies in Natural Populations Can Be Differentiated on the Basis of rRNA Gene Internal Transcribed Spacer Diversity, Appl. Environ. Microbiol. 70 (2004), 3979–3987.

173) Welker, M., Brunke, M., Preussel, K., Lippert, I. and von Döhren, H., Diversity and Distribution of *Microcystis* (*Cyanobacteria*) Oligopeptide Chemotypes from Natural Communities Studied by Single-Colony Mass Spectrometry, Microbiology 150 (2004), 1785–1796.

174) Haande, S., Ballot, A., Rohrlack, T., Fastner, J., Wiedner, C. and Edvardsen, B., Diversity of *Microcystis aeruginosa* Isolates (Chroococcales, Cyanobacteria) from East-African Water Bodies, Arch. Microbiol. 188 (2007), 15–25.

175) Ryzhov, V. and Fenselau, C., Characterization of the Protein Subset Desorbed by MALDI from Whole Bacterial Cells, Anal. Chem. 73 (2001), 746–750.

176) Demirev, P., Ho, Y.P., Ryzhov, V. and Fenselau, C., Microorganism Identification by Mass Spectrometry and Protein Database Searches, Anal. Chem. 71 (1999), 2732–2738.

177) Demirev, P., Pineda, F., Lin, J. and Fenselau, C., Bioinformatics and Mass spectrometry for Microorganism Identification: Proteome-Wide Post-Translational Modifications and Database Search Algorithms for Characterization of Intact *H. pylori*, Anal. Chem. 73 (2001), 4566–4573.

178) Pineda, F.J., Antoine, M.D., Demirev, P.A., Feldman, A.B., Jackman, J., Longenecker, M. and Lin, J.S., Microorganism Identification by Matrix-Assisted Laser/Desorption Ionization Mass Spectrometry and Model-Derived Ribosomal Protein Biomarkers, Anal. Chem. 75 (2003), 3817–3822.

179) Currently accessible at www.rmidb.org.

180) Teramoto, K., Sato, H., Sun, L., Torimura, M. and Tao, H., A Simple Intact Protein Analysis by MALDI-MS for Characterization of Ribosomal Proteins of Two Genome-Sequenced Lactic Acid Bacteria and Verification of Their Amino Acid Sequences, J. Proteome Res. 6 (2007), 3899–3907.

181) Teramoto, K., Sato, H., Sun, L., Torimura, M., Tao, H., Yoshikawa, H., Hotta, Y., Hosoda, A. and Tamura, H., Phylogenetic Classification of *Pseudomonas putida* Strains by MALDI-MS Using Ribosomal Subunit Proteins as Biomarkers, Anal. Chem. 79 (2007), 8712–8719.

182) Stump, M.J., Jones, J.J., Fleming, R.C., Lay Jr, J.O. and Wilkins, C.L., Use of Double-Depleted 13C and 15N Culture Media for Analysis of Whole Cell Bacteria by MALDI Time-of-Flight and Fourier Transform Mass Spectrometry, J. Am. Soc. Mass Spectrom. 14 (2003), 1306–1314.

183) Wilcox, S.K., Cavey, G.S. and Pearson, J.D., Single Ribosomal Protein Mutations in Antibiotic-Resistant Bacteria Analyzed by Mass Spectrometry, Antimicrob. Agents Chemother. 45 (2001), 3046–3055.

184) Clauser, K.R., Baker, P. and Burlingame, A.L., Role of Accurate Mass Measurement (+/−10 ppm) in Protein Identification Strategies Employing MS or MS/MS and Database Searching, Anal. Chem. 71 (1999), 2871–2882.

185) Demirev, P.A., Feldman, A.B., Kowalski, P. and Lin, J.S., Top-Down Proteomics for Rapid Identification of Intact Microorganisms, Anal. Chem. 77 (2005), 7455–7461.

186) Harris, W.A. and Reilly, J.P., On-Probe Digestion of Bacterial Proteins for MALDI-MS, Anal. Chem. 74 (2002), 4410–4416.

187) English, R.D., Warscheid, B., Fenselau, C. and Cotter, R.J., *Bacillus* Spore Identification via Proteolytic Peptide Mapping with a Miniaturized MALDI TOF Mass Spectrometer, Anal. Chem. 75 (2003), 6886–6893.

188) Warscheid, B., Jackson, K., Sutton, C. and Fenselau, C., MALDI Analysis of *Bacilli* in Spore Mixtures by Applying a Quadrupole Ion Trap Time-of-Flight Tandem Mass Spectrometer, Anal. Chem. 75 (2003), 5608–5617.

189) Warscheid, B. and Fenselau, C., Characterization of *Bacillus* Spore Species and Their Mixtures Using Postsource Decay with a Curved-Field Reflectron, Anal. Chem. 75 (2003), 5618–5627.

190) Warscheid, B. and Fenselau, C., A Targeted Proteomics Approach to the Rapid Identification of Bacterial Cell Mixtures by Matrix-Assisted Laser Desorption/Ionization Mass Spectrometry, Proteomics 4 (2004), 2877–2892.

191) Holland, R.D., Duffy, C.R., Rafii, F., Sutherland, J.B., Heinze, T.M., Holder, C.L., Voorhees, K.J. and Lay Jr, J.O., Identification of Bacterial Proteins Observed in MALDI TOF Mass Spectra From Whole Cells. Anal. Chem. 71 (1999), 3226–3230.

192) Wunschel, D., Wahl, J., Willse, A., Valentine, N. and Wahl, K., Small Protein Biomarkers of Culture in *Bacillus* Spores Detected Using Capillary Liquid Chromatography Coupled with Matrix Assisted Laser Desorption/Ionization Mass Spectrometry, J. Chromatogr., B 843 (2006) 25–33.

193) Vaidyanathan, S., Rowland, J.J., Kell, D.B. and Goodacre, R., Discrimination of Aerobic Endospore-Forming Bacteria via Electrospray-Ionization Mass Spectrometry of Whole Cell Suspensions, Anal. Chem. 73 (2001), 4134–4144.

194) Li, T.Y., Liu, B.H. and Chen, Y.C., Characterization of *Aspergillus* Spores by Matrix-Assisted Laser Desorption/Ionization Time-of-Flight Mass Spectrometry, Rapid Commun. Mass Spectrom. 14 (2000), 2393–2400.

195) Valentine, N.B., Wahl, J.H., Kingsley, M.T. and Wahl, K.L., Direct Surface Analysis of Fungal Species by Matrix-Assisted Laser Desorption/Ionization Mass Spectrometry, Rapid. Commun Mass Spectrom. 16 (2002), 1352–1357.

196) Chen, H.Y. and Chen, Y.C., Characterization of Intact *Penicillium* Spores by Matrix-Assisted Laser Desorption/Ionization Mass Spectrometry, Rapid Commun. Mass Spectrom. 19 (2005), 3564–3568.

197) Hettick, J.M., Green, B.J., Buskirk, A.D., Kashon, M.L., Slaven, J.E., Janotka, E., Blachere, F.M., Schmechel, D. and Beezhold, D.H., Discrimination of *Aspergillus* Isolates at the Species and Strain Level by Matrix-Assisted Laser Desorption/Ionization Time-of-Flight Mass Spectrometry Fingerprinting, Anal. Biochem. 380 (2008), 276–281.

198) Qian, J., Cutler, J.E., Cole, R.B. and Cai, Y., MALDI-TOF Mass Signatures for Differentiation of Yeast Species, Strain Grouping and Monitoring of Morphogenesis Markers, Anal. Bioanal. Chem. 392 (2008), 439–449.

199) Krader, P. and Emerson, D., Identification of Archaea and some Extremophilic Bacteria Using Matrix-Assisted Laser Desorption/Ionization Time-of-Flight (MALDI-TOF) Mass Spectrometry, Extremophiles 8 (2004), 259–268.

200) Zhang, X., Scalf, M., Berggren, T.W., Westphall, M.S. and Smith, L.M., Identification of Mammalian Cell Lines Using MALDI-TOF and LC-ESI-MS/MS Mass Spectrometry, J. Am. Soc. Mass Spectrom. 17 (2006), 490–499.

14

The Proteomic Road Map to Explore Novel Mechanisms of Bacterial Physiology

Haike Antelmann and Michael Hecker
Institute of Microbiology, Ernst-Moritz-Arndt-University of Greifswald, Greifswald, Germany

14.1 Introduction

Bacillus subtilis is one of the best-studied model organisms of the low-GC content Gram-positive bacteria. There is a long-standing interest in bacilli as hosts for biotechnological production processes. *Bacillus* is also a simple model for cellular differentiation and there is excellent genetic and biochemical characterization. During the last 10 years detailed functional analyses programmes (e.g. in Europe and Japan) have studied systematically the regulation and functions of genes with unknown functions in *B. subtilis*. Several stress and starvation-responsive stimulons have been characterized in *B. subtilis* using functional genomic approaches such as proteomics and transcriptomics and a complex dataset of gene expression profiling is now available. At the second stage stimulons were dissected into specific regulons by the comparative proteome and transcriptome analyses of wild-type strains and mutants in central regulatory genes. However, *B. subtilis* encodes more than 286 transcription factors and 34 two-component systems, many of which remain functionally unknown. There might be many unidentified stimuli which induce unknown regulatory mechanisms that remain to be elucidated in future studies; this is a challenge for functional genomics in the post-genomic era.

From a physiological point of view there are two major proteomes of bacteria: vegetative proteomes of growing cells with mainly house-keeping functions; and proteomes

Mass Spectrometry for Microbial Proteomics   Edited by Haroun N. Shah and Saheer E. Gharbia
© 2010 John Wiley & Sons, Ltd

required for adaptation to environmental changes such as stress, starvation or exposure to toxic or antimicrobial compounds [1–3]. The first step for physiological proteomics is to define the expression profile and to study the regulation of all proteins under defined physiological conditions using mutants in regulatory genes. However, the alternative approaches of proteomics in the postgenomic era are also to analyse protein targeting and localization, interaction, aggregation, posttranslational modification, protein damage, repair and finally degradation on a global scale.

In this chapter we demonstrate the power of gel-based proteomics to get new insights into mechanisms of cell physiology in *B. subtilis*. We review three of our outstanding proteomic studies which include (1) novel mechanisms of protein secretion in *B. subtilis,* (2) the wide range of irreversible and reversible thiol modifications and novel regulatory pathways for thiol-reactive electrophiles and (3) the discovery of new targets of the Fur-regulated small RNA FsrA and three basic proteins (FbpA, B, C). The studies on protein secretion mechanisms were performed in collaboration with the Groningen secretion groups of Jan Maarten van Dijl and Sierd Bron, with Vesa Kontinen and Matti Sarvas (Helsinki) and Colin Harwood (Newcastle). In addition, the regulatory pathways which control the electrophile responsive mechanisms in *B. subtilis* were studied in collaboration with the groups of Peter Zuber (Oregon) and John D. Helmann (Ithaca). Finally, the latest example concerning the target identification for the Fur-controlled small RNA FsrA was also peformed in collaboration with John D. Helmann. These examples should highlight how the power of gel-based proteomics can be exploited to understand bacterial physiology on a genome-wide scale.

14.2   Proteomics of Protein Secretion Mechanisms in *B. subtilis*

14.2.1   Protein Export Machineries of *B. subtilis*

The completion of the *B. subtilis* genome sequence in 1997 [4] allowed the prediction of extracellular and surface-associated proteins according to the presence of N-terminal signal peptides, cell wall binding and membrane-spanning domains [5, 6]. The secretome of *B. subtilis* includes 300 predicted secretory proteins which are translocated mainly via the 'General secretion pathway' (Sec pathway). In contrast, the 'Twin arginine translocation pathway' (Tat pathway) can be regarded as a special purpose pathway via which only a small subset of often folded, complex cofactor-containing and oligomeric proteins are transported. Most of these secreted proteins are synthesized with N-terminal signal peptides and secreted into the extracellular medium as these lack retention signals. Other exported proteins are involved in processes at the membrane–cell wall interface (e.g. cell wall turnover, substrate binding, folding or modification of exported proteins). These proteins are retained in the cytoplasmic membrane or cell wall after their translocation [5, 6]. Cytoplasmic chaperones and targeting factors, such as Ffh/ FtsY or CsaA are required for signal peptide recognition, keep these precursors in a translocation-competent conformation and facilitate targeting to the Sec-translocase in the membrane [7]. During or shortly after translocation of the (pre-)proteins across the membrane, the N-terminal signal peptide is cleaved by the signal peptidases (SPases). Secretory and twin arginine signal peptides are cleaved by the multiple type-I signal peptidases (SipSTUVW) [5]. Lipoprotein signal peptides (type-II signal peptides) are present at the N-terminus of pre-lipoproteins in *B. subtilis* which are Sec-dependently translocated, lipid-modified at the conserved cysteine residue of the lipobox by the diacyl glycerol transferase (Lgt) and subsequently cleaved by the type-II signal peptidase (LspA) [8]. Furthermore, quality control factors constitute additional components of the *B. subtilis* protein export machinery that are involved in post-translocational processes, such as folding or modification. The most important folding catalyst involved in protein secretion in *B. subtilis* is the chaperone and peptidyl-prolyl-cis/trans-isomerase PrsA which is essential for protein secretion and cell viability [9–11]. Secreted proteins that are not correctly folded by the PrsA foldase are substrates for the numerous proteases of *B. subtilis* located in the extracellular medium, cell wall and membrane. Quality control factors, such as the HtrA proteases (HtrA and HtrB) are induced in response to secretion stress in a CssSR-dependent manner in *B. subtilis* by overproduction of the heterologous secreted AmyQ protein of *B. amyloque-faciens* [12, 13].

In the first part of this chapter we demonstrate how the power of gel-based proteomics was used to analyse the mechanisms of protein secretion in *B. subtilis*. These analyses were performed in collaboration with the 'European Bacillus Secretion Group' (EBSG), most importantly with the secretion groups of Jan Maarten van Dijl and Sierd Bron (Groningen) and Matti Sarvas and Vesa Kontinen (Helsinki). Based on the extracellular and cell wall proteome analyses of *B. subtilis* mutants lacking components of the protein export machinery as well as quality control factors new proteomic insights into the mechanisms of protein secretion in *B. subtilis* were obtained. Some proteomic highlights of these outstanding collaborations over the last 8 years are summarized in the first part of this chapter.

14.2.2 The Extracellular Proteome of *B. subtilis*

As a Gram-positive bacterium *B. subtilis* lacks an outer membrane and is able to secrete large amounts of extracellular enzymes directly into the growth medium. Thus, we can define four subproteomes in *B. subtilis*: the cytoplasmic, membrane, cell wall and extracellular proteomes. The first proteome analyses of secreted proteins of *B. subtilis* were performed in minimal medium containing glucose or alternative carbon sources [14] as well as under the conditions of phosphate starvation [15]. The PhoPR-controlled extracellular phosphatases, phosphodiesterases and nucleotidases which are involved in the uptake and utilization of alternative phosphate sources are major parts of the phosphate starvation extracellular proteome [Figure 14.1(a)] [15].

The highest level of protein secretion is observed in rich LB-medium, in particular during the stationary phase. Thus, the master gel for the extracellular proteome was defined in LB during the stationary phase [Figure 14.1(b)] [16]. In the extracellular proteome of *B. subtilis* 113 proteins were identified including 50% predicted secreted proteins which have signal peptides but no additional retention signals. The remaining proteins are cytoplasmic, flagella- and prophage-related proteins that lack signal peptides, or lipoproteins and cell wall proteins that possess retention signals in addition to the signal peptides.

(a) (b)

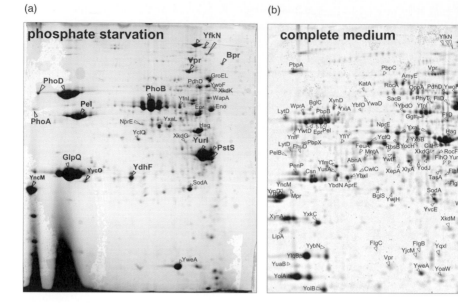

Figure 14.1 *The extracellular proteome of B. subtilis 168 under conditions of phosphate starvation (a) and in complete medium (b) [15, 16]*

14.2.3   The Cell Wall Proteome of *B. subtilis*

Noncovalently linked cell wall-binding proteins (CWBPs) can be extracted using the LiCl extraction procedure [17]. This cell wall proteome was also defined in LB medium during the stationary phase which consists of the WapA processing products (CWBP105 and CWBP62), the WprA cell wall protease processing products (CWBP52 and CWBP23), autolysins and autolysin modifier proteins (LytC and LytB), flagellin (Hag), YwsB and YqgA [Figure 14.2(a)] [17]. Interestingly, different cell wall autolysins such as LytD, YocH, YvcE and YwtD are present in the extracellular fraction but not in the cell wall proteome. This indicates that the presence of cell wall binding repeats is not a guarantee for retention in the cell wall [18].

Recently, we identified one remarkable novel cell wall protein, the oxalate decarboxylase OxdC which accumulates as the most abundant protein in the presence of acidic conditions (e.g. phytate) in the cell wall proteome [Figure 14.2(b)] [19]. Since OxdC has no signal peptide, the mechanism of OxdC targeting the cell wall is completely unknown. It was shown by John Helmann`s group that OxdC expression is regulated by a novel sigma factor (YvrI) and two accessory proteins (YvrHa and YvrL) that affect YvrI activity [20]. Our recent collaborative proteomic results showed that acid induction of OxdC and cell wall sorting both depends on YvrI and the negative factor YvrL [21]. Also in this case, proteomics has discovered a completely unexpected physiological phenomenon which could be based on a novel protein secretion mechanism.

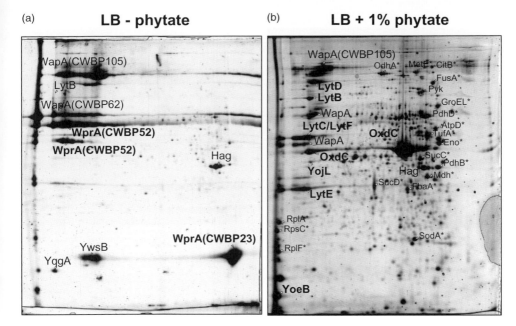

Figure 14.2   *The cell wall proteome of B.* subtilis *168 grown in LB at neutral pH (a) and under acidic conditions provoked by 1% acidic phytate (pH 5.4) (b). Cell wall proteins were extracted using a LiCl extraction procedure [17]. The WprA processing products are absent and several autolysins and autolysin modifier proteins are increased in the cell wall proteome of B.* subtilis *grown with phytate. The OxdC protein is most strongly induced in the cell wall proteome under acidic conditions. Induction and secretion of OxdC by acid stress depends on the novel sigma factor YvrI [19–21]. Cytoplasmic proteins that indicate increased cell lysis with phytate are marked by asterisks*

14.2.4   The Membrane Attached Lipoproteome of *B. subtilis*

To analyse the 'lipoproteome' fraction of *B. subtilis*, a washed cell membrane fraction was prepared that was extracted with the detergent ß-D-maltoside as described previously [22, 23]. This lipoproteome fraction shows an enrichment of 10 lipoproteins [Figure 14.3(b)] [23]. These include OppA, YfmC and MntA as the most abundant lipoproteins, which are also part of the extracellular proteome due to proteolytical shedding. In addition, the lipoproteins FeuA, PbpC, RbsB, YfiY, YusA and YxeB could be enriched in the ß-D-maltoside-extracted lipoproteome fraction of wild type cells. These are redistributed from the membrane to the medium fraction in the *lgt* mutant which is impaired in lipidmodication of pre-lipoproteins (Figure 14.3). Lipoproteins are in most cases substrate binding components of high-affinity ABC transporters. For example, the highly expressed lipoproteins FeuA, YfiY, YxeB and YfmC are under negative control of the iron-uptake respressor Fur and mediate uptake of the catecholate siderohores enterobactin and bacillibactin (FeuA), the hydroxamate siderophores schizokinen, arthrobactin, and corprogen (YfiY), ferrichrome (FhuD) and ferrioxamine (YxeB), respectively [24].

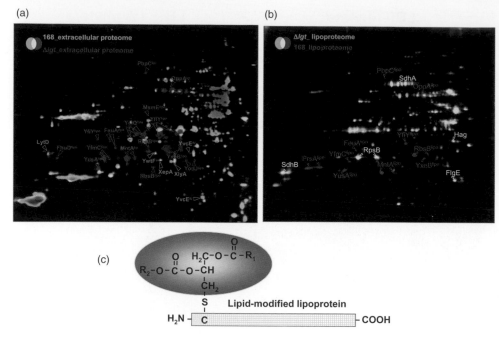

Figure 14.3 *The extracellular proteome (a) and the membrane-attached 'lipoproteome' (b) of the diacylglyceryl transferase (Δlgt) mutant (red image, a; green image, b) in comparison with the wild type (red image, b; green image, a). B. subtilis wild type and Δlgt mutant strains were grown in complete medium into the stationary phase and extracellular proteins were separated using 2D PAGE [16]. The lipoproteome fraction was obtained after stepwise extraction with ß-D-maltoside as described previously [22, 23] (b). Lipoproteins that are released into the medium of the Δlgt strain (a) and absent from the lipoproteome membrane fraction of the Δlgt strain (b) are indicated by red labels and marked with 'lipo'. The structure of lipoproteins and modification by the diacyl glyceryl moiety at the conserved Cys residue of the 'lipo box' which is catalysed by the diacyl glyceryl transferase (Lgt) in B. subtilis is shown in (c)*

14.2.5   The Proteome Analysis of Protein Secretion Mechanisms in *B. subtilis*

The comparative proteome analysis of *B. subtilis* mutants in different components of the protein export machinery provided insights into the protein secretion mechanisms for predicted and unpredicted secreted proteins (Figure 14.4). The transport of most predicted secreted proteins with signal peptides is dependent on both components of the bacterial signal recognition particle (Ffh and FtsY) in *B. subtilis*, since they were present at decreased levels in the extracellular proteomes after depletion of FtsY or Ffh (Figure 14.5) [25]. In contrast, the Ffh- homologous FlhF protein was shown to be dispensable for protein secretion in *B. subtilis* since no difference was detected in the extracellular proteome comparison between wild-type and *flhF* mutant cells [26].

In the extracellular proteome of *B. subtilis* 14 secreted proteins with twin arginine signal peptides have been identified which could be directed into the Sec-independent alternative

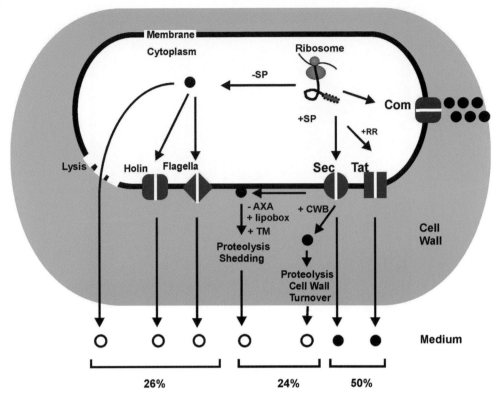

Figure 14.4 *Mechanisms for 'secretion' of 113 extracellular proteins. A total of 113 different extracellular proteins was identified in the extracellular proteomes of cells grown in complete or phosphate starvation media. Extracellular proteins were classified as 'predicted' or 'unpredicted' secretory proteins based on the presence or absence of signal peptides [5]. Accordingly, 54 extracellular proteins were predicted to be secreted because these possess N-terminal signal peptides (SPs) with cleavage sites for type I SPases and lack retention signals. The remaining 59 are unpredicted because these either lack an SP (17 cytoplasmic proteins, 5 phage-related proteins and 7 flagella- related proteins) or have retention signals in addition to the SP (18 lipoproteins, 6 transmembrane proteins and 6 cell wall proteins) [16]*

Tat pathway [27, 28]. Using extracellular proteome analysis of the *tatAdCd* mutant in response to phosphate starvation we showed that the phosphate-starvation-induced PhoPR-dependent alkaline phosphodiesterase PhoD is specifically transported via the TatAdCd translocase (Figure 14.6) [27, 28]. In addition, *B. subtilis* encodes a second minimal Tat translocase TatAyCy via which the iron-starvation induced Fur-controlled iron peroxidase YwbN is specifically translocated [29].

These proteomic efforts had also a great impact on identification of substrates for the Tat pathway in pathogenic bacteria, such as *Streptomyces scabies*, *Pseudomonas aeruginosa* and *Shigella flexneri* which were performed in recent collaborations within the Tat machine

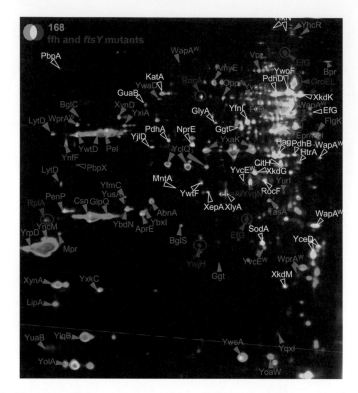

Figure 14.5   *The extracellular proteomes of B. subtilis strain IftsY grown in the absence of IPTG (FtsY depletion, red image) compared with the wild type (green image). Cells were grown in complete medium into the stationary phase and extracellular proteins were separated using 2D PAGE [16, 25]. Proteins with signal peptides which are decreased after depletion of FtsY are marked in blue [25]. Cytoplasmic proteins are more abundant in the extracellular proteome in the absence of FtsY (labelled in red) indicating cell lysis*

consortium together with the groups of Tracy Palmer (Dundee), Long-Fei Wu and Rome Voulhoux (Marseille).

The extracellular proteome of *B. subtilis* also includes cytoplasmic, prophage, flagella, cell wall-associated, and membrane-associated proteins which are unpredicted secretory proteins. Our proteomic results revealed that the release of cytoplasmic proteins is most probably mediated by partial cell lysis. The five prophage-related proteins identified in the extracellular proteome of *B. subtilis* could be secreted via prophage-encoded holins [16, 18]. Flagella-related proteins could be exported via type-III secretion machinery for assembly of flagella similar to in Gram-negative bacteria. Lipoproteins are normally retained in the cytoplasmic membrane via the lipid anchor. Since the lipoproteins in the extracellular proteome lack the N-terminal lipid-modified cysteine residue, these are probably released from the membrane by proteolytic shedding.

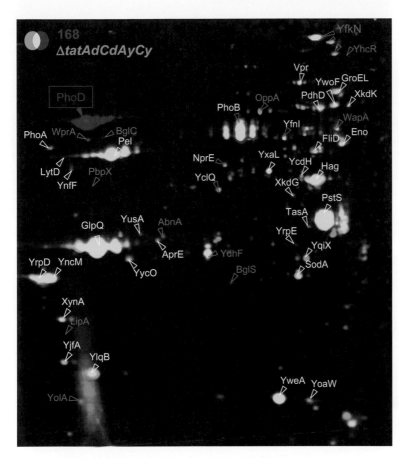

Figure 14.6   *The phosphate-starvation extracellular proteome of the B. subtilis Δtotal-tat mutant (green image) compared with the wild type (red image). B. subtilis strains were grown under phosphate starvation conditions and extracellular proteins were separated using 2D PAGE [15, 16]. PhoD was identified as the specific substrate of the TatAdCd pathway. Other blue-labelled secreted proteins possess also RR/KR-motifs in their N-terminal signal peptides but these are no substrates for the Tat pathway*

Cell wall proteins are also present in the extracellular proteome. We found that cell wall proteins such as processing products of WapA and WprA and also some autolysins are secreted predominantly during the exponential growth and switched off during the transition phase [17]. Our proteome analysis showed stabilization of some cell wall proteins (e.g. WapA, YvcE) during the stationary phase in multiple extracellular protease and *sigD* mutants (Figure 14.7). These results indicate that cell wall proteins are substrates for degradation by extracellular proteases during the stationary phase in *B. subtilis*. In addition, cell wall proteins might be released due to high cell wall turnover predominantly during growth in *B. subtilis* which is mediated by the activity of the autolysins. We were

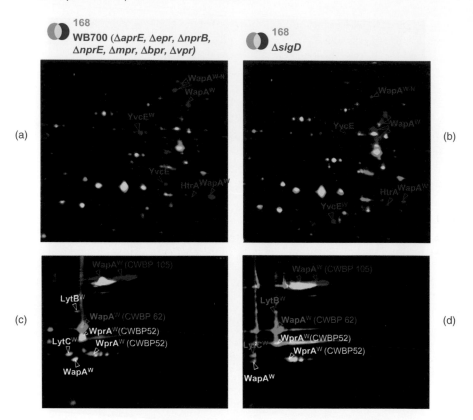

Figure 14.7   *The extracellular proteomes (a, b) and cell wall proteomes (c, d) of the B. subtilis multiple protease mutant WB700 (left) and of the sigD mutant (right) (red images) in comparison with the wild type (green images). Secreted cell wall proteins including the large WapA processing products, unprocessed YvcE, HtrA and YqxI are stabilized in the absence of extracellular proteases (a) and in the sigD mutant (b) in the extracellular proteome [17]. The WapA processing product CWBP105 is increased and the WapA processing product CWBP62 is decreased in the cell wall proteomes of the protease (c) and sigD mutants (d)*

interested in the effect of protease inhibitors on the composition of the extracellular proteome. Thus, we added the protease-inhibitor cocktail 'Complete' during the exponential growth phase at a concentration which had no effect on the growth rate of *B. subtilis*. The proteome analysis showed that WapA and YvcE protein were stabilized in the extracellular proteome after the addition of Complete similar to in the protease depletion strain. Thus, it seems that the released forms of the cell wall proteins WapA and YvcE are targets for extracellular proteases which can be chemically inhibited with protease inhibitors [30].

Extracellular proteome analysis of a *prsA* conditional mutant revealed that the major folding catalyst PrsA of *B. subtilis* is required for secretion and folding of most of the predicted secreted proteins with signal peptides [31]. Detailed proteome analysis using different *B. subtilis prsA* point mutants showed that the N-terminal domain is essential for protein secretion but not the PPIase active centre in the parvulin domain [31]. Besides the proteases and the PrsA foldase, the HtrA proteases are quality control factors for malfolded secreted proteins. Extracellular proteome analysis showed that two proteins are induced in response to secretion stress, HtrA and YqxI. Induction of HtrA is transcriptionally regulated by the CssRS two-component system [12, 13]. In contrast, induction of YqxI is mediated by a post-transcriptional mechanism in a HtrA-dependent manner suggesting a chaperone-like activity of HtrA for stabilization of YqxI [32].

14.3 Definition of Proteomic Signatures to Study Cell Physiology

As soil bacterium, *B. subtilis* is exposed to different stress and starvations conditions as well as toxic and antimicrobial compounds. Several stress, starvation and antibiotic responsive stimulons have been studied in *B. subtilis* using functional genomic approaches and a complex dataset of gene expression profiling is now available. At the second stage stimulons were dissected into regulons by the comparative proteome and transcriptome analyses of wild type strains and mutants in central regulatory genes. Based on these large sets of available expression profiles and the genetic understanding of the regulatory mechanisms, *B. subtilis* is also an excellent model bacterium for physiological studies. In the next part of this chapter we summarize our progress on the definition of proteomic signatures in response to different stress and starvation conditions in *B. subtilis*.

14.3.1 Proteomic Signatures of *B. subtilis* in Response to Stress and Starvation

From a physiological point of view, there are two sub-proteomes in bacterial cells: the proteome of growing cells; and the proteome under growth-restricted conditions (stress, starvation, antibiotics). The proteome of growing cells includes a core-set of highly conserved and abundant housekeeping proteins such as central metabolic enzymes or the translation machinery which are required for cellular growth [1]. In the vegetative proteome of *B. subtilis* 745 proteins were identified using the 2D gel-based approach. Most of these vegetative proteins could be separated in the standard pH range 4–7 [33]. Gel-free proteomic approaches have added 473 proteins to the list of vegetative proteins resulting in the identification of 814 proteins [34, 35].

The proteome under growth-restricted conditions is highly dynamic and is influenced by a complex adaptative regulatory network of transcription factors. Proteomic signatures are defined as marker proteins which are induced in response to environmental stimuli that contribute to specific stress or starvation conditions [1, 36, 37]. These stimulus-specific marker proteins can be dissected into regulon-specific marker proteins. Proteomic signatures of *B. subtilis* were defined at first in response to different stress and starvation conditions using the colour-coding approach of the Decodon Delta 2D software (Figure 14.8) [2, 38, 39]. In total, 201 stress or starvation induced proteins were identified including 79 marker proteins for heat, salt, oxidative stress and 155 marker proteins indicative of glucose, phosphate, ammonium or tryptophan starvation [2]. Following this, a catalogue of proteomic signatures for heat, salt, and oxidative stress (H_2O_2, paraquat) and starvation of glucose, phosphate, ammonium and tryptophan was established [2, 3]. The heat shock

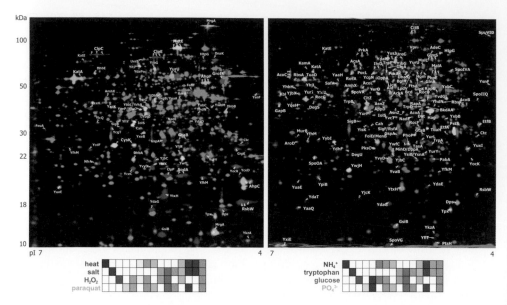

Figure 14.8 *Fused proteome maps of B. subtilis exposed to heat, salt, hydrogen peroxide and paraquat stress (a) or starvation for ammonium, tryptophan, glucose and phosphate (b). The protein synthesis patterns (autoradiograms) of B. subtilis exposed to different stress (a) or starvation (b) conditions were combined to generate a fused stress or starvation proteome map, respectively [2]. Induced marker proteins were colour coded according to their expression profile. Proteins can be classified as specifically or generally induced stress or starvation proteins using the respective colour codes*

proteomic signature is indicated by the induction of the heat-specific HrcA-dependent chaperones, the protein-damage responsive CtsR and Spx regulons and the σ^B-dependent general stress response. Hydrogen peroxide and paraquat caused inductions of the oxidative stress specific PerR and Fur regulons and the Spx regulon as indicators for irreversible and reversible thiol modifications in proteins [40–44]. Furthermore, the SOS regulon induction is specific for DNA damage caused by peroxides [45, 46]. The cysteine metabolism (CymR) regulon and the S-box regulon involved in methionine biosynthesis were specifically induced by paraquat [47–49]. The specific adaptative functions of the stress specific regulons (e.g. HrcA, PerR) include detoxification of the stressor or repair of the damage [1].

Proteomic signatures reflect also the changes in response to different starvation conditions which are encountered by microorganisms. The induction of the TnrA- and σ^L/BkdR-dependent catabolic enzymes for alternative nitrogen sources indicates an ammonium starvation specific proteome signature [50–52]. The TRAP-regulated tryptophan biosynthesis enzymes are the tryptophan starvation specific proteomic signature [53, 54]. The CcpA, CcpN and AcoR-dependent carbon catabolite controlled marker proteins are indicative for the glucose starvation proteomic signature [55–57]. The PhoPR regulon induction is the specific proteome signature for phosphate starvation [15, 58, 59]. The starvation-specific TnrA, TRAP, CcpA or PhoPR regulons are involved in the uptake and utilization of alternative nutrient sources upon nutrient depletion.

Besides these starvation-specific regulons, many starvation-induced proteins respond to more than one nutrient starvation condition. These general starvation proteins belong to the CodY, σ^B and σ^H transition phase regulons which are required for the adaptation of the cell to stationary phase processes such as survival, competence or sporulation [60–62]. Interestingly, a subset of six σ^H-dependent proteins YvyD, YtxH, YisK, YpiB, Spo0A and YuxI and the CodY-dependent RapA were identified as general starvation proteins induced under all starvation conditions. Of these general starvation proteins, YvyD, YtxH, Spo0A, YpiB and RapA proteins are under positive stringent control, indicating the transition from exponential growth to stationary phase [63].

14.3.2   Proteomic Signatures of *B. subtilis* in Response to Thiol-Reactive Electrophiles Uncovered the Novel MarR-Type Regulators MhqR and YodB

This catalogue of stress and starvation proteome signatures is the basis to understand the mode of action of unknown antimicrobials, antibiotics and xenobiotics in *B. subtilis*. Julia Bandow established the first catalogue of proteomic signatures for *B. subtilis* in response to 30 different antibiotics [64, 65]. The aim of our study was to analyse the physiological response of *B. subtilis* to different phenolic antimicrobial compounds to identify novel resistance mechanisms [3]. Specifically, proteomic and transcriptomic signatures of *B. subtilis* were defined after exposure to phenol, salicylic acid, chromanon, catechol, 2-MHQ and diamide [66–69]. These genome-wide expression profilings discovered novel thiol-stress specific electrophile resistance mechanisms that are conserved in *B. subtilis* and other Gram-positive bacteria [3]. The results of these studies which are based on physiological proteomics and novel thiol-specific post-translational modifications are summarized here.

Basically, our proteome and transcriptome analyses showed a strong overlap in the response of *B. subtilis* to catechol, 2-MHQ and diamide. Diamide is an electrophilic azo-compound that reacts with protein thiolates leading to the formation of disulfides or S-thiolated proteins [69–71]. Catechol and hydroquinone-like compounds are auto-oxidized to toxic quinones. Quinones are electrophiles which form S-adducts with cellular thiols via the thiol-(S)-alkylation reaction that leads to aggregation of thiol-containing proteins *in vivo* as irreversible thiol modification [72–74]. Thus, quinone compounds and diamide have the same mode of action but react via different thiol chemistries and lead to depletion of protein and nonprotein low molecular weight (LMW) thiols. This depletion of the thiol redox buffer leads to induction of the complex thiol-specific stress response to electrophilic compounds in *B. subtilis*, which is regulated by Spx, CtsR, PerR and CymR [43, 69, 75–78]. Using the transcription factor/transformation array technology the novel thiol-stress responsive MarR-type repressors MhqR and YodB were discovered in cooperation with Kazuo Kobayashi (Nara) and Peter Zuber (Oregon) [3, 77–80]. We further used proteomic and transcriptomic approaches to identify the genes and proteins regulated by YodB and MhqR. All YodB and MhqR regulon members are strongly induced by

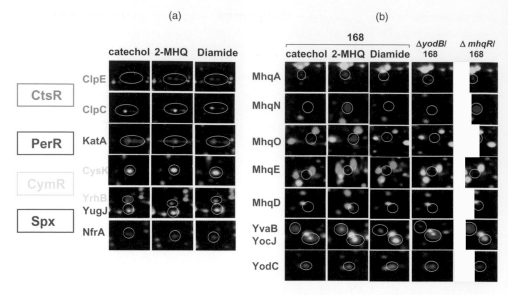

Figure 14.9 *Proteome analysis of the thiol-specific stress response in* B. subtilis *in response to diamide and quinone-like electrophiles. Induction of CtsR-, PerR-, CymR- and Spx-dependent marker proteins by electrophiles indicates depletion of the thiol redox buffer in* B. subtilis *(a) [3]. The MhqR-dependent proteins (AzoR2, MhqA, MhqD, MhqE, MhqN, MhqO) and YodB-dependent proteins (AzoR1, YodC) are involved in specific detoxification of diamide and quinones (b).* B. subtilis *wild type cells grown in minimal medium to an OD500 of 0.4 were pulse-labelled for 5 min each with L-(35S)methionine at control conditions (green image) and in the wild type 10 min after MHQ, catechol and diamide challenge or in the ΔyodB and ΔmhqR mutants at control conditions (red images)*

the thiol-reactive compounds diamide, MHQ and catechol at the proteome level (Figure 14.9). This proteomic signature was in fact the basis to get interested in the regulation of these novel detoxification genes. The YodB and MhqR regulons include paralogous azoreductases (AzoR1 and AzoR2), nitroreductases (YodC and MhqN) and thiol-dependent dioxygenases (MhqA, MhqE and MhqO) that confer resistance to quinone-like electrophiles and diamide. Azoreductases and nitroreductases function in the reduction of quinones and diamide. Dioxygenases are involved in the thiol-dependent ring-cleavage of the quinone-S-adducts which are formed in the detoxification reaction of the LMW thiol buffer with quinones (Figure 14.10) [3, 80]. Recently, we have identified in collaboration with Kazuo Kobayashi the YodB-paralogous thiol-stress sensing MarR/DUF24-type regulator, YvaP (CatR) as repressor of the quinone-inducible *yfiDE* (*catDE*) operon which encodes a DoxX-like oxidoreductase CatD and a catechol-2,3-dioxygenase (CatE) (our unpublished data). Together, the MarR-type repressors MhqR, YodB and CatR control paralogous oxidoreductases and thiol-dependent dioxygenases and constitute a quinone resistance network of *B. subtilis* [3, 79, 80]. This network could have evolved to detoxify quinone-like electrophiles which are naturally produced during electron transfer in the respiratory chain [3].

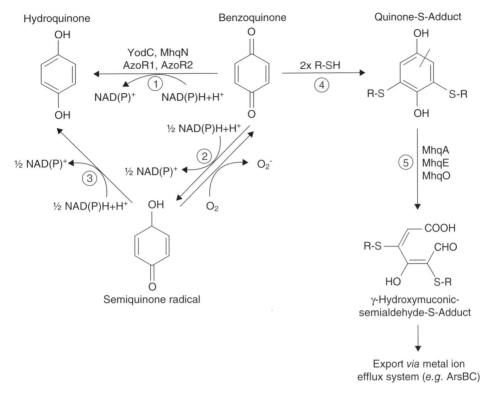

Figure 14.10 *Thiol-dependent detoxification of quinones in B. subtilis. (1) Benzoquinone (BQ) is detoxified to hydroquinone in the two-electron reduction by NAD(P)H-dependent quinone reductases (AzoR1, AzoR2, YodC and MhqN) which are controlled by YodB and MhqR [3, 74]. (2) The one-electron reduction of BQ generates the semiquinone radical which reduces molecular oxygen to the superoxide anion. (3) Conversion of accumulated semiquinone radical to hydroquinone. (4) BQ reacts primarily electrophilic with cellular protein thiols via the Michael addition chemistry [thiol-(S)-alkylation] leading to S-adduct formation. Since BQ has two reactive sites for the nucleophilic addition, it can cross-link different protein thiols in vivo (5) The paralogous thiol-dependent dioxygenases (MhqA, MhqE, MhqO) which are regulated by MhqR might be involved in the dioxygenolytic cleavage of the thiol-(S)-adducts that are probably exported via metal ion efflux systems*

14.3.3   The MarR/DUF24-Family YodB Repressor is Directly Sensing Thiol-Reactive Electrophiles via the Conserved Cys6 Residue

The novel MarR/DUF24-family YodB-repressor is highly conserved among low-GC Gram-positive bacteria. Moreover, *B. subtilis* encodes eight MarR/DUF24-family regulators (YybR, YdeP, YkvN, YtcD, YdzF, HxlR, YodB and YvaP). The first described MarR/DUF24-family regulator of *B. subtilis* was HxlR, which responds to formaldehyde and activates transcription of the *hxlAB* operon encoding the ribulose monophosphate pathway [81]. Our recent genome-wide studies revealed that also formaldehyde triggers a thiol-specific stress response [82].

YodB paralogues have in common a conserved cysteine residue around position 6 in the N-terminal part. We have further proposed that DNA binding activity of the thiol-stress sensing YodB repressor could be inhibited via thiol-(S)-alkylation at the conserved Cys6 residue in response to quinone-like electrophiles *in vitro* which remains to be shown *in vivo* [3]. This alkylation of regulatory thiols is an irreversible thiol modification that cannot be reduced using the thiol-disulfide exchange reaction [80]. In addition, the conserved Cys6 residue of YodB was essential for repression using site directed mutagenesis *in vivo* and *in vitro*. Cys residues have been shown to be involved in the redox control of other oxidative stress transcription factors, such as OxyR, Hsp33 of *E. coli* and OhrR of *B. subtilis* [76, 83–86]. Furthermore, S-thiolation has been shown as one regulatory mechanism for OhrR and OxyR [86, 87]. Our recent data indicate that YodB is redox-controlled via intersubunit disulfide formation in response to diamide *in vivo* (our unpublished data). This YodB mechanism is similar to the two-Cys sensing mechanism for OhrR of *Xanthomonas campestris* which also involves an intersubunit disulfide bond between one N-terminal Cys22 of one subunit and one C-terminal Cys127 of the other subunit [88, 89]. Detailed future work is required to elucidate the functions and regulatory mechanisms of all MarR/DUF24 family regulators and their regulons in this thiol-stress responsive regulatory network.

14.4   Proteomics as a Tool to Visualize Reversible and Irreversible Thiol Modifications

The state-of-the-art of proteomics in the twenty-first century is the global analysis of post-translational modifications of proteins such as protein phosphorylation, glycosylation, oxidation or aggregation. These post-translational modifications can trigger their activation, inactivation or even degradation. As examples for protein oxidation and aggregation we will review our recent proteome-wide results of reversible and irreversible thiol modifications in cytoplasmic proteins in response to the different thiol-reactive electrophiles, diamide and quinones. Proteomic data reflect the different thiol chemistries of diamide and quinones. Diamide leads to reversible disulfide formations and quinones react with thiol-containing proteins largely via irreversible thiol-(S)-alkylation which lead to protein aggregation.

14.4.1   The Thiol-Redox Proteome of *B. subtilis* in Response to Diamide and Quinones

The recently developed 2D gel-based fluorescense alkylation of reversible oxidized thiols assay (FALKO assay) was developed to analyse reversible thiol modifications (e.g. disulfide bonds and S-thiolated proteins) which are formed *in vivo* in response to oxidative or thiol-specific stress [70]. First, all reduced protein thiols are chemically blocked with iodoacetamide (IAM) and subsequently all proteins with reversible thiol oxidations are reduced with tris(2-carboxyethyl)phosphine (TCEP) followed by labelling with the fluorescence dye BODIPY FL C_1-IA.

The dual channel images of the fluorescence image (thiol-redox proteome, false-coloured in red) and the Sypro ruby-stained images (false-coloured in green) of exponentially

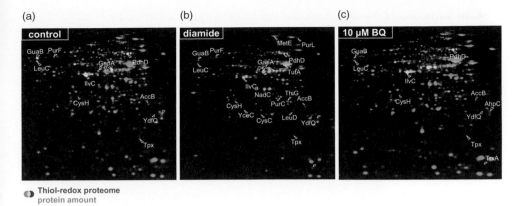

Figure 14.11   *The Sypro-ruby-stained proteome (green) compared with the thiol-redox proteome (red) at control conditions (a) and after exposure to 1 mM diamide (b) and 10 μM BQ (c). Cell extracts were alkylated with IAM to block reduced protein thiols and subsequently treated with TCEP and BODIPY FL C₁-IA to reduce and fluorescence-label oxidized protein thiols [70, 74]. Proteins with reversible thiol oxidations are marked in the thiol-redox proteomes (red images). Diamide causes strongly increased disulfide formations in numerous cytoplasmic proteins. GapA is the only new target for reversible thiol oxidation in response to quinones*

growing cells is shown in Figure 14.11(a). Only a few proteins with oxidized thiols were detected under control conditions during exponential growth which reflects the reduced state of the cytoplasm [70]. These include proteins that are known to have catalytic or redox-active disulfide bonds, such as the thiol-dependent peroxidases AhpC and Tpx. However, there is a strongly increased formation of reversible disulfide bonds in the thiol-redox proteome upon diamide stress [Figure 14.11(b)] [70]. Mass spectrometry revealed the formation of mixed disulfides of protein thiols with the LMW thiol cysteine (S-cysteinylated proteins) in response to diamide stress [71]. This mechanism of S-cysteinylation is thought to protect catalytic cysteine residues against overoxidation to irreversible sulfinic acids and sulfonic acids. This formation of mixed disulfides with cysteine and a novel 398 Da thiol was shown as a regulatory mechanism for the organic hydroperoxide repressor OhrR in *B. subtilis in vivo* after exposure to cumene hydroperoxides [86].

Exposure of cells to sublethal quinone concentrations results in a thiol-modification image that resembles mostly that of untreated control cells [Figure 14.11(c)] [74]. This indicates that in contrast to diamide sublethal quinone concentrations do not lead to increased disulfide bond formations in cellular protein thiols *in vivo*. However, the glyceraldehyde 3-phosphate dehydrogenase (GapA) was strongly oxidized after exposure to toxic quinones and represents the only target for reversible thiol modifications in response to toxic quinones [Figure 14.11(c)]. Consistent with these results, the GapDH activity is decreased by growth-inhibitory concentrations of quinone-like electrophiles *in vivo* and *in vitro*.

14.4.2 Depletion of Thiol-Containing Proteins by Quinones Due to Thiol-(S)-Alkylation

We also analysed the changes in the proteome which are caused by toxic concentrations of BQ in *B. subtilis*. Surprisingly, numerous proteins are depleted in the BQ-proteome compared with the proteome of untreated controls cells (Figure 14.12). These BQ-depleted proteins represent all Cys-rich proteins with 4–10 cysteine residues. These include for example proteins involved in leucine and isoleucine biosynthesis (IlvA, IlvB, IlvD, LeuA, LeuC and LeuD) and the TCA cycle enzyme OdhA, SucC and SucD that are most strongly

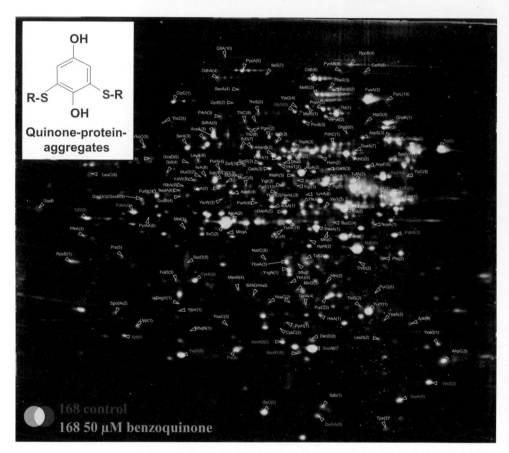

Figure 14.12 *Benzoquinone causes irreversible aggregation of thiol-containing proteins which are depleted in the proteome. Dual channel-images of Coomassie-stained cytoplasmic proteins before (red image) and after exposure to 50 μM BQ in vivo (green image) [74]. Red proteins are absent or decreased in the proteome after BQ stress. Those proteins that do not contain Cys residues are unchanged or increased in the proteome after BQ stress. The number of Cys residues in the labelled cytoplasmic proteins is shown in parentheses. (The subsequent depletion of thiol-containing proteins in the proteome by increasing concentrations of BQ is shown at: http://microbio1.biologie.uni-greifswald.de/ publications.html)*

depleted from the BQ proteome. In contrast, proteins lacking Cys residues are present at similar amounts in the BQ proteome.

We hypothesized that these BQ-depleted proteins are thiol-(S)-alkylated proteins that escape the 2D gel separation. Western blot analyses showed that the strongly BQ-depleted proteins SecA and IlvB are present in the BQ-protein extracts but appear as higher molecular mass BQ-aggregated proteins [74]. The same BQ-protein aggregates and protein thiol depletion can be generated *in vitro* upon modification of the control protein extract with 1 mM BQ. Finally, we were able to identify 90 proteins in the upper gel part of the BQ proteome which represent the BQ-aggregated and depleted protein thiols.

Using GC-MS metabolome analysis we could further measure a 30- to 100-fold decreased level of cysteine after BQ exposure [74]. The role of cysteine in electrophile detoxification was also verified by medium supplementation with extracellular cysteine before quinone exposure which was protective against lethal quinone concentrations. In conclusion, these proteomic and metabolomic results indicate that thiol-(S)-alkylation is the mechanism for cysteine and protein thiol deletion in response to quinones in *B. subtilis* and most likely also in other bacteria.

14.5   Proteomics as a Tool to Define Regulons and Targets for Noncoding RNAs

This chapter demonstrates that gel-based proteomics is a powerful approach to discover novel mechanisms of bacterial physiology. However, there are still limitations of gel-based proteomic approaches since only about 20–30% of the expressed proteins are covered by gel-based studies. Thus, monitoring of gene regulation on the proteome level leads only to a basic understanding of general cell physiology and metabolism because most of the metabolic pathways are covered by gel-based proteomics. To define regulons, protein expression profiling must be complemented by more genome-wide transcriptome analyses which allow a complete view of the changes in gene expression at the mRNA level. Regulons are defined as groups of genes distributed over the chromosome but controlled by a unique transcription factor (e.g. repressor, activator, sigma factor or two-component system response regulator). Several stress, starvation or antibiotic responsive regulons have been successfully defined in *B. subtilis* using comparative proteome and transcriptome analyses of wild type strains and mutants in central regulatory genes (e.g. HrcA, CtsR, Spx, PerR, Fur, MntR, Zur, CymR, CodY, RelA, CcpA, PhoPR, TnrA, σ^B, σ^H, σ^M, σ^W and σ^X). We used proteomic approaches for example to define the cytoplasmic and extracellular PhoPR, DegSU and ScoC regulons of *B. subtilis* [15, 16, 90, 91]. All phosphate starvation-specific secreted extracellular phosphatases/phosphodiesterases and lipoproteins (PhoA, PhoB, PhoD, GlpQ, YfkN, YurI and PstS) were shown to be regulated by the PhoPR two-component system since these are absent in the extracellular proteome of the *phoR* mutant [15, 58, 59]. To analyse the extracellular DegSU regulons, the pleiotropic *degU32(hy)* mutant was used which displays the phenotype hyperphosphorylated constitutively active DegU response regulator. This stable Asp-56 phosphorylation of DegU results in an acid-shifted DegU~P spot in the cytoplasmic proteome [91]. Several degradative enzymes were upregulated in the mutant such as proteases, amylases, glucanases, xylanases, pectate lyases and nucleotidases [16, 91]. Furthermore, the levels of cell wall proteins and flagella-related proteins which are regulated by the motility sigma factor SigmaD are repressed in the *degU32(hy)* mutant.

As reviewed in detail in the last section, proteomics was also used to define the electrophile-stress responsive cytoplasmic YodB and MhqR regulons [3]. Recently, our proteomic approaches were also successfully applied to analyse more complex regulatory mechanisms, such as the roles of noncoding RNAs. Small noncoding RNAs are untranslated RNAs that often act as antisense RNAs of trans-encoded mRNAs, thus causing translational inhibition [92–94]. Several strategies are available to identify targets for the many newly discovered sRNAs, such as bioinformatic, genetic, microarray (particulary based on tiling arrays) and biochemical fishing approaches [95]. Proteomics is especially suited to identify sRNA targets compared with microarray approaches if the translation of the target mRNA is blocked without causing mRNA decay [95].

As the last part of this chapter we will summarize our recent proteomic results from identifying sRNA targets in *B. subtilis*. In collaboration with the group of John Helmann (Ithaca), positive and negative transcriptional and post-transcriptional control mechanisms of the ferric uptake repressor Fur and the Fur-regulated small RNA FsrA were investigated at the proteome level [96]. The Fur repressor responds to iron starvation conditions and controls negatively siderophore biosynthesis and uptake systems for bacillibactin, ferric citrate or elemental Fe. All these Fe uptake pathways are required to increase iron assimilation. Proteome analyses revealed the derepression of siderophore biosynthesis proteins (DhbA, DhbB, DhbE) in the cytoplasmic proteome of the *Δfur* mutant (Figure 14.13) [24]. In addition, most of the Fe-transporter binding proteins can be enriched in the lipoproteome membrane fraction which is accessible to 2D PAGE in contrast to integral membrane proteins. All lipoproteins of the Fur-regulated Fe uptake systems are strongly up-regulated in the lipoproteome of the *Δfur* mutant (FeuA, YxeB, YfiY, YclQ, YfmC, YhfQ and FhuD) (Figure 14.14) [96].

In addition, Fur also positively controls an iron-sparing response which acts to repress iron-rich proteins such as aconitase (CitB), succinate dehydrogenase (SdhA) or biosynthesis enzymes for branched chain amino acid (LeuCD) when iron is limited [96]. This iron-sparing response is mediated by the Fur-controlled small RNA (FsrA) and three basic proteins (FbpA, FbpB and FbpC) that are postulated to function as RNA chaperones. Using proteomics we have discovered at least some target genes of this complex iron-sparing response which are affected by FsrA and/or the FbpA, FbpB and FbpC proteins. For example, repression of the aconitase CitB under iron starvation depends on FsrA only (Figures 14.13 and 14.14). The negative regulation of SdhA is mediated by FsrA and requires in part the FbpAB proteins. Regulation of SdhA could be also visualized in the lipoproteome since SdhA is one of the most prominent proteins in the membrane fraction (Figure 14.14). In contrast the isopropyl malate dehydrogenase LeuCD which is required for branched chain amino acids requires FsrA and FbpC for repression. Repression of the ferredoxin-like protein YvfW and YvbY was shown to be dependent on FsrA and the FbpAB proteins (Figure 14.13). Moreover, the proteome analysis revealed also many other potential targets for this FsrA or FbpAB/FbpC-mediated iron-sparing response which requires further experiments. A further interesting result was that there is probably a

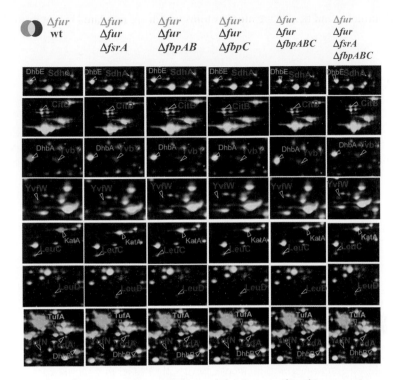

Figure 14.13 *Cytoplasmic proteome analysis of the Fur-regulated iron-sparing response in* B. subtilis. *Cytoplasmic proteins including SdhA, CitB, YvfW, YvbY, LeuC, LeuD, YufN and YcdA are repressed in the cytoplasmic proteome of the Δfur mutant (green images) and restored in the ΔfurΔfsrA and/or ΔfurΔfbpAB, ΔfurΔfhpC, ΔfurΔfbpΛBC,ΔfurΔfsrA ΔfbpABC mutant strains (red images). This figure is adapted from reference [96]*

decreased proteolytic processing of EF-Tu in the *Δfur* mutant since the right processed TufA spot disappeared in the *Δfur* mutant and the left main EF-Tu spot is shifted to the acidic region (Figure 14.13). This decreased TufA processing and acidic TufA shift depends on the FbpAB or FbpC proteins suggesting that these small basic proteins might have also different functions.

In summary, we review here some recent proteomic results which have led to new mechanisms of bacterial physiology. We demonstrate that gel-based proteomics is the most powerful tool to analyse protein targeting, localization and export mechanisms, reversible and irreversible thiol modifications as examples for protein damage or protein expression regulation on a proteome-wide scale. This is essentially the roadmap for physiological proteomics: Proteomic results should be the basis for exciting new ideas for future research, followed by detailed genetic and biochemical experiments that result in novel mechanisms of cell physiology.

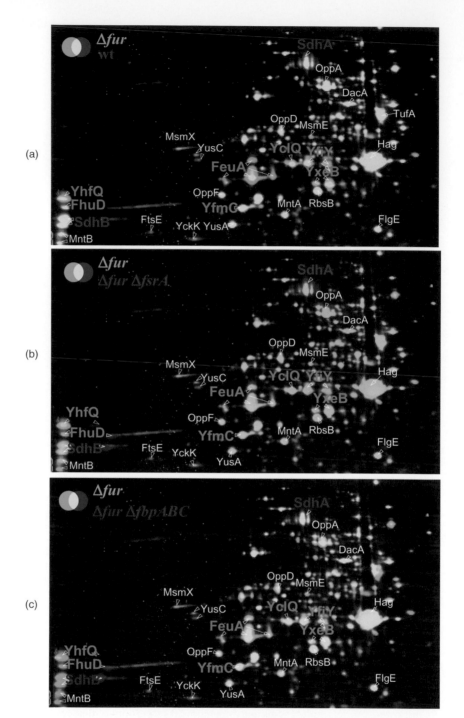

Figure 14.14 *Lipoproteome analysis of the Fur-regulated iron-sparing response in B. subtilis. Lipoproteins were enriched in the ß-D-maltoside-soluble membrane fraction [22, 23]. The abundant Fe-S-membrane protein SdhA is downregulated in the Δfur mutant (green images) compared with the wild type (a), fully restored in the ΔfurΔfsrA mutant (b) but only partly restored in the ΔfurΔfbpABC mutant (c) (red images). The ABC transporter binding proteins of siderophore uptake systems FeuA, YxeB, YfiY, YclQ, YfmC, YhfQ and FhuD are derepressed in the Δfur mutant and marked by green labels. This figure is adapted from reference [96]*

Acknowledgements

We thank the Decodon Company for support with the Decodon Delta 2D software and Sebastian Grund for excellent technical assistance. We are especially grateful to John D. Helmann, Peter Zuber, Kazuo Kobayashi, Jan Maarten van Dijl and Sierd Bron for the very successful and long-standing collaborations which are summarized in this chapter. We further thank the members of the European Bacillus secretion group and the Tat-Machine consortium for stimulating discussions about the work on secretion mechanisms. This work was supported by grants from the Deutsche Forschungsgemeinschaft (AN746/2-1) to Haike Antelmann, the Bundesministerium für Bildung und Forschung (BACELL-SysMo 031397A), the Fonds der Chemischen Industrie, the Bildungsministerium of the country Mecklenburg-Vorpommern, Genencor International and European Union grants BACELL-Factory (QLK3-CT-1999-00413), BACELL-Health (LSHG-CT-2004-503468), Tat-Machine (LSGH-CT-2004-005257), BACELL-BaSysBio (LSHG-CT-2006-037469) and BACELL-BaSysMo (LSHG-CT-2006-031397A) to Michael Hecker.

References

1. Hecker, M.; Völker, U., Towards a comprehensive understanding of *Bacillus subtilis* cell physiology by physiological proteomics; Proteomics 2004, 4, 3727–3750.
2. Tam, L.T.; Antelmann, H.; Eymann, C.; Albrecht, D.; Bernhardt, J.; Hecker, M., Proteome signatures for stress and starvation in *Bacillus subtilis* as revealed by a 2-D gel image color coding approach; Proteomics 2006a, 6, 4565–4585.
3. Antelmann, H.; Hecker, M.; Zuber, P., Proteomic signatures uncover thiol-specific electrophile resistance mechanisms in *Bacillus subtilis*; Expert. Rev. Proteomics 2008, 5, 77–90.
4. Kunst, F.; Ogasawara, N.; Moszer, I. *et al.*, The complete genome sequence of the Gram-positive bacterium *Bacillus subtilis*; Nature 1997, 390, 249–256.
5. Tjalsma, H.; Bolhuis, A.; Jongbloed, J.D.H.; Bron, S.; van Dijl, J.M., Signal peptide-dependent protein transport in *Bacillus subtilis*: a genome-based survey of the secretome; Microbiol. Mol. Biol. Rev. 2000, 64, 515–547.
6. van Dijl J.M.; Bolhuis, A.; Tjalsma, H.; Jongbloed, J.D.H.; de Jong, A.; Bron, S., Protein transport pathways in *Bacillus subtilis*: a genome-based road map; in Sonenshein, A.L.; Hoch, J.A.; Losick, R. (eds), *Bacillus subtilis* and its Closest Relatives: From Genes to Cells, ASM Press, Washington, DC, 2002, pp. 337–355.
7. Honda, K.; Nakamura, K.; Nishiguchi, M.; Yamane, K., Cloning and characterization of a *Bacillus subtilis* gene encoding a homolog of the 54-kilodalton subunit of mammalian signal recognition particle and *Escherichia coli* Ffh; J. Bacteriol. 1993, 175, 4885–4894.
8. Leskelä, S.; Wahlström, E.; Kontinen, V.P.; Sarvas, M., Lipid modification of prelipoproteins is dispensable for growth but essential for efficient protein secretion in *Bacillus subtilis*: characterization of the *lgt* gene; Mol. Microbiol. 1999, 31, 1075–1085.
9. Kontinen, V.P.; Saris, P.; Sarvas, M., A gene (*prsA*) of *Bacillus subtilis* involved in a novel, late stage of protein export; Mol. Microbiol. 1991, 5, 1273–1283.
10. Kontinen V.P; Sarvas, M., The PrsA lipoprotein is essential for protein secretion in *Bacillus subtilis* and sets a limit for high-level secretion; Mol. Microbiol. 1993, 8, 727–737.
11. Jacobs, M.; Andersen, J.B.; Kontinen, V.P.; Sarvas, M., *Bacillus subtilis* PrsA is required *in vivo* as an extracytoplasmic chaperone for secretion of active enzymes synthesized either with or without pro-sequences; Mol. Microbiol. 1993, 8, 957–966.
12. Darmon, E.; Noone, D.; Masson, A.; Bron, S.; Kuipers, O.P.; Devine, K.M.; van Dijl, J.M., A novel class of heat and secretion stress-responsive genes is controlled by the autoregulated CssRS two-component system of *Bacillus subtilis*; J. Bacteriol. 2002, 184, 5661–5671.

13. Hyyrylainen, H.L.; Bolhuis, A.; Darmon, E.; Muukkonen, L.; Koski, P.; Vitikainen, M.; Sarvas, M.; Pragai, Z.; Bron S.; van Dijl, J.M.; Kontinen, V.P., A novel two-component regulatory system in *Bacillus subtilis* for the survival of severe secretion stress; Mol. Microbiol. 2001, 41, 1159–1172.

14. Hirose, I.; Sano, K.; Shioda, I.; Kumano, M.; Nalamura, K.; Yamane, K., Proteome analysis of *Bacillus subtilis* extracellular proteins: a two-dimensional protein electrophoretic study; Microbiology 2000, 146, 65–75.

15. Antelmann, H.; Scharf, C.; Hecker, M., Phosphate starvation-inducible proteins in *Bacillus subtilis*: proteomics and transcriptional analysis; J. Bacteriol. 2000, 182, 4478–4490.

16. Antelmann, H.; Tjalsma, H.; Voigt, B.; Ohlmeier, S.; Bron, S.; van Dijl, J.M.; Hecker, M., A proteomic view on genome-based signal peptide predictions; Gen. Res. 2001, 11, 1484–1502.

17. Antelmann, H.; Yamamoto, H.; Sekiguchi, J.; Hecker, M., Stabilization of cell wall proteins in *Bacillus subtilis*: a proteomic approach; Proteomics 2002, 2, 591–602.

18. Tjalsma, H.; Antelmann, H.; Jongbloed, J.D.; Braun, P.G.; Darmon, E.; Dorenbos, R.; Dubois, J.Y.; Westers, H.; Zanen, G.; Quax, W.J.; Kuipers, O.P.; Bron, S.; Hecker, M.; van Dijl, J.M., Proteomics of protein secretion by *Bacillus subtilis*: separating the 'secrets' of the secretome; Microbiol. Mol. Biol. Rev. 2004, 68, 207–233.

19. Antelmann, H.; Töwe, S.; Albrecht, D.; Hecker, M., The phosphorous source phytate changes the composition of the cell wall proteome in *Bacillus subtilis*; J. Proteome Res. 2007, 6, 897–903.

20. MacLellan, S.R.; Wecke, T.; Helmann, J.D., A previously unidentified sigma factor and two accessory proteins regulate oxalate decarboxylase expression in *Bacillus subtilis*; Mol. Microbiol. 2008, 69, 954–967.

21. MacLellan, S.R.; Helmann, J.D.; Antelmann, H., The YvrI alternative sigma factor is essential for acid-stress induction of oxalate decarboxylase in *Bacillus subtilis*; J Bacteriol. 2009, 191, 931–939.

22. Bunai, K.; Ariga, M.; Inoue, T.; Nozaki, M.; Ogane, S.; Kakeshita, H.; Nemoto, T.; Nakanishi, H.; Yamane, K., Profiling and comprehensive expression analysis of ABC transporter solute-binding proteins of *Bacillus subtilis* membrane based on a proteomic approach; Electrophoresis 2004, 25, 141–155.

23. Antelmann, H.; van Dijl, J.M.; Bron, S.; Hecker, M., Proteomic survey through the secretome of *Bacillus subtilis*; Methods Biochem. Anal. 2006, 49, 179–208.

24. Ollinger, J.; Song, K.B.; Antelmann, H.; Hecker, M.; Helmann, J.D., Role of the Fur regulon in iron transport in *Bacillus subtilis*; J. Bacteriol. 2006, 188, 3664–3673.

25. Zanen, G.; Antelmann, H.; Meima, R.; Jongbloed, J.D.; Kolkman, M.; Hecker, M.; van Dijl, J.M.; Quax, W.J., Proteomic dissection of potential signal recognition particle dependence in protein secretion by *Bacillus subtilis*; Proteomics 2006, 6, 3636–3648.

26. Zanen, G.; Antelmann, H.; Westers, H.; Hecker, M.; van Dijl, J.M.; Quax, W.J., FlhF, the third signal recognition particle-GTPase of *Bacillus subtilis*, is dispensable for protein secretion; J. Bacteriol. 2004, 186, 5956–5960.

27. Jongbloed, J.D.; Martin, U.; Antelmann, H.; Hecker, M.; Tjalsma, H.; Venema, G.; Bron, S.; van Dijl, J.M.; Müller, J., TatC is a specificity determinant for protein secretion *via* the twin-arginine translocation pathway; J. Biol. Chem. 2000, 275, 41350–41357.

28. Jongbloed, J.D.; Antelmann, H.; Hecker, M.; Nijland, R.; Bron, S.; Airaksinen, U.; Pries, F.; Quax, W.J.; van Dijl, J.M., Selective contribution of the twin-arginine translocation pathway to protein secretion in *Bacillus subtilis*; J. Biol. Chem. 2002, 277, 44068–44078.

29. Jongbloed, J.D.; Grieger, U.; Antelmann, H.; Hecker, M.; Nijland, R.; Bron, S.; van Dijl, J.M., Two minimal Tat translocases in *Bacillus*; Mol. Microbiol. 2004, 54, 1319–1325.

30. Westers, L.; Westers, H.; Zanen, G.; Antelmann, H.; Hecker, M.; Noone, D.; Devine, K.M.; van Dijl, J.M.; Quax, W., Genetic or chemical protease inhibition causes significant changes in the *Bacillus subtilis* exoproteome; Proteomics 2008, 8, 2704–2713.

31. Vitikainen, M.; Lappalainen, I.; Seppala, R.; Antelmann, H.; Boer, H.; Taira, S.; Savilahti, H.; Hecker, M.; Vihinen, M.; Sarvas, M.; Kontinen, V.P., Structure-function analysis of PrsA reveals roles for the parvulin-like and flanking N- and C-terminal domains in protein folding and secretion in *Bacillus subtilis*; J. Biol. Chem. 2004, 279, 19302–19314.

32. Antelmann, H.; Darmon, E.; Noone, D.; Veening, J.W.; Westers, H.; Bron, S.; Kuipers, O.; Devine, K.; Hecker, M.; van Dijl, J.M., The extracellular proteome of *Bacillus subtilis* under secretion stress conditions; Mol. Microbiol. 2003, 49, 143–156.

33. Eymann, C.; Dreisbach, A.; Albrecht, D.; Bernhardt, J.; Becher, D.; Gentner, S.; Tam le, T.; Büttner, K.; Buurman, G.; Scharf, C.; Venz, S.; Völker, U.; Hecker, M., A comprehensive proteome map of growing *Bacillus subtilis* cells; Proteomics 2004, 4, 2849–2876.

34. Wolff, S.; Otto, A.; Albrecht, D.; Zeng, J.S.; Büttner, K.; Glückmann, M.; Hecker, M.; Becher, D., Gel-free and gel-based proteomics in *Bacillus subtilis*: A comparative study; Mol. Cell. Proteomics 2006, 5, 1183–1192.

35. Wolff, S.; Antelmann, H.; Albrecht, D.; Becher, D.; Bernhardt, J.; Bron, S.; Buttner, K.; van Dijl, J.M.; Eymann, C.; Otto, A.; Tam, L.T.; Hecker, M., Towards the entire proteome of the model bacterium *Bacillus subtilis* by gel-based and gel-free approaches; J. Chromatogr., B 2008, 849, 129–140.

36. Van Bogelen, R.A.; Schiller, E.E.; Thomas, J.D.; Neidhardt, F.C., Diagnosis of cellular states of microbial organisms using proteomics; Electrophoresis 1999, 20, 2149–2159.

37. Neidhardt, F.C.; VanBogelen, R.A., In Storz, G., Hengge-Aronis, R. (eds), Bacterial Stress Responses, ASM Press, Washington, DC, 2000, p. 445.

38. Bernhardt, J.; Büttner, K.; Scharf, C.; Hecker, M., Dual channel imaging of two-dimensional electropherograms in *Bacillus subtilis*; Electrophoresis 1999, 20, 2225–2240.

39. Berth, M.; Moser, F.M.; Kolbe, M.; Bernhardt, J., The state of the art in the analysis of two-dimensional gel electrophoresis images; Appl. Microbiol. Biotechnol. 2007, 76, 1223–1243.

40. Helmann, J.D.; Wu, M.F.; Gaballa, A.; Kobel, P.A.; Morshedi, M.M.; Fawcett, P.; Paddon, C., The global transcriptional response of *Bacillus subtilis* to peroxide stress is coordinated by three transcription factors; J. Bacteriol. 2003, 185, 243–253.

41. Mongkolsuk, S.; Helmann, J.D., Regulation of inducible peroxide stress responses; Mol. Microbiol. 2002, 45, 9–15.

42. Lee, J.W.; Helmann, J.D., Functional specialization within the Fur family of metalloregulators; Biometals 2007, 20, 485–499.

43. Zuber, P., Spx-RNA polymerase interaction and global transcriptional control during oxidative stress; J. Bacteriol. 2004, 186, 1911–1918.

44. Mostertz, J.; Scharf, C.; Hecker, M.; Homuth, G., Transcriptome and proteome analysis of *Bacillus subtilis* gene expression in response to superoxide and peroxide stress; Microbiology 2004, 150, 497–512.

45. Fernandez, S.; Ayora, S.; Alonso, J.C., *Bacillus subtilis* homologous recombination: genes and products; Res. Microbiol. 2000, 151, 481–486.

46. Au, N.; Kuester-Schoeck, E.; Mandava, V.; Bothwell, L.E.; Canny, S.P.; Chachu, K.; Colavito, S.A.; Fuller, S.N.; Groban, E.S.; Hensley, L.A.; O'Brien, T.C.; Shah, A.; Tierney, J.T.; Tomm, L.L.; O'Gara, T.M.; Goranov, A.I.; Grossman, A.D.; Lovett, C.M., Genetic composition of the *Bacillus subtilis* SOS system; J. Bacteriol. 2005, 187, 7655–7666.

47. Even, S.; Burguiere, P.; Auger, S.; Soutourina, O.; Danchin, A.; Martin-Verstraete, I., Global control of cysteine metabolism by CymR in *Bacillus subtilis*; J. Bacteriol. 2006, 188, 2184–2197.

48. Grundy, F.J.; Henkin, T.M., The T box and S box transcription termination control systems; Front Biosci. 2003, 8, 20–31.

49. Henkin, T.M.; Grundy, F.J., Sensing metabolic signals with nascent RNA transcripts: the T box and S box riboswitches as paradigms; Cold Spring Harbor Symp. Quant. Biol. 2006, 71, 231–237.

50. Fisher, S.H., Regulation of nitrogen metabolism in *Bacillus subtilis*: vive la difference!; Mol. Microbiol. 1999, 32, 223–232.

51. Wray, L.V. Jr; Ferson, A.E.; Rohrer, K.; Fisher, S.H., TnrA, a transcription factor required for global nitrogen regulation in *Bacillus subtilis*; Proc. Natl Acad. Sci. USA 1996, 93, 8841–8845.

52. Yoshida, K.; Yamaguchi, H.; Kinehara, M.; Ohki, Y.H.; Nakaura, Y.; Fujita, Y., Identification of additional TnrA-regulated genes of *Bacillus subtilis* associated with a TnrA box; Mol. Microbiol. 2003, 49, 157–165.

53. Babitzke, P.; Gollnick, P., Posttranscription initiation control of tryptophan metabolism in Bacillus subtilis by the trp RNA-binding attenuation protein (TRAP), anti-TRAP, and RNA structure; J. Bacteriol. 2001, 183, 5795–5802.
54. Gollnick, P.; Babitzke, P.; Antson, A.; Yanofsky, C., Complexity in regulation of tryptophan biosynthesis in *Bacillus subtilis*; Annu. Rev. Genet. 2005, 39, 47–68.
55. Bernhardt, J.; Weibezahn, J.; Scharf, C.; Hecker, M., *Bacillus subtilis* during feast and famine: visualization of the overall regulation of protein synthesis during glucose starvation by proteome analysis; Genome. Res. 2003, 13, 224–237.
56. Koburger, T.; Weibezahn, J.; Bernhardt, J.; Homuth, G.; Hecker M., Genome-wide mRNA profiling in glucose starved *Bacillus subtilis* cells; Mol. Genet. Genomics 2005, 274, 1–12.
57. Yoshida, K.; Kobayashi, K.; Miwa, Y.; Kang, C.M.; Matsunaga, M.; Yamaguchi, H.; Tojo, S.; Yamamoto, M.; Nishi, R.; Ogasawara, N.; Nakayama, T.; Fujita, Y., Combined transcriptome and proteome analysis as a powerful approach to study genes under glucose repression in *Bacillus subtilis*; Nucleic Acids Res. 2001, 29, 683–692.
58. Allenby, N.E.; O'Connor, N.; Pragai, Z.; Ward, A.C.; Wipat, A.; Harwood, C.R., Genome-wide transcriptional analysis of the phosphate starvation stimulon of *Bacillus subtilis*; J. Bacteriol. 2005, 187, 8063–8080.
59. Hulett, F.M., The Pho regulon; in Sonenshein, A.L.; Hoch, J.A.; Losick, R. (eds), *Bacillus subtilis* and its Closest Relatives: From Genes to Cells, ASM Press, Washington, DC, 2002, pp. 193–201.
60. Ratnayake-Lecamwasam, M.; Serror, P.; Wong, K.W.; Sonenshein, A.L., *Bacillus subtilis* CodY represses early-stationary-phase genes by sensing GTP levels; Genes Dev. 2001, 15, 1093–1103.
61. Molle, V.; Nakaura, Y.; Shivers, R.P.; Yamaguchi, H.; Losick, R.; Fujita, Y.; Sonenshein, A.L., Additional targets of the *Bacillus subtilis* global regulator CodY identified by chromatin immunoprecipitation and genome-wide transcript analysis; J. Bacteriol. 2003, 185, 1911–1922.
62. Britton, R.A.; Eichenberger, P.; Gonzalez-Pastor, J.E.; Fawcett, P.; Monson, R.; Losick, R.; Grossman, A.D., Genome-wide analysis of the stationary-phase sigma factor (sigma-H) regulon of *Bacillus subtilis*; J. Bacteriol. 2002, 184, 4881–4890.
63. Eymann, C.; Homuth, G.; Scharf, C.; Hecker, M., *Bacillus subtilis* functional genomics: global characterization of the stringent response by proteome and transcriptome analysis; J. Bacteriol. 2002, 184, 2500–2520.
64. Bandow, J.E.; Brötz, H.; Leichert, L.I.; Labischinski, H.; Hecker M., Proteomic approach to understanding antibiotic action; Antimicrob. Agents Chemother. 2003, 47, 948–955.
65. Bandow, J.E.; Hecker, M., Proteomic profiling of cellular stresses in *Bacillus subtilis* reveals cellular networks and assists in elucidating antibiotic mechanisms of action; Prog. Drug Res. 2007, 64, 81–101.
66. Tam, L.T.; Eymann, C.; Albrecht, D.; Sietmann, R.; Schauer, F.; Hecker, M.; Antelmann, H., Differential gene expression in response to phenol and catechol reveals different metabolic activities for the degradation of aromatic compounds in *Bacillus subtilis*; Environ. Microbiol. 2006, 8, 1408–1427.
67. Duy, N.V.; Mäder, U.; Tran, N.P.; Cavin, J.F.; Tam le, T.; Albrecht, D.; Hecker, M.; Antelmann, H., The proteome and transcriptome analysis of *Bacillus subtilis* in response to salicylic acid; Proteomics 2007, 7, 698–710.
68. Duy, N.V.; Wolf, C.; Mader, U.; Lalk, M.; Langer, P.; Lindequist, U.; Hecker, M.; Antelmann, H., Transcriptome and proteome analyses in response to 2-methylhydroquinone and 6-brom-2-vinyl-chroman-4-on reveal different degradation systems involved in the catabolism of aromatic compounds in *Bacillus subtilis*; Proteomics 2007, 7, 1391–1408.
69. Leichert, L.I.; Scharf, C.; Hecker, M., Global characterization of disulfide stress in *Bacillus subtilis*; J. Bacteriol. 2003, 185, 1967–1975.
70. Hochgräfe, F.; Mostertz, J.; Albrecht, D.; Hecker, M., Fluorescence thiol modification assay: oxidatively modified proteins in *Bacillus subtilis*; Mol. Microbiol. 2005, 58, 409–425.
71. Hochgräfe, F.; Mostertz, J.; Pöther, D.C.; Becher, D.; Helmann, J.D.; Hecker, M., S-cysteinylation is a general mechanism for thiol protection of *Bacillus subtilis* proteins after oxidative stress; J. Biol. Chem. 2007, 282, 25981–25985.

72. O'Brian, P.J., The molecular mechanisms of quinone toxicity; Chem. Biol. Interact. 1991, 80, 1–41.

73. Monks, T.J.; Hanzlik, R.P.; Cohen, G.M.; Ross, D.; Graham, D.G., Quinone chemistry and toxicity; Toxicol. Appl. Pharmacol. 1992, 112, 2–16.

74. Liebeke, M.; Pöther, D.C.; van Duy, N.; Albrecht, D.; Becher, D.; Hochgräfe, F.; Lalk, M.; Hecker, M.; Antelmann, H., Depletion of thiol-containing proteins in response to quinones in *Bacillus subtilis*; Mol. Microbiol. 2008, 69, 1513–1529.

75. Nakano, S.; Kuster-Schock, E.; Grossman, A.D.; Zuber, P., Spx-dependent global transcriptional control is induced by thiol-specific oxidative stress in *Bacillus subtilis*; Proc. Natl Acad. Sci. USA 2003, 100, 13603–13608.

76. Nakano, S.; Erwin, K.N.; Ralle, M.; Zuber, P., Redox-sensitive transcriptional control by a thiol/disulphide switch in the global regulator, Spx; Mol. Microbiol. 2005, 55, 498–510.

77. Leelakriangsak, M.; Kobayashi, K.; Zuber, P., Dual negative control of *spx* transcription initiation from the P3 promoter by repressors PerR and YodB in *Bacillus subtilis*; J. Bacteriol. 2007, 189, 1736–1744.

78. Leelakriangsak, M.; Zuber, P., Transcription from the P3 promoter of the *Bacillus subtilis spx* gene is induced in response to disulfide stress; J. Bacteriol. 2007, 189, 1727–1735.

79. Töwe, S.; Leelakriangsak, M.; Kobayashi, K.; Duy, N.Y.; Hecker, M.; Zuber, P.; Antelmann, H., The MarR-type repressor MhqR (YkvE) regulates multiple dioxygenases/glyoxalases and an azoreductase which confer resistance to 2-methylhydroquinone and catechol in *Bacillus subtilis*; Mol. Microbiol. 2007, 65, 40–56.

80. Leelakriangsak, M.; Huyen, N.T.T.; Töwe, S.; van Duy, N.; Becher, D.; Hecker, M.; Antelmann, H.; Zuber, P., Regulation of quinone detoxification by the thiol-stress sensing DUF24/MarR-like repressor, YodB in *Bacillus subtilis*; Mol. Microbiol. 2008, 67, 1108–1124.

81. Yurimoto, H.; Hirai, R.; Matsuno, N.; Yasueda, H.; Kato, N.; Sakai, Y., HxlR, a member of the DUF24 protein family, is a DNA-binding protein that acts as a positive regulator of the formaldehyde-inducible *hxlAB* operon in *Bacillus subtilis*; Mol. Microbiol. 2005, 57(2), 511–519.

82. Huyen, N.T.T., Eiamphungporn, W., Mäder, U.; Hecker, M.; Helmann, J.D.; Antelmann, H., Genome wide responses to carbonyl electrophiles in *Bacillus subtilis*-control of the thiol-dependent formaldehyde dehydrogenase AdhA and cysteine proteinase YraA by the novel MerR-family regulator YraB (AdhR); Mol. Microbiol. 2009, 71, 876–894.

83. Barbirz, S.; Jakob, U.; Glocker, M.O., Mass spectrometry unravels disulfide bond formation as the mechanism that activates a molecular chaperone; J. Biol. Chem. 2000, 275, 18759–18766.

84. Graumann, J., Lilie, H.; Tang, X.; Tucker, K.A.; Hoffmann, J.H.; Vijayalakshmi, J.; Saper, M.; Bardwell, J.C.; Jakob, U., Activation of the redox-regulated molecular chaperone Hsp33–a two-step mechanism; Structure 2001, 9, 377–387.

85. Hong, M.; Fuangthong, M.; Helmann, J.D.; Brennan, R.G., Structure of an OhrR-*ohrA* operator complex reveals the DNA binding mechanism of the MarR family; Mol. Cell. 2005, 20, 131–141.

86. Lee, J.W.; Soonsanga, S.; Helmann, J.D., A complex thiolate switch regulates the *Bacillus subtilis* organic peroxide repressor OhrR; Proc. Natl Acad. Sci. USA 2007, 104, 8743–8748.

87. Kim, S.O.; Merchant, K.; Nudelman, R.; Beyer Jr, W.F.; Keng, T.; DeAngelo, J.; Hausladen, A.; Stamler, J.S., OxyR: a molecular code for redox-related signaling; Cell 2002, 109, 383–396.

88. Panmanee, W.; Vattanaviboon, P.; Poole, L.B.; Mongkolsuk, S., Novel organic hydroperoxide-sensing and responding mechanisms for OhrR, a major bacterial sensor and regulator of organic hydroperoxide stress; J. Bacteriol. 2006, 188, 1389–1395.

89. Soonsanga, S.; Lee, J.W.; Helmann, J.D., Oxidant-dependent switching between reversible and sacrificial oxidation pathways for *Bacillus subtilis* OhrR; Mol. Microbiol. 2008, 68, 978–986.

90. Antelmann, H.; Sapolsky, R.; Miller, B.; Ferrari, E.;. Chotani, G.; Weyler, W.; Gaertner, A.; Hecker, M., Quantitative proteome profiling during the fermentation process of pleiotropic *Bacillus subtilis* mutants; Proteomics 2004, 4, 2408–2424.

91. Mäder, U.; Antelmann, H.; Buder, T.; Dahl, M.K., Hecker, M.; Homuth, G., *Bacillus subtilis* functional genomics: genome wide analysis of the DegS-DegU regulon by transcriptomics and proteomics; Mol. Genet. Genomics 2003, 268, 455–467.

92. Romby, P.; Vandenesch, F.; Wagner, E.G., The role of RNAs in the regulation of virulence-gene expression; Curr. Opin. Microbiol. 2006, 9, 229–236.
93. Majdalani, N.; Vanderpool, C.K.; Gottesman, S., Bacterial small RNA regulators; Crit. Rev. Biochem. Mol. Biol. 2005, 40, 93–113.
94. Storz, G.; Altuvia, S.; Wassarman, K. M., An abundance of RNA regulators. Annu. Rev. Biochem. 2005, 74, 199–217.
95. Vogel, J.; Wagner, E.G., Target identification of small noncoding RNAs in bacteria; Curr. Opin. Microbiol. 2007, 10, 262–270.
96. Gaballa, A.; Antelmann, H.; Aguilar, C.; Khakh, S.-K., Song, K.-B., Smaldone, G.T.; Helmann, J.D., The *Bacillus subtilis* iron-sparing response is mediated by a Fur-regulated small RNA and three small basic proteins; Proc. Natl. Acad.Sci. USA 2008, 105, 11927–11932.

15

The Proteome of *Francisella tularensis*; Methodology and Mass Spectral Analysis for Studying One of the Most Dangerous Human Pathogens

Jiri Stulik, Juraj Lenco, Jiri Dresler, Jana Klimentova, Lenka Hernychova,
Lucie Balonova, Alena Fucikova and Marek Link
Institute of Molecular Pathology, Faculty of Military Health Science UO,
Hradec Kralove, Czech Republic

15.1 Introduction to Molecular Pathogenesis of *Francisella tularensis* Infection

The facultative Gram-negative intracellular bacterium *Francisella tularensis* is a causative agent of the zoonotic disease tularemia (1–3). This microbe is widely distributed only in the Northern hemisphere and there are five subspecies of *F. tularensis*: *tularensis* (also called *F. tularensis* type A), *novicida*, *mediaasiatica*, *holarctica* (*F. tularensis* type B) and a variant of *holarctica* found in Japan. Of them only *F. tularensis* subsp. *tularensis* and subsp. *holarctica* cause disease in humans (4). Subtypes *holarctica* and *tularensis* differ in their geographic distribution and also in their virulence for humans. Subtype *tularensis* that prevails in Northern America is considered the most virulent for humans. The minimal infectious dose is less than 10 CFU and mortality is 5–6% in untreated cases of cutaneous disease (5). In contrast, subtype *holarctica* predominates in Europe and Asia, and it does

Mass Spectrometry for Microbial Proteomics   Edited by Haroun N. Shah and Saheer E. Gharbia
© 2010 John Wiley & Sons, Ltd

not trigger fatal disease. Interest in *F. tularensis* substantially increased recently due to serious concerns of misuse of *F. tularensis* subtype A as a biological weapon.

The molecular mechanism of *F. tularensis* pathogenesis is still poorly understood. The genome sequencing revealed according to aberrant GC content the existence of a *F. tularensis* pathogenicity island (FPI) that includes genes required for intramacrophage growth and survival, and additionally, genes encoding the components that are homologues of proteins assembling the new type VI secretion system-T6SS (5). Most of these genes, however, do not exhibit any homology to other known bacterial virulence factors. The expression of all FPI genes is regulated by MglA protein, a homologue of the *Escherichia coli* starvation transcriptional regulator. The MglA protein controls transcription of a wide array of *F. tularensis* genes, including genes originating from the FPI and its deletion mutant is attenuated for virulence in mice (6).

The progress in the identification of new virulence factors is being sped up now by the development of new genetic tools like new plasmids and transposons (7). Based on these, great attention is being paid to *F. tularensis* virulence outer membrane proteins. Up to now the role of several outer membrane proteins in molecular mechanisms of *F. tularensis* pathogenesis has been described. For example in-frame deletion of the *pdpD* gene that occurs within the FPI but interestingly only in more virulent type A strains and in *F. tularensis* subsp. *novicida* was found to lead to a highly attenuated phenotype (8). The product of this gene, the 140 kDa protein, was detected in the outer membrane leaflet and that location was dependent on the presence of the T6SS homologues. Similarly, respiratory-burst-inhibiting acid phosphatase, whose activity is mostly associated with the outer membrane is required for *F. tularensis* intramacrophage survival and virulence in mice (9). Additionally, the interaction of two outer membrane lipoproteins, the TUL4 and the DsbA homologue, with the host cell Toll-like receptor 2 is the prerequisite for the induction of the innate immune response to *F. tularensis* (10). This is in contrast to the *F. tularensis* lipopolysaccharide that exhibits low toxicity and elicits only negligible proinflammatory response in comparison with other Gram-negative bacteria.

The modulation of gene expression of Gram-negative microbes in response to a hostile environment is usually regulated by a two-component signal transduction system (TCS). The existence of a fully functional *F. tularensis* TCS is not clear. On the one hand there are reports claiming the lack of classical two-component systems in *F. tularensis*, while on the other hand *F. tularensis* genes encoding the homologues of orphaned response regulators were identified in the published *F. tularensis* genome. One of them exhibits the sequence similar to *Salmonella typhimurium* protein PmrA that is a component of the PmrAB TCS. The *F. tularensis* PmrA homologue deletion mutant showed a replication defect in macrophages, *in vivo* attenuation and its microarray analysis then revealed alteration in the transcription of more than 60 different genes, some of them also originating from the FPI (5).

Globally, it is assumed that new components of the *F. tularensis* virulence have to be discovered in order to construct better diagnostic, prophylactic and therapeutic tools. The current data indicate that components of protein origin might be the most important ones; hence, advanced proteomic techniques could significantly contribute to this research area.

15.2   *Francisella tularensis* LVS Proteome Alterations Induced by Different Temperatures and Stationary Phase of Growth

15.2.1   Introduction

During the infectious cycles, pathogenic bacteria alternate between different environments. In order to succeed, the bacteria must sense local conditions, and modify their physiological state and virulence phenotype accordingly. Therefore, elucidation of global impact of specific environmental factors on *F. tularensis* proteome might afford deeper insight into the nature of its virulence. For decades, the gel-based approach has been used in quantitative proteomic studies. However, despite the improvements involving introduction of IPG strips, fluorescent dyes, and difference gel electrophoresis (DIGE), this approach still suffers from several disadvantages, limiting the analysis. Enormous development in the field of mass spectrometry (MS) has made gel-free techniques a real alternative to two-dimensional polyacrylamide gel electrophoresis (2D-PAGE). In particular, dynamic development may be seen in the area of quantitative shotgun proteomics. Several years ago, Gygi *et al.* evolved the first quantitative technique for applications of shotgun proteomics that denoted isotope-coded affinity tags (ICAT) (11). Since that time, several other methods have been developed and successfully introduced into proteomics. Among them, iTRAQ technology has gained an exceptional position because of its ability to simultaneously quantify up to four different samples in one analysis (12); in the latest version, even eight samples (citace) have been used. Due to the isobaric design of the tags, differently labelled peptides originating from four samples appear as single ions. Generation of iTRAQ reporter ions is achieved in the MS/MS mode via cleavage between the balancer and the reporter group. Four different reporter ions in the range of 114–117 amu provide quantitative information on proteins (Figure 15.1). Differently labelled peptides appear as a single precursor; therefore the analysis of multiplex samples influences neither the speed nor the quality of MS analysis. Furthermore, peptides originating from low abundance proteins are more likely to reach the intensity threshold for fragmentation. Finally, since the protein quantification is performed using a number of MS/MS spectra, data suitable for statistical evaluation are obtained. Indeed, iTRAQ shotgun analysis has become a widely popular technique able to generate an enormous amount of significant data. These, however, require complex statistical evaluation. In order to find important information in the multiplex expression profiles of hundreds of proteins, the exploratory analysis should not be omitted (13). Through acceptance of the hypothesis that proteins similarly produced under different environmental conditions might participate in the same biological process, a combination of iTRAQ technique with multidimensional statistical data mining can lead to outlines of biological functions of uncharacterized proteins.

   The aim of the study presented below was to take advantage of the iTRAQ shotgun proteomic analysis followed by comprehensive statistical evaluation to search for variations in the proteome of *F. tularensis* LVS associated with cultivation at different temperatures or in the stationary phase. To our knowledge, this is the first proteomic comprehensive investigation of temperature- and growth-phase-related proteomes performed with the dangerous human pathogen *F. tularensis*.

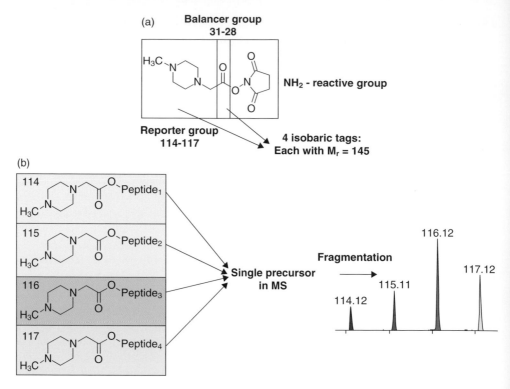

Figure 15.1   *(a) Diagram showing the components of the iTRAQ chemistry. (b) A mixture of four identical peptides each labelled with one member of the iTRAQ kit appears as a single precursor ion in MS. Upon fragmentation, the four reporter group ions appear as distinct masses (114–117 Da)*

15.2.2 Experimental Procedure

The bacterial cells were grown in a chemically defined medium prepared according to Chamberlain at three different temperatures (25, 37 and 42 °C). After 5 h of cultivation, the cultivation of cultures grown at 37 and 42 °C was stopped (OD 600 nm reached 0.75). As the proliferation rate of cells grown at 25 °C was significantly lower, the cultivation was prolonged up to 18 h (OD 600 nm reached 0.75). The fourth culture was grown at 37 °C for 24 h in order to evoke the stationary phase. After spin down, the bacterial cells were resuspended in the lysis buffer composed of 0.3% sodium dodecyl sulfate (SDS) and the sample tubes were immersed in a boiling waterbath for 3 min. Protein concentration in the supernatant was determined using a BCA Kit (Sigma, St Louis, USA). Samples containing 100 μg of protein were diluted 10 times using 50 mM triethylammonium bicarbonate buffer pH 8.5 (Sigma, St Louis, USA) and were concentrated to 30 μl on spin filters (Microcon MWCO 3000; Millipore, Bedford, USA). This step led to a reduction in SDS concentration. Disulfide bonds were reduced [tris(2-carboxyethyl) phosphine hydrochloride] and blocked (methyl methanethiosulfonate) according to the protocol of the iTRAQ manufacturer (Applied Biosystems, Framingham, USA). The proteins were digested overnight at 37 °C using trypsin in the ratio 1 : 50 (Promega, Madison, USA). Peptides were dried out in a vacuum centrifuge and solubilized in 30 μl of 500 mM triethylammonium bicarbonate buffer pH 8.5. Peptides were covalently modified using iTRAQ labels dissolved in 70 μl of absolute ethanol according to the following schema: iTRAQ 114 (25 °C, 5 h), iTRAQ 115 (37 °C, 5 h), iTRAQ 116 (42 °C, 5 h), and iTRAQ 117 (37 °C, 24 h). The reaction was stopped by adding 300 μl of water. After 60 min of incubation, the samples were mixed in the ratio 1 : 1 : 1 : 1 and dried out in a vacuum centrifuge. Peptides were reconstituted with 100 μl of mobile phase A for strong cation exchange (SCX) chromatography (50 mM KH_2PO_4, pH 3.0 in water), and SCX chromatography was used to fractionate the bacterial digest into 26 peptide-containing fractions. Separation was performed on a 2.1 mm × 100 mm PolySulfoethyl A column packed with 5 μm, 300 Å resins (The Nest Group, Southborough, USA). The fractionation was performed at a flow rate of 150 μl min^{-1} using an Alliance HPLC system (Waters, Milford, USA). After loading 90 μl of sample onto the column, the gradient was maintained at 100% of mobile phase SCX-A for 5 min. Peptides were then separated using a linear gradient of 0–100% of mobile phase SCX-B (50 mM KH_2PO_4, pH 3.5, 250 mM KCl in water) over 45 min at a flow rate of 150 μl min^{-1}. Fractions were dried in a vacuum centrifuge and peptides were dissolved in 30–50 μl of mobile phase A for reversed phase (RP) chromatography (2% acetonitrile, 98% water, and 0.1% formic acid). For nano liquid chromatography (nanoLC)-MS/MS, the peptides contained in 1-min SCX fractions were injected into the CapLC system (Waters, Manchester, UK) and trapped on μ PrecolumnTM 300 μm × 5 mm filled with PepMapTM C18, 5 μm, 100 Å (LC-Packings, Sunnyvale, USA). The peptides were eluted and separated on the analytical column NanoEaseTM 75 μm × 150 mm filled with AtlantisTM dC18, 3 μm (Waters, Milford, USA) by a gradient formed by mobile phase A and mobile phase B (98% acetonitrile, 2% water and 0.1% formic acid) over 75 min at a flow rate of 300 nl min^{-1}. The peptides eluted from the RP column were electrospray ionized via a PicoTip needle (New Objective, Woburn, MA, USA) and were analysed online on a quadrupole-TOF tandem mass spectrometer Q – TOF UltimaTM API (Waters, Manchester, UK). To identify the eluting peptides, the mass spectrometer was operated in a data-dependent acquisition mode, in which a 1.0 s full MS scan (*m/z* 400–1600) was followed by 0.7 s MS/MS scans (*m/z* 50–1800) of three precursor ions. The three most intense precursor ions were dynamically selected and subjected to collision-induced dissociation. Dynamic exclusion duration of 180 s was used. For each LC-MS/MS run, a peak list specifying peptide and associated fragment masses was exported into a pkl file using ProteinLynx module of MassLynx 4.0 software (Waters, Manchester, UK). During this process, the acquired spectra were smoothed by the Savitzky–Golay method twice, deisotoped, and centroided at 80% of peak high. Peak lists obtained from all fractions were submitted to the Phenyx 2.1 search platform (GeneBio, Geneva, Switzerland). The software enabled extracting and exporting intensities of iTRAQ reporter ions with associated peptide sequences. For the database search, the following parameters were set up: database NCBInr (r. 20070111); taxonomy, other bacteria; iTRAQ fixed modification on N-termini and Lys; fixed methyl methanethiosulfonate modification of Cys; variable oxidation of Met; one missed cleavage; and parent error tolerance of 0.08 Da. For alignment of peptide sequence to the MS/MS spectra, the following acceptance parameters were selected: minimum peptide length of six amino acids; minimum peptide z-Score 5.0; and maximum peptide p-value 1.0×10^{-5}. The results together with iTRAQ reporter ions intensities were exported into STATISTICA cz 7.0 software (StatSoft, Inc., Tulsa, USA) for further evaluation. Exported intensities of reporter ions were corrected using factors supplied with the iTRAQ kit. In order to correct any experimental or systematic bias, normalization using ratio histograms was further performed. All the statistical analyses were done using relative (IR) values calculated for each reporter intensity, for example, $IR_{114} = I_{114}/\Sigma_{(I114,\ I115,\ I116,\ I117)}$. Outlier values for each protein were eliminated by visual inspection of two types of graphs: $\log_{10}$ scatterplots of corrected values and scatterplots of relative values. For exploratory analyses, means of relative values for each protein were used. At first, clustering using k-means was carried out. The method produces exactly the adjusted number of different clusters of greatest possible distinction. Another exploratory analysis involved in the study was tree clustering. This method groups the proteins in a hierarchical structure. Proteins with the most related quantification profile within the mixed sample appear together on the shortest branch.

15.2.3   Results and Discussion

In order to characterize the proteome of *F. tularensis* LVS grown under four different conditions, progressive shotgun quantitative iTRAQ analysis was carried out. Altogether, off-line two-dimensional chromatography combined with MS/MS provided approximately 10 000 MS/MS spectra. To 3712 of them, a peptide sequence from the *F. tularensis* subsp. *holarctica* protein database was aligned with high significance, resulting in 518 unique identified proteins. Of the 521 proteins, 421 were identified using two or more peptides. After filtering out the outlier iTRAQ values, 291 unique proteins were identified and quantified using at least three peptides. Only these proteins were admitted to statistical analysis. After data processing, the quantitative values were evaluated via clustering using k-means. This method simplified the evaluation of data, allowing rapid focus on proteins with specific production profile. Globally, five different trends in production profiles were observed, and thus the proteins were arranged into five clusters (Figure 15.2). Cluster 1 covered proteins with slightly reduced abundance in response to increased temperature and without any change in synthesis in stationary phase compared with culture grown exponentially at 37 °C. Proteins, whose levels positively correlated with increased temperature, fell into Cluster 2. In this cluster, only proteins corresponding to COG functional category O (posttranslational modification, protein turnover, chaperones) (14) were found: six heat shock proteins and a hypothetical protein FTL 0617 that showed good homology to bacterioferritin/ferritin. Proteins on average not influenced by any of these conditions were arranged in Cluster 3. Proteins grouped in Cluster 4 were strongly upregulated during the stationary phase, and on average, their levels slightly decreased at 42 °C compared with 37 °C. In this cluster, the proteins encoded by the intracellular growth locus genes iglA, iglB, and iglC occurred. Moreover, using the BLASTp algorithm it was found that an additional three hypothetical proteins grouped together with Igl proteins in the same cluster (FTL 0126, FTL 0125, FTL 0116) are homologues of PdpA, PdpB, and PdpC. The genes encoding both Igl and Pdp proteins lie within a DNA region designated the Francisella Pathogenicity Island (FPI). Currently we are not able to discern the crucial aspect which is responsible for growth in the stationary phase (high density of bacterial cells or scarcity of essential nutrients piecemeal depleted during long-term cultivation) and transcription factors implicated in triggering elevated production of virulence factors. However,

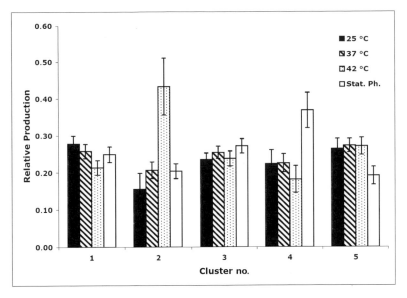

Figure 15.2 *Expression profiles of five clusters obtained by k-means clustering. Means of relative production including standard deviation are depicted for each cluster*

it has been recently demonstrated that expression of virulence factors encoded by igl (15, 16) and pdp (15) genes is significantly increased in *F. tularensis* cells starved for iron, resulting also in the stationary phase. This finding has also guaranteed a connection with the recently described regulation of genes localized in FPI by MglA (6) that is a homologue of SspA from *Escherichia coli*, which is, together with SspB, a global regulator of the stringent starvation response. In a very recent work, MglA was shown to control production of proteins involved in response to starvation and oxidative stress in *F. tularensis* subsp. *novicida* (17). Cluster 5 contained proteins whose abundances were reduced during the stationary phase. Most of them were proteins involved in translation and transcription, for example, ribosomal proteins. Decreased production of these proteins was an expected consequence of reduced bacterial metabolism resulting in growth arrest. In order to find proteins with similarly regulated synthesis under four different conditions of cultivation, tree clustering was carried out. In several cases, proteins certainly known to participate in the same biological process were arranged in one cluster (e.g. ribosomal proteins) or were even arranged together on the same branch (GroEL and GroES, DNA directed RNA polymerase beta chain and beta subunit, ATP synthase beta chain and gamma chain). Although these observations could be expected, they were of great significance because they further document the significance of acquired data and the high credibility of data evaluation using the tree clustering method. Even more interesting was a cluster characterized by the presence of proteins whose production was strongly induced during the stationary phase [Figure 15.3(a)]. The expression profile and list of clustered proteins were very similar to the expression profile of proteins grouped in Cluster 4 obtained using *k*-means [Figure 15.3(b)]. In this tree cluster, IglA, IglB, IglC and homologues of PdpA, PdpB, and PdpC (FTL 0126, FTL 0125 and FTL 0116) were present. Earlier studies have already

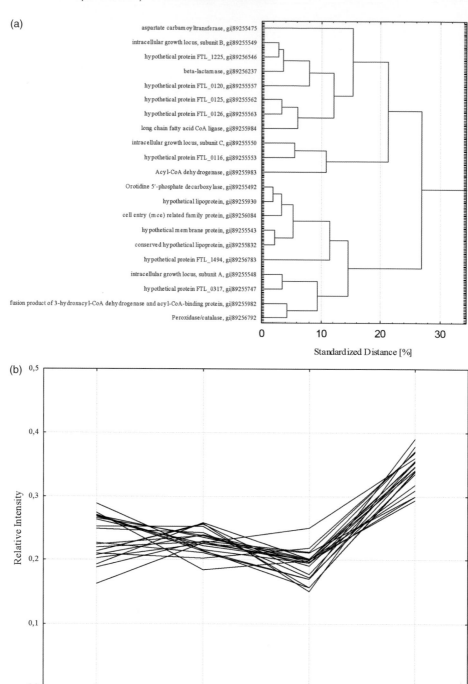

Figure 15.3 *Results from tree clustering. (a) Cut-out of tree diagram for 291 proteins showing a cluster characterized by the presence of several known F. tularensis virulence factors. (b) Expression profiles of proteins grouped in the tree cluster characterized by the presence of several known F. tularensis virulence factors*

established that IglC and IglA proteins and members of the pdp operon as well are required for full intracellular growth of the bacterium in macrophages (2, 18–20). Furthermore, in the case of PdpB protein, knockout mutant strains were shown to be significantly attenuated in virulence (21). This finding supports the notion that additional proteins occurring in the same cluster might also participate in the Francisella pathogenicity. There is additional evidence and earlier published coherent data that are in accordance with this hypothesis. One protein from this cluster is related to the Mycobacterium cell entry (Mce) protein family. Six putative invasin/adhesin-like proteins MceA-F are encoded in four homologous mce operons in *M. tuberculosis*. Mutations in mce operons resulted in significant attenuation of virulence of *M. tuberculosis* (22). Peroxidase/catalase, affording protection against hydrogen peroxide, was also classified in this cluster. This protein was earlier shown to be important for full virulence of *F. tularensis* LVS strain in mice (but not for SCHU S4 strain) (23). Seven other hypothetical proteins were arranged in this cluster. For any of them BLASTp algorithm did not reveal any clear homology to proteins with known functions (it was found that only the conserved hypothetical lipoprotein gi|89255832 corresponds to Tul4 protein). Using InterProScan database search the authors revealed that four proteins, namely, gi|89255832, gi|89255543, gi|89255930, and gi|89256546, have a membrane lipoprotein lipid attachment site and this prediction was further strengthened by the detection of signal peptidase II motifs in protein structures using LipoP 1.0 algorithm. The arrangement of these, typically highly immunogenic, membrane-anchored proteins together with known or potential virulence factors may be associated with accumulating evidence that lipoproteins play a significant role in bacterial pathogenesis (24, 25). However, in the same cluster, two enzymes that are involved in catabolism of fatty acids were observed, which are essential components of lipoproteins, namely, acyl-CoA dehydrogenase and long chain fatty acid-CoA ligase. Lipoproteins of *F. tularensis* seem to play meaningful roles under conditions leading to increased expression of virulence genes from FPI, as they are produced more while the catabolism of fatty acids is activated simultaneously.

15.2.4   Conclusion

To conclude, using exploratory data analysis it has been found that under tested conditions the expression of some hypothetical proteins were coregulated with known *F. tularensis* virulence factors, suggesting their role in pathogenesis of tularemia. In order to confirm this hypothesis, construction of knockout mutants for hypothetical proteins grouped in the tree cluster discussed is currently underway.

15.3   Analysis of Membrane Protein Complexes of *F. tularensis*

15.3.1   Introduction

One of the basic properties of proteins is their ability to form noncovalent complexes with other proteins. Such protein–protein interactions play a crucial role in a major part of biological processes and functions. Identifying and characterizing protein–protein interaction and the interaction networks (so-called 'interactomes', if discussed on a whole cell level) are therefore prerequisites to understanding these processes on a molecular and biological level (26).

Membranes are one of the most important interfaces in biological systems. A membrane system contains many kinds of receptors, transporters and channel proteins with a critical role for the biological activity in every cell (27). In the case of bacteria, membrane proteins often also provide the interaction between bacterial and host cells which can finally determine the pathogenesis on a molecular level. The analysis of membrane proteins and protein complexes is thus highly relevant to our understanding of the interaction of bacteria with their environment including interaction with their hosts.

The problems in the investigation and separation of membrane protein complexes (MPCs) originate from their hydrophobicity (usually they have several transmembrane domains with high affinity for membrane lipids). The problem of isolation and separation of MPCs thus leads the investigator to find a compromise between effective and gentle solubilization of membranes to (a) release the MPCs and (b) retain their tertiary structure.

Membrane proteins were found to be underrepresented when employing the 'classical' isoelectric focusing and sodium dodecyl sulfate polyacylamide gel electrophoresis (SDS-PAGE). In contrast, the technology of blue native polyacrylamide gel electrophoresis (BN-PAGE) afforded excellent results for membrane proteins and especially MPCs in a molecular mass range between 100 kDa and 10 000 kDa (28). Furthermore, the combination of BN-PAGE in the first dimension with SDS-PAGE in the second provides more detailed information about subunit composition.

15.3.1.1   Blue Native Polyacrylamide Gel Electrophoresis: an Effective Tool for Studying Membrane Protein Complexes

BN-PAGE was developed by Shägger and von Jagow in 1991 for the separation of MPCs from the respiratory chain of human mitochondrial membranes (29, 30) and nowadays it is widely used for a number of applications, including the analysis of bacterial MPCs (31–36).

Sample preparation and further workflow follows several common principles:

- Nonionic detergents (e.g. digitonin, *n*-dodecyl-*β*-D-maltoside or triton X-100) are used for solubilization of biological membranes. The ideal detergent for BN-PAGE should effectively dissociate protein complexes from other membrane components, should not unfold or denature the complex and should be readily removable following solubilization. One of the mildest and most frequently used detergents is digitonin. Dodecylmaltoside and triton X-100 have stronger delipidating properties but can dissociate labile hydrophobic interactions (37). The detergent type as well as its concentration has to be determined experimentally depending on the sample (28).
- Nondenaturing conditions must be preserved during the whole sample preparation and electrophoresis. This includes: neutral pH, low salt concentration, no reducing and denaturing agent (SDS or urea), no heating and ideally also no freezing because repeated freezing and thawing of the sample may cause formation of artificial aggregates and/or break-up of existing MPCs leading to false results.
- Following solubilization of membranes anionic dye Coomassie blue G-250 is added to the sample and into the cathode buffer. This dye binds nonspecifically to MPCs imposing negative charge on them. This enables even basic proteins to migrate to the anode at pH 7.5 during electrophoresis. Because the proteins are negatively charged by binding Coomassie, they do not tend to aggregate (37).

Native MPCs during the BN-PAGE migrate in a gradient gel until they reach their specific size-dependent pore-size limit (38) according to their size (molecular weight and shape). The resulting native gel can be used for different applications. The complexes can be visualized both by Coomassie and silver staining and by antibody detection after immunoblot (39). MS analysis of the visualized bands is possible as well but usually requires a preseparation step because the bands correspond to whole MPCs and thus contain a mixture of several proteins.

15.3.1.2   *Two-Dimensional BN/SDS-PAGE Separation of Membrane Protein Complexes*

The subunits of protein complexes resolved in the first dimension by BN-PAGE can be further resolved in a subsequent electrophoretic step. In this 2D approach, the protein complexes are in-gel denatured with SDS, reduced with a reducing agent (e.g. *β*-mercaptoethanol or dithiothreitol) and optionally alkylated (with iodoacetamide or *N,N'*-dimethyl acrylamide). This results in the decomposition of protein complexes into individual subunits. A native 1D gel strip equilibrated in this manner is then transferred to a standard SDS gel in a horizontal position to separate the released protein subunits according to their molecular weight. After this SDS-PAGE step the protein subunits corresponding to one MPC are aligned vertically (Figure 15.4).

15.3.2   Analysis of Oxidative Phosphorylation Membrane Protein Complexes of *F. tularensis* by 2D BN/SDS-PAGE

15.3.2.1   *Methods*

Membrane Fraction Isolation and Solubilization.   Francisella tularensis LVS was cultured in chemically defined Chamberlain's medium at 36.8 °C under constant agitation.

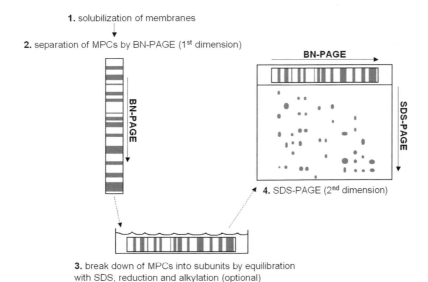

Figure 15.4   *Workflow of BN-PAGE and BN/SDS-PAGE of membrane protein complexes*

The overnight culture was centrifuged and the pellet was diluted with fresh Chamberlain's medium to OD_{600nm} 0.1 and cultivated until OD_{600nm} 0.8 (ca. 6 h, late exponential phase of growth). The bacteria were pelleted and washed three times by ice-cold PBS. The pellet was then resuspended in ice-cold BN-lysis buffer (20 mM BisTris, 500 mM 6-aminohexanoic acid, 20 mM NaCl, 2 mM EDTA, 10% (v/v) glycerol, supplemented by protease inhibitor cocktail Complete EDTA-free (Roche Diagnostics, Germany). The bacteria were disrupted in a French pressure cell (Thermo IEC, Needham Heights, USA), two passages at 16 000 psi, and unbroken cells were removed by centrifugation. The supernatant was treated by benzonase (150 U ml^{-1}) for 1 h on ice.

The membranes were pelleted by centrifugation at 100 000 g for 30 min at 4 °C (Beckman Optima MAX ultracentrifuge, Palo Alto, USA), washed once with fresh BN-lysis buffer and pelleted again. The pellet was solubilized in ice-cold BN-lysis buffer supplemented with 2% (w/v) digitonin and without freezing immediately submitted to BN-PAGE.

BN-PAGE (1st Dimension).   BN-PAGE was performed using XCell SureLock™ Mini-Cell (Invitrogen, Carlsbad, USA). The gradient gels (4–15%) and electrophoresis buffers were prepared as described previously (40). The samples were supplemented with Coomassie G-250 to a final concentration of 0.5% (w/v) and volume corresponding to 30 μg of protein was loaded. Electrophoresis was performed at 4 °C at a constant voltage of 100 V for ca. 30 min (until the samples reached the separating gel), then the voltage was increased to 200 V and the run continued for ca. 2 h. Then ca. 1 h before the end of the run, the blue cathode buffer was replaced with the colourless one (without Coomassie). The native gels were either subdued to Coomassie G-250 staining and MS analysis or further treated for 2D SDS-PAGE.

SDS-PAGE (2nd Dimension).   Immediately after the first dimension, the lines of interest were cut off and equilibrated for 15 min in NuPAGE® LDS Sample Buffer (Invitrogen) with 50 mM dithiothreitol (5 ml per gel strip), 15 min in NuPAGE® LDS Sample Buffer with 50 mM *N,N'*-dimethyl acrylamide and finally 15 min in NuPAGE® LDS Sample Buffer with 5 mM dithiothreitol and 20% (v/v) ethanol. The equilibrated gel strips were placed horizontally on the top of 12% SDS gels, carefully gel-to-gel, and surrounded by agarose. The electrophoresis was performed at room temperature at 100 V until the dye reached the separating gel (ca. 20 min) and then at 200 V until it reached the bottom of the gel (ca. 80 min). The proteins were visualized by Coomassie G-250 staining for MS identification.

In Gel Digestion.   The pretreatment of gel pieces set for trypsin digestion and MS analysis differed in native and SDS gels, because the proteins in the native gels had to be reduced and alkylated first.

SDS gels: Selected spots were excised and destained in 100 mM Tris/HCl (pH 8.5) in 50% (v/v) acetonitrile for 30 min at 30 °C. Then they were treated with equilibration buffer (50 mM ammonium bicarbonate, pH 7.8 in 5% acetonitrile) for 30 min at 30 °C and subsequently vacuum dried. The gel pieces were swollen at 4 °C in a mixture containing 0.1 μg of sequencing grade trypsin in 0.5 μl of 50 mM acetic acid (Promega, Madison, USA) and 4.5 μl of equilibration buffer. After 20 min, 15–20 μl of equilibration buffer was added to cover the gel pieces. The samples were mildly shaken overnight at 37 °C. Supernatants were collected and gels were extracted by adding 2% trifluoroacetic acid, shaking for 20 min at room temperature followed by adding 30% acetonitrile and shaking for another

20 min. The extracts were combined with the first supernatants, dried in vacuum and subjected to MS analysis.

BN gels: The bands were excised and treated according to Fandiño *et al.* (41). Briefly, the bands were dehydrated in acetonitrile and denatured in 8 M urea for 1 h at 8 °C. Then they were destained by 100 mM ammonium bicarbonate (pH 8) at room temperature for 30 min and in acetonitrile under sonication until the blue colour was completely removed. The proteins were reduced by 10 mM dithiothreitol in 100 mM ammonium bicarbonate (pH 8) for 1 h at 56 °C and then alkylated by 55 mM iodoacetamide in 100 mM ammonium bicarbonate (pH 8) for 45 min at room temperature in the dark. The digestion and subsequent extraction then followed the same procedure as in the SDS gels; a mixture containing 0.2 μg of sequencing grade trypsin in 1 μl of 50 mM acetic acid and 4 μl of equilibration buffer was used.

MS Analysis.   The digests of 2D gels were analysed by a 4800 MALDI-TOF/TOF™ Analyser (Applied Biosystems, Foster City, USA). The samples were dissolved in 10 μl of 50% acetonitrile and 0.1% trifluoroacetic acid. Then 2 μl of this solution was mixed with 2 μl of matrix solution (0.5% α-cyano-4-hydroxy cinnamic acid in 50% acetonitrile and 0.1% trifluoroacetic acid) and 0.8–1.0 μl was spotted on the MALDI plate in triplicate. After drying, the spots were promptly rinsed with 1.5 μl of 10 mM ammonium citrate dibasic.

The digests of 1D native gels were analysed by nanoLC-MS/MS. The peptide extracts were dissolved in 10–15 μl of 2% acetonitrile with 0.1% formic acid. The samples were injected into a CapLC system (Waters, Manchester, UK) that was connected to a Q-TOF Ultima™ API (Waters).

The database search was performed using Phenyx GeneBio and GPS Explorer software platform.

15.3.2.2   Results and Discussion

MPCs of *F. tularensis* separated by BN-PAGE and by BN/SDS-PAGE are shown in Figure 15.5(a) (b), respectively. The proteins, regarding oxidative phosphorylation (ATP synthase and complexes I and II of the respiratory chain), in the corresponding bands and spots are listed in Tables 15.1 and 15.2, respectively. It was evident that many proteins belonging to one MPC co-migrated in the first dimension and the corresponding spots were found in the second dimension aligned vertically and exhibiting similar shape. In the bottom part of the 1D gel, most proteins were detected as released subunits.

ATP Synthase.   The principle of BN-PAGE is well documented in the example of ATP synthase. ATP synthase (F_1F_0-ATP synthase) is an MPC that performs ATP synthesis/hydrolysis involving the transport of protons across the membrane. The complex consists of two discrete sectors (subcomplexes), designated F_0 and F_1. F_0 is membrane-embedded and provides a pathway for the passage of protons through the membrane, down the electrochemical gradient. F_1 is membrane-extrinsic and contains catalytic sites for ATP synthesis. In bacteria F_1 consists of five subunits (α, β, γ, δ and ε) in a stoichiometry of $\alpha_3\beta_3\gamma\delta\varepsilon$; F_0 consists of three subunits (A, B and C) in a stoichiometry of AB_2C_{9-12} (42).

In Figure 15.5(a) (1D BN-PAGE) F_1F_0-ATP synthase represents one of the most intensive bands (band 1, MW ca. 680–700 kDa according to MW standards) in which all of the subunits were detected by MS analysis except for one – the δ subunit, see Table 15.1. A

Table 15.1 *Oxidative phosphorylation MPCs detected on 1DBN-PAGE membrane fraction of F. tularensis LVS*

Band[a]	Protein homology	ORF number	Theoretical molecular weight (Da)	
ATP synthase				
1	ATP synthase alpha chain	FTL_1797	55 405	F_1F_0 ATP synthase[b]
	ATP synthase beta chain	FTL_1795	49 734	
	ATP synthase gamma chain	FTL_1796	33 104	
	ATP synthase epsilon chain	FTL_1794	15 606	
	ATP synthase A chain	FTL_1801	29 869	
	ATP synthase B chain	FTL_1799	17 252	
	ATP synthase C chain	FTL_1800	10 084	
2	ATP synthase alpha chain			F_1 ATP synthase
	ATP synthase beta chain			
	ATP synthase gamma chain			
	ATP synthase epsilon chain			
3	ATP synthase alpha chain			F_1 ATP synthase
	ATP synthase beta chain			– fragment
4	ATP synthase alpha chain			Released subunits
5	ATP synthase beta chain			
NADH dehydrogenase (complex I)				
6	NADH dehydrogenase I, C subunit	FTL_1827	24 857	Fragment
	NADH dehydrogenase I, D subunit	FTL_1827	47 455	
	NADH dehydrogenase I, G subunit	FTL_1824	24 857	
7	NADH dehydrogenase I, C subunit			Fragment
	NADH dehydrogenase I, D subunit			
8	NADH dehydrogenase I, G subunit			Released subunit
Succinate dehydrogenase/fumarate reductase (complex II)				
9 and 10	Succinate dehydrogenase iron-sulfur protein (SdhB)	FTL_1785	26 435	Succinate dehydrogenase[c]
	Succinate dehydrogenase, catalytic and NAD/flavoprotein subunit (SdhA)	FTL_1786	65 730	
	Succinate dehydrogenase, cytochrome b556 (SdhC)	FTL_1788	14 274	
11	Succinate dehydrogenase iron-sulfur protein (SdhB)			Released subunit

[a] Numbers of bands correspond to those in Figure 15.5.
[b] The whole machinery of F_1F_0 ATP synthase complex was detected except for the subunit delta (FTL_ 1798).
[c] Whole complex detected except for the subunit succinate dehydrogenase hydrophobic membrane anchor protein (FTL_1787).

Table 15.2   *Oxidative phosphorylation MPCs detected on 2DBN/SDS-PAGE of membrane fraction of F. tularensis LVS*

Protein homology	2D spot[a]
ATP synthase	
ATP synthase alpha chain	14, 15
ATP synthase beta chain	16, 17
ATP synthase gamma chain	18, 19
ATP synthase B chain	20
NADH dehydrogenase	
NADH dehydrogenase I, G subunit	21
NADH dehydrogenase I, D subunit	22, 23
NADH dehydrogenase I, L subunit[b]	24
Succinate dehydrogenase (fumarate reductase)	
Succinate dehydrogenase iron-sulfur protein	25
Succinate dehydrogenase, catalytic and NAD/flavoprotein subunit	26, 27
Succinate dehydrogenase, cytochrome b556	28

[a] Numbers of spots correspond to those in Figure 15.5.
[b] NADH dehydrogenase I, L subunit [FTL_1819, calculated molecular weight. (MW) 73 725 Da].

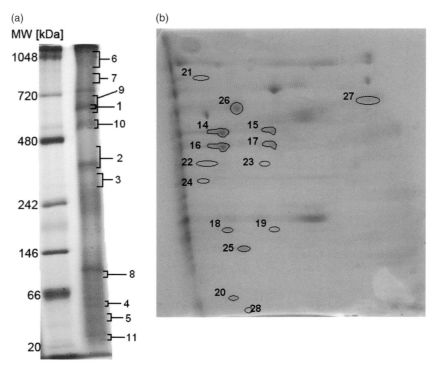

Figure 15.5   *(a) 1D BN-PAGE of membrane protein complexes of* F. tularensis. *Proteins detected in the marked bands are listed in Table 15.1. (b) 2D BN/SDS-PAGE. Proteins detected in the marked spots are listed in Table 15.2*

wide band (band 2, MW ca. 400–450 kDa) seems to be the F_1 subcomplex for it contains only subunits α–ε (except for δ again). Another band (band 3, MW ca. 290–330 kDa) comprises only subunits α and β and probably resulted from partial disruption of the complex caused artificially by solubilization or electrophoresis conditions. On the other hand, bands 4 and 5 contain subunits α and β, respectively, which migrated as free subunits (their position corresponded well with their respective MW).

In Figure 15.5(b) (2D BN/SDS-PAGE) spots corresponding to bands 1 and 2 from Figure 15.5(a) are aligned vertically: spots 14, 16, 18 and 20 resulted from band 1 and spots 15, 17 and 19 from band 2. On 2D gel subunits A, C and δ were not detected.

NADH Dehydrogenase.   Respiratory chain NADH dehydrogenase is also known as complex I or NADH:ubiquinone oxidoreductase. It is a multisubunit (at least 14 subunits) complex that catalyses the first step in the respiratory electron-transport chain that provides electron transfer of NADH to quinone coupled with the transfer of protons across the membrane (43).

In Figure 15.5(a) bands 6 and 7 (MW ca. 1000–1050 and 850 kDa, respectively) contained probably fragments of this complex; subunits C, D and G were detected in band 6, and only C and D in band 7. In addition, 2D BN/SDS-PAGE revealed subunit L.

Succinate Dehydrogenase/Fumarate Reductase.   Succinate dehydrogenase is part of the respiratory chain (complex II) that couples interconversion of succinate and fumarate with quinone and quinol. In *E. coli* the complex is composed of four nonidentical subunits organized into two domains. The membrane extrinsic subcomplex exhibits succinate dehydrogenase activity and is composed of flavoprotein (SdhA) and iron-sulfur subunit (SdhB); the membrane embedded subcomplex (SdhC and D) forms cytochrome b556 (44). The whole enzyme forms a trimeric complex (trimer of heterotetramers).

In bands 9 and 10 (MW ca. 650–690 and 540–580 kDa, respectively) subunits SdhA, B and C were detected. Band 10 is further resolved by 2D BN/SDS-PAGE (spots 25, 26 and 28).

15.3.2.3   Conclusion

The presented method of BN-PAGE coupled with subsequent SDS-PAGE offers advantageous possibilities in the study of membrane protein complexes of *F. tularensis*. Various complexes were separated and their subunit composition was studied. This technology can be used not only to study the MPCs connected with the energy metabolism but also with other cell functions.

15.4   Analysis of *F. tularensis* Glycoproteins and Phosphoproteins

15.4.1   Introduction

Post-translational modifications (PTMs) of proteins play crucial roles in the assembly, degradation, structure, and function of expressed genes. PTMs of bacterial pathogenic proteins strongly influence the nature of interaction with the host cell system. However, little is known about the character and function of such modifications in intracellular bacteria.

Glycosylation, together with phosphorylation, represent the most common PTMs of proteins. Over 50% of today's known proteins, as well as 80% of membrane proteins, are estimated to be modified with glycans (45). The commonly accepted theory limited the ability of organisms to glycosylate their proteins to eukaryotes. This false conclusion originated from the fact that the most frequently studied prokaryotes, such as *E. coli*, *Bacillus subtilis* and *Salmonella* species were identified as nonglycosylated (46–48). However, advances in analytical technologies and genome sequencing enabled the discovery of glycosylation in prokaryotes, such as bacteria. The existence of prokaryotic glycoproteins is no longer considered to be novel and is well-documented, as rapid progress has been made in the last few years (49–51). The first report on the occurrence of glycosylation in prokaryotes appeared in 1976 from the discovery of surface-layer glycoproteins in the Gram-negative *Halobacterium salinarum* (52). Current enhanced interest in glycoprotein discovery in bacteria can be explained by a proven correlation between the presence of glycosylation and bacterial pathogenicity (53–58). To date, a noticeable number of membrane-associated, surface-associated, exoenzymes and even secreted glycoproteins from diverse bacterial species have been characterized.

Protein phosphorylation is a reversible PTM that has tremendous regulatory and signalling potential (59). Modification of serine, threonine and tyrosine residues results in the formation of O-phosphates, similar to some unusual amino acids, e.g. hydroxyproline N-, S- and acyl phosphorylation which are far less common. In this case, histidine, lysine, cysteine, and aspartic and glutamic acid residues are modified. Initial studies have revealed that bacteria use histidine/aspartate phosphorylation, mainly in their two-component systems, which represent a paradigm of bacterial signal transduction (60). Interestingly signalling via serine/threonine/tyrosine phosphorylation is often implicated in the regulation of bacterial virulence (61), and in some cases, it is also known to interfere with eukaryotic signal transduction, thereby rendering the host more prone to infection.

Despite their importance, these PTMs have not been studied in the Gram-negative intracellular pathogenic bacterium *F. tularensis* so far. Up to now, PilA is the only *F. tularensis* protein that has been found to be probably glycosylated (62). In this study, the presence of glycoproteins and phosphoproteins in *F. tularensis* was investigated using basic proteomic methods.

15.4.2 Methods

15.4.2.1 Glycoproteins

F. tularensis subsp. *holarctica* strain LVS was cultivated in Chamberlain medium (36.6 °C, 200 rpm) until the late logarithmic growth phase of bacteria. Bacteria were harvested by centrifugation and pellets were washed three times with PBS pH 7.5. The bacteria were lysed using a FrenchPress (16 000 psi). Undisrupted microbes were eliminated by centrifugation. The fraction enriched in membrane proteins was prepared by carbonate extraction (pH 11) and ultracentrifugation (115 000 g) according to the method described by Molloy *et al.* (63). The final membrane protein-containing pellet was solubilized in rehydration buffer for 2D electrophoresis [7 M urea, 2 M thiourea, 1% ASB14, 4% (w/v) CHAPS, 1% (w/v) DTT, 1% Ampholytes pH 3–10 and 0.5% Pharmalytes pH 8–10.5]. Proteins were separated by 2D mini gel electrophoresis using PAGE gel strips with an immobilized pH gradient of 3–10 in the first dimension. In the second dimension, the IPG strips were embedded onto 12% homogeneous PAGE gels, thus allowing protein separation according to their molecular weight. The separated proteins were transferred onto a BioTrace NT 0.45 μm nitrocellulose membrane, and stained with the Ponceau S solution to vizualise successful protein transfer.

The lectin Peanut agglutinin (PNA) was utilized in this study for detection of glycoproteins according to the manufacturer's instructions for the Dig glycan differentiation kit (Roche). Asialofetuin was used as a positive control for PNA lectin. As a negative control, recombinant *F. tularensis* FTT Igl C protein was used. To avoid nonspecific binding, the nitrocellulose membrane was incubated in blocking solution overnight. After washing, the membrane was incubated in the digoxigenin-labelled lectin solution for 1 h. Unbound lectin was removed by washing. The lectin-retained membrane glycoproteins were then incubated with anti-digoxigenin-labelled alkaline phosphatase for 1 h. Following repeated washes, a staining solution containing the substrate NBT/BCIP was used to visualize the presence of glycoproteins.

Protein spots corresponding to on-blot detected putative glycoproteins were excised from the preparative gel stained with the CBB. The pieces of gel were subjected to in-gel tryptic digestion. The mass spectra were recorded in reflectron mode on a 4800 MALDI-TOF/TOF mass spectrometer (Applied Biosystems) with CHCA as the matrix. Acquired data were processed using GPS Explorer™ Software version 3.6 (Applied Biosystems). Database searching was performed using the same software platform against the *F. tularensis* genome databases.

15.4.2.2   *Phosphoproteins*

F. tularensis subsp. *tularensis* strain Schu S4 was cultivated (until the same conditions as strain LVS), harvested, and lysed within a BioSafety Level 3 containment facility. The whole cell lysate was prepared by lysis of bacteria in buffer (28 mM Tris-HCl, 22 mM Tris-Base, 0.3% SDS, protease and phosphatase inhibitors). The suspension was doused in boiling water for 3 min, immediately cooled on ice and treated with benzonase. The undisrupted microbes were removed by centrifugation. The protein-containing pellet was solubilized in rehydration buffer for 2D electrophoresis (6 M urea, 2 M thiourea, 4% CHAPS, 40 mM Tris-Base, De Streak Reagent, 1%). Proteins were separated by 2D mini gel electrophoresis using PAGE gel strips with an immobilized pH gradient 4–7 in the first dimension. In the second dimension, the IPG strips were embedded onto 12% homogeneous PAGE gels, thus allowing protein separation according to their molecular weight. To detect the phosphorylated proteins, gel was first stained with Pro-Q Diamond Phosphoprotein Gel Stain according to the manufacturer's instructions (Molecular Probes) followed by staining with SYPRO Ruby. The 2D electrophoresis gel images were digitalized using a charge-coupled device camera Image station 2000R (Estman Kodak).

Protein spots visualized with Pro-Q Diamond Phosphoprotein Gel Stain corresponding to on-gel detected putative phosphoproteins were excised from the preparative gel stained with the SYPRO Ruby and subjected to in-gel tryptic digestion. The mass spectra were recorded in reflectron mode on a 4800 MALDI-TOF/TOF mass spectrometer (Applied Biosystems) with CHCA as the matrix. Acquired data were processed using GPS Explorer™ Software version 3.6 (Applied Biosystems). Database searching was performed using the same software platform against the *F. tularensis* genome databases.

15.4.3   Results

15.4.3.1   Glycoproteins

The presence of membrane localized *F. tularensis* strain LVS glycoprotein was investigated using basic proteomic methods: 2D electrophoresis, Dig glycan differentiation kit (Roche) and MS analysis. The selected lectins can provide insight into the type of sugar residues likely to be present in the glycan moiety. For example, PNA recognizes the core disaccharide galactose (1–3) *N*-acetylgalactosamine. Figure 15.6 shows the 2D electrophoresis representative nitrocellulose membrane with the detected PNA-lectin specific putative glycoproteins. Table 15.3 summarizes the glycoproteins identified by the MALDI-TOF/TOF mass spectrometer.

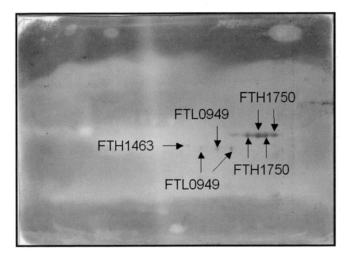

Figure 15.6   *The representative 2D nitrocellulose membrane (pH 3–10) of proteins from the membrane enriched fraction of* F. tularensis *strain LVS. The PNA-specific putative glycoproteins were detected by Dig glycan differentiation kit*

Table 15.3   *PNA-specific* F. tularensis *strain LVS glycoproteins identified by MS*

Protein name	Gene name	Molecular weight (kDa)	pI
Glycerophosphodiester phosphodiesterase	FTH1463	38.9	5.39
Ribose-phosphate pyrophosphokinase	FTL0949	34.8	5.68
Recombinase A protein	FTH1750	38.7	5.97
Fatty acid/phospholipid synthesis protein IsX	FTH1117	37.7	9.17
Membrane protease subunit HflC	FTH0887	34.5	9.35
Succinate dehydrogenase iron-sulfur protein	FTL1785	26.4	8.42
50S ribosomal protein L5	FTL0248	19.9	9.75

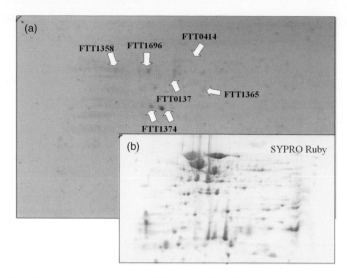

Figure 15.7   *(a) The representative mini 2D gel (pH 4–7) of whole cell lysates prepared from F.* tularensis *strain Schu S4 with phosphoproteins detected by Pro-Q Diamond phosphoprotein gel stain. (b) The representative mini 2D gel stained with SYPRO Ruby protein gel stain after detection of phosphoproteins*

Table 15.4   F. tularensis *strain Schu S4 phosphoproteins identified by MS*

Protein name	Gene name	Molecular weight (kDa)	pI
Elongation factor Tu	FTT0137	43.4	5.12
Phosphoglucomutase	FTT0414	59.6	5.39
Intracellular growth locus, subunit B	FTT1358	58.9	4.69
Fructose-1,6-bisphosphate aldolase	FTT1365	38.1	5.34
Chaperonin GroEL	FTT1696	57.4	4.96
Malonyl CoA-acyl carrier protein transacylase	FTT1374	33.5	5.09

15.4.3.2   *Phosphoproteins*

Pro-Q Diamond phosphoprotein gel stain detects only a subset of the *Francisella* phosphoproteome. The first results of *F. tularensis* strain Schu S4 phosphoproteins are displayed on the representative mini 2D gel (Figure 15.7) and Table 15.4 summarizes the phosphoproteins identified by the MALDI-TOF/TOF mass spectrometer.

15.4.4   Discussion

Our results demonstrate post-translational modified proteins that could be connected with factors of virulence. For example protein D, having a glycerol-3-phosphodiester phosphodiesterase activity, is found at the surface of all *Haemophilus influenzae* strains and is a possible virulence factor. The involvement of protein D in the entry of microbes into human monocytic cells has been reported (64). *H. influenzae* adhered to and entered monocytic cells up to four times more efficiently compared with the mutant lacking the gene for protein D. The genome sequence of *Listeria monocytogenes* has revealed genes for 16 putative two-component systems, four putative tyrosine phosphatases, three putative serine-threonine kinases and two putative serine-threonine phosphatases. Archambaud *et al.* (65) found that one of the latter genes, *stp*, encodes a functional Mn^{2+}-dependent serine-threonine phosphatase similar to PPM eukaryotic phosphatases (Mg^{2+}- or Mn^{2+}-dependent protein phosphatase) and is required for growth of *L. monocytogenes* in a murine model of infection. They identified that the first target for Stp was the elongation factor EF-Tu. Post-translational phosphorylation of EF-Tu was shown to prevent its binding to amino-acylated transfer RNA as well as to kirromycin, an antibiotic known to inhibit EF-Tu function. Moreover EF-Tu was located at the surface of *Lactobacillus johnsonii* NCC533 (La1). *In vitro* results indicated that EF-Tu, through its binding to the intestinal mucosa, might participate in gut homeostasis (66).

15.4.5 Conclusion

Studying bacterial glycoproteins has gained importance due to the recently revealed role of these proteins in host–pathogen interactions. Nowadays is well known that prokaryotic glycoproteins rely on a wider variety of monosaccharide constituents than do those of eukaryotes. Innovative proteomic and glycomic technologies have provided sufficient capabilities for identification and characterization of putative bacterial glycosylated proteins. Substantial progress in describing the enzymes involved in bacterial and archaeal glycosylation pathways has been made. It is becoming clear that enhanced knowledge of bacterial glycosylation enzymes may be of therapeutic value.

Likewise new knowledge in bacterial phosphoproteomics is expected to tackle some global questions about bacterial protein phosphorylation at the level of system biology. Modelling of signal transduction cascades in bacteria was previously performed for two component systems (67). New data show the feasibility of bacterial serine/threonine/ tyrosine phosphorylation, which is why further work is needed to assign the newly found phosphorylated proteins to their cognate kinases and phosphatases and quantify the phosphorylation events.

15.5 Identification of *F. tularensis* Transcription Factors Potentially Involved in Its Virulence

15.5.1 Introduction

Virulence phenotype of microbes is characterized by production of virulence factors, which interact with the host cells in order to promote the infection and/or protect the bacterium against the destructive tools of both, innate and acquired immunity. The expression of the virulence factors comes under the control of transcription factors. The prokaryotic transcription factors can be divided into two groups, activators and repressors. If the repressor binds the operator sequence, this sequence will not be accessible for binding of RNA polymerase and the appropriate operon is not transcribed. However, the activator makes the bond of weak binding RNA polymerase stronger and promotors become fully active. It would be uneconomical to have one specific transcription factor for each gene or small group of genes, thereby transcription factors cooperate together, alternatively they have an activating function on one promotor and inhibiting effect on another (68).

The importance of transcription factors for molecular pathogenesis of infection is reflected by the effort to develop new antimicrobial chemotherapeutics causing the pharmacological targeting of function of some selected transcription factors (12, 69).

Currently, there is a wide range of molecular approaches suitable for studying transcription factors involved in bacterial virulence. However, since the transcription factors are present in cells in low number of copies, they might be hardly identified in whole cell protein extract without certain enrichment or a purification step. One of the possible approaches as to how to enrich DNA binding proteins is affinity chromatography that exploits specific interactions between the protein and DNA molecule when the protein selectively binds the specific nucleotide sequence. This mechanism is a guarantee of good yield and high purity of the sample (70, 71). It can be used for isolation of DNA binding proteins on specific, nonspecific, single- or double-stranded DNA, with dependence on binding properties of protein and DNA.

The other alternative approaches are based on chromatographic techniques used to obtain the specific group of proteins, e.g. transcription factors, telomerases, polymerases, etc. Widely exploited methods for purification of transcription factors are e.g. heparin-agarose or phosphocellulose chromatography or concatemer columns (72). Many reviews and papers concerning DNA affinity chromatography and other techniques have been already published (70, 73–75).

15.5.2 Methods

Francisella tularensis LVS was grown in liquid Chamberlain medium at 37 °C. Pelleted cells were resuspended in lysis buffer (100 mM Tris, 1 mM EDTA, 1% CHAPS, pH 7.0), and homogenized using a French pressure cell press (Thermo Electron Corp., USA). One gram of cellulose particles with immobilized single-stranded (ss-DNA) or double-stranded (ds-DNA) calf thymus DNA (Sigma, St Louis, USA) was added to the samples. The suspensions were equilibrated at ambient temperature for 15 min. After spin down, pellets were washed with washing buffer (20 mM Tris, 1 mM EDTA, 10% acetonitrile, pH 7.4). Proteins bound to the particles were eluted using an incresing concentration of NaCl in three steps: 0.2 M NaCl; 0.5 M NaCl, and 1.0 M NaCl. Proteins in each fraction were desalted and subsequently digested by trypsin overnight. Proteins were identified by means of LC-MS/MS (CapLC and Q-TOF Ultima API, Waters, UK). MS/MS spectra were assigned to peptide sequences using Phenyx 2.1 software (Genbio, Switzerland). The DNA binding motifs of identified proteins were searched against publicly available algorithms – Panther, Pfam, Prodom, TIGR and Prosite.

15.5.3 Results and Discussion

We attempted to enrich and analysed DNA binding proteins of *F. tularensis* in order to identify transcription factors, which may potentially control some of its virulence factors. The sequences of identified proteins were screened for putative DNA binding domains using available prediction algorithms (http://npsa-pbil.ibcp.fr/cgi-bin/npsa_automat. pl?page=/NPSA/npsa_hth.html). Furthermore, identified proteins were searched against readily available bioinformatics databases in order to select transcription factors that might be potentially involved in regulation of its virulence phenotype representing possible targets for further functional studies. Globally, 98 different proteins were identified using ss-DNA sorbent and 214 different proteins using ds-DNA sorbent, respectively. Of them, 133 proteins with possible DNA binding properties were predicated. They represent ribosomal proteins, tRNA syntethases, and transcription factors predominantly. Among the identified proteins, 18 different proteins with helix-turn-helix DNA-binding motif (probability >50%) were found. The rest of the proteins may contain another binding motif but there are no available prediction algorithms currently. Three identified proteins showing homology to transcription factors involved in virulence in other pathogenic bacteria were found.

15.5.3.1   Ferric Uptake Regulation Protein (gi 118498242)

The Fur family includes metal ion uptake regulator proteins and this Fur protein is a 17 kDa, iron binding protein, which represses promoters of operons determining siderophore biosynthesis or virulence genes in response to the intracellular levels of Fe^{2+}. Iron deprivation is a signal of entrance into the iron-scarce medium of a mammalian host and this process is under the control of *fur* gene family products. Fur proteins are associated in processes such as acid-shock response, defense against oxygen radicals and chemotaxis. Fur seems to be the global regulator of many metabolic pathways rather than a specific transcription factor (76).

15.5.3.2   Transcription Regulator (gi 56708441)

This is a family of proteins found in a single copy in at least 10 different early completed bacterial genomes. It is a homologue of the Bvg accessory factor, a protein required, in addition to the regulatory operon *bvgAS*, for transcription of the *Bordetella pertussis* toxin operon in *E. coli*. The BvgAS two-component system controls the expression of pertussis toxin and a number of other *B. pertussis* virulence factors. The HTH motif was not found.

15.5.3.3   Conserved Hypothetical Protein (gi 56604555)

This uncharacterized bacterial protein has been shown to be evolutionarily related to the *E. coli* hypothetical proteins YfhP and YjeB, and *Mycobacterium tuberculosis* hypothetical protein Rv1287, and *S. typhimurium*. The protein has been shown to play a role in the regulation of virulence factors in both *S. typhimurium* and *E. coli*. Its functions include inhibition of the initiation of DNA replication from the OriC site and promotion of Hin-mediated DNA inversion (77).

15.5.4   Conclusion

Although DNA binding proteins form only a small part of pathogenic bacteria research, they represent one of the main control mechanisms of the cell. Studies focused on this issue can provide valuable information about virulence mechanisms and their possible influences. The method described is the first basic approach tried in our laboratory. Despite low specificity, we successfully identified several proteins with DNA binding properties. Moreover, the literature search revealed that three of them might be involved in *F. tularensis* virulence. Further work will lead to construction of the deletion mutant strain and subsequent research in this area.

Acknowledgements

This study was supported by the Czech Science Foundation GACR 310/06/P266, by Ministry of Defence OVUOFVZ200808, MO0FVZ0000501 and by Ministry of Education, Youth and Sports OC151.

We thank Maria Safarova, Lenka Luksikova, Alena Firychova and Jana Michalickova for their excellent technical assistance.

References

1. Fortier AH, Green SJ, Polsinelli T, *et al*. Life and death of an intracellular pathogen: *Francisella tularensis* and the macrophage. Immunol. Ser. 1994; 60: 349–61.
2. Golovliov I, Baranov V, Krocova Z, Kovarova H, Sjostedt A. An attenuated strain of the facultative intracellular bacterium *Francisella tularensis* can escape the phagosome of monocytic cells. Infect. Immun. 2003; 71(10): 5940–50.
3. Larsson P, Oyston PC, Chain P, *et al*. The complete genome sequence of Francisella tularensis, the causative agent of tularemia. Nat. Genet. 2005; 37(2): 153–9.
4. McLendon MK, Apicella MA, Allen LA. *Francisella tularensis*: taxonomy, genetics, and immunopathogenesis of a potential agent of biowarfare. Annu. Rev. Microbiol. 2006; 60: 167–85.
5. Barker JR, Klose KE. Molecular and genetic basis of pathogenesis in *Francisella tularensis*. Ann. NY Acad. Sci. 2007; 1105: 138–59.
6. Brotcke A, Weiss DS, Kim CC, *et al*. Identification of MglA-regulated genes reveals novel virulence factors in *Francisella tularensis*. Infect. Immun. 2006; 74(12): 6642–55.
7. Weiss DS, Brotcke A, Henry T, Margolis JJ, Chan K, Monack DM. In vivo negative selection screen identifies genes required for Francisella virulence. Proc. Natl Acad. Sci. U S A 2007; 104(14): 6037–42.
8. Ludu JS, de Bruin OM, Duplantis BN, *et al*. The Francisella pathogenicity island protein PdpD is required for full virulence and associates with homologues of the type VI secretion system. J. Bacteriol. 2008; 190(13): 4584–95.
9. Mohapatra NP, Balagopal A, Soni S, Schlesinger LS, Gunn JS. AcpA is a Francisella acid phosphatase that affects intramacrophage survival and virulence. Infect. Immun. 2007; 75(1): 390–6.
10. Thakran S, Li H, Lavine CL, *et al*. Identification of *Francisella tularensis* lipoproteins that stimulate the toll-like receptor (TLR) 2/TLR1 heterodimer. J. Biol. Chem. 2008; 283(7): 3751–60.
11. Gygi SP, Rist B, Gerber SA, Turecek F, Gelb MH, Aebersold R. Quantitative analysis of complex protein mixtures using isotope-coded affinity tags. Nat. Biotechnol. 1999; 17(10): 994–9.
12. Ross PL, Huang YN, Marchese JN, *et al*. Multiplexed protein quantitation in *Saccharomyces cerevisiae* using amine-reactive isobaric tagging reagents. Mol. Cell Proteomics 2004; 3(12): 1154–69.
13. Duncan MW, Hunsucker SW. Comments on standards in proteomics and the concept of fitness-for-purpose. Proteomics 2006; 6(S2): 45–7.
14. Tatusov RL, Natale DA, Garkavtsev IV, *et al*. The COG database: new developments in phylogenetic classification of proteins from complete genomes. Nucleic Acids Res. 2001; 29(1): 22–8.
15. Deng K, Blick RJ, Liu W, Hansen EJ. Identification of *Francisella tularensis* genes affected by iron limitation. Infect. Immun. 2006; 74(7): 4224–36.
16. Lenco J, Hubalek M, Larsson P, *et al*. Proteomics analysis of the *Francisella tularensis* LVS response to iron restriction: induction of the *F. tularensis* pathogenicity island proteins IglABC. FEMS Microbiol. Lett. 2007; 269(1): 11–21.
17. Guina T, Radulovic D, Bahrami AJ, *et al*. MglA regulates *Francisella tularensis* subsp. *novicida* response to starvation and oxidative stress. J. Bacteriol. 2007; 189(18): 6580–6.

18. Lindgren H, Golovliov I, Baranov V, Ernst RK, Telepnev M, Sjostedt A. Factors affecting the escape of *Francisella tularensis* from the phagolysosome. J. Med. Microbiol. 2004; 53(Pt 10): 953–8.

19. Gray CG, Cowley SC, Cheung KK, Nano FE. The identification of five genetic loci of Francisella novicida associated with intracellular growth. FEMS Microbiol. Lett. 2002; 215(1): 53–6.

20. de Bruin OM, Ludu JS, Nano FE. The Francisella pathogenicity island protein IglA localizes to the bacterial cytoplasm and is needed for intracellular growth. BMC Microbiol. 2007; 7: 1.

21. Tempel R, Lai XH, Crosa L, Kozlowicz B, Heffron F. Attenuated *Francisella novicida* transposon mutants protect mice against wild-type challenge. Infect. Immun. 2006; 74(9): 5095–105.

22. Gioffre A, Infante E, Aguilar D, *et al.* Mutation in mce operons attenuates *Mycobacterium tuberculosis* virulence. Microbes Infect. 2005; 7(3): 325–34.

23. Lindgren H, Shen H, Zingmark C, Golovliov I, Conlan W, Sjostedt A. Resistance of *Francisella tularensis* strains against reactive nitrogen and oxygen species with special reference to the role of KatG. Infect. Immun. 2007; 75(3): 1303–9.

24. Zhang H, Niesel DW, Peterson JW, Klimpel GR. Lipoprotein release by bacteria: potential factor in bacterial pathogenesis. Infect. Immun. 1998; 66(11): 5196–201.

25. Aliprantis AO, Yang RB, Mark MR, *et al.* Cell activation and apoptosis by bacterial lipoproteins through toll-like receptor-2. Science 1999; 285(5428): 736–9.

26. Piehler J. New methodologies for measuring protein interactions in vivo and in vitro. Curr. Opin. Struct. Biol. 2005; 15(1): 4–14.

27. Kashino Y. Separation methods in the analysis of protein membrane complexes. J. Chromatogr., B 2003; 797(1–2): 191–216.

28. Reisinger V, Eichacker LA. How to analyze protein complexes by 2D blue native SDS-PAGE. Proteomics 2007; 7 (Suppl. 1): 6–16.

29. Schagger H, von Jagow G. Blue native electrophoresis for isolation of membrane protein complexes in enzymatically active form. Anal Biochem. 1991; 199(2): 223–31.

30. Schagger H, Cramer WA, von Jagow G. Analysis of molecular masses and oligomeric states of protein complexes by blue native electrophoresis and isolation of membrane protein complexes by two-dimensional native electrophoresis. Anal. Biochem. 1994; 217(2): 220–30.

31. Stenberg F, Chovanec P, Maslen SL, *et al.* Protein complexes of the *Escherichia coli* cell envelope. J. Biol. Chem. 2005; 280(41): 34409–19.

32. Lasserre JP, Beyne E, Pyndiah S, Lapaillerie D, Claverol S, Bonneu M. A complexomic study of *Escherichia coli* using two-dimensional blue native/SDS polyacrylamide gel electrophoresis. Electrophoresis 2006; 27(16): 3306–21.

33. Pyndiah S, Lasserre JP, Menard A, *et al.* Two-dimensional blue native/SDS gel electrophoresis of multiprotein complexes from *Helicobacter pylori*. Mol. Cell Proteomics 2007; 6(2): 193–206.

34. Farhoud MH, Wessels HJ, Steenbakkers PJ, *et al.* Protein complexes in the archaeon *Methanothermobacter thermautotrophicus* analyzed by blue native/SDS-PAGE and mass spectrometry. Mol Cell Proteomics 2005; 4(11): 1653–63.

35. Stroh A, Anderka O, Pfeiffer K, *et al.* Assembly of respiratory complexes I, III, and IV into NADH oxidase supercomplex stabilizes complex I in *Paracoccus denitrificans*. J. Biol. Chem. 2004; 279(6): 5000–7.

36. Heuberger EH, Veenhoff LM, Duurkens RH, Friesen RH, Poolman B. Oligomeric state of membrane transport proteins analyzed with blue native electrophoresis and analytical ultracentrifugation. J. Mol. Biol. 2002; 317(4): 591–600.

37. Wittig I, Braun HP, Schagger H. Blue native PAGE. Nat. Protoc. 2006; 1(1): 418–28.

38. Braun RJ, Kinkl N, Beer M, Ueffing M. Two-dimensional electrophoresis of membrane proteins. Anal. Bioanal. Chem. 2007; 389(4): 1033–45.

39. Reisinger V, Eichacker LA. Analysis of membrane protein complexes by blue native PAGE. Proteomics 2006; 6 (Suppl. 2): 6–15.

40. Swamy M, Siegers GM, Minguet S, Wollscheid B, Schamel WW. Blue native polyacrylamide gel electrophoresis (BN-PAGE) for the identification and analysis of multiprotein complexes. Sci STKE 2006; 2006(345): pl4.

41. Fandiño AS, Rais I, Vollmer M, Elgass H, Schagger H, Karas M. LC-nanospray-MS/MS analysis of hydrophobic proteins from membrane protein complexes isolated by blue-native electrophoresis. J. Mass Spectrom. 2005; 40(9): 1223–31.
42. Weber J, Senior AE. Catalytic mechanism of F1-ATPase. Biochim. Biophys. Acta 1997; 1319(1): 19–58.
43. Hirst J. Energy transduction by respiratory complex I – an evaluation of current knowledge. Biochem. Soc. Trans. 2005; 33(Pt 3): 525–9.
44. Maklashina E, Iverson TM, Sher Y, *et al.* Fumarate reductase and succinate oxidase activity of *Escherichia coli* complex II homologs are perturbed differently by mutation of the flavin binding domain. J. Biol. Chem. 2006; 281(16): 11 357–65.
45. Apweiler R, Hermjakob H, Sharon N. On the frequency of protein glycosylation, as deduced from analysis of the SWISS-PROT database. Biochim. Biophys. Acta 1999; 1473(1): 4–8.
46. Messner P. Prokaryotic glycoproteins: unexplored but important. J. Bacteriol. 2004; 186(9): 2517–9.
47. Eichler J. Facing extremes: archaeal surface-layer (glyco)proteins. Microbiology 2003; 149(Pt 12): 3347–51.
48. Messner P, Schaffer C. Prokaryotic Glycoproteins. Progress in the Chemistry of Organic Natural Products. Vienna: Springer-Verlag, 2001, pp. 51–124.
49. Virji M. Post-translational modifications of meningococcal pili. Identification of common substituents: glycans and alpha-glycerophosphate – a review. Gene 1997; 192(1): 141–7.
50. Parge HE, Forest KT, Hickey MJ, Christensen DA, Getzoff ED, Tainer JA. Structure of the fibre-forming protein pilin at 2.6 A resolution. Nature 1995; 378(6552): 32–8.
51. Castric P. pilO, a gene required for glycosylation of *Pseudomonas aeruginosa* 1244 pilin. Microbiology 1995; 141 (Pt 5): 1247–54.
52. Mescher MF, Strominger JL. Purification and characterization of a prokaryotic glucoprotein from the cell envelope of *Halobacterium salinarium*. J. Biol. Chem. 1976; 251(7): 2005–14.
53. Power PM, Roddam LF, Rutter K, Fitzpatrick SZ, Srikhanta YN, Jennings MP. Genetic characterization of pilin glycosylation and phase variation in *Neisseria meningitidis*. Mol. Microbiol. 2003; 49(3): 833–47.
54. Schmidt MA, Riley LW, Benz I. Sweet new world: glycoproteins in bacterial pathogens. Trends Microbiol. 2003; 11(12): 554–61.
55. Dobos KM, Khoo KH, Swiderek KM, Brennan PJ, Belisle JT. Definition of the full extent of glycosylation of the 45-kilodalton glycoprotein of Mycobacterium tuberculosis. J. Bacteriol. 1996; 178(9): 2498–506.
56. Benz I, Schmidt MA. Never say never again: protein glycosylation in pathogenic bacteria. Mol. Microbiol. 2002; 45(2): 267–76.
57. Kuo C, Takahashi N, Swanson AF, Ozeki Y, Hakomori S. An N-linked high-mannose type oligosaccharide, expressed at the major outer membrane protein of *Chlamydia trachomatis*, mediates attachment and infectivity of the microorganism to HeLa cells. J. Clin. Invest. 1996; 98(12): 2813–8.
58. Aas FE, Vik A, Vedde J, Koomey M, Egge-Jacobsen W. *Neisseria gonorrhoeae* O-linked pilin glycosylation: functional analyses define both the biosynthetic pathway and glycan structure. Mol. Microbiol. 2007; 65(3): 607–24.
59. Pawson T, Scott JD. Protein phosphorylation in signaling – 50 years and counting. Trends Biochem. Sci. 2005; 30(6): 286–90.
60. Deutscher J, Saier MH, Jr. Ser/Thr/Tyr protein phosphorylation in bacteria – for long time neglected, now well established. J. Mol. Microbiol. Biotechnol. 2005; 9(3–4): 125–31.
61. Hoch JA. Two-component and phosphorelay signal transduction. Curr. Opin. Microbiol. 2000; 3(2): 165–70.
62. Forslund AL, Kuoppa K, Svensson K, *et al.* Direct repeat-mediated deletion of a type IV pilin gene results in major virulence attenuation of *Francisella tularensis*. Mol. Microbiol. 2006; 59(6): 1818–30.
63. Molloy MP, Herbert BR, Slade MB, *et al.* Proteomic analysis of the *Escherichia coli* outer membrane. Eur. J. Biochem. 2000; 267(10): 2871–81.

64. Ahren IL, Janson H, Forsgren A, Riesbeck K. Protein D expression promotes the adherence and internalization of non-typeable *Haemophilus influenzae* into human monocytic cells. Microb. Pathog. 2001; 31(3): 151–8.

65. Archambaud C, Gouin E, Pizarro-Cerda J, Cossart P, Dussurget O. Translation elongation factor EF-Tu is a target for Stp, a serine-threonine phosphatase involved in virulence of Listeria monocytogenes. Mol. Microbiol. 2005; 56(2): 383–96.

66. Granato D, Bergonzelli GE, Pridmore RD, Marvin L, Rouvet M, Corthesy-Theulaz IE. Cell surface-associated elongation factor Tu mediates the attachment of *Lactobacillus johnsonii* NCC533 (La1) to human intestinal cells and mucins. Infect. Immun. 2004; 72(4): 2160–9.

67. Kollmann M, Lovdok L, Bartholome K, Timmer J, Sourjik V. Design principles of a bacterial signalling network. Nature 2005; 438(7067): 504–7.

68. Monsalve M, Calles B, Mencia M, Salas M, Rojo F. Transcription activation or repression by phage psi 29 protein p4 depends on the strength of the RNA polymerase-promoter interactions. Mol. Cell. 1997; 1(1): 99–107.

69. Pennypacker KR, Hong JS, McMillian MK. Pharmacological regulation of AP-1 transcription factor DNA binding activity. Faseb J. 1994; 8(8): 475–8.

70. Gadgil H, Jurado LA, Jarrett HW. DNA affinity chromatography of transcription factors. Anal. Biochem. 2001; 290(2): 147–78.

71. Gadgil H, Oak SA, Jarrett HW. Affinity purification of DNA-binding proteins. J. Biochem. Biophys. Methods. 2001; 49(1–3): 607–24.

72. Briggs MR, Kadonaga JT, Bell SP, Tjian R. Purification and biochemical characterization of the promoter-specific transcription factor, Sp1. Science 1986; 234(4772): 47–52.

73. Chockalingam PS, Jurado LA, Robinson FD, Jarrett HW. DNA affinity chromatography. Methods Mol Biol. 2000; 147: 141–53.

74. Jarrett HW. Affinity chromatography with nucleic acid polymers. J. Chromatogr. 1993; 618(1–2): 315–39.

75. Kadonaga JT. Purification of sequence-specific binding proteins by DNA affinity chromatography. Methods Enzymol. 1991; 208: 10–23.

76. Escolar L, Perez-Martin J, de Lorenzo V. Binding of the fur (ferric uptake regulator) repressor of *Escherichia coli* to arrays of the GATAAT sequence. J. Mol. Biol. 1998; 283(3): 537–47.

77. Goldberg MD, Johnson M, Hinton JC, Williams PH. Role of the nucleoid-associated protein His in the regulation of virulence properties of enteropathogenic *Escherichia coli*. Mol Microbiol. 2001; 41(3): 549–59.

16

Bacterial Post-Genomics Approaches for Vaccine Development

Giulia Bernardini, Daniela Braconi and Annalisa Santucci
Department of Molecular Biology, University of Siena, Siena, Italy

16.1   Introduction

In spite of advances made in the treatment and prevention of infections at the beginning of the twenty-first century, bacterial infections remain a major cause of disease and mortality in human and animals worldwide and still pose a major threat to global public health. New and emerging infections, together with the development of microbial resistance to antibiotics and antibacterial drugs has made the control of infections more difficult and expensive.

The introduction of vaccination more than 200 years ago was a watershed in science and prevented illness and death of millions of individuals. Vaccines generated immunological protection for a specific disease upon subsequent contact with a given pathogen. However, traditional approaches to vaccine development are time consuming and yield a limited number of candidates following extensive trial and error methods of testing to eliminate of candidate antigens (1). Moreover, the conventional methods are not applicable to noncultivable organisms such as hepatitis C virus and *Treponema pallidum* (2).

In the context of protection induced by humoral immunity, a vaccine candidate must meet the following criteria: it must retain antigenic conservation among clinical isolates; elicit functional antibodies; protect in animal models; and, ultimately be safe and effective for human use.

The genomic era has given the scientific community insights into the disease process for many organisms and offers exciting opportunities for vaccines research, in particular for those infectious diseases that are still waiting for an effective vaccine to be developed, or for which traditional approaches to vaccines discovery have failed due to the high

Mass Spectrometry for Microbial Proteomics   Edited by Haroun N. Shah and Saheer E. Gharbia
© 2010 John Wiley & Sons, Ltd

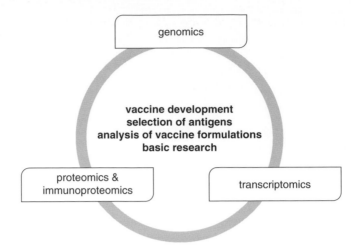

Figure 16.1   *The virtuous cycle: how post-genomics technologies can contribute to vaccine development*

standards of safety and for the chemical-physical characterization that regulatory authorities now require. However, modern approaches have so far had a limited impact on vaccine production considering the fact that only a couple of vaccines have been developed using recombinant DNA technology. Some examples of microbial pathogens against which subunit recombinant vaccines are commercially available include *Bordetella pertussis* (3), hepatitis B virus (4), *Vibrio cholerae* (5) and *Borrelia burgdoferi* (6).

Notwithstanding the advent of the various post-genomics technologies, new high-throughput technologies and bioinformatics can expedite the discovery process by allowing the rapid generation of a set of potential vaccine candidates that may normally be produced by a microbial pathogen in limited amounts, and possibly missed or underrepresented using traditional antigen discovery methodologies. These techniques, increasingly applied in laboratories, offer tremendous opportunities for vaccines research (Figure 16.1).

Sequencing of the first bacterial genome in 1995 (7) marked the beginning of a 'genomic era', changing our understanding of microbial life and the landscape of modern biology, leading to a new approach in vaccine design. Because of the rapid advances in DNA sequencing technology and sequence annotation, a wealth of new data have become available; at the beginning of 2008, according to the Genome On Line Database (GOLD, www.genomeonline.org) 866 completed genomes were published and 2654 are ongoing (incomplete) sequencing projects.

In silico whole-genome analysis has the potential to provide the basis for the complete understanding of genetics, biochemistry, physiology and pathogenesis of microorganisms. For example, the availability of genome sequence of closely related pathogenic and non-pathogenic bacteria can help in identifying virulence proteins that could be potential vaccine candidates and in predicting possible functions of newly sequenced proteins, or predicting cell surface localized or secreted bacterial proteins. These types of cellular proteins are important targets for the host immune system conferring protective immunity in a range of infection models, in particular in extracellular pathogen infections where the predominant immunological response is antibody mediated (8).

Moreover, the development of post-genomic approaches, whose real power arose from the combination between comparative genomics, molecular biology, transcriptomics, proteomics, biophysics and metabolomics, has accelerated the discovery of protein functional information, and the selection of virulence factors and potential antigen candidates, in particular for those pathogens that are still waiting for an effective vaccine to be available.

16.2 Comparative Genomics

Analysis of single genomes is no longer satisfactory. As the number of complete genome sequences increases, comparative genomics became a popular tool to compare a significant number of genomes from bacteria belonging to closely related species, to identify virulence or pathogenesis related factors, such as genes involved in environmental persistence of a pathogen. The microbial genome is a dynamic entity, and the virulence of many pathogens often correlates with the presence of DNA traits encoding disease-related factors, usually acquired by genetic horizontal transfer (GHT) and absent in nonpathogenic species (9). In particular, the analysis of the genetic variability between pathogens and closely related nonpathogenic microorganisms leads to the rapid identification of the complete set of genes potentially responsible for bacterial colonization, invasion, and infection in the human host. Although DNA acquisition, the main cause of inter- and intra-species genetic differences, is not the only determinant of pathogenicity, the association between GHT and the acquisition of virulence has a practical implication in vaccine discovery and therapeutic intervention. Genome sequence analysis of disease related genes provides crucial insights into the evolution of bacterial pathogenesis deciphering different mechanisms evolved by microorganisms to alter their antigenic appearance: phase variation (10), gene duplication (11) and potential loss of rudimentary functions (12). Proteins encoded by these genes form a promising class of candidates to be used as purified antigens in subunit-based vaccines and can provide the rational basis for a safe and stable attenuation of live vaccine candidates or vectors for vaccine delivery.

The comparative genomic hybridization (CGH) approach, by means of DNA microarray technology, is a powerful tool to investigate genome diversity and relatedness of bacteria, virus and parasites, circumventing the need for multiple closely related genome sequences. CGH involves the use of microarrays containing DNA from a sequenced reference microorganism that can be used to compare genomes of different unsequenced isolates by detecting genes that are conserved between them. However, CHG has intrinsic technical limitations; detection is restricted to the DNA spotted on the array and its inability to detect acquisition events with respect to the reference strain. On the other hand, CGH analyses permit a more accurate evaluation of the genetic stability of a great number of bacterial pathogens (13). Typically, such experiments produce large data sets that are difficult to process. Only recently, Carter and collaborators (14) developed a simple and robust CGH microarray data analysis process to overcome the bottle neck persisting in accurate processing and mathematical analysis of data.

The technique of comparative genomics has been applied to *Neisseria meningitidis* (15) to compare its genome with that of *Neisseria gonorreae* and *Neisseria lactamica*. Meningococcal specific and pathogenicity related genes present in meningococci and gonococci but absent in *N. lactamica* have been identified. The same technique has been applied for epidemiological genotyping of *Campylobacter jejuni* (16) with a better resolution than MLST and *N. meningitidis* serogroup C (17,18).

Comparative genomics strategies include the examination of genome variability, evolution and plasticity in pathogenic bacteria such as *Cryptococcus neoformans* (19) or *Escherichia coli* (20), to detect and investigate plasmid-mediated drug resistance (21), to define genomic island and pathogenic specific genes (22), to determine invasiveness and to differentiate clones of *Streptococcus pneumoniae* (23).

The comparison of multiple strains of a single species provides a pan-genome, as a measure of the total gene repertoire that pertains to a given microorganism from which important practical information for vaccine design may be deduced. Although the core genes represent the most desirable source for the selection of conserved and, therefore potentially universally applicable vaccine candidates, they are also more likely to be immunologically silent in a successful pathogen (24). The first application of a pan-genome approach to vaccine design was performed on *Streptococcus agalactiae* (GBS) (25). Bioinformatics was used to select genes from two sub-genomes (core and dispensable genomes) encoding proteins as potential vaccine candidates. This approach led to a vaccine formulation comprising a combination of four antigens which provided overall almost universal strain coverage, with high levels of protection (25). Recently, a new approach known as metagenomics, which involves the use of genome sequencing and other genomics technologies to study microorganisms directly in their habitats, has emerged (26). In such environmental niches, e.g. the female reproductive tract, the skin, the oral cavity or the gut, complex microbial communities interact with each other and have co-evolved with their human host and therefore play an important part in human health and disease (27). Metagenomics applied to specific niches of human host could produce a comprehensive picture of the microbiome. These can be viewed as a human 'accessory' genome, to assess the functional repertoire of bacterial communities as shaped by host–pathogen interactions and facilitate our understanding of the molecular basis of specific interaction with the host to better define the boundaries between pathogenic and commensal sub-populations of the same species.

16.3 Transcriptomics

Although comparative genomic approaches help reveal the molecular differences between phenotypes among bacterial species or strains, they are not sufficient to uncover the complexities of the interaction between pathogen and host, the underlying basis of the infectious disease. To this end, post-genomics technologies such as transcriptomics and proteomics, enable monitoring of the global changes in gene and protein expression in both the pathogen and host during the infectious process.

Metabolic profiling using DNA microarrays has been applied to determine the global effects that changes in growth conditions, gene mutations, antimicrobial agents and environmental stimuli have on the transcriptome of an organism, and to possibly identify new regulatory associations between genes (28). However, for vaccine antigen discovery research, it is important to know which genes are up-regulated during infection as they represent the most likely protective vaccine candidates. Furthermore, understanding of the molecular and cellular details of these host–microbe interactions may lead to the identification of virulence-associated microbial genes and host–defence strategies. This information is invaluable for the design of a new generation of medical investigative tools. In this regard, whole-genome microarrays are fundamental platforms for the analysis of gene expression that occurs during infection (29). They help in the identification of genes that are essential for virulence and pathogenesis, defining the challenge of host–pathogen interactions (30,31), and in understanding how a pathogen orchestrates its responses to the host environment.

Unfortunately, because of the lack of efficient extraction methods of different bacterial mRNAs from tissues, most microarray studies have been initiated to organisms grown *in vitro*. Since environmental parameters influence microbial growth and therefore gene expression, care should be taken in extrapolating such data to *in vivo* conditions. Thus, traditional biological studies remain important (32) and to date published studies have primarily described the host response to infection. By contrast, the pathogen's responses within infected host cells are still poorly documented. In general, DNA microarray-based gene expression profiling for the study of bacterial infections has faced many challenges (33), including both the quantity and the quality of recovered bacterial mRNA but these will undoubtedly improve with the development of new technologies.

16.4 Proteomics and Immmunoproteomics

Proteomics (high-throughput analysis of the complete repertoire of proteins of an organism) theoretically bridges the gap between the genome sequence information of a cell and its dynamic protein profile; moreover, in spite of the fact that gene expression can be elegantly assessed using gene chips, mRNA levels do not necessarily correlate with the level of protein (26). Thus for many bacterial pathogens, whose genomes have been sequenced, proteomic techniques are increasingly being applied for identification of potential targets for vaccine development.

Compared with genomics and transcriptomics, proteomics also has the advantage of defining proteins that are differentially located or secreted (34). A key step in this process is the analysis of membrane proteins as it provides experimental support to *in silico* prediction of protein compartmentalization and allows investigation of new aspects of the topological organization of prokaryotic membranes (see Chapter 8). For vaccines, a reliable list of surface exposed proteins offers the possibility to identify and screen additional antigens for their capacity to elicit a protective immune response. These proteins are crucial in pathogenic processes such as motility, colonization of and adherence to host cells, and in providing channels for the removal of antibiotics and injection of toxins and extracellular proteases. As such, they are regarded as being of central importance as potential drug targets and as the basis for novel vaccines (35).

Genomic mining and proteomics have been used to characterize the surface proteins of *Chlamydia pneumoniae* (36). Integrated proteomic strategies have been also been successfully applied in the discovery of antigens from *Helicobacter pylori* (37), *Legionella pneumophila* (38), *S. pneumoniae* (39), *Bacillus anthracis* (40) and *Clostridium difficile* (41).

When the goal of proteomics or, more generally, post-genomics is to probe deeper into the physiology of an organism, careful consideration should be given towards establishing standardized and systematic methodologies. This is true for studies of both microbial and multicellular organisms. Most pathogenic bacteria cross very different environments but as far as possible they should be studied in their natural state and environment. Currently, this is not possible and *in vitro* model systems of bacterial infection are in urgent need of development.

The interaction of a pathogenic microorganism with its host to produce clinical diseases is determined by multiple factors acting individually or synergically at different stages during infection. An attractive application of proteomics is its use in combination with serological analysis (SERPA: serological proteome analysis) to screen and select novel *in vivo* immunogens that may be valuable as vaccine candidates (42). A combination of two-dimensional eletrophoresis with immunoblotting is widely used as a tool to investigate the humoral immune response against microbial pathogens during natural infections (43–45), since a large number of antigenic proteins that are recognized by polyclonal patient's sera, can be simultaneously detected.

The analytical limitations of two-dimensional electrophoresis, as a standard tool for proteomic research is driving the development of alternative strategies, mainly mass spectrometry-based technologies (46). 'Shotgun proteomics' or multidimensional protein identification (mudPIT) is a gel-free approach based on multidimensional liquid chromatography separation of complex peptide mixtures (i.e. from chemical or enzymatic digestion of cellular proteins, via different chemical or physical properties) coupled with mass spectrometry. Due to the increased complexity of the sample, the mechanisms of separation should be orthogonal so that the resolution gained in the first dimension can be maximized in the subsequent dimension. For pathogenic bacteria, this approach has proven to be highly successful for membrane protein analysis (47) aimed at vaccine development (e.g. *Plasmodium falciparum*) (48). The use of multidimensional separation techniques has greatly enhanced the number of routinely identified proteins, even if they do not usually provide quantitative information. One of the first described quantitative proteomic strategies was isotope-coded affinity tag (ICAT) and its application to the study of enteropathogenic *E. coli* and *Methanococcoides burtonii* is well documented (49,50). Other early methods for quantitative proteomics include stable isotope labelling with amino acids in cell culture (SILAC) and iTRAQ. In the SILAC method, cells subjected to different biological conditions are grown in culture in the presence of an essential amino acid with a stable isotopic nucleus. Therefore, one sample can be incubated with an unlabelled amino acid and the test sample with a deuterated form of the same amino acid (51). The iTRAQ technique is a new approach that makes differentially labelled peptide masses indistinguishable but produces diagnostic fragment peaks providing relative quantitative information on proteins. The iTRAQ reagents are a set of amine reactive, isobaric, isotope tags that can be used to simultaneously track up to four samples in a single experiment. *E. coli, Saccharomyces cerevisiae, Solfolobus solfataricus* and *Synechocystis* spp. are examples of microorganisms that have been investigated using these techniques (52–54).

An additional advantage of proteomics over genomics and transcriptomics is its capacity to analyse post-translational modifications (PTMs) that may not be apparent from the analysis of nucleotide sequence data. PTMs have been demonstrated to play an important role in many aspects of bacterial pathogenesis. Moreover, PTMs provide an effective means to generate diversity and influence antigenicity.

16.5   Other High-Throughput Technologies

In vivo gene expression has been exploited in order to identify virulence genes, which is a key step in vaccine design. A recent technology which does not strictly depend on but is facilitated by genome sequencing is signature-tagged mutagenesis (STM) (55), successfully used to discover virulence genes from a variety of bacterial species including *Salmonella enterica* (56) and *Salmonella gallinarum* (57). *In vivo* expression technology (IVET), developed by John Mekalanos (58), and its variant recombination-based *in vivo* expression technology (RIVET) (59), are high-throughput gene expression techniques that have been exploited in vaccine discovery research to identify bacterial genes for which expression occurs specifically when bacteria infect their host (59,60).

Another high-throughput technology that offers an interesting contribution to the use of genomics for vaccine-candidate identification is expression library immunization (ELI). This approach consists of making an expression library of whole-genomic DNA or cDNA from the pathogen in genetic immunization vectors, followed by immunization with DNA of the library and screening using an animal disease model (61). Yero and collaborators applied ELI and genetic immunization for generating a functional immune response against *N. meningitidis* (62), demonstrating that immunization with a meningococcal genomic library led to the induction of a protective humoral immune response. ELI technology has been applied to vaccine discovery of other pathogens such as *C. pneumoniae* (63), *Chlamydia muridarum* (64) and *Coccidioides immitis* (65).

16.6   Meningococcal Vaccines and Reverse Vaccinology

Neisseria meningitidis, the meningococcus, is a Gram-negative bacterium with a coccoid shape, mostly associated with meningitis and severe sepsis with often fatal outcome. On the basis of the immunochemistry of the capsular polysaccharides, at least 13 different serogroups have been defined but serogroups A, B, C, Y and W135 account for almost all cases of disease worldwide (66).

Vaccination is the only way to control this disease. Conjugate vaccines against serogroups A, C, Y and W135 are presently in clinical development and are expected to be licensed shortly (67). On the other hand, serogroup B capsular polysaccharide-based vaccines are poorly immunogenic and may induce autoimmune responses. An alternative approach to an antimeningococcus B vaccine focused on the use of subcapsular antigens, in particular on surface-exposed proteins contained in outer membrane vesicles (OMVs) (68). These have been shown in clinical trials to elicit both a serum bactericidal antibody response while also protecting against developing meningococcal diseases. Bacterial outer membrane preparations are characterized by the presence of five major outer membrane proteins: PorA, PorB, Opa, Omp85 and RmpM, and high lipid content. On account of these features proteomics should be valuable for biochemical characterization of membrane and surface-associated proteins in OMV-based vaccines (69). Systematic proteome analyses of commercial OMV-based vaccines were performed to fully identify the protein components of the preparations (70–75). Apart from the classical major antigens, a number of periplasmic, membrane-associated and cytoplasmic proteins have so far been found whose presence should be carefully considered for in-depth investigations for the development of future meningococcal vaccines. Moreover, OMV preparative procedures should be carefully standardized to include the quantity and quality of protein patterns, since differences can have important outcomes for the immune response and influence the efficacy and safety of the OMVs. The opportunity provided by proteomics to characterize these preparations, allows differences between preclinical development of vaccines from different batches/manufacturers (76) to be assessed and helps to reduce cost. However, a drawback of these vaccines is the high variability of their main components, PorA, PorB and opacity proteins, that are strain specific and can only be used against clonal disease outbreaks (77).

Therefore, the biggest challenge in the development of a protein-based meningococcus B vaccine is the identification of highly conserved antigens capable of inducing protective immunity.

In an attempt to identify such vaccine candidates, two parallel approaches were followed, one using bioinformatics algorithms to screen the complete MenB genome in order to predict putative antigens (78), the other exploiting DNA microarray technology to identify antigens specifically induced during bacterial adhesion (79). In the first approach, the availability of the complete meningoccal genome made it possible to choose *a priori* potential surface-exposed proteins in a reverse manner, starting from the genomic sequence rather than from the microorganism; a strategy referred to as 'reverse vaccinology' (80). Once applied to serogroup B meningococcus, this approach allowed the rapid identification of 29 antigens that were able to elicit an antibody response that could kill the bacteria *in vitro* in the presence of complement, and the selection of five highly conserved antigens with potential for a universal meningococcal vaccine (81). Soon, reverse vaccinology approaches were described for *S. pneumoniae* (82), group A streptococcus (83), group B streptococcus (84), *Porphyromonas gengivalis* (85) and *Chlamydia* spp. (36). Although the method has been successful in generating innovative vaccines that are protective in animal models, the real practical value of reverse vaccinology and the extent to which it is effective in providing new-generation vaccines (2) must await the results of ongoing clinical trials.

Another approach utilized transcriptomic analyses of *N. meningitidis* B-based MC58 strain during the adhesion to human epithelial or endothelial cells (79,86,87) or after treatment with human serum (87). Of the adhesion-regulated genes, about 40% potentially encode membrane proteins, suggesting a remodelling of MenB membrane components during contact between the bacterium and its host. Since these antigens are of particular importance in the infection processes, they are expected to be good candidates for vaccine development.

16.7   *Helicobacter pylori* Vaccines

H. pylori is a Gram-negative bacterium that establishes a chronic infection in the human stomach (see Chapter 1). *H. pylori* colonizes the gastrointestinal mucosa of 50% of the population in developed countries and approximately 80% in the developing world. It is linked with a diverse spectrum of gastrointestinal clinical disorders including peptic ulcer, gastric atrophy, gastric cancer and MALT lymphoma in the stomach (88). The genomes of three *H. pylori* isolates have been sequenced (strain 26695, J99 and HPAG1) and represent powerful resources for the study of the biology of this microorganism (89–91).

Development of a vaccine against *H. pylori* could be cost-effective and a desirable alternative to the current antibiotic therapy. Vaccination in experimental animal models with orally administered urease is feasible but clinical trials in patients with chronic infection have failed, mainly because of the poor immunogenicity. Numerous studies have investigated the immunological basis of *H. pylori* immunization, and despite significant advances, the mechanisms of protection against this pathogen are poorly understood (92,93).

H. pylori is a highly adapted bacterial pathogen, and understanding the mechanisms of interactions between this microorganism and its host is essential to explain the host's response to this infectious agent and the diverse clinical manifestations. Despite a vigorous and cellular immune response to *H. pylori* at the local and systemic level, the bacterium persists throughout the lifetime of its host. The key to avoidance of the immune system by the bacterium most likely lies in the fact that it has evolved together with its host, achieving a balance in which the immune system is stimulated sufficiently to cause inflammation and epithelial cell damage, while modulating the response to prevent elimination of bacteria (94).

Humoral immune response to *H. pylori* infection can be used for diagnostic purposes, as well as a basis for vaccine development. Immunoproteomics can be particularly useful to identify subgroups of proteins consistently recognized by patient sera but not in non-infected controls, that may be used as serological markers for detecting and monitoring *H. pylori* infection. Moreover, immunoproteomics would be particularly helpful in identifying antigens associated with different pathologies or different stages of bacterial infection and can be adopted to discriminate asymptomatic carriers from patients that are at high risk of developing severe diseases (95–98).

Several groups have used this approach to detect candidate antigens of *H. pylori* for diagnosis, therapy and vaccine development, as well as to investigate potential associations between specific immune responses and manifestations of disease, and reveal a high variability of humoral recognition pattern among different patients and *H. pylori* isolates (99–101).

In each study, there was a considerable inter-individual variation that may be partially explained since strains derived from patients affected by different pathologies have shown different protein and antigenic repertoires. In this regard, Mini and collaborators (100) highlighted the importance of the strain to be used in immunoproteomics studies. Strain specificity emerged from the difference in antigenic patterns of the same sera tested against three different strains and a strain immunogenicity/antigenicity profile could be deduced, with gastric carcinoma being more immunogenic than duodenal ulcer, and this in turn being more immunogenic compared with chronic gastritis.

The DNA microarray approach has also been used to study host–gene expression in response to *H. pylori* infection or immunization, to shed light on the clarification of immune protection, and the design of vaccines targeting specific aspects of the host immune system (102).

Recently, a Phase I study was carried out to assess the safety and immunogenicity of a *H. pylori* vaccine in noninfected volunteers. Vaccination with three conserved antigens (CagA, VacA and NAP) was well-tolerated and strongly immunogenic and generated specific antibody and T-cell response (103).

16.8   Conclusions

In the field of vaccinology, trial and error experimentation is more likely to succeed than rational design approach. However, current development in systems biology will eventually provide a better understanding of the immune system which in turn will lead to the development of improved vaccines.

The immune system involves the dynamic interaction of a wide array of tissues, cells, and molecules and leads to a variety of different biological outcomes. This complexity makes the study of immune responses particularly apt to systems biology approaches by high-throughput new approaches (104).

Recent advances in global genome, mRNA and protein analysis provide access to certain aspects of the biological complexity but much work still needs to be done to link such composite data and the dynamic metabolic and physiological aspects of cellular system. Historically, the portfolio of technologies that are now generally referred to as proteomics has been studied mainly by academic groups with a clear focus on protein analysis and driven by the technologies. Regulatory administrators such as the FDA, however, place increasing pressure on drug companies involved in the development and marketing of therapeutic proteins which has created a huge demand for new technologies in protein chemistry. Hence, the developments described here were paralleled by the efforts of pharmaceutical companies to get the approval of drug regulatory administrations for the marketing of genetically engineered therapeutic proteins and a steadily increasing demand for product quality and drug safety. At present many antibacterial new generation vaccines are protein-based and in this regard post-genomics, and particularly proteomics, can be crucial technologies for quality control/assurance and GMP (Good Manufacturing Practice).

References

1 Chakravarti DN, Fiske MJ, Fletcher LD, Zagursky RJ. Application of genomics and proteomics for identification of bacterial gene products as potential vaccine candidates. Vaccines 2001, 19, 601–612.

2 Fraser CM, Rappuoli R. Application of microbial genomic science to advanced therapeutics. Annu. Rev. Med. 2005, 56, 459–474.

3 Pizza M, Covacci A, Bartoloni A, *et al.* Mutants of pertussis toxin suitable for vaccine development. Science 1989, 246, 497–500.

4 Assad S, Francis A. Over a decade of experience with a yeast recombinant hepatitis B vaccine. Vaccine 1999, 18(1–2), 57–67.

5 Ryan ET, Calderwood SB. Cholera vaccines. Clin. Infect. Dis. 2000, 31, 561–565.

6 Luft BJ, Dunn JJ, Lawson CL. Approaches toward the directed design of a vaccine against *Borrelia burgdorferi*. J. Infect. Dis. 2002, 185 (Suppl. 1), S46-S51.

7 Fleischmann RD, Adams MD, White O, *et al.* Whole genome random sequencing and assembly of *Haemophilus influenzae* Rd. Science 1995, 269, 496–512.

8 Rappuoli R. Bridging the knowledge gaps in vaccine design. Nat. Biotechnol. 2007, 25, 1361–1366.

9 Hacker J, Kaper JB, Pathogenicity islands and the evolution of microbes. Annu. Rev. Microbiol. 2000, 54, 641–679.

10 van der Woude MW, Bäumler AJ. Phase and antigenic variation in bacteria. Clin. Microbiol. Rev. 2004, 17, 581–611.

11 Alm RA, Bina J, Andrews BM, Doig P, Hancock RE, Trust TJ. Comparative genomics of *Helicobacter pylori*: analysis of the outer membrane protein families. Infect. Immun. 2000, 68, 4155–4168.

12 Preston A, Parkhill J, Maskell DJ. The bordetellae: lessons from genomics. Nat. Rev. Microbiol. 2004, 2, 379–390.

13 Bryant PA, Venter D, Robins-Browne R, Curtis N. Chips with everything: DNA microarrays in infectious diseases. Lancet Infect. Dis. 2004, 4, 100–111.

14 Carter B, Wu G, Woodward MJ, Anjum MF. A process for analysis of microarray comparative genomics hybridisation studies for bacterial genomes. BMC Genomics 2008, 9, 53.

15 Perrin A, Bonacorsi S, Carbonnelle E, *et al.* Comparative genomics identifies the genetic islands that distinguish *Neisseria meningitidis*, the agent of cerebrospinal meningitis, from other Neisseria species. Infect. Immun. 2002, 70, 7063–7072.

16 Rodin S, Andersson AF, Wirta V, *et al.* Performance of a 70-mer oligonucleotide microarray for genotyping of *Campylobacter jejuni*. BMC Microbiol. 2008, 8, 73.

17 Peng J, Zhang X, Shao Z, Yang L, Jin Q. Characterization of a new *Neisseria meningitidis* serogroup C clone from China. Scan. J. Infect. Dis. 2008, 40, 63–66.

18 Peng J, Zhang X, Yang E, *et al.* Characterization of serogroup C meningococci isolated from 14 provinces of China during 1966–2005 using comparative genomic hybridization. Sci. China C Life Sci. 2007, 50, 1–6.

19 Hu G, Liu I, Sham A, Stajich JE, Dietrich FS, Kronstad JW. Comparative hybridization reveals extensive genome variation in the AIDS-associated pathogen *Cryptococcus neoformans*. Genome Biol. 2008, 9, R41.

20 Zhang Y, Laing C, Steele M, *et al.* Genome evolution in major *Escherichia coli* O157:H7 lineages. BMC Genomics 2007, 16, 121.

21 Tan TY, Ng SY, Teo L, Koh Y, Teok CH. Detection of plasmid-mediated AmpC in *Escherichia coli*, *Klebsiella pneumoniae* and *Proteus mirabilis*. J. Clin. Pathol. 2008, 61, 642–644.

22 Lloyd AL, Rasko DA, Mobley HL. Defining genomic islands and uropathogen-specific genes in uropathogenic *Escherichia coli*. J. Bacteriol. 2007, 189, 3532–3546.

23 Obert CA, Gao G, Sublett J, Tuomanen EI, Orihuela CJ. Assessment of molecular typing methods to determine invasiveness and to differentiate clones of *Streptococcus pneumoniae*. Infect. Genet. Evol. 2007, 7, 708–716.

24 Muzzi A, Masignani V, Rappuoli R. The pan-genome: towards a knowledge-based discovery of novel targets for vaccines and antibacterials. Drug Discov. Today 2007, 12, 429–439.

25 Maione D, Margarit I, Rinaudo CD, *et al.* Identification of a universal Group B streptococcus vaccine by multiple genome screen. Science 2005, 309, 148–150.

26 Medini D, Serruto D, Parkhill J, *et al.* Microbiology in the post-genomic era. Nat. Rev. Microbiol. 2008, 6, 419–430.

27 Dethlefsen L, McFall-Ngai M, Relman DA. An ecological and evolutionary perspective on human-microbe mutualism and disease. Nature 2007, 449, 811–818.

28 Suen G, Arshinoff BI, Taylor RG, Welch RD. Practical applications of bacterial functional genomics. Biotechnol. Genet. Eng. Rev. 2007, 24, 213–242.

29 Fleischmann RD, Alland D, Eisen JA, *et al.* Whole-genome comparison of *Mycobacterium tuberculosis* clinical and laboratory strains. J. Bacteriol. 2002, 184, 5479–5490.

30 Walduck A, Rudel T, Meyer TF. Proteomic and gene profiling approaches to study host responses to bacterial infection. Curr. Opin. Microbiol. 2004, 7, 33–38.

31 Rappuoli R, Pushing the limits of cellular microbiology: microarrays to study bacteria-host cell intimate contacts. Proc. Natl Acad. Sci. USA 2000, 97, 13.467–13.469.

32 Scarselli M, Giuliani MM, Adu-Bobie J, Pizza M, Rappuoli R. The impact of genomics on vaccine design. Trends Biotechnol. 2005, 23, 84–91.

33 La MV, Raoult D, Renesto P. Regulation of whole bacterial pathogen transcription within infected hosts. FEMS Microbiol. Rev. 2008, 32, 440–460.

34 Galka F, Wai SN, Kusch H, *et al.* Proteomic characterization of the whole secretome of *Legionella pneumophila* and functional analysis of outer membrane vesicles. Infect. Immun. 2008, 76, 1825–1836.

35 Cordwell SJ. Technologies for bacterial surface proteomics. Curr. Opin. Microbiol. 2006, 9320–9329.

36 Montigiani S, Falugi F, Scarselli M, *et al.* Genomic approach for analysis of surface proteins in *Chlamydia pneumoniae*. Infect. Immun. 2002, 70, 368–379.

37 Bernardini G, Braconi D, Lusini P, Santucci A. *Helicobacter pylori*: immunoproteomics related to different pathologies. Expert Rev. Proteomics. 2007, 4, 679–689.

38 Khemiri A, Galland A, Vaudry D, *et al.* Outer-membrane proteomic maps and surface-exposed proteins of *Legionella pneumophila* using cellular fractionation and fluorescent labelling. Anal. Bioanal. Chem. 2008, 390, 1861–1871.

39 Morsczeck C, Prokhorova T, Sigh J, *et al. Streptococcus pneumoniae*: proteomics of surface proteins for vaccine development. Clin. Microbiol. Infect. 2008, 14, 74–81.

40 Liu YT, Lin SB, Huang CP, Huang CM. A novel immunogenic spore coat-associated protein in *Bacillus anthracis*: characterization via proteomics approaches and a vector-based vaccine system. Protein Expr Purif. 2008, 57, 72–80.

41 Wright A, Drudy D, Kyne L, Brown K, Fairweather NF. Immunoreactive cell wall proteins of *Clostridium difficile* identified by human sera. J. Med. Microbiol. 2008, 57, 750–756.

42 Klade CS. Proteomics approaches towards antigen discovery and vaccine development. Curr. Opin. Mol. Ther. 2002, 4, 216–223.

43 Nowalk AJ, Gilmore RD Jr, Carroll JA. Serologic proteome analysis of *Borrelia burgdorferi* membrane-associated proteins. Infect. Immun. 2006, 74, 3864–3873.

44 Vytvytska O, Nagy E, Blüggel M, *et al.*, Identification of vaccine candidate antigens of *Staphylococcus aureus* by serological proteome analysis. Proteomics 2002, 2, 580–590.

45 Jungblut PR. Proteome analysis of bacterial pathogens. Microbes Infect. 2001, 3, 831–840.

46 Roe MR, Griffin TJ. Gel-free mass spectrometry-based high throughput proteomics: tools for studying biological response of proteins and proteomes. Proteomics 2006, 6, 4678–4687.

47 Fischer F, Wolters D, Rögner M, Poetsch A. Toward the complete membrane proteome: high coverage of integral membrane proteins through transmembrane peptide detection. Mol. Cell. Proteomics 2006, 5, 444–453.

48 Florens L, Washburn MP, Raine JD, *et al.* A proteomic view of the *Plasmodium falciparum* life cycle. Nature 2002, 419, 520–526.

49 Hardwidge PR, Donohoe S, Aebersold R, Finlay BB. Proteomic analysis of the binding partners to enteropathogenic *Escherichia coli* virulence proteins expressed in *Saccharomyces cerevisiae*. Proteomics 2006, 6, 2174–2179.

50 Goodchild A, Raftery M, Saunders NF, Guilhaus M, Cavicchioli R. Cold adaptation of the Antarctic archaeon, *Methanococcoides burtonii* assessed by proteomics using ICAT. J. Proteome Res. 2005, 4, 473–480.

51 Ong SE, Blagoev B, Kratchmarova I, *et al.* Stable isotope labeling by amino acids in cell culture, SILAC, as a simple and accurate approach to expression proteomics. Mol Cell Proteomics 2002, 1, 376–386.

52 Neher SB, Villén J, Oakes EC, *et al.* Proteomic profiling of ClpXP substrates after DNA damage reveals extensive instability within SOS regulon. Mol. Cell. 2006, 22, 193–204.

53 Choe LH, Aggarwal K, Franck Z, Lee KH. A comparison of the consistency of proteome quantitation using two-dimensional electrophoresis and shotgun isobaric tagging in *Escherichia coli* cells. Electrophoresis 2005, 26, 2437–2449.

54 Chong PK, Gan CS, Pham TK, Wright PC. Isobaric tags for relative and absolute quantitation (iTRAQ) reproducibility: Implication of multiple injections. J. Proteome Res. 2006, 5, 1232–1240.

55 Hensel M, Shea JE, Gleeson C, Jones MD, Dalton E, Holden DW. Simultaneous identification of bacterial virulence genes by negative selection. Science 1995, 269, 400–403.

56 Carnell SC, Bowen A, Morgan E, Maskell DJ, Wallis TS, Stevens MP. Role in virulence and protective efficacy in pigs of *Salmonella enterica* serovar Typhimurium secreted components identified by signature-tagged mutagenesis. Microbiology 2007, 153, 1940–1952.

57 Shah DH, Lee MJ, Park JH, *et al.* Identification of *Salmonella gallinarum* virulence genes in a chicken infection model using PCR-based signature-tagged mutagenesis. Microbiology 2005, 151, 3957–3968.

58 Mahan MJ, Slauch JM, Mekalanos JJ. Selection of bacterial virulence genes that are specifically induced in host tissues. Science 1993, 259, 686–688.

59 Slauch JM, Camilli A. IVET and RIVET: use of gene fusions to identify bacterial virulence factors specifically induced in host tissues. Methods Enzymol. 2000, 326, 73–96.

60 Lombardo MJ, Michalski J, Martinez-Wilson H, *et al.* An *in vivo* expression technology screen for *Vibrio cholerae* genes expressed in human volunteers. Proc. Natl Acad. Sci. USA 2007, 104, 18 229–18 234.

61 Barry MA, Lai WC, Johnston SA. Protection against mycoplasma infection using expression-library immunization. Nature 1995, 377, 632–635.

62 Yero D, Pajón R, Pérez Y, *et al.* Identification by genomic immunization of a pool of DNA vaccine candidates that confer protective immunity in mice against *Neisseria meningitidis* serogroup B. Vaccine 2007, 25, 5175–5188.

63 Li D, Borovkov A, Vaglenov A, *et al.* Mouse model of respiratory *Chlamydia pneumoniae* infection for a genomic screen of subunit vaccine candidates. Vaccine 2006, 24, 2917–2927.

64 McNeilly CL, Beagley KW, Moore RJ, Haring V, Timms P, Hafner LM. Expression library immunization confers partial protection against *Chlamydia muridarum* genital infection. Vaccine 2007, 25, 2643–2655.

65 Ivey FD, Magee DM, Woitaske MD, Johnston SA, Cox RA. Identification of a protective antigen of *Coccidioides immitis* by expression library immunization. Vaccine 2003, 21, 4359–4367.

66 Taha MK, Deghmane AE, Antignac A, Zarantonelli ML, Larribe M, Alonso JM. The duality of virulence and transmissibility in *Neisseria meningitidis*. Trends Microbiol. 2002, 10, 376–382.

67 Pichichero M, Casey J, Blatter M, *et al.* Comparative trial of the safety and immunogenicity of quadrivalent (A, C, Y, W-135) meningococcal polysaccharide-diphtheria conjugate vaccine versus quadrivalent polysaccharide vaccine in two- to ten-year-old children. Pediatr. Infect. Dis. J. 2005, 24, 57–62.

68 de Moraes JC, Perkins BA, Camargo MC, *et al.* Protective efficacy of a serogroup B meningococcal vaccine in Sao Paulo, Brazil. Lancet 1992, 340, 1074–1078.

69 Bernardini G, Braconi D, Martelli P, Santucci A. Postgenomics of *Neisseria meningitidis* for vaccines development. Expert Rev. Proteomics 2007, 4, 667–677.

70 Post DM, Zhang D, Eastvold JS, Teghanemt A, Gibson BW, Weiss JP. Biochemical and functional characterization of membrane blebs purified from *Neisseria meningitidis* serogroup B. J. Biol Chem. 2005, 280, 38 383–38 394.

71 Ferrari G, Garaguso I, Adu-Bobie J, *et al.* Outer membrane vesicles from group B *Neisseria meningitidis* delta gna33 mutant: proteomic and immunological comparison with detergent-derived outer membrane vesicles. Proteomics 2006, 6, 1856–1866.

72 Uli L, Castellanos-Serra L, Betancourt L, *et al.* Outer membrane vesicles of the VA-MENGOC-BC vaccine against serogroup B of *Neisseria meningitidis*. Analysis of protein components by two-dimensional gel electrophoresis and mass spectrometry. Proteomics 2006, 6, 3389–3399.

73 Vipond C, Suker J, Jones C, Tang C, Feavers IM, Wheeler JX. Proteomic analysis of a meningococcal outer membrane vesicle vaccine prepared from the group B strain NZ98/254. Proteomics 2006, 6, 3400–3413.

74 Vipond C, Wheeler JX, Jones C, Feavers IM, Suker J. Characterization of the protein content of a meningococcal outer membrane vesicle vaccine by polyacrylamide gel electrophoresis and mass spectrometry. Hum Vaccine. 2005, 1, 80–84.

75 Vaughan TE, Skipp PJ, O'Connor CD, *et al.* Proteomic analysis of *Neisseria lactamica* and *Neisseria meningitidis* outer membrane vesicle vaccine antigens. Vaccine 2006, 24, 5277–5293.

76 Bernardini G, Braconi D, Santucci A. The analysis of *Neisseria meningitidis* proteomes: Reference maps and their applications. Proteomics 2007, 7, 2933–2946.

77 Sierra GV, Campa HC, Varcacel NM, *et al.* Vaccine against group B *Neisseria meningitidis*: protection trial and mass vaccination results in Cuba. NIPH Ann. 1991, 14, 195–207.

78 Pizza M, Scarlato V, Masignani V, *et al.* Identification of vaccine candidates against serogroup B meningococcus by whole-genome sequencing. Science 2000, 287, 1816–1820.

79 Grifantini R, Bartolini E, Muzzi A, *et al.* Previously unrecognized vaccine candidates against group B meningococcus identified by DNA microarrays. Nat. Biotechnol. 2002, 20, 914–921.

80 Mora M, Veggi D, Santini L, Pizza M, Rappuoli R. Reverse vaccinology. Drug Discov. Today 2003, 8, 459–464.

81 Giuliani MM, Adu-Bobie J, Comanducci M, *et al.* A universal vaccine for serogroup B meningococcus. Proc. Natl Acad. Sci. USA 2006, 103, 10 834–10 839.

82 Wizemann TM, Heinrichs JH, Adamou JE, *et al.* Use of a whole genome approach to identify vaccine molecules affording protection against *Streptococcus pneumoniae* infection. Infect. Immun. 2001, 69, 1593–1598.

83 McMillan DJ, Batzloff MR, Browning CL, *et al.* Identification and assessment of new vaccine candidates for group A streptococcal infections. Vaccine 2004, 22, 2783–2790.

84 Tettelin H, Masignani V, Cieslewicz MJ, *et al.* Complete genome sequence and comparative genomic analysis of an emerging human pathogen, serotype V *Streptococcus agalactiae.* Proc. Natl Acad. Sci. USA 2002, 99, 12391–12396.

85 Ross BC, Czajkowski L, Hocking D, *et al.* Identification of vaccine candidate antigens from a genomic analysis of *Porphyromonas gingivalis.* Vaccine 2001, 19, 4135–4142.

86 Dietrich G, Kurz S, Hübner C, *et al.* Transcriptome analysis of *Neisseria meningitidis* during infection. J. Bacteriol. 2003, 185, 155–164.

87 Kurz S, Hübner C, Aepinus C, *et al.* Transcriptome-based antigen identification for *Neisseria meningitidis.* Vaccine 2003, 21, 768–775.

88 Portal-Celhay C, Perez-Perez GI. Immune responses to *Helicobacter pylori* colonization: mechanisms and clinical outcomes. Clin. Sci. 2006, 110, 305–314.

89 Tomb JF, White O, Kerlavage AR, *et al.* The complete genome sequence of the gastric pathogen *Helicobacter pylori.* Nature 1997, 388, 539–547.

90 Alm RA, Ling LS, Moir DT, *et al.* Genomic-sequence comparison of two unrelated isolates of the human gastric pathogen *Helicobacter pylori.* Nature 1999, 397, 176–180.

91 Oh JD, Kling-Bäckhed H, Giannakis M, *et al.* The complete genome sequence of a chronic atrophic gastritis *Helicobacter pylori* strain: evolution during disease progression. Proc. Natl Acad. Sci. USA 2006, 103, 9999–10 004.

92 Pappo J, Torrey D, Castriotta L, Savinainen A, Kabok Z, Ibraghimov A. *Helicobacter pylori* infection in immunized mice lacking major histocompatibility complex class I and class II functions. Infect. Immun. 1999, 67, 337–341.

93 Lucas B, Bumann D, Walduck A, *et al.* Adoptive transfer of CD4+ T cells specific for subunit A of *Helicobacter pylori* urease reduces *H. pylori* stomach colonization in mice in the absence of interleukin-4 (IL-4)/IL-13 receptor signaling. Infect. Immun. 2001, 69, 1714–1721.

94 Linz B, Balloux F, Moodley Y, *et al.* An African origin for the intimate association between humans and *Helicobacter pylori.* Nature 2007, 445, 915–918.

95 Kimmel B, Bosserhoff A, Frank R, Gross R, Goebel W, Beier D. Identification of immunodominant antigens from *Helicobacter pylori* and evaluation of their reactivities with sera from patients with different gastroduodenal pathologies. Infect. Immun. 2000, 68, 915–920.

96 Krah A, Miehlke S, Pleissner KP, *et al.* Identification of candidate antigens for serologic detection of *Helicobacter pylori*-infected patients with gastric carcinoma. Int. J. Cancer 2004, 108, 456–463.

97 Bumann D, Jungblut PR, Meyer TF. *Helicobacter pylori* vaccine development based on combined subproteome analysis. Proteomics 2004, 4, 2843–2848.

98 Mini R, Annibale B, Lahner E, Bernardini G, Figura N, Santucci A. Western blotting of total lysate of *Helicobacter pylori* in cases of atrophic body gastritis. Clin. Chem. 2006, 52, 220–226.

99 Haas G, Karaali G, Ebermayer K, *et al.* Immunoproteomics of *Helicobacter pylori* infection and relation to gastric disease. Proteomics 2002, 2, 313–324.

100 Mini R, Bernardini G, Salzano AM, Renzone G, Scaloni A, Figura N, Santucci A. Comparative proteomics and immunoproteomics of *Helicobacter pylori* related to different gastric pathologies. J. Chromatogr., B 2006, 833, 63–79.

101 Lin YF, Wu MS, Chang CC, *et al.* Comparative immunoproteomics of identification and characterization of virulence factors from *Helicobacter pylori* related to gastric cancer. Mol. Cell Proteomics 2006, 5, 1484–1496.

102 Walduck A, Schmitt A, Lucas B, Aebischer T, Meyer TF. Transcription profiling analysis of the mechanisms of vaccine-induced protection against *H. pylori.* FASEB J. 2004, 18, 1955–1957.

103 Malfertheiner P, Schultze V, Rosenkranz B, *et al.*, Safety and immunogenicity of an intramuscular *Helicobacter pylori* vaccine in noninfected volunteers: A phase I study. Gastroenterology 2008, 135, 787–795.

104 Van Regenmortel MHV. The rational design of biological complexity: a deceptive metaphor. Proteomics 2007, 7, 965–975.

Part VI

Statistical Analysis of 2D Gels and Analysis of Mass Spectral Data

17

Data Mining for Predictive Proteomics

Graham Ball[1] and Ali Al-Shahib[2]

[1] *School of Biomedical and Natural Sciences, Nottingham Trent University, Nottingham, UK*
[2] *Health Protection Agency Centre for Infections, London, UK*

17.1 Introduction

The analysis of biological samples using mass spectrometry (MS) based methods poses a number of challenges. One of the most significant challenges in the analysis of data produced by MS is the complexity of the data produced for intact biological samples. This complexity has led commonly to the use of MS based analysis in the later stages of the projects, where following extensive sample clean-up and processing are used to identify and validate biological components. Analysis of this type of data is much simpler as the approach is being used to confirm an earlier proposed hypothesis and a lot is known about a few specific target components.

In more recent years, more groups have undertaken analysis of intact proteomic samples [1,2] using MS based techniques. These characteristically have limited sample preprocessing and are used primarily in the discovery phase rather than the validation phase. Frequently, the sample is cleaned up by chromatographic techniques that eliminate the nonprotein and peptide components of the sample and seek to preserve a large component of the proteome. Thus, by the nature of the sample being profiled by the mass spectrometer, the data generated are complicated and highly dimensional. Commonly a sample is represented by thousands of peaks across a wide mass range, each representing a potential protein component of the biological sample.

These complications will therefore lead to problems in defining and interpreting the spectra which will ultimately result in some degree of ambiguity at each peptide identification. Further problems will arise when comparing the expected MS/MS spectra with the

Mass Spectrometry for Microbial Proteomics   Edited by Haroun N. Shah and Saheer E. Gharbia
© 2010 John Wiley & Sons, Ltd

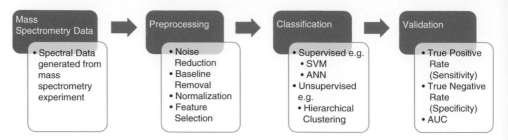

Figure 17.1   *Summary of the proteomics data mining workflow. Some examples of the methods involved in each process are given*

observed spectra using search algorithms which can lead to high false discovery rate (FDR) and low confidence in biomarker discovery. Several attempts to reduce the FDR have been outlined in the literature [3–5]. Many involve statistical scoring systems where a score is set after the theoretical and observed spectra are matched to represent the degree of correlation for each peptide. Others argue that applying a target-decoy (shuffled database) search strategy provides a substantial improvement over the scoring systems. Whatever the options, there is a paramount need to reduce the FDR and increase peptide coverage for accurate biomarker discovery.

An alternative approach to reduce the FDR is to follow the data mining process (Figure 17.1). In the classification stage for example, machine learning algorithms such as Support Vector Machines (SVMs) and Neural Networks have been used to distinguish between correctly and incorrectly identified peptides [4,6]. The classification models are generated by learning on training data represented by a set of features in the hope of discriminating between the correct and incorrect peptides.

In this chapter, we outline the workflow in a typical proteomic data mining process. The chapter will be divided into three sections each representing a stage in the data mining process. Prior to classification, a pre-processing stage is needed to reduce errors and noise. At the classification stage machine learning classification models are trained to classify between correct and incorrect peptides according to certain discriminatory features. This model is assessed and evaluated in the final stage of the process. The data mining workflow in proteomics is shown in Figure 17.1.

17.2   Pre-Processing MS Data

Performing a MS experiment will inevitably produce many data that are affected by redundancy, noise and errors. This is mainly due to sample preparation, the insertion of the sample into the mass spectrometer and the functionality of the mass spectrometer itself. There is therefore a need for a thorough pre-processing stage prior to the classification stage.

17.2.1   Noise Reduction, Baseline Removal and Normalization

There are two different types of noise that result from MS data: chemical and electrical noise. As the name suggests, the chemical noise is caused by chemical contaminants (e.g.

ion over loading, clusters or ionized matrix molecules hitting the detector) in the sample or matrix and the noise usually shows up as a baseline along the spectrum [7]. Several statistical methods have been proposed to remove the baseline [8]. Generally, we compute the local minimum value from the spectrum, draw a baseline as the background noise and subtract the baseline from the spectrum [9].

Another type of noise that may damage the classification results is known as the electrical noise. This is caused during the mass spectrum acquisition by the mass spectrometer, when the instrument randomly changes the intensity. A common way to reduce this noise is to apply wavelet transformation that removes the low value wavelet coefficients [10].

After the de-noising process, the data must be normalized to correct for experimental variations (e.g. degradation in the protein sample) and to provide some degree of comparability within different spectra.

17.2.2   Feature Selection

Once the peaks have been detected in the mass spectrometer and the mass spectra produced, the next step is to identify the peptides by matching the experimental spectra with the theoretical spectra using database search algorithms. Associated with each peptide spectrum match (PSM) is a collection of statistics (or features) that reflects the quality of the match. The statistical scores are used to select the closely matched peptides and the peptides with low scores are discarded.

To generate highly sensitive and specific classification models, we therefore need to select the features that are able to discriminate between correct and incorrect peptides with high accuracy. If this is not considered appropriately, the data can lead to false representation of samples, false detection of biomarkers or determination of biomarkers that actually represent noise in the sample. This curse of dimensionality problem is commonly tackled using various feature selection methods.

One of the common methods used is principal component analysis (PCA). PCA is a multivariate technique used for visualization and dimensionality reduction in MS datasets [11–13]. This technique appears widely as a tool on MS associated instrument software that provides a useful approach for the visualization of structures within a population of MS runs and can be used to identify the main component ions associated with that structure.

PCA transforms the input space into a new space described by what are known as principal components, which are expressed as linear combinations of the original variables, essentially drawing a line through the data that explains most variation (Figure 17.2). These principal components lie orthogonal to one another and are ranked according to an eigenvalue and thus the amount of variation in the data space that they explain. By selecting these vectors based on the eigenvalues, the vectors which explain the variation in the data space are determined. Thus, the ultimate aim of PCA is to capture those vectors, or principal components which explain the most variation in the data, thus reducing the dimensionality of the data space [14]. The main limitation of using PCA for proteomic and gene expression data is the inability to verify the association of a principal component vector with the known experimental variables. This often makes it difficult to identify the importance of the protein ions in a given system as multiple ions can explain a given principal component.

Marengo *et al.* [15] applied PCA analysis to proteomic data generated from neuroblastoma tumour samples and identified two groups of samples in the data set. By analysing the loadings of the principal components, they could identify the discriminatory variables

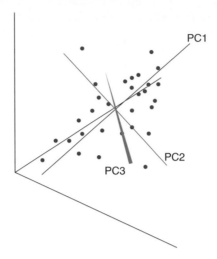

Figure 17.2   *The use of vectors to explain the variation in data using principal component analysis. A three-dimensional example is shown for presentation purposes with three orthogonal principal components. In reality more dimensions are used but cannot be illustrated here*

and by following this up with MS they identified proteins responsible for the differences occurring between healthy and diseased samples.

Another feature selection method proposed in the literature is Support Vector Machines Recursive Feature Elimination (SVMRFE) [16]. Initially SVMRFE was used to select microarray expression genes but has been used recently in MS [4]. The features are selected using a backward elimination search approach in a wrapper feature selection method. SVMRFE uses (or 'wraps' around) the support vector machine algorithm in a recursive procedure to select a subset of discriminatory features for classification. It starts with all the features and eliminates one feature at a time (recursively) until all features are ranked according to a score computed as a result of the SVM classification. If the SVM's separating hyper-plane is orthogonal to a feature's dimensions then this feature is given a high ranking score and vice versa. From the implementation of SVMRFE in the classification of matrix assisted laser desorption/ionization (MALDI)-MS data, Liu *et al.* [9] showed that SVMRFE is more sensitive than other feature selection methods (GLGS) and concluded that SVM outperforms other classification algorithms in classifying MALDI-MS data.

17.3   Classification of MS Data

In general terms, a method is said to be learning if it incorporates information or knowledge from a training sample. In machine learning, this learning is of two types: supervised and unsupervised.

In supervised learning, a 'teacher' instructs the learner to generate a classifier or a model according to the information provided to the learner. This information includes defining the various classes involved in the observed data, as well as providing the examples belonging to these classes. The output (a classification model or classifier) is used to predict the unseen data.

In contrast, there is no 'teacher' in unsupervised learning where no information regarding the classes is provided to the learner. The learning algorithm therefore finds a way of clustering the examples into classes as well as finding descriptions for these classes.

17.3.1 Clustering

Clustering techniques are an example of an unsupervised learning. One very common method is known as hierarchical clustering.

Hierarchical clustering is routinely used as the method of choice when analysing gene expression data [17,18] and functions by arranging the profiles of samples into a tree-like structure so that the most similar profiles lie close together, and profiles very different to one another lie farther apart, allowing for the rapid visual assessment of patterns within the data. The methodology is based on the construction of a distance matrix which enables the two samples with the most similar profiles to be determined. These are then placed together in the tree to form a cluster, and the distance between this newly defined cluster and the remaining samples is calculated. A new cluster is then determined and this process is repeated until all of the samples have been placed in a cluster. There are various linkage methods used for calculating distance, such as single linkage, complete linkage and average linkage. Single linkage computes the distance as the distance between the two nearest points in the clusters being compared. Complete linkage computes the distance between the two farthest points, whilst average linkage averages all distances across all the points in the clusters being compared. Similarly, there are also several distance metrics which can be used to compute this value, such as Pearson correlation and Euclidean distance. Different linkage methods and methods of calculating distances often lead to very different dendrograms, so it is recommended that many methods are applied before drawing conclusions regarding the relationships in the data [19].

The one major problem concerning clustering is that it suffers from the curse of dimensionality when analysing complex datasets. In a high dimensional space, it is likely that for any given pair of points within a cluster, there exist dimensions on which these points are far apart from one another. Therefore distance functions using all input features equally may not be truly effective [20]. Furthermore, clustering methods will often fail to identify coherent clusters due to the presence of many irrelevant and redundant features [21].

Relatively few studies have employed hierarchical clustering to MS analysis. This may be because it does not simplify the complex data into a manageable set of features or identify biomarkers associated with a particular class of interest unless associated with another methodology. One common use of hierarchical clustering is to create phylogenetic trees or identify structures in populations. Seibold *et al.* [22] used hierarchical clustering coupled with PCA to differentiate *Francisella tularensis* at the level of subspecies and individual strains.

Clustering has also been utilized to identify the relationship between ions where samples have been trypically digested prior to MS being run in reflectron mode. Dekker *et al.* [23] utilized clustering to group tryptic ions prior to development of a classifier for lepto-meningeal metastases in patients with breast cancer. Such clustering can also be used to potentially optimize protein identification strategies.

17.3.2 Support Vector Machines

SVMs are a relatively powerful supervised learning algorithm that has been widely implemented in MS. It is a binary classification algorithm that separates the positive and negative examples by constructing a straight line or hyper plane (Figure 17.3) that is rotated in multidimensional space. The advantage SVMs have over other linear separators is that the data can be first projected into a higher dimensional space (by a kernel function, for example polynomial or radial basis functions) before being separated by a linear method, which allows for discrimination of nonlinear regions of space, and therefore separation of nonlinear data. The class of the unknown sample is then determined by the side of the 'maximal marginal hyper plane' on which it lies [24]. This is in contrast to artificial neural networks (ANNs) where a function that separates the classes is defined rather than data transformation to achieve separation. The major disadvantages associated with SVMs are that they are affected by speed and size, both in training and testing, and can be extremely slow when in the test phase [25]. Furthermore from a practical point of view, for large scale tasks extensive memory is required due to the high complexity of the data [26].

One of the most recent and successful implementations of the SVM algorithm in classification of correct and incorrect peptides is by Brosch *et al.* [5]. They propose a method called Percolator that trains SVMs to discriminate between correct (PSMs derived from searching the target database) and decoy (PSMs derived from searching the shuffled protein sequence database) spectrum identifications. This discrimination depends on a set of features provided by the database search algorithm together with an extended feature set calculated by Percolator. They have shown that Percolator, when used as a postprocessor to a collection of target and decoy PSMs produced by SEQUEST and MASCOT database search algorithms, produces more true positive (unique) peptides than other machine learning methods. This clearly indicates the power of SVMs in correctly identifying peptides in MS.

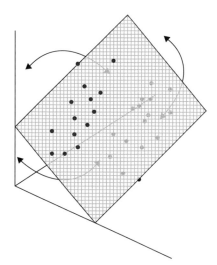

Figure 17.3 *Separation of two classes in multidimensional space by use of a linear hyper plane in the construction of a support vector machine*

17.3.3 Decision Trees

Decision trees are essentially an extension of clustering approaches where rules are applied to the clustering dendrogram to try and separate individuals in the population into meaningful classes. Thus, a decision is made using a rule (logical or numerical) that separates classes (one branch of the cluster dendrogram from another) within the population (Figure 17.4). This approach is used for both classification and regression problems. Through the process of analysis (as in decision trees) multiple if-then logical splits are produced which allow classification or regression rules to be derived.

Adam *et al.* [27] attempted to identify biomarkers for prostate cancer discovery. They used proteins derived from protein biochip surface enhanced laser desorption/ionization (SELDI) MS and decision trees to discriminate prostate cancer proteins and benign prostate hyperplasia/healthy men. They show that peak detection by the SELDI approach coupled with decision trees is a very sensitive method for increase in peptide coverage and eventually biomarker discovery.

An extension of decision trees that has been also used in the classification of MS data is Classification and Regression Trees (CART) [28]. CART allows for the prediction of either continuous dependant variables (regression) or discreet dependant classes (classification). This together with their simplicity of results has been useful in classification problems in cancer data. For example, Wu *et al.* [29] compared CART with other classification approaches for the classification of ovarian cancer from MALDI data. They reported that CART performed moderately well against other methods including Random Forests.

Regression trees may be further enhanced by a classification combination method called boosting. Boosting is a process which combines a number of weak classifiers with a strong

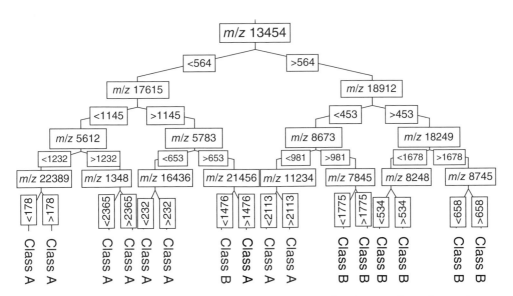

Figure 17.4 *Example of a decision tree in a proteomic study. The tree shows how rules based on the intensity of ions at given m/z values can be used to discriminate between classes*

classifier to attempt to improve overall classifier performance after combination and recalculation of weights. The boosting process essentially adds an adaptive step to the decision tree process. A number of boosting algorithms have been proposed, a range of examples being described by Robert E. Schapire [30].

17.3.4   Random Forest

Another extension of tree based classification methods is the random forest classifier proposed by Breiman [31]. Here, multiple trees are linked by taking the mode of the output class prediction. Breiman showed this approach was very good at making generalized classifications. The approach essentially derives each tree from a random vector with equivalent distribution from within the dataset, essentially an extensive form of cross-validation. As with previous decision tree based methodologies the main disadvantage of random forests is the complexity of the tree produced for high dimensional datasets. This is particularly true where there are large numbers of irrelevant features in the data such as for MS data. This in part can be overcome by the examination of consensus markers across multiple trees.

One example of the use of random forests in MS was reported by Izmirlian [32] who used the approach in the analysis of SELDI data within a cancer prevention trial. He concludes that the Random Forest algorithm is robust to noisy data, does not depend on tuning parameters and is very fast.

17.3.5   Artificial Neural Networks

ANNs are algorithms that determine a solution to a particular problem iteratively fitting a complex function (model) to data. The magnitude of change for a given iteration is proportional to the error of the previous iteration. In this way ANNs, a form of artificial intelligence, 'learn' a solution to a given problem. Learning is achieved by updating the weights that exist between the processing elements that constitute the network topology (Figure 17.5). The most important aspect of the ANN function is that it is built from a potentially large number of interconnected processing elements, also known as nodes, which work together in a network. One of the most common forms of network used is the multilayer perceptron (MLP). The algorithm fits multiple logistic functions to the data to define a given class in an iterative fashion, essentially an extension of logistic regression. This algorithm has been used to solve many types of problems such as pattern recognition and classification, function approximation, and prediction. Once trained, ANNs can be used to predict the class of an unknown sample of interest. Additionally, the variables of the trained ANN model may be extracted to assess their importance in the system of interest. The major disadvantage usually associated with ANNs is that the ability to interpret how they reach an optimal solution is often perceived as difficult, and as such they have been referred to as 'black boxes' [33,34]. Additionally if used to solve complex problems using large numbers of parameters they suffer from the curse of dimensionality. This has been overcome by the use of a stepwise approach to model development [35]. A review of the use of ANNs in a clinical setting is presented by Lisboa and Taktak [36].

BP-MLP ANNs were first proposed for use in the identification of biomarkers from SELDI MS data by Ball *et al.* [1]. This work has been developed further for melanoma where biomarkers have been derived by MALDI MS for late stage disease [37].

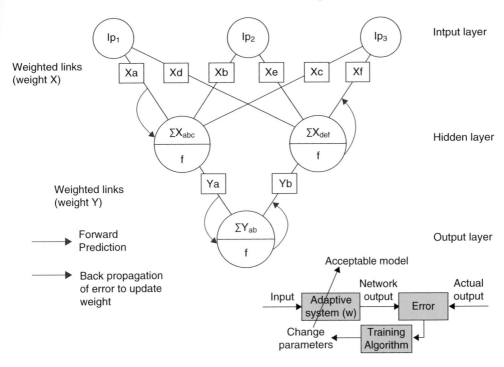

Figure 17.5   *Learning an artificial neural network. The figure shows a three-layer multilayer perception architecture with forward prediction and back propagation of error*

17.4   Evaluation of Classification Models

To simulate the accuracy of the classifier on future proteomics data, we can estimate its discriminating ability by using the following equation:

$$D = \Pr(x_0 > x_1) + \frac{1}{2}\Pr(x_0 = x_1)$$

where x_0 is the test score of a random class member and x_1 the score of a random nonclass member. D is the probability that the classifier is able to make the correct assignment when presented with a random member and a random nonmember of the class. In the case of a binary classifier with scores $\{0,1\}$, D can be simply estimated as:

$$\hat{D} = TPR \times TNR + \frac{1}{2}(TPR \times FPR + TNR \times FNR)\frac{1}{2}TPR + \frac{1}{2}TNR$$

where TPR is the true positive rate $TP/(TP + FN)$ and TNR is the true negative rate $TN/(TN + FP)$, and correspondingly for the false positive rate and false negative rate.

For a classifier which produces continuous scores for each decision, a nonparametric estimate of $\hat{D}$ is provided by the area under the receiver operating characteristic curve (AUC):

$$\hat{D} = \text{AUC} = \int_{-\infty}^{\infty} F_0(x)\,dF_1(x) = \text{Pr}(x_0 \geq x_1)$$

where F_0 is the distribution function of test scores in the class members, and F_1 the distribution for nonclass members.

A receiver operating characteristic (ROC) graph is a powerful technique for visualizing classifier performances that is widely used in evaluating models generated from MS data. It is a two-dimensional graph with its x-axis as the false positive rate and its y-axis is the true positive rate. A perfect classifier will have a true positive rate of 100% and a false positive rate of 0%. However, when multiple classifier performances are plotted in the ROC space and two curves cross each other, it is then hard to determine which classifier is better than the other. However, the AUC is used to measure the discriminability of a set of classifiers. A trivial classifier will have an AUC of less than 0.5, a random classifier will have an AUC score of 0.5 and a perfect classifier will have an AUC score of 1. Various studies in the literature have shown that using AUCs as a classifier evaluation measurement is much more powerful than other measurements [38,39].

Using the discriminating ability D (or the AUC) as a descriptor of classifier performance has the important advantage that it is independent of the class distribution of the test set (although the most accurate estimate will be achieved when the test set is fully balanced).

To conclude, it is worthwhile mentioning the sensitivity at every stage of the data mining protocol for MS data analysis. In the pre-processing stage, normalization of the signal intensities for example is essential for the correctness of any experimental variations that might have occurred. Dimensionality reduction is also vital but there is a risk of neglecting relevant peak information that emphasizes the importance of applying the correct feature selection algorithm. In the next stage, the choice of the learning algorithm and the tuning of the parameters for classification are important for optimal model construction. Furthermore, the risk of overfitting the training data must be addressed at classification, and finally assessing the quality of the model for prediction purposes must be performed in a statistically sound way.

References

1. Ball G, Mian S, Holding F, Allibone RO, Lowe J, Ali S, Li G, McCardle S, Ellis IO, Creaser C, Rees RC. An integrated approach utilizing artificial neural networks and SELDI mass spectrometry for the classification of human tumours and rapid identification of potential biomarkers. *Bioinformatics* 18, 395–404 (2002).
2. Petricoin III EF, Ardekani AM, Hitt BA, Levine PJ, Fusaro VA, Steinberg SM, Mills GB, Simone C, Fishman DA, Kohn EC, Liotta LA. Use of proteomic patterns in serum to identify ovarian cancer. *The Lancet* 359, 572–577 (2002).
3. Huttlin EL, Hegeman AD, Harms AC, Sussman MR. Prediction of error associated with false positive rate determination for peptide identification in large-scale proteomics experiments using a combined reverse and forward peptide sequence database strategy. *J. Proteome Res.* (2008).
4. Duan K, Rajapakse J. SVM-RFE peak selection for cancer classification with mass spectrometry data. *Proc. 3rd Asia-Pacific Bioinformatics Conf.* 191–200 (2004).
5. Brosch M, Yu L, Hubbard T, Choudhary J. Accurate and sensitive peptide identification with Mascot Percolator. *J. Proteome Res.* 8, 3176–3181 (2009).
6. Lancashire LJ, Powe DG, Reis-Filho JS, Rakha E, Lemetre C, Weigelt B, Abdel-Fatah TM, Green AR, Mukta R, Blamey R, Paish EC, Rees RC, Ellis IO, Ball GR. A validated gene

expression profile for detecting clinical outcome in breast cancer using artificial neural networks. *Breast Cancer Res. Treat.* (2009).

7. Barla A, Jurman G, Riccadonna S, Merler S, Chierici M, Furlanello C. Machine learning methods for predictive proteomics. *Brief Bioinform.* 2, 119–128 (2008).

8. Williams B, Cornett S, Dawant B, Crecelius A, Bodenheimer B, Caprioli R. An algorithm for baseline correction of MALDI mass spectra. *Proc. 43rd Annual Southeast Regional Conf.* 1, 137–142 (2005).

9. Liu Q, Sung AH, Qiao M, Chen Z, Yang JY, Yang MQ, Huang X, Deng Y. Comparison of feature selection and classification for MALDI-MS data. *BMC Genomics* 10 (2009).

10. Chen S, Hong D, Shyr Y. Wavelet-based procedures for proteomic mass spectrometry data processing. *Comput. Stat. Data Anal.* 52, 211–220 (2007).

11. Arnold RJ, Karty JA, Ellington AD, Reilly JP. Monitoring the growth of a bacteria culture by MALDI-MS of whole cells. *Anal. Chem.* 71(10), 1990–1996 (1999).

12. de B Harrington P, Vieira NE, Espinoza J, Nien JK, Romero R, Yergey AL. Analysis of variance–principal component analysis: A soft tool for proteomic discovery. *Anal. Chim. Acta* 544(1–2), 118–127 (2005).

13. Polley ACJ, Mulholland F, Pin C, Williams EA, Bradburn DM, Mills SJ, Mathers JC, Johnson IT. Proteomic analysis reveals field-wide changes in protein expression in the morphologically normal mucosa of patients with colorectal neoplasia. *Cancer Res.* 66, 6553–6562 (2006).

14. Haykin S. *Neural Networks: A Comprehensive Foundation.* Prentice-Hall, Upper Saddle River, NJ (1999).

15. Marengo E, Robotti E, Righetti PG, Campostrini N, Pascali J, Ponzoni M, Hamdan M, Astner H. Study of proteomic changes associated with healthy and tumoral murine samples in neuroblastoma by principal component analysis and classification methods. *Clin. Chim. Acta* 345, 55–67 (2004).

16. Guyon I, Weston J, Barnhill S, Vapnik V. Gene selection for cancer classification using support vector machines. *Machine Learning* 389–422 (2002).

17. Alon U, Barkai N, Notterman DA, Gish K, Ybarra S, Mack D, Levine AJ. Broad patterns of gene expression revealed by clustering analysis of tumor and normal colon tissues probed by oligonucleotide arrays. *Proc. Natl Acad. Sci. USA* 96, 6745–6750 (1999).

18. Pomeroy SL, Tamayo P, Gaasenbeek M, Sturla LM, Angelo M, McLaughlin ME, Kim JY, Goumnerova LC, Black PM, Lau C, Allen JC, Zagzag D, Olson JM, Curran T, Wetmore C, Biegel JA, Poggio T, Mukherjee S, Rifkin R, Califano A, Stolovitzky G, Louis DN, Mesirov JP, Lander ES, Golub TR. Prediction of central nervous system embryonal tumour outcome based on gene expression. *Nature* 415, 436–442 (2003).

19. Stekel D. *Microarray. Bioinformatics.* Cambridge University Press, Cambridge (2003).

20. Domeniconi C, Papadopoulos D, Gunopulos D, Ma S. Subspace clustering of high dimensional data. In: *Subspace Clustering of High Dimensional Data*, pp. 517–521 (2004).

21. Greene D, Cunningham P. Producing accurate interpretable clusters from high-dimensional data. In: *Knowledge Discovery in Databases: PKDD*, pp. 486–494 (2005).

22. Seibold E, Bogumil R, Vorderwülbecke S, Al Dahouk S, Buckendahl A, Tomaso H, Splettstoesser W. Optimized application of surface-enhanced laser desorption/ionization time-of-flight MS to differentiate *Francisella tularensis* at the level of subspecies and individual strains. *FEMS Immunol. Med. Microbiol.* 49, 364–373 (2007).

23. Dekker LJ, Boogerd W, Stockhammer G, Dalebout JC, Siccama I, Zheng P, Bonfrer JM, Verschuuren JJ, Jenster G, Verbeek MM, Luider TM, Sillevis Smitt PA. MALDI-TOF mass spectrometry analysis of cerebrospinal fluid tryptic peptide profiles to diagnose leptomeningeal metastases in patients with breast cancer. *Mol. Cell. Proteomics* 4, 1341–1349 (2005).

24. Crisianini N, Shawe-Taylor J. *An Introduction to Support Vector Machines (and Other Kernel-Based Learning Methods).* Cambridge University Press, Cambridge (2000).

25. Burges CJC. A tutorial on support vector machines for pattern recognition. *Data Mining Knowledge Discov.* 2, 121–167 (1998).

26. Osuna E, Girosi F. Reducing run-time complexity in support vector machines. In: Schölkopf B, Burges CJC, Smola AJ (eds), *Advances in Kernel Methods: Support Vector Learning.* The MIT Press, Cambridge, MA, p. 392 (1999).

27. Adam BL, Qu Y, Davis JW, Ward MD, Clements MA, Cazares LH, Semmes OJ, Schellhammer PF, Yasui Y, Feng Z, Wright GLJ. Serum protein fingerprinting coupled with a pattern-matching

algorithm distinguishes prostate cancer from benign prostate hyperplasia and healthy men. *Cancer Res.* 214–219 (2002).

28. Breiman L, Friedman JH, Stone C, Olshen RA. *Classification and Regression Trees.* Chapman & Hall, Monterey, CA (1984).
29. Wu B, Abbott T, Fishman D, McMurray W, Mor G, Stone K, Ward D, Williams K, Zhao H. Comparison of statistical methods for classification of ovarian cancer using mass spectrometry data. *Bioinformatics* 19, 1636–1643 (2003).
30. Schapire RE. The boosting approach to machine learning: An overview. In: Denison DD, Hansen MH, Holmes C, Mallick B, Yu B. (eds), *Nonlinear Estimation and Classification.* Springer (2003).
31. Breiman L. Random Forests. *Machine Learning* 45, 5–32 (2001).
32. Izmirlian G. Application of the random forest classification algorithm to a SELDI-TOF proteomics study in the setting of a cancer prevention trial. *Ann. NY Acad. Sci.* 1020, 154–174 (2004).
33. Duh MS, Walker AM, Ayanian JZ. Epidemiologic interpretation of artificial neural networks. *Am. J. Epidemiol.* 147, 1112–1122 (1998).
34. Spining MT, Darsey JA, Sumpter BG, Noid DW. Opening up the black box of artifical neural networks. *J. Chem. Educ.* 71, 406–411 (1994).
35. Lancashire LJ, Powe DG, Reis-Filho JS, Rakha E, Lemetre C, Weigelt B, Abdel-Fatah TM, Green AR, Mukta R, Blamey R, Paish EC, Rees RC, Ellis IO, Ball GR. A validated gene expression profile for detecting clinical outcome in breast cancer using artificial neural networks. *Breast Cancer Res. Treat.* (2009).
36. Lisboa PJ, Taktak AF. The use of artificial neural networks in decision support in cancer: A systematic review. *Neural Networks* 19, 408–415 (2006).
37. Matharoo-Ball B, Ratcliffe L, Lancashire L, Ugurel S, Miles AK, Weston DJ, Rees R, Schadendorf D, Ball G, Creaser CS. Diagnostic biomarkers differentiating metastatic melanoma patients from healthy controls identified by an integrated MALDI-TOF mass spectrometry/ bioinformatic approach. *Proteomics Clin, Applicat.* 1, 605–620 (2007).
38. Fawcett T. ROC graphs: notes and practical considerations for data mining researchers. *Technical Report HPL-2003-4* (2004).
39. Ling CX, Huang J, Zhang H. AUC: a statistically consistent and more discriminating measure than accuracy. *Proc. Int. Joint Conf. Artificial Int.* (2003).

18

Mass Spectrometry for Microbial Proteomics: Issues in Data Analysis with Electrophoretic or Mass Spectrometric Expression Proteomic Data

Natasha A. Karp

Wellcome Trust Sanger Institute, Wellcome Trust Genome Campus, Hinxton, Cambridge, UK

18.1   Introduction

Expression proteomics is the comparison of distinct proteomes (e.g. control versus treatment or control versus disease) to identify protein species with changes in expression or post-translational state. There are now numerous methods to interrogate the proteome in a quantitative manner. Simplistically, these fall into two major categories; methods based around two-dimensional gel electrophoresis (2DE) with post-staining [1, 2] or pre-labelling [3, 4], or methodologies where quantification occurs using mass spectrometric measurement, for example stable isotope labelling, both *in vitro* [5–7] or *in vivo* [8], and label-free approaches [9, 10]. There are some technology specific issues, for example the normalization methods needed to allow comparison between collection runs, which are beyond the scope of this chapter. All these technologies have in common the generation of datasets which are long and lean with many variables (proteins) being quantified simultaneously with few samples that lead to some significant mathematical issues which need consideration in the data analysis applied. Furthermore, the design of the experiment also influences the type of analysis that can be applied, the type of questions that can be answered and the robustness of the results obtained. Consequently, it is important to plan

Mass Spectrometry for Microbial Proteomics   Edited by Haroun N. Shah and Saheer E. Gharbia
© 2010 John Wiley & Sons, Ltd

the entire process from defining the experimental objective, to designing the experiment, and finally planning the data analysis. This will ensure the experiment will truly answer the researcher's question with robust conclusions.

Data analysis can be divided into univariate or multivariate in approach, depending on whether variables are considered in isolation or simultaneously. Classical univariate statistical tools, such as the Student's *t*-test and the analysis of variance (ANOVA) test, treat each individual variable as being independent, and thus, cannot easily capture information about correlated trends. Multivariate statistics are methods that consider a group of variables together rather than focusing on only one variable at a time. These include approaches such as principal component analysis (PCA), partial least squares (PLS), hierarchical cluster analysis (HCA), linear discriminate analysis (LDA), self-organizing maps (SOMs) and multivariate analysis of variance (MANOVA). Univariate data analysis is the first port of call for the majority of researchers and hence this chapter will consider the design and data analysis issue with direct reference to univariate techniques. It should be noted, that in general design issues that improve quality of data for univariate analysis will also improve analysis at the multivariate level. The univariate method, however, is the simplest to interpret conceptually and most commonly used.

This chapter will focus on some of the issues surrounding the design of expression proteomic experiments. It will also give insight into some of the statistical tests commonly employed and discuss their appropriateness for use with different types of quantitative proteomic datasets. As awareness of these issues in the community has grown, to ensure robust results are reported, minimum expectations for good practice on experimental design and data analysis have been published by many journals issuing guidelines for the submission of supporting proteomics data [11, 12].

18.2   Experimental Design

Good experimental design is essential to extract meaningful information from an experiment. From randomization to replicate type, various aspects of experimental design relevant to expression proteomic studies are discussed in more detail.

18.2.1   How Many Measurements/Observations are Needed?

Power can be defined as the ability of a univariate statistical test to detect change and depends on the variance (noise), the effect size (the change in expression), the number of replicates, the nominal significance the researcher sets (typically 0.05 which is related to confidence desired e.g. 95%) and the statistical test employed. In designing an exploratory experiment, a target power of 0.8 is optimal, meaning if the experiment was repeated five times then on average that change in expression would be detected four times. Whilst for confirmation experiments, where the objective is to test for a planned hypothesis, a higher target power (e.g. 0.95) is optimal.

For the majority of studies, the researcher only has control over the number of observations taken. Whilst increasing the number of measurements is essential to distinguish between true difference in expression and random fluctuations and hence improves power, increasing the number of replicates beyond a certain point has little impact on the power

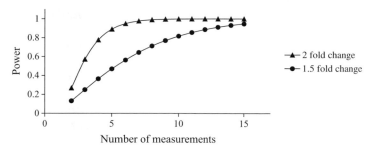

Figure 18.1 *Example power curves showing how the number of measurement influences the resulting power for two typical fold-changes that interest researchers. Looking at a graphical display of the power calculation, the number of replicates needed to achieve a target power of 0.8 can be estimated. For a 2-fold change, five replicates are needed whilst for a 1.5-fold change 10 replicates are needed. The effect size selected by the researcher will depend on the objectives of the study and the anticipated size of change expected from the treatment*

(Figure 18.1). An under-sized study will be a waste of resources, as it will not have the capacity to detect scientifically important changes as statistically significant, whilst an over-sized study will use more resources than are necessary.

Noise or variation can be defined as fluctuation in data measurements that obscures or reduces the clarity of a signal and consequently reduces the sensitivity of an experiment.

By considering the various sources of variation, the researcher can maximize the sensitivity of the study for the research in question by reducing this variation where possible; for example, using fluorescence difference gel electrophoresis (DIGE) instead of post-stained 2DE [13] or using cultured samples rather than collected samples. Attention can be focused on the reduction of technical variation as this can be a significant source of variation. In two post-stained 2DE studies looking at depleted human plasma and established cell lines it was found that technical variation contributed up to 50% of the total noise [14, 15]. Understanding the sources of technical variation can lead to improvements in experimental design. In metabolic labelling studies, samples to be compared are combined very early on in the experimental schema where a standard sample is labelled with ^{15}N-enriched media and is added in equal amounts to the sample of interest labelled with ^{14}N-enriched media at the stage of the sample preparation to ensure that any differences reflect quantitation differences [16]. In a label-free mass spectrometry (MS) study, where sample combining only occurs *in silico*, most of the technical variability could be attributed to the liquid chromatography(LC)-MS injection [17]. The importance of careful standardization in sample preparation and quantitative methodology is essential to reduce technical variation [18–21]. For example, a study looking for early post mortem changes in bovine muscles found metabolic and heat shock proteins with significant changes [22].

Exactly how many measurements are needed are thus a function of the sample type, sample collection, the technology used, the size of change expected from the treatment and consequently there is not a simple 'one-fits-all' answer. This was recently demonstrated in a publication which completed two 2DE experiments with the same design of four biological replicates per group [23]. One of the experiments studying a *DsbA*

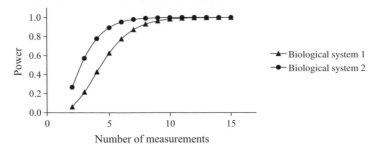

Figure 18.2   *Example power curves showing how the variance influences the power for various numbers of measurements in detecting a 1.5-fold change with a 95% confidence. Biological system 1 is the soluble proteome from mouse liver whilst biological system 2 is an Erwinia soluble proteome extract from wildtype cells. The variance measure was one that encompassed 75% of the spots in a DIGE experiment [23]*

knockout with cultured *Erwinia* cells identified 139 protein species with a significant change in expression whilst the second experiment working with *Per2* knockout in mice liver failed to identify a single protein species with any changes. The knockout of *Per2* as a key negative regulator should have significant impact on the proteome. The failure to detect any changes within this experiment compared with the *Erwinia* experiment was due to low power arising from greater biological variance and smaller and fewer changes being triggered by the *Per2* knockout (Figure 18.2).

To estimate the power for a given experimental design, a measure of variance is needed for the proposed sample type with the quantitation technology intended to be used. This measure can either come from a preliminary or similar study and a number of recent studies have measured the variance for various proteomic techniques and sample type [14, 15, 24, 25]. Power can be calculated for each individual species or an experiment as a whole. Approaches for a whole experiment can include using an average noise value or a value that encompasses 75% of the protein species. Various power analysis software programs exist, both commercial (e.g. Power and Precision by Biostat, USA) and free (e.g. Lenth power tool [26]).

18.2.2   What Type of Repeat Measurement Should be Used?

It is important to consider the type of repeat measurement, as the type of replicates used affects the type of statistical analysis that can be carried out and the conclusions that can be drawn. Repeat measurement, replicates, can be divided into two main types and are related to the variation source they encompass. Technical replicates, also called repeat measures, are repeat measures from the same biological sample and reduce the uncertainty around the true reading for a given sample. Biological replicates are different samples from the same treatment group. Biological variability is intrinsic to all organisms and can arise from genetic or environmental factors. When only technical replicates are used all that can be concluded for species identified as significant is that the individuals sampled are different from one another and that difference is larger than the technical variance. Biological replicates also encompass the technical noise in a given system. Without biological replicates, no inferences can be made about the differences within populations,

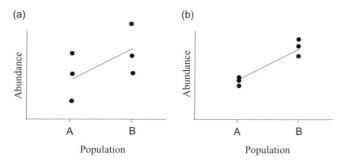

Figure 18.3 *Effect of replicate type and demonstration that for the same size fold change technical replicates (b) apparently increase power due to the lower variance compared with biological replicates (a)*

unless the unrealistic assumption is made that there is no biological variability. The random sampling of biological replicates from populations subjected to different treatments allows inferences to be made about the effect of the treatments relative to the biological noise of the system. Biological variation arises both from natural genetic variation and environment differences. In a study looking at the natural variability in *Arabidopsis* with post-stained 2DE it was found that 95% of proteins varied in spot quantity (2- to 53-fold) and thus biological variability can be substantial [27].

It often is tempting to only include technical replicates within an experimental design, but the apparent increase in the power of the study is only illusionary and significant findings may simply reflect change fluctuation between the particular sample chosen for the study [28] (Figure 18.3). For 2DE, whilst technical replicates reduce measurement error for a sample, as a function of cost and resources the general consensus is that biological replicates provide considerably more scientific information than analytical replicates [29]. It can, however, be argued that in situations of high technical noise there is a role for technical replicates [30, 31]. This situation could arise with alternate quantitative technology or in a situation where pre-fraction is used to prepare a focused sample.

For microbiology studies, where interest lies not on the individual but rather on characteristics of the population from which the individuals are obtained an alternative approach to biological replicates is to use pooled samples. This may be necessary when insufficient material is obtained from an individual sample preparation, or when the number of possible replicates is limited (e.g. cost limitations) and the biological variance is high compared with technical variation. Pooling from randomly selected biological replicates reduces the biological variance by forming an average sample. If pooling is required, it is better to use several small pools such that the variance among pooled samples within treatments can be estimated and outliers can be identified [32]. The pooling process relies on an assumption of biological averaging, where protein abundances average out when pooled. This may not be true as an average on the scale of raw signal will not necessarily correspond to an average after the data processing occurs. This assumption has been tested for DIGE technology and found no evidence for systematic bias being introduced with sampling pooling [33]. Furthermore the authors have investigated with power analysis when sampling pooling would beneficial. Further research is needed in the use of pooling in proteomic studies with alternate technologies.

If multiple replicate types, i.e. both technical and biological, are to be used within an experiment, this influences the data analysis. The basic statistical tests, such as the Student's *t*-test, assume independence which will not be met when replicate types are mixed as technical replicates from the same biological sample will be more similar than technical replicates from an independent biological sample. Instead, a more complex test which considers the hierarchy in the data is needed. In a DIGE study, a nested ANOVA was used to account for the hierarchical structure arising from mixing biological and technical replicates [30]. It was shown that the use of a Student's *t*-test led to an inflation of false positives as it overestimated the significance of the difference between groups highlighting the risks if an inappropriate statistical test is applied for the experimental design employed.

18.2.3   Sampling Depth

Fundamentally samples are collected from a population to make inferences about those populations. The population chosen will limit the inferences that can be made, as it is important to avoid overgeneralization and applying the conclusions to a larger population than that studied. Depending on the objective of the study, careful consideration is needed to ensure the right subjects are selected in the process. Frequent microbial proteomic experiment will consider the influence of a treatment by repeat measures on one cultural strain [34]. As a consequence, the treatment effects can only be concluded to be arising for that strain of that species. A 2DE comparative analysis of two *Salmonella enterica* subspecies 1 serovars revealed a high degree of variation among isolates obtained from different sources [35, 36]. Furthermore, the authors found that 'in some cases, the variation was greater between isolates of the same serovar than between isolates with different sero-specificity' and they also found several serovar-specific proteins [35]. The implications of this are that the conclusions are limited when only one cultural strain is used and instead where possible a large number of strains of a species should be used to account for the intrinsic diversity of a bacterial population.

A further example of sampling depth consideration is in the emerging field of metaproteomics. Metaproteomics has been defined as the 'large-scale characterization of the entire protein complement of environmental microbiota at a given point in time' [37]. As microbial derived commodities or processes can often be a result of mixed populations or organisms the composition of which can be varied rather than a single axenic culture, the role of metaproteomics will grow. In an example metaproteomics study, Wilmes and Bond examined protein expression in a mixed microbial community from two activated sludges from a laboratory-scale sequencing batch reactor with dissimilar phosphorus removal performances with 2DE and compared the anaerobic and aerobic phases of the sludge [38]. For the sludge with good enhanced biological phosphorus removal performance (EBPR), dominated by Betaproteobacteria, 9.4% of all protein spots were statistically different between the two phases. For the non-EBPR sludge, dominated by tetrad-forming Alphaproteobacteria, 14.7% of protein species exhibited statistical differences between the two phases. The comparison of proteome in the two sludges found that their metaproteomes were substantially different and this was reflected in their microbial community structures and metabolic transformations.

When thinking about sampling depth, this is not only in regards to the populations studied but also to the levels of the treatment applied; as conclusions can only be made regarding those treatment tested. For example, a study looking at the effect of pH on growth can only make statements relative to the pH range tested. In particular, time courses benefit from greater sampling depth as this will allow better modelling of behaviour with time.

18.2.4   Randomization within an Experimental Design

With the use of hypothesis testing, the central idea is to design the experiment such that only one factor that is relevant to the outcome varies. This makes it possible to observe the effect of that factor on the outcome i.e. keep all variables equal between groups except for that which is being examined. Two methods are used to ensure these assumptions are met; 'controlling' and 'randomizing'. In controlling, the environment for the study is made as uniform and homogeneous as possible. For example, each culture flask is the same size and held at the same temperature. Randomization reduces bias by equalizing other factors that have not been explicitly accounted for in the experimental design. Randomization is a core principle in experimental design and it is essential throughout the design of an experiment to avoid hidden bias leading to systematic errors in the data. Examples of randomization include the random assignment of culture flask positions in a shaker to a treatment group, and the random order of which samples are run in time. Randomization can practically be achieved by assigning a number and then using a random number generator or with a raffle system. The ease at which systematic bias can arise has been demonstrated and examples have arisen in 2DE from dye effects, gel casters, data analysis, and immobilized pH gradient (IPG) stripes from different packets where samples have not been fully randomized across these groups [39, 40]. In summary, to address bias all variables must be kept equal between groups at every stage and this includes design, conduct and interpretation and then reporting these steps in an explicit and transparent way.

18.3   Data Analysis

The following focuses on the data analysis issues for expression proteomics. Technology specific issues, such as the use of normalization approaches to address systematic bias or methods to handle missing values are not included as this is beyond the scope of this chapter.

18.3.1   Understanding the Statistical Tests: Is It Statistically Significant?

18.3.1.1   Pair-Wise Comparison with Fold-Change Threshold

One approach in expression studies is to compare one sample with another by the calculation of a ratio. Identical samples with no experimental variation would lead to a ratio of one. With different samples, any ratios above a threshold of significance are selected as significant changes in expression [6, 8, 41]. The significant threshold is determined by assessing the system experimental noise by running an experiment initially with the same sample to assess the natural occurring variation (Figure 18.4). This method of analysis limits the sensitivity of the study, as biologically relevant changes smaller than the threshold cannot be detected. For example, with DIGE in a pair-wise comparison of *Erwinia* soluble protein extract the sensitivity is limited to changes above a 2-fold threshold as this threshold encompassed approximately 90% of the experimental variability [42].

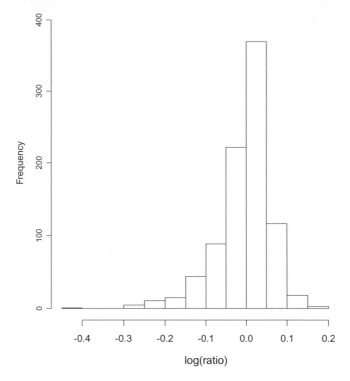

Figure 18.4   *A same-same frequency histogram of the log(ratio) from a pair-wise comparison of wildtype* Erwinia *on a DIGE gel. Even with identical samples, due to technical variation, the log ratios has a spread in values centred around zero (equivalent to the ratio equalling one). For a pair-wise comparison a threshold has to be chosen that encompasses the natural variation such that the difference being seen indicates a biological difference between the samples rather than a technical effect*

Furthermore, any changes in expression identified are changes in the two samples compared and the assumption is made that these samples are representative of the populations they represent. A methodology that does not consider the variability of the protein will run the risk of selecting variable proteins due to sample selection (Figure 18.5). To address these issues, expression proteomic experiments have been moving towards multi-sample studies. Use of a fold-change approach has a potential role in preliminary experiments but the limitation of this method must always be considered when interpreting the data.

18.3.1.2   *Hypothesis Tests*

Hypothesis tests, e.g. Student's *t*-test, assess whether the differences between groups is an effect of chance arising from a sampling effect or reflects a real statistically significant difference between the groups. Hypothesis tests are usually stated in terms of both a condition that is doubted, which is called the null hypothesis (H_0) and a condition that is believed, which is called the alternative hypothesis (H_1). When we are comparing two groups the null hypothesis will state that no differences exist between the group means whilst the alternative is that the group means are different. The aim of the test is to deter-

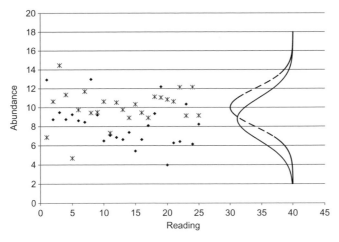

Figure 18.5 *Is the sample representative? The graph display two populations which at a population level have a difference in mean. The two populations are shown with different symbols showing the measured abundance (y-axis) for each data point drawn at random from the population (x-axis). Shown on the right are the two frequency distribution profiles indicating how the populations have a difference in mean. The figure highlights how different data points (samples) drawn from a population can give very different results depending on where they are drawn from their population*

mine whether there is enough evidence to reject the null hypothesis and accept the alternative hypothesis. However, to accept a null hypothesis does not mean that it is true, only that there is no evidence to reject the hypothesis. The tests calculate a *p*-value, which is the probability of obtaining these results assuming the null hypothesis is correct. A significance level is set, typically 0.05 or 0.01, and defines the sensitivity of the test. A significance level of 0.05 means that the test incorrectly rejects the null hypothesis 5% of the time when it is in fact true, i.e. the differences seen are just sampling effects but are classed as statistically significant. This is also called a type I error. The significance level chosen is somewhat arbitrary, although in practice values of 0.05 and 0.01 are commonly used depending on the confidence the researcher requires.

18.3.2 Understanding the Statistical Tests: The Hypothesis Test Outcomes

In expression proteomics, the null hypothesis is typically that the mean of the groups is the same i.e. the treatment has no effect on expression. The calculated probability (*p*) is thus the chance of observing a test statistic more extreme than the one calculated from the data given that the samples are from the same population (i.e. that any apparent change in expression occurs by chance alone). Typically an expression change is considered significant if the calculated *p*-value falls below a prescribed significance threshold, typically 0.01 (the per comparison error rate, PCER). Two types of errors are possible (Table 18.1). A false positive (type I error, α) occurs when a protein species is declared to be differentially expressed erroneously. A false negative (type II error, β) occurs when the test fails to detect a differentially expressed species. Power is the probability that the test

Table 18.1   *Possible outcomes for a univariate statistical test*

		Biological event	
		No change in expression	Change in expression
Test outcome	Statistically significant (*p*-value < threshold)	False positive Type I error	Correct acceptance
	Statistically not significant (*p*-value > threshold)	Correct rejection	False negative Type II error

will reject a false null hypothesis, that it will not make a type II error, and thus is equal to $1-\beta$.

18.3.3   Understanding the Statistical Tests: Is It Biologically Significant?

The test of significance is asking whether the difference between samples is big enough to signify a real difference between two populations. It is important to highlight that 'statistical significance' does not necessarily imply interesting or biologically relevant. If sufficient readings are taken or the variance is very low then a very small change in mean can be considered statistically significant. Consequently, even when you reject a null hypothesis, effect sizes should be taken into consideration. 'Effect size' is simply a way of measuring the size of the difference between two groups, i.e. the observed effect of the treatment. Typically in proteomic experiments, researchers have been considering effect size by looking at the average ratio change between the two groups being compared and applying both a fold-change threshold and a *p*-value threshold in selecting significant species [43–45].

An alternative approach to fold change is to calculate an effect size measure that considers the variation in the protein, for example Cohen's *d* effect size. Mathematically it is a standardized mean difference and is calculated as the difference between two means divided by the pooled standard deviation for those means. The effect size can be classified as small, medium or large using standard thresholds. Particularly in ranking species for decisions for downstream analysis, methods that consider the variation are useful as protein species with a large change relative to the variation are more significant than species with large changes within a noisy background. An effect size method that considered variation was used in an iTRAQ study on a mouse model of cerebellar dysfunction [45]. Proteins were classified as significantly changed when the *p*-value was less than 0.05 and an absolute expression ratio fell at least one standard deviation beyond the population mean.

18.3.4   Understanding the Statistical Tests: The Multiple Testing Problem

In expression studies many thousands of statistical tests are conducted, one for each protein species. A substantial number of false positives may accumulate at the 0.01 confidence level as 1% of sample differences will be significant even if no changes in protein expression exist between the two populations being tested. This accumulation of false positives is termed the multiple testing problem and is a general property of a confidence based statistical test when applied multiple times. Consider a typical 2DE experiment with 1000 spots matched across an eight-gel series; at a simplistic level, in a comparison of two groups across these 1000 spots, 10 false positives could arise with the significance threshold currently used in the field ($p < 0.01$). Frequently, a PCER of 0.01 was applied and choosing this low threshold was an attempt to manage the multiple testing problem [23]. However, the PCER was developed for a single test and the situation becomes quite complex when applied to multiple tests simultaneously. The complexity arises as some species are changing in expression, different species have different effect size and not all species are independent (e.g. charge trains or components of a common pathway). When the PCER method is applied the error rate control is lost.

One approach to address the multiple testing problem is to control the family wise error rate (FWER), which controls the probability of one or more false rejections (type I error) among all tests conducted. The simplest and most conservative FWER approach is the Bonferroni correction, which adjusts the threshold of significance by dividing the PCER by the number of comparisons being completed [46]. The use of these conservative methods to address the multiple testing problem, results in the expression studies having very little power. This can result in no species being detected as having significant changes in expression [23]. However, confidence is high for those changes detected as significant when using a FWER approach. It can also be argued that in the context of exploratory experiments, where later confirmatory investigations are utilized, allowing a few false leads would not present a serious problem if the majority of significant species were correctly chosen [47–49]. This has led to the application of methodologies to control the false discovery rate (FDR) where the focus is on achieving an acceptable ratio of true and false positives. Benjamini and Hochberg originally defined the FDR as a proportion of changes identified as significant that are false [50]. For example, a FDR rate of 5% means that on average 5% of changes identified as significant would be expected to have arisen from type I errors. This method was used in a study looking for early cancer biomarkers and the authors estimated that 20% of their 26 significant spots would be false calls [29].

An extension to the FDR was developed by Storey and calculates a q-value for each tested feature [51, 52] (Figure 18.6). The q-value is the expected proportion of false positives incurred when making a call that this feature and all other more significant species have a significant change in expression (Figure 18.6). It is important to note that this methodology estimates the error rate; it does not however tell you which species are the errors. The idea here is that in the downstream analysis the error rate forms a caveat to the analysis, where the researcher always has to bear in mind the risk that this protein could be the 'red herring'.

The risks of using a PCER versus a q-value was demonstrated in a DIGE study on the effect of *per2* knockout on murine liver and highlighted that using a PCER threshold of 0.01 would have led to six protein species being identified as significant; however, the q-value estimated that 80% of these would have been false positives [23]. In a second study, looking at the effect of a mutation on the soluble protein extract from *Erwinia*, the q-value allowed the researchers to select a FDR acceptable to their study [23]. With a 1% FDR, 58 protein species were classed as significant with an estimated one false positive. Alternatively, with a 5% FDR, 139 species were significant with an estimated eight false positives. These numbers demonstrate how allowing a few false positives in the species identified as statistically significant considerably increased the power of the experiment to detect change.

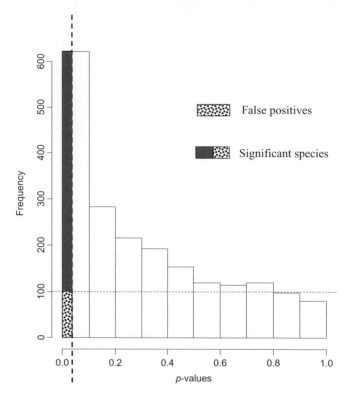

Figure 18.6   *A typical p-value frequency distribution that is obtained in a quantitative experiment when 23% of the species tested are changing. The profile arises from a mixture of species with no change in expression contributing to a uniform frequency distribution whilst species that change in expression will tend to have a low p-value. To calculate the q-value, the process estimates the background distribution from the unchanging distribution using the flat portion of the frequency histogram assuming most p-values near one will be background events (horizontal dashed line). The q-value is calculated for each species, by using that species p-value as a significance threshold (vertical dashed line) and estimating the proportion of false positives within the total number of species called significant*

18.3.5   Understanding the Statistical Tests: The Assumptions

Underlying statistical tests are assumptions about the data. For example a Student's *t*-test assumes normality, independent sampling, and homogeneity of variance (Figure 18.7). If these assumptions are not met then the tests will not behave as anticipated and the false positive and false negative rates will not be controlled as anticipated and the results will not be robust. Whether the assumptions are met will be a function of the experimental design and the technology used.

For example, consider the use of DIGE to compare two sample types with a Student's *t*-test. Here samples are labelled with spectrally resolvable fluorescent CyDyes (Cy2, Cy3, and Cy5; GE Healthcare) prior to electrophoresis. For each gel in a multi-gel study three labelled samples are mixed, one of which is the internal standard sample. This internal

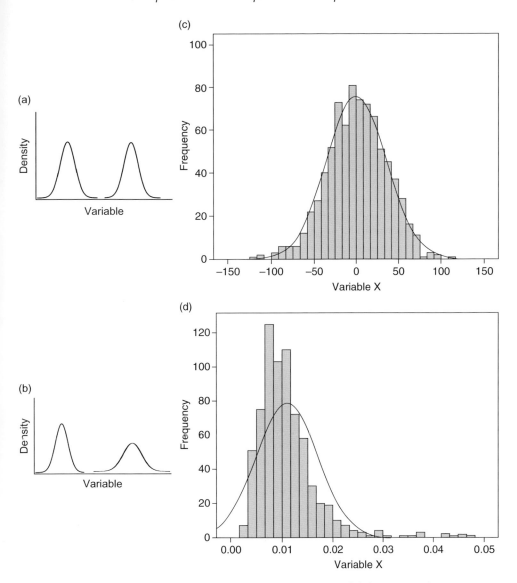

Figure 18.7 *Graphs showing data behaviour which meets and fails to meet the assumptions of normality and homogeneity of variance (homoscedasticity). (a) Data exhibiting homoscedasticity, even though the mean has changed the variance is the same. (b) Data exhibiting heteroscedasticity, the complement to homoscedasticity. Here the mean has changed and the variance. (c) Frequency distribution data exhibiting a normal distribution profile as the data have a characteristic frequency distribution (black line is the normal probability curve). (d) Frequency distribution data exhibiting a non-normal distribution profile as the data show significant skew (black line is the normal probability curve)*

standard sample allows both inter- and intra-gel matching, and is used in the standardization of spot volumes in different gels. This methodology greatly reduces variation in spot intensities due to gel-specific experimental factors, including protein loss during sample entry into the IPG strip, as the effects will be the same for each sample within a single DIGE gel. Studies have found that a log transformation of the standardized spot volume is essential for the assumption of normality and homogeneity of variance to be met [24]. As independent biological replicates were being used for each sample, it was initially thought that the assumption of independent sampling was met allowing the use of the Student's *t*-test [24]. Later studies found that with the three-dye experimental design sampling was not independent because the three-dye system had a correlated error structure from the use of a common internal standard for two samples in a gel [23]. To address this, a more complex statistical test or a two-dye methodology which avoids this correlation could be used. This example highlights how both the technology and experimental design need consideration to ensure the assumptions are met and as a consequence the appropriate statistical test is applied.

Based on distribution assumptions, hypothesis tests can be divided into two groups: parametric and nonparametric. Parametric tests assume that the distribution of the variables being assessed belong to known probability distributions. For example, the Student's *t*-test assumes the variable comes from a normal distribution. Nonparametric tests do not rely on a known underlying distribution as they do not involve estimation of parameters (such as the mean). An example of a nonparametric test is the Mann–Whitney test, which ranks all the values from low to high and then compares the mean rank in the two groups. It is important to note that nonparametric tests are not assumption free, for example the Mann–Whitney test relies on the groups only differing in parameter location and the distributions have a similar shape. The price for the robustness of the nonparametric test is less power. The majority of published proteomic studies utilize a parametric test (author observation). In a discussion of the pitfalls of proteomic experiments, Biron *et al.* recommended assessing the normality for each protein species and then either selecting a parametric or nonparametric test [53]. This approach was used in a direct injection LC-MS quantitation experiment looking for biomarkers in a study comparing healthy subjects with rheumatoid arthritis patients [25].

Whether the assumptions will be met depends on the design, the technology and the data analysis being utilized. Consequently, consideration of the assumptions underlying the data analysis utilized needs to be an integral stage of the experimental planning.

18.4   Validation

Many researchers have called for downstream validation to confirm the findings from expression studies. For example, Mehta *et al.* argue the need for constructive replication with orthogonal validation, where a method most different from those that resulted in the original finding is used to look for the difference [54]. This approach was used in a study where western blots were used to validate the proteins identified as induced in *Salmonella enterica* serovar Typhimurium with LC-MS based quantitation during infection of macrophages [9]. It has been highlighted that unless systematic bias exists in the system and correlates with sample group, validation by studying the same samples is a flawed practice as measurement error does not lead to false positives [55, 56]. Instead, attention should focus on validation with new samples. This would validate against sampling variability leading to a difference when no population variability exists.

The issue of power arises in thinking about validation. The probability of two events occurring sequentially is the probabilities of the individual events multiplied. Consequently, if the probability of observing an event is 0.8 the probability of seeing it twice in succession is 0.64. Hence when a change is not observed the second time, unless the power is high the researcher cannot distinguish whether the difference is a false negative issue in the second experiment or a false positive in the first experiment.

In general, validation has focused on false positives ensuring the downstream leads are appropriate. However, it is also possible to consider false negatives, those whose test statistics were almost significant. In these cases, the concern is that measurement error is leading to a missed lead. For these, an additional study could be completed by a more precise technology and/or larger sample size.

The call for validation has occurred; however as yet little debate has occurred as to the criteria for a researcher accepting a validated finding. For example, do you need a similar fold change or what *p*-value score threshold should be used? Duncan and Hunsucker argue against the need for a 'full picture' on the first pass publication instead they argue for the need for impartial review of the experiment highlighting the strengths and weaknesses of a study [57]. They pointed out that the laboratories that discover are often not set-up for high throughput validation and a delay for validation prevents the dissemination of the information.

18.5   Conclusions

The advancements made in expression proteomics technology combined with the advancement made in tackling the mathematical issues arising from these large datasets has led to powerful techniques that can give robust answers in elucidating biological questions. By researchers understanding the mathematical issues that arise from the use of these technologies and the necessity to embrace experimental design and data analysis issues, then experiments can be designed that deliver. In advance of practical work, time needs to be spent thinking through the whole procedure from defining the objectives, selecting the samples, addressing the technology and data/sample collection issues and finally how to analyse the data. Initially this will increase the work load and potentially the cost of these expression studies but it will improve the understanding of the biological mechanism in question and ensure resources are used effectively.

With the use of hypothesis-driven statistical tests, researchers have to become aware that all hypothesis tests have unavoidable but quantifiable risks of making the wrong conclusion which is essential in interpreting the results of these studies. This is a particular issue when working with these long and lean datasets with many variables but few observations and the resulting problem of multiple testing. For exploratory experiments, a significant advance has come with the concept of allowing a proportion of false discoveries as this approach increases the power of these experiments whilst controlling the error rate. The knowledge will provide a caveat in the later interpretation and downstream analysis of protein species that are being identified as significant.

Improving the experimental design, the subsequent quality of data and the understanding of the results will significantly improve the value that expression proteomic experiments bring to biological studies.

References

[1] Fievet J, Dillmann C, Lagniel G, Davanture M, Negroni L, Labarre J, de Vienne D. Assessing factors for reliable quantitative proteomics based on two-dimensional gel electrophoresis. Proteomics 2004; 4(7): 1939–49.

[2] Smejkal GB, Robinson MH, Lazarev A. Comparison of fluorescent stains: relative photostability and differential staining of proteins in two-dimensional gels. Electrophoresis 2004; 25(15): 2511–9.

[3] Coulthurst SJ, Lilley LK, Salmond GPC. Genetic and proteomic analysis of the role of luxS in the enteric phytopathogen, *Erwinia carotovora*. Mol. Plant Pathol. 2006; 7(1): 31–45.

[4] Vipond C, Suker J, Jones C, Tang C, Feavers IM, Wheeler JX. Proteomic analysis of a meningococcal outer membrane vesicle vaccine prepared from the group B strain NZ98/254. Proteomics 2006; 6(11): 3400–13.

[5] Gygi SP, Rist B, Gerber SA, Turecek F, Gelb MH, Aebersold R. Quantitative analysis of complex protein mixtures using isotope-coded affinity tags. Nat. Biotechnol. 1999; 17(10): 994–9.

[6] Zhou H, Ranish JA, Watts JD, Aebersold R. Quantitative proteome analysis by solid-phase isotope tagging and mass spectrometry. Nat. Biotechnol. 2002; 20(5): 512–5.

[7] Yao X, Freas A, Ramirez J, Demirev PA, Fenselau C. Proteolytic 18O labeling for comparative proteomics: model studies with two serotypes of adenovirus. Anal. Chem. 2001; 73(13): 2836–42.

[8] Everley P, Krijgsveld J, Zetter B, Gygi S. Quantitative cancer proteomics: stable isotope labeling with amino acids in cell culture (SILAC) as a tool for prostate cancer research. Mol. Cell Proteomics 2004; 3(7): 729–35.

[9] Shi L, Adkins JN, Coleman JR, Schepmoes AA, Dohnkova A, Mottaz HM, Norbeck AD, Purvine SO, Manes NP, Smallwood HS, Wang H, Forbes J, Gros P, Uzzau S, Rodland KD, Heffron F, Smith RD, Squier TC. Proteomic analysis of *Salmonella enterica* serovar typhimurium isolated from RAW 264.7 macrophages: identification of a novel protein that contributes to the replication of serovar typhimurium inside macrophages. J. Biol. Chem. 2006; 281(39): 29131–40.

[10] Zhang B, VerBerkmoes NC, Langston MA, Uberbacher E, Hettich RL, Samatova NF. Detecting differential and correlated protein expression in label-free shotgun proteomics. J. Proteome Res. 2006; 5(11): 2909–18.

[11] Wilkins MR, Appel RD, Van Eyk JE, Chung MC, Gorg A, Hecker M, Huber LA, Langen H, Link AJ, Paik YK, Patterson SD, Pennington SR, Rabilloud T, Simpson RJ, Weiss W, Dunn MJ. Guidelines for the next 10 years of proteomics. Proteomics. 2006; 6(1): 4–8.

[12] Carr S, Aebersold R, Baldwin M, Burlingame A, Clauser K, Nesvizhskii A. The need for guidelines in publication of peptide and protein identification data: Working Group on Publication Guidelines for Peptide and Protein Identification Data. Mol. Cell. Proteomics 2004; 3(6): 531–3.

[13] Karp N, Feret R, Rubtsov D, Lilley K. Comparison of DIGE and post-stained gel electrophoresis with both traditional and SameSpots analysis for quantiative proteomics. Proteomics 2007; 8(5): 948–60.

[14] Hunt SM, Thomas MR, Sebastian LT, Pedersen SK, Harcourt RL, Sloane AJ, Wilkins MR. Optimal replication and the importance of experimental design for gel-based quantitative proteomics. J. Proteome Res. 2005; 4(3): 809–19.

[15] Molloy M, Brzezinski E, Hang J, McDowell M, Van Bogelen R. Overcoming technical variation and biological variation in quantitative proteomics. Proteomics 2003; 3: 1912–19.

[16] MacCoss M, Wu C, Liu H, Sadygov R, Yates J. A correlation algorithm for the automated quantitative analysis of shotgun proteomics data. Anal. Chem. 2003; 75(24): 6912–21.

[17] Higgs RE, Knierman MD, Gelfanova V, Butler JP, Hale JE. Comprehensive label-free method for the relative quantification of proteins from biological samples. J. Proteome Res.; 4(4): 1442–50.

[18] Challapalli KK, Zabel C, Schuchhardt J, Kaindl AM, Klose J, Herzel H. High reproducibility of large-gel two-dimensional electrophoresis. Electrophoresis 2004; 25(17): 3040–7.

[19] Hynd MR, Lewohl JM, Scott HL, Dodd PR. Biochemical and molecular studies using human autopsy brain tissue. J. Neurochem. 2003; 85(3): 543–62.

[20] Veenstra TD, Conrads TP, Hood BL, Avellino AM, Ellenbogen RG, Morrison RS. Biomarkers: mining the biofluid proteome. Mol. Cell Proteomics 2005; 4(4): 409–18.

[21] Thongboonkerd V, Chutipongtanate S, Kanlaya R. Systematic evaluation of sample preparation methods for gel-based human urinary proteomics: quantity, quality, and variability. J. Proteome Res. 2006; 5(1): 183–91.

[22] Jia X, Hildrum KI, Westad F, Kummen E, Aass L, Hollung K. Changes in enzymes associated with energy metabolism during the early post mortem period in longissimus thoracis bovine muscle analyzed by proteomics. J. Proteome Res. 2006; 5(7): 1763–9.

[23] Karp NA, McCormick PS, Russell MR, Lilley KS. Experimental and statistical considerations to avoid false conclusions in proteomic studies using differential in-gel electrophoresis. Mol. Cell Proteomics 2007; 6(8): 1354–64.

[24] Karp NA, Lilley KS. Maximising sensitivity for detecting changes in protein expression: experimental design using minimal CyDyes. Proteomics 2005; 5(12): 3105–15.

[25] Mimi Roy S, Anderle M, Lin H, Becker C. Differential expression profiling of serum proteins and metabolites for biomarker discovery. Int. J. Mass Spectrom. 2004; 238: 163–71.

[26] Lenth R. Some practical guidelines for effective sample size determination. Am. Stat. 2001; 55: 187–93.

[27] Ruebelt MC, Lipp M, Reynolds TL, Astwood JD, Engel KH, Jany KD. Application of two-dimensional gel electrophoresis to interrogate alterations in the proteome of genetically modified crops. 2. Assessing natural variability. J. Agric. Food Chem. 2006; 54(6): 2162–8.

[28] Churchill GA. Fundamentals of experimental design for cDNA microarrays. Nat. Genet. 2002; 32 (Suppl.): 490–5.

[29] Rowell C, Carpenter M, Lamartiniere CA. Modeling biological variability in 2-D gel proteomic carcinogenesis experiments. J. Proteome Res. 2005; 4(5): 1619–27.

[30] Karp NA, Spencer M, Lindsay H, O'Dell K, Lilley KS. Impact of replicate types on proteomic expression analysis. J. Proteome Res. 2005; 4(5): 1867–71.

[31] Asirvatham VS, Watson BS, Sumner LW. Analytical and biological variances associated with proteomic studies of *Medicago truncatula* by two-dimensional polyacrylamide gel electrophoresis. Proteomics 2002; 2(8): 960–8.

[32] Kendziorski C, Irizarry RA, Chen KS, Haag JD, Gould MN. On the utility of pooling biological samples in microarray experiments. Proc. Natl Acad. Sci. USA 2005; 102(12): 4252–7.

[33] Karp NA, Lilley KS. Investigating sample pooling strategies for DIGE experiments to address biological variability. Proteomics 2009; 9(2): 388–97.

[34] Graham R, Graham C, McMullan G. Microbial proteomics: a mass spectrometry primer for biologists. Microb. Cell Fact. 2007; 6: 26.

[35] Encheva V, Wait R, Gharbia SE, Begum S, Shah HN. Proteome analysis of serovars Typhimurium and Pullorum of *Salmonella enterica* subspecies I. BMC Microbiol. 2005; 5(1): 42.

[36] Encheva V, Wait R, Begum S, Gharbia SE, Shah HN. Protein expression diversity amongst serovars of *Salmonella enterica*. Microbiology 2007; 153(Pt 12): 4183–93.

[37] Wilmes P, Bond PL. The application of two-dimensional polyacrylamide gel electrophoresis and downstream analyses to a mixed community of prokaryotic microorganisms. Environ. Microbiol. 2004; 6(9): 911–20.

[38] Wilmes P, Bond PL. Towards exposure of elusive metabolic mixed-culture processes: the application of metaproteomic analyses to activated sludge. Water Sci. Technol. 2006; 54(1): 217–26.

[39] Fuxius S, Eravci M, Broedel O, Weist S, Mansmann U, Eravci S, Baumgartner A. Technical strategies to reduce the amount of 'false significant' results in quantitative proteomics. Proteomics 2008; 8(9): 1780–4.

[40] Karp N, Griffin J, Lilley K. Application of partial least squares discriminant analysis to two dimensional difference gel studies in expression proteomics. Proteomics 2005; 5(1): 81–90.

[41] Hu Y, Wang G, Chen GY, Fu X, Yao SQ. Proteome analysis of *Saccharomyces cerevisiae* under metal stress by two-dimensional differential gel electrophoresis. Electrophoresis 2003; 24(9): 1458–70.

[42] Karp N, Kreil D, Lilley K. Determining a significant change in protein expression with DeCyderTM during a pair-wise comparison using two-dimensional difference gel electrophoresis. Proteomics 2004; 4(5): 1421–32.

[43] Bi X, Lin Q, Foo TW, Joshi S, You T, Shen HM, Ong CN, Cheah PY, Eu KW, Hew CL. Proteomic analysis of colorectal cancer reveals alterations in metabolic pathways: mechanism of tumorigenesis. Mol. Cell Proteomics 2006; 5(6): 1119–30.

[44] Gade D, Thiermann J, Markowsky D, Rabus R. Evaluation of two-dimensional difference gel electrophoresis for protein profiling. Soluble protiens of the marine bacterium Pirellula sp. strain 1. J. Mol. Microbiol. Biotechnol. 2003; 5(4): 240–51.

[45] Hu J, Qian J, Borisov O, Pan S, Li Y, Liu T, Deng L, Wannemacher K, Kurnellas M, Patterson C, Elkabes S, Li H. Optimized proteomic analysis of a mouse model of cerebellar dysfunction using amine-specific isobaric tags. Proteomics 2006; 6(15): 4321–34.

[46] Bland JM, Altman DG. Multiple significance tests: the Bonferroni method. Br. Med. J. 1995; 310(6973): 170.

[47] Smyth GK, Yang YH, Speed T. Statistical issues in cDNA microarray data analysis. Methods Mol. Biol. 2003; 224: 111–36.

[48] Draghici S. Statistical intelligence: effective analysis of high-density microarray data. Drug Discov. Today 2002; 7(11): S55–63.

[49] Cui X, Churchill GA. Statistical tests for differential expression in cDNA microarray experiments. Genome Biol. 2003; 4(4): 210.

[50] Benjamini Y, Hochberg Y. Controlling the false discovery rate – A practical and powerful approach to multiple testing. J. R. Stat. Soc. B 1995; 57(1): 289–300.

[51] Storey JD, Tibshirani R. Statistical significance for genomewide studies. Proc. Natl Acad. Sci. USA 2003; 100(16): 9440–5. Epub 2003 Jul 25.

[52] Storey JD. A direct approach to false discovery rates. J. R. Stat. Soc. B 2002; 64(3): 479–98.

[53] Biron DG, Brun C, Lefevre T, Lebarbenchon C, Loxdale HD, Chevenet F, Brizard JP, Thomas F. The pitfalls of proteomics experiments without the correct use of bioinformatics tools. Proteomics 2006; 6(20): 5577–96.

[54] Mehta TS, Zakharkin SO, Gadbury GL, Allison DB. Epistemological issues in omics and high-dimensional biology: give the people what they want. Physiol. Genomics 2006; 28(1): 24–32.

[55] Allison DB, Cui X, Page GP, Sabripour M. Microarray data analysis: from disarray to consolidation and consensus. Nat. Rev. Genet. 2006; 7(1): 55–65.

[56] Ransohoff DF. Rules of evidence for cancer molecular-marker discovery and validation. Nat. Rev. Cancer 2004; 4(4): 309–14.

[57] Duncan MW, Hunsucker SW. Comments on standards in proteomics and the concept of fitness-for-purpose. Proteomics 2006; 6(S2): 45–7.

Part VII

DNA Resequencing by MALDI-TOF-Mass Spectrometry and Its Application to Traditional Microbiological Problems

19

Comparative DNA Sequence Analysis and Typing Using Mass Spectrometry

Christiane Honisch[1], Yong Chen[1] and Franz Hillenkamp[2]
[1] *Sequenom, Inc., San Diego, CA, USA*
[2] *Institute for Medical Physics and Biophysics, University of Muenster, Muenster, Germany*

19.1   Introduction

Over the past decade, mass spectrometry (MS) has played a central role in the study of biological molecules and is a cornerstone in proteomics. Twenty years after the discovery of the soft ionization techniques electrospray ionization (ESI) and matrix assisted laser desorption/ionization (MALDI) both have found widespread applications in research as well as routine analysis in the life sciences and in medicine (1–3). Applications in lipid-omics and metabolomics follow the suit of the wide field of protein analysis.

The use of MALDI for the analysis of nucleic acids was demonstrated soon after its introduction for the analysis of proteins and peptides (4). In contrast to proteins, ions of nucleic acids are less stable in the vacuum of a mass spectrometer, limiting the maximum size of the analytes. In MALDI-MS with UV-laser excitation the maximum size for DNA is limited to about a length of 50 nucleotides, i.e. a mass of approximately 15 kDa. Above this molecular mass limit increasing fragmentation of the ions in the mass spectrometer will complicate interpretation of the spectra and decrease detection sensitivity. Mainly for this reason MS is mostly suited for DNA resequencing rather than *de novo* sequencing, where one analyses limited modifications of an otherwise known sequence such as single nucleotide polymorphisms (SNPs). Even in the field of DNA resequencing, MS can be used to advantage only if its specific strengths are taken into account, mainly high accuracy, high specificity and information content, high sensitivity and high adaptability to varying analytical tasks and samples. Key to any such application is the development of

Mass Spectrometry for Microbial Proteomics   Edited by Haroun N. Shah and Saheer E. Gharbia
© 2010 John Wiley & Sons, Ltd

suitable assays and the associated software, which allow access to the analytical information despite the limited accessible mass range. A combination of clever experimental designs that connect molecular biology methods with the detection of nucleic acids by matrix assisted laser desorption/ionization time-of-flight mass specrtometry (MALDI-TOF MS) makes the technology a user friendly method for the characterization of DNA and RNA.

Analytical tasks are different for nucleic acid versus protein MS. Sample preparation for nucleic acid analysis is usually based on polymerase chain reactions (PCRs). Unique primer sets can be used when targeted analysis is done to increase the amount of material that is available for MALDI-TOF analysis. Very small amounts of DNA down to a single copy can be amplified. Thus, in contrast to proteomics, sensitivity and sample amount are usually not limiting problems, even in expression analysis where the RNA concentration spans many orders of magnitude. In the actual analysis the dynamic range of two to three orders of magnitude and the signal-to-noise ratio (SNR) are usually limiting, not the absolute amount. For the same reason pre-fractionation of samples by electrophoresis or chromatography is rarely required.

The primary structure of nucleic acids with only four basic building blocks (adenine, A; cytosine, C; guanine, G; thymine, T; or uracile, U in RNA molecules) compared with the 21+ different amino acids sets different boundary conditions for analysis by MS. Higher order MS of analyte fragments after successive collision-induced dissociation (CID) or other suitable methods (MS/MS or MS^n), commonly used in peptide analysis for sequence elucidation, is of very limited use for nucleic acids and used only rarely. Because of the four building blocks nucleic acids have a limited variation in desorption and ionization efficiencies between similar sequences. This makes quantification based on signal intensities in the spectra amenable. Though not trivial, quantification is much simpler for nucleic acids than for proteins, because of their uniform constituents.

Enzymatic digests as applied to proteins in peptide mapping or peptide mass fingerprinting can be applied and are similarly useful for nucleic acid analysis, as will be shown later in this chapter.

The MALDI-TOF MS data analysis for nucleic acid stretches (oligonucleotides) is similar to that of intact protein mass spectra. Nucleic acid spectra are dominated by peaks corresponding to $z = +1$. The peak areas or heights can be used as a measure of the quantity of the analyte and the centroids can be used to determine the mass of the analyte. The masses can be compared with models of the analyte and these models can be ranked according to their fit, which allows the determination of identity and modification of the analytes.

Several algorithms and bioinformatics search tools for protein analysis were developed e.g. to match an observed set of peptide masses with the known sequence of the full-length protein being studied. Nucleic acid analysis tools specific for a given analytical task have been developed as discussed later.

Whereas in proteomics of e.g. a cell lysate one is dealing with a very complex mixture of *a priori* unknown sequences and structures, PCR products are much more homogeneous. At the same time the correct identification and localization of SNPs in a sequence of only four building blocks requires more sophistication than e. g. that of a glycosylation site in a peptide.

The lower stability of DNA as compared with proteins in the vacuum of the mass spectrometer may at first be surprising, because it is the other way around in the biological environment of the cell. The reason for this instability is rather basic and related to the structure of the molecule. For the typically singly charged ions in MALDI-MS the fragmentation mechanisms have been well clarified (5). The phosphate proton on the 5′-side of any really 'basic' base (A, C, G) is apparently somewhat delocalized between the phosphate and the base. When temporarily localized at the base, it weakens the N-glycosidic bond facilitating a loss of the base. This 1′-2′ trans-elimination leaves an unstable ribose with a single nonsymmetric double bond, which stabilizes itself to a furan ring by a backbone cleavage at the 3′-position. This sequence of fragmentation reactions can happen at many positions sequentially or even in parallel, dissociating the full length sequence into many small fragments of little information content.

This basic structural feature can be counteracted by placing a suitable ligand at the 2′-position of the ribose to increase the electron density in the N-glycosidic bond. Unfortunately, most of the potential ligands, such as fluorine (6), interfere with the various steps of enzymatic sample preparation and are therefore not acceptable in any analytical assay. This also holds for the alkylation of the phosphate (7). The only natural exception is the hydroxyl group of RNA in the 2′-position. In the vacuum of the mass spectrometer RNA is indeed substantially more stable than DNA; this is used to advantage in some of the assays, as will be shown later in the chapter.

Traditionally MALDI ion sources have been coupled with TOF mass spectrometers and this is still the analyzer of choice for the analysis of oligonucleotides. In the higher mass range the limited mass resolution and accuracy of simple TOF instruments may not suffice to differentiate e. g. a ribocytidine from a ribouridine which differ by only 1 Da in mass. However, in such cases it is usually simpler to modify the assay to increase the mass difference, rather than using a more sophisticated and more expensive mass analyzer.

The biochemistries and the dispensing of resulting products onto a matrix carrier chip for data acquisition by MALDI-TOF MS can be automated by liquid handling robotics for high-throughput application.

One of the most successful nucleic acid assays on MALDI-TOF MS so far is SNP genotyping by primer extension (iPLEX™). PCR is used to amplify the templates containing the SNPs of interest and to provide sensitivity and specificity. In the assay, a primer complementary to the sequence one base upstream of the target SNP is extended by a single nucleotide. The reaction gives rise to primers and products of different extensions and thus different resulting mass in subsequent MALDI-TOF MS analysis. An overview of the assay format (Sequenom) is shown in Figure 19.1. The generated mass spectra are analysed by determining the mass and intensity of the peaks. Masses are used to identify which of the extended primers are present in the sample, and the intensities of the peaks are used as a measure of their quanitity. Within one mass spectrum many SNPs can be detected simultaneously in a few seconds. This means assays can be multiplexed with currently up to 60 SNPs per individual reaction. Quantitative applications gain from the dynamic range and the detection down to a single molecule (8). MS outperforms most other methods in the limit of detection, quantitative precision, and the ability to measure many different analytes simultaneously (9).

Additional applications include comparative sequence analysis by MALDI-TOF MS (iSEQ™, Sequenom), which will be the focus of the remainder of the chapter and methylation analysis (EpiTYPER™, Sequenom), the detection of the methylation status of CpG dinucleotides, especially in the human genome. In methylation analysis candidate gene amplification and a base-specific cleavage assay in combination with MALDI-TOF MS

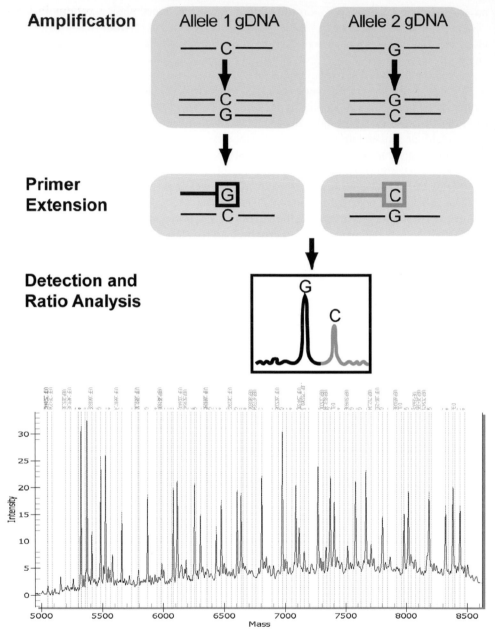

Figure 19.1   *PCR and primer extension assay (Sequenom) in a SNP region with two alleles. MALDI-TOF MS detection of resulting extension products. Example spectrum of a multiplex of 29 SNP regions in a* Streptococcus *species. One hundred per cent homozygous (single allele) genotyping calls*

are used for the quantitative characterization of cytosine methylation pattern in CpG islands (10, 11). Changes in DNA methylation are a dynamic process and resulting patterns are tightly associated with diseases like cancer and their progression (12).

19.2   Comparative Sequence Analysis by MALDI-TOF MS

For many applications, including typing of microorganisms and viruses, it is desirable to go beyond SNP based genotyping and analyse 200–1000 bp genomic regions. This clearly exceeds the range of directly measurable ions by MALDI-TOF MS, which is limited to a length of 50 nucleotides as described earlier. A clever assay design solution is required, which breaks down genomic regions into measurable DNA or favourably more stable RNA units. Similar to the information obtained by the protease digests in protein fingerprinting, where the information on how the protein is digested and the size and composition of the digested fragments are used, base-specific endonuclease digests of RNA followed by MALDI-TOF MS provide a solution for nucleic acid mass fingerprinting and comparative sequence analysis (13, 14).

As shown in Figure 19.2, amplicons of the desired genomic region are generated by PCR utilizing target specific primers tagged with RNA polymerase promoter sequences. PCR amplified sequence regions are subject to *in vitro* transcription and base-specific RNA cleavage. For this step one PCR reaction is split into four subsequent reaction wells, each containing a different transcription and cleavage cocktail (MassCLEAVE™). In two wells a genetically modified T7 polymerase facilitates transcription of the forward DNA strand into RNA, whereas in the remaining two wells a genetically modified SP6 polymerase allows for the transcription of the complementary reverse strand. This transcription step might at first sight look like an added complication of the assay. It must be kept in mind though that oligoribonucleotides are substancially more stable in a mass spectrometer and that this step not only adds another level of up to 200-fold amplification but more importantly generates single stranded copies of the forward and the reverse strand, a prerequisite for easy spectra interpretation.

Reaction mixes for the transcription reactions include a ribonuclease for base-specific cleavage. In general, base-specific riboendonucleases are described in the literature but they either exhibit rather low specificity, or lack robustness or both (15). RNase A evolved as a suitable candidate but cleaves at all pyrimidines, i.e. at ribo-C and ribo-U(T). To resolve this multiplicity one of the cleavage sites is blocked during the *in vitro* transcription process by incorporating dCTP for a U(T) specific cleavage and dTTP for a C specific cleavage reaction. In total, four different cleavages are obtained, which represent cleavage at virtually all four nucleotides.

Specific mass signal patterns of the four resulting cleavage reactions, mixtures of RNA fragments, are acquired and provide fingerprints of the sample. Since the exact masses of each of the bases within the RNA fragments are known, base compositions of each signal can be derived. The list of possible base compositions is constrained by the single representation of the known cleavage base at the 3'-end of the fragment.

One early concern of this approach was the specifity of the assay. The mass of each cleavage product is defined by its nucleotide composition but not by the order of nucleotides in the cleavage fragments. Fragments with identical nucleotide composition have

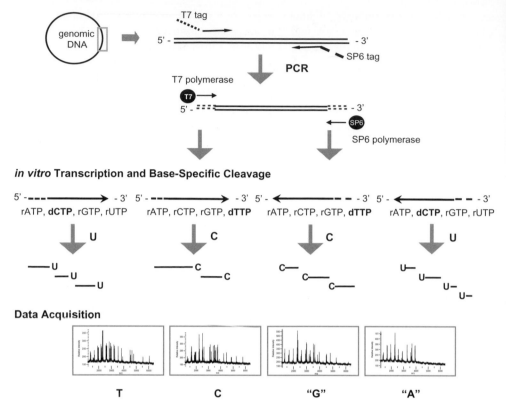

Figure 19.2   *PCR amplification of the genomic target region of interest with T7- and SP6-promoter tagged primers. In vitro transcription of the PCR product. Base-specific cleavage at virtually all four bases facilitated by cleavage of the transcribed forward as well as the reverse strand at U and C*

identical mass. If only four different nucleotides and cleavage sites are available (compared with 21 different standard peptides and cleavages at a few sites), one could expect a very high abundance of small cleavage products, 4 mers on average. In addition, many different cleavage products with identical mass were anticipated to overlap in the low mass range, or with mass differences too small to be separated with a simple linear TOF mass spectrometer.

A simulation was performed which provided results for *in silico* digests of approximately 250 000 coding reference sequences of the human genome (http://www.ensembl.org/index.html) with 62 million overall nucleotide counts and amplicon lengths between 200 bp and 1500 bp (16, 17). Simulation results for all possible SNPs over a nucleotide range between 200 nt and 1500 nt are shown in Figure 19.3. As can be seen, the rate of correctly identified SNPs is surprisingly high if all four cleavage reactions are run. Detection rates between 95% and 100% are achieved for homozygous SNPs in this range, an indication for the applicability of the technology in microbial and viral genotyping with a discriminatory power down to a single nucleotide. The detection range of 95–100% is maintained in the heterozygous SNP context for a nucleotide mass range between 200 nt

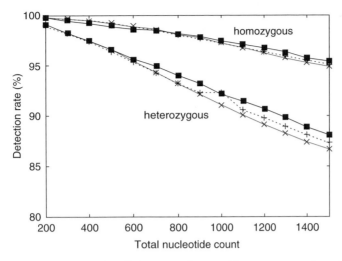

Figure 19.3 *Simulation results for all theoretically possible SNPs using the reference sequence for all exons of Ensembl (build 34b). The y-axis provides the percentage of sequence change events that can be detected with base-specific cleavage using four cleavage reactions. The x-axis provides the simulation results for overall amplicon length of 200 nt up to 1500 nt (16)*

and 800 nt, a sequence read length comparable with today's standard dideoxy sequencing equipment. The success rate decreases somewhat, if only one or two cleavage reactions are run. However, as shown below (Figure 19.4), for many practical resequencing applications based on known reference sets an *in silico* digest prior to the experiment can tell how much information will be lost if the number of reactions is limited. The up to four cleavage reactions will introduce substantial redundancy which increases the reliability of a clinical application.

   The discrimination between bacterial and viral isolates is central to many aspects of clinical microbiology and the vastly growing genomic reference information in the public domain enables the identification of deviations between genomes. Comparative sequencing projects play a key role in epidemiology and population studies. They can be applied to the development of outbreak control strategies and provide insight into pathogenesis and surveillance of infectious diseases. They further help in elucidating the significance of the genetic diversity within a population for the behaviour of certain genotypes, for example epidemicity, virulence and antibiotic resistance. The specific advantage of sequence-based comparative typing schemes in clinical microbiology and outbreak monitoring is the availability and accessibility of sequence databases, the comparability of results and the intercentre data exchange via computer networks.

   Automated comparative sequence analysis by MALDI-TOF MS (iSEQ™, Sequenom) takes advantage of the accessibility of these sequence databases to identify the experimental sample-specific patterns of the four base-specific cleavage reactions and ultimately compare them with the best matching *in silico* reference mass signal patterns. This is a task similar to a central step in proteomic data processing, where peptide sequences are identified by correlating acquired fragment ion spectra with theoretical spectra predicted for each peptide contained in a protein sequence database (18).

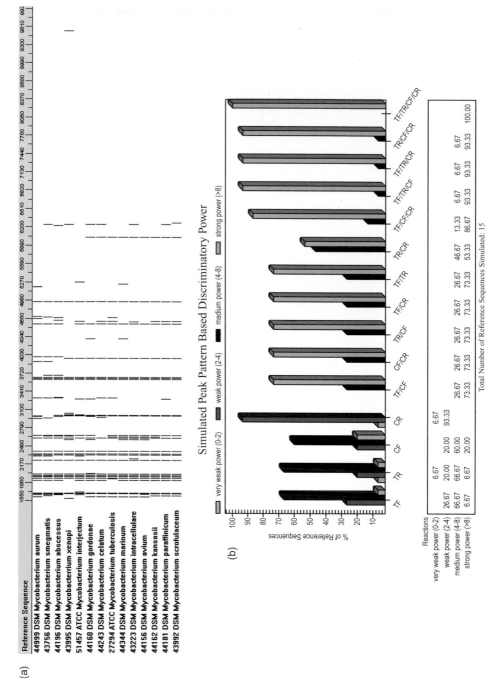

Figure 19.4 *Base-specific cleavage and MS simulation results for a set of 15 mycobacterial reference sequences. (a) Banding pattern of the C forward cleavage reaction. Reference sequences are clustered according to their pattern similarity. (b) Bar graph of discriminatory power for 4, 3, 2 and for individual cleavage reactions. The y-axis displays the % of sequences falling into the individual power catagories. A threshold of 2 (equivalent to 2 full peak changes) is required to distinguish one*

The total nucleic acid comparative sequence analysis process can be divided into analysing reference sequence sets *in silico* to generate reference patterns, measuring spectra quality and extracting sample peaks, identifying the best matching reference sequence out of a set of sequences and finally evaluating the identified best matching reference sequence for any sequence variations.

The purpose of analysing reference sequence sets *in silico* is to create reference peak patterns based on a known reference sequence set. Reference peak patterns for each reference sequence and cleavage reaction are obtained by *in silico* base-specific RNaseA cleavage. Resulting peak patterns or peak lists can be displayed as banding patterns, as shown in Figure 19.4. For each reaction, the peaks of all reference sequences in the set are aligned by mass and form a data array. In extension, the combination of data for all four cleavage reactions forms an even larger data array representative of the whole reference sequence set and all cleavage reactions. Distance calculations between peak patterns of different reference sequences allow for peak pattern based clustering and relatedness analysis.

In addition, the reference peak lists and aligned peak patterns can be used to assess if all references in the set can be discriminated by base-specific cleavage and MALDI-TOF MS and what the minimal set of reactions is to discriminate all references in a set.

In this simulation, references are first grouped into clusters based on discriminating features by finding peaks present in one set of references but absent in others. Clusters are then grouped into subclusters until each cluster has only one sequence or a set of indistinguishable sequences. Values for discriminating power are calculated by summing up intensities of all the discriminating features. These are the unique peaks present only in the cluster as well as peaks with changes of peak intensities between clusters. A value above the threshold of discriminating power, typically set to 2, is required to distinguish one reference from another with good confidence.

If references in a set are substantially different from one another, one reaction can be enough to discriminate them all. Figure 19.4 shows results of the iSEQ™ Simulation Software (Sequenom, Inc.), which utilizes the presented analysis concept and displays results for a set of mycobacterial 16 S rDNA reference sequences (19). In this example, two individual cleavage reactions (the T forward or the C forward reactions) are sufficient to discriminate the given set of reference sequences.

Measuring the spectra quality and extracting sample peaks are the next steps in the comparative sequence analysis process. To ensure that quality spectra are acquired, spectra are evaluated during MS acquisition by comparing the detected peak patterns with a set of anchor peaks selected from the *in silico* reference peak patterns. Anchor peak sets contain peaks present in all references or at least large groups of references within the *in silico* reference set. They are identified in such a way that all the references are represented by one or more peaks in each anchor peak set and that they are likely to be found in the acquired spectra. Typically, 10–20 anchor peak sets are used.

Once spectra are acquired, meaningful peaks need to be extracted and differentiated from the noise. After initial filtering, peak positions are identified by finding local maxima. Depending on peak separation, which is determined by the MS instrument resolution, one or a set of peaks are grouped together and a common baseline in the original spectrum is determined for the group. Subsequently, these baseline corrected data points are fitted to Gaussian peak shape.

Peak intensities and SNRs are then calculated from the heights and widths. Peaks with low SNRs are evaluated to obtain the cut-off for chemical noise peaks, which are filtered from the final peak list. Afterwards peak intensities are normalized against internal references.

Raw peak intensities vary substantially over the mass range depending on the mass spectrometer and the tuning of the instrument. For the mass spectrometer used in our studies (MassARRAY® Compact Analyzer, Sequenom, Inc.) spectral profiles are acquired in a mass range of 1100–11 000 Da using delayed ion extraction. On average, peaks have the highest intensities between 2000 Da and 4000 Da. Mass dependent variations are corrected by a mass scaling curve, which is calculated for each spectrum. Depending on spectrometers, alternative fittings may apply.

The resulting profile is then used to normalize detected intensities and calculate revised intensities for all detected peaks. This becomes very important for the comparison, clustering and relatedness analysis of samples or sample populations based on their mass spectrometric fingerprints as shown below.

The resulting detected peak lists are then screened for contaminant and side product peaks, such as salt adduct peaks and signals of doubly charged ions, which will be excluded from score calculations during the matching process.

Salt adduct peaks are the results of ubiquitous cations such as alkali salt ions like sodium (Na^+) and potassium (K^+) from the buffer system of the biochemical assay binding to the negative nucleic acid phosphodiester backbone. This can cause a multiplicity of signals in the spectrum, mostly peaks at +22 Da for Na^+ and +38 Da for K^+. Salt adducts can be minimized by suitable treatment of the reaction products with ion-exchange resin to exchange alkali ions for NH_4^+. NH_4^+ is quantitatively lost upon laser desorption, leaving only the signals of the free acid in the spectra.

It has been observed that the flight behaviour and the resulting peak intensities of nucleic acid fragments are dependent on their nucleotide compositions. In particular, when the C-specific RNAse A cleavage is applied, T-rich fragments of the C-cleavage reaction show decreasing signal intensities with increasing T content, as shown in Figure 19.5. Intensities of T-rich signals can sometimes be lower than the intensity of an adduct peak. To better distinguish real T-rich signals from adduct or other noise peaks, an empirical relationship between adjusted peak intensity and base composition for C-cleavage products was implemented. As a result, intensities for the T-rich reference peaks are adjusted using the empirical model.

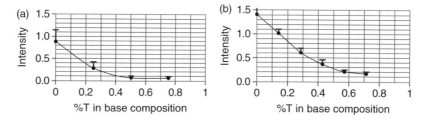

Figure 19.5 *Intensity distribution of T-rich fragments of C-specific cleavage reactions in MALDI-TOF MS. %T base composition and corresponding intensities in MS for fragments of (a) 4 (4 mer) and (b) 7 (7 mer) nucleotides are plotted. Each compomer plot averages up to 9000 datapoints. Signal intensities decrease with increasing T content*

Once detected peaks for a sample are extracted, normalized and scaled, the next step is to identify the best matching reference or references. During this identification process, an overall score is calculated for each reference in comparison with the sample peak list by combining three different subscores. The first subscore is calculated by comparing all reference peaks generated in the simulation with the detected peaks. A value of zero applies if for a reference peak there is no matching detected peak. Otherwise, values are calculated by evaluating the intensity ratio of reference versus detected. The total subscore is then calculated by averaging the values for all the reference peaks weighted by reference intensities and the mass scaling curve as described earlier. Thus, peaks with T-rich nucleic acid compositions or peaks in the low or high mass range with lower intensities have less impact on the score.

The second subscore is calculated in a similar fashion, except that only a subset of peaks that can discriminate one reference from another or one set of references from another set is taken into consideration. It is more sensitive in picking up minor differences between the peak intensities and crucial for differentiation of similar references. Finally, the third subscore, a distance score, is calculated based on the Euclidian distance of detected peaks for each sample compared with the detected peaks of the other samples in the set. The overall score is the dynamic combination of all three scores.

During identification, all the references are sorted by the overall score. The subset of sequences with the highest scores is then used to refine the intensities of the detected peak lists. The overall score is calculated again for this subset of sequences and a new subset is defined. As illustrated in Figure 19.6, this iterative process continues until one sequence

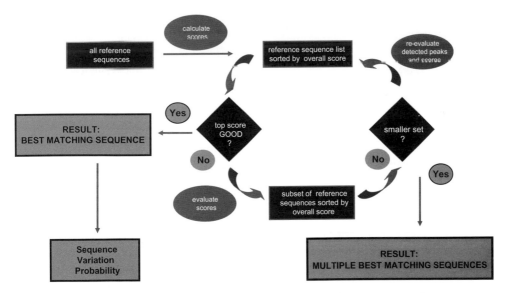

Figure 19.6 *Schematic representation of the iteratives identification process for comparative sequence analysis by MS. All references are sorted by score. If the top score is significantly better than the rest of the scores, the reference sequence is considered the best match and subject to sequence variation analysis and probability evaluation. If multiple equal scores are at the top, the subset is reevaluated by refining the intensities of the detected peak list. The iterative process continues until one sequence or several best matching sequences with close scores are identified*

or several sequences with close scores that are considerably better than the rest are identified. After the best matching reference or references are found, they are further evaluated for final scores, sequence variations and confidence.

Final scores need to be comparable among different samples. They ignore minor differences between peak intensities but include contributions from new peaks that are not expected for the reference.

The identified best matching reference pattern and the detected sample pattern might differ from each other if SNPs are present. Differences between the expected and the observed mass signal pattern can be interpreted and enable identification of microheterogeneities. Single base deviations affect one or more cleavage reactions and generate additional as well as missing signals. Time-efficient algorithms utilize the redundancy of the four cleavage reactions and detected deviations to identify and localize sequence differences down to single base pair change and identify novel sequences (20).

With a comparative sequence analysis process in place comprised of a matching and a single nucleotide detection module, the final question is how confident one can be in the results. In other words, how well does the selected reference match the sample and how likely are additional variations?

The common approach for confidence evaluation is to calculate the probability value (*P*-value), which estimates the probability of a random reference having a better score than the selected one. However, to get reasonably accurate *P*-values, the sampling space has to be large, which would be computationally prohibitive to do.

Thus, an empirical model was applied assuming that at least one sample in the analysis data set matches the top matching reference (with or without resolved variations). The model was built based on training data sets with known reference sequences. First, all the samples in the training sets were identified. Then for each sample, single base mutations in the top match reference were simulated at random and scores were calculated for all mutated sequences in comparison with the sample spectra. Finally, the density distributions for the resulting scores were plotted. For all the samples simulated, density distributions for scores can be described by a Gaussian distribution. Alternatively, other distributions such as Poisson can also be used to describe the density distribution. Density contributions for scores from two or more mutations are usually significantly weaker than those from single mutations, and can be ignored. Thus the density distributions identified from single mutations are used to estimate the probability of additional mutations in the best matching reference sequence of the sample. We have established a threshold of 0.05, which means that samples with sequence variation probabilities > 0.05 are likely to have additional sequence variations. Figure 19.7 provides an overview of the sample identification and data output.

As applied for the analysis of the reference sequence set, where Euclidean distance matrices of the simulated cleavage pattern can be calculated, Euclidean distance matrices can be obtained for the detected peak pattern taking peak intensities into account.

The detected peak based distance matrix provides a fast and efficient way to group samples. Applications of this distance matrix approach go from clustering under experimental conditions where samples do not match the known references to clustering of mixed populations without resolving them to the individual sequence level. The presence of mixtures can often be tolerated and utilized as exemplified in the next section.

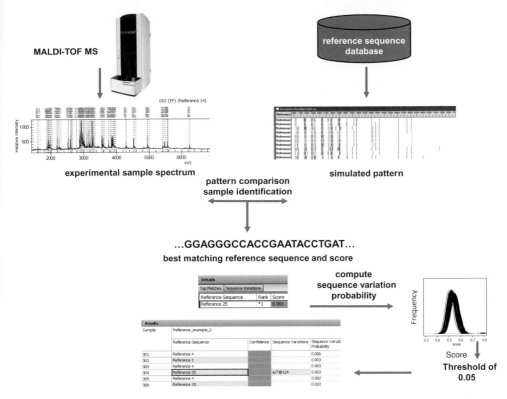

Figure 19.7 *Sample identification and data output of the comparative sequence analysis process by MS. Acquired spectra (up to four per reaction) are correlated against the theoretical peak pattern derived from an input reference sequence set as defined by the user. A scoring scheme is used to measure the degree of similarity. Matching reference sequences are ranked according to the computed score. The reference sequence with the highest score is selected for further statistical analysis. The sequence variation probability assesses the quality of the match between the top matching reference pattern and the sample pattern and expresses the likelihood of any unexplained sequence variation in the selected best matching reference sequence*

19.3 Applications of Nucleic Acid Analysis by MALDI-TOF MS in Clinical Microbiology

In a routine clinical laboratory, species identification of cultured isolates usually still relies on phenotypic methods, such as panels of biochemical pathways, antibiotic resistance reactions, antigens presented on the cell's surface and fatty acid patterns. As most of these methods rely on gene expression, these tests all have a tendency to vary, based on environmental influences such as changes in growth conditions, growth phase, and the frequency of occurrence of spontaneous mutations.

About a decade ago MALDI-TOF MS began to be used to record spectra of small proteins in whole cells, cell lysates or crude bacterial extracts and to classify bacteria by a quantitative fingerprint comparison of whole spectra or selected signals. Again, quality and reproducibility of results were dependent on cultivation conditions and on the availability of reliable reference data sets (21–23).

To overcome these drawbacks genotypic DNA-based technologies have been introduced in clinical laboratories over the last two decades. These include, but are not restricted to SNP detection, ribotyping, insertion sequence (IS) profiling, variable number of tandem repeat (VNTR) analysis, gel electrophoresis or fingerprinting based on electrophoretic mobility like pulse-field gel electrophoresis (PFGE) (24), or a combination of these.

Next generation sequencing technologies which enable rapid whole genome sequencing are still prohibitively expensive and lack ease of use to allow for an easy comparison of data from multiple isolates in an automated fashion.

The same applies for whole genome DNA microarrays and their routine application. In addition, PCR product microarrays generally do not have the resolution to detect minor deletions and point mutations (25).

In particular, PCR-based typing systems, which provide the ability to exponentially amplify marker regions in a given template genome even for trace amount of material or uncultured species, are increasingly being used in clinical laboratories. Dideoxy sequencing of the 16S rDNA gene is a reference method for species identification and studies have shown its superiority to phenotypic methods for the identification of various groups of bacteria (26, 27).

The challenge for sequence-based typing is to identify combinations of informative SNPs or sequence regions within microbial or viral genomes which exhibit the right amount of variation. The availability of genome sequences provide a target source to rationalize the choice of DNA signatures that may serve as PCR targets, according to the degree of specificity, the epidemiological question and the population structure of the species under study. This also provides a means to place an unknown organism in its phylogenetic context allowing for the rapid identification of new organisms.

Sequenced reference genomes available in the public domain today include nearly all clinically important human bacterial pathogens, covering all the phylogenetic domains of bacteria (28), an excellent resource for automated comparative nucleic acid analysis tools.

Based on the specific strength of MS and the available assay formats DNA methods for the detection and identification of microbes and viruses fall into two categories – multiplexed SNP genotyping and comparative sequence based typing.

SNPs are easy to assay and provide a powerful strategy for large-scale molecular population-genetic studies examining phylogenetic frameworks, which provide new insights into the worldwide evolution of large samples of strains (29). With MS plexing levels up to at least 60 SNPs per reaction the analysis of hundreds of SNPs in thousands of isolates is feasible and economically realistic.

One example, which has shown the feasibility of this approach for clinical application is the MALDI-TOF MS based SNP genotyping of human papilloma virus (HPV). HPV is the major cause of invasive cervical cancer and genotyping of HPV infections is important for cervical screens as different genotypes confer distinctly different risks for the development of disease. Knowledge of the prevalence and associated risks for each HPV genotype has also been important for the development of vaccines against HPV and monitoring of the effectiveness of such vaccination programmes. Fourteen specific high-risk oncogenic HPV genotypes were multiplexed into a single reaction for MALDI-TOF MS analysis, which performed as well or better than one of the current standard methods based on reverse dot blot hybridization (30). This exemplifies that pathogenic strains in a given environment can be differentiated by SNPs and measured simultaneously. In extension, a competitive MALDI-TOF MS PCR approach was developed to detect HPV (31). Competitors are synthesized templates that match the target sequences except for a single base mutation, which can then be discriminated from the target allele during primer extension. If added at known quantities competitors can be used to determine target DNA quantities and can serve as positive controls for a working PCR. Here, the specificity of the MALDI-TOF MS SNP genotyping as opposed to e.g. fluorescent methods derives from the requirement that backgrounds have to have the same molecular weight as the detected extended primers.

Experimental evolution has become a powerful tool for studying biological processes, principles and systems in bacteria and viruses. Another MALDI-TOF MS application example demonstrates the versatility of SNP genotyping for monitoring of mutations in parallel laboratory evolutions of *Escherichia coli* populations under specific growth constrains. *E. coli*-K12 evolutions were studied on the genotypic level by a combination of a whole-genome comparative hybridization array and comparative sequence analysis by MALDI-TOF MS resulting in a small number of mutations describing the genetic basis for adaption (32). Strains were following slightly divergent adaptive paths towards the same improved phenotype manifested in slightly different combinations of mutations. During the course of adaption and in subsequent competition experiments the ratios of wild-type versus mutant genotypes were measured by multiplexed primer extension assays and MALDI-TOF MS (33, 34). Peak intensities of the alleles were used to determine the ratios of wild type versus mutants. The approach provides significantly increased sensitivity to detect fitness differences compared with optical-density-based growth rate measurements that were previously used to study the evolved strains (35). Other methods for competition monitoring like the use of selectable markers in one of the competed strains to allow for discriminatory colony counting, fluorescent protein expression (36) and a plasmid sequence-based code to trace lineages (37) are labour intensive and provide less sensitivity than primer extension and the subsequent MALDI-TOF MS analysis.

Alternatively, one is interested in detecting and classifying a new isolate by establishing its sequence proximity to previously known isolates. In this case comparative sequence analysis by MALDI-TOF MS (iSEQ™, Sequenom) enables automated reference sequence based identification of DNA or RNA sequences and is suited to screen multiple loci in parallel. Compared with traditional methods for analysing PCR amplicons, including gel electrophoresis and dideoxy sequencing, automated data analysis of the comparative sequence analysis approach by MALDI-TOF MS avoids time consuming trace analysis and sequence alignments. As opposed to dideoxy sequencing, band compression artifacts by repeats of single nucleotides in a sequence are not an issue and do not cause misreading of the sequence.

The concept of base-specific cleavage and MALDI-TOF MS was successfully applied to 16S rDNA gene based typing to the discrimination of Mycobacteria (19) and *Bordetella* species (13, 38) and to multi-locus sequence typing (MLST) to identify lineages of the bacterial pathogen *Neisseria meningitidis* (39). MLST is an elegant strategy for the inter- and intraspecies classification of bacterial isolates on the basis of sequences of internal fragments of so-called housekeeping genes (40). Reference sequence information is publicly available for over 25 organisms (http://www.mlst.net) and signature regions can

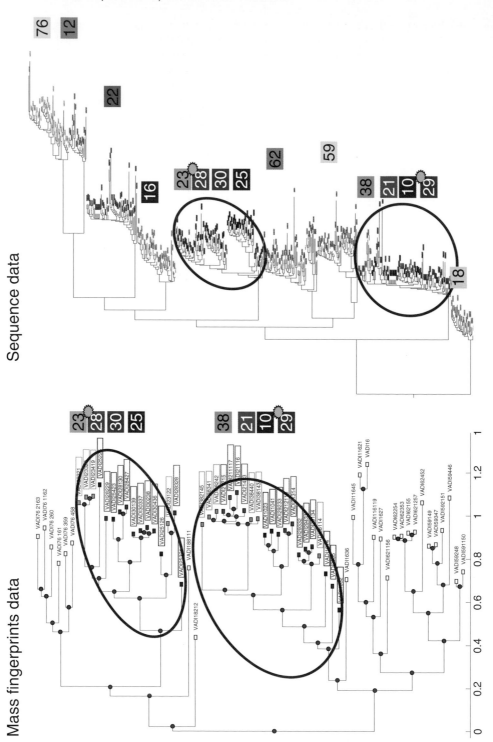

Figure 19.8 Phylogenetic analysis of an outbreak investigation of hepatitis C (HCV) based on fragments from the HVR1 region. Circled are clusters of acute cases and the respective chronic sources. Chronic sources are labelled with stars. Sequence data and mass fingerprints provide concordant results. (Data provided by the CDC, Atlanta, CA, USA).

easily be configured into base-specific cleavage assays (MassCLEAVE™) for the MALDI-TOF MS platform. Validation for a set of 100 clinical *N. meningitidis* samples showed concordance with dideoxy sequencing; 98.9% of all sequences were identified identically and four samples revealed mixed populations (34). Sequence changes down to one nucleotide difference and thus new sequences not previously listed in the database were detectable for three samples. The throughput and speed of MALDI-TOF MS allows for processing of all 100 samples over seven MLST housekeeping genes within one working day, which is sufficient to track an ongoing epidemic. Overall 97.6% of all typeable sample alleles were automatically assigned to the correct sequence by the software algorithm described in the previous paragraph, 1.8% were among a group of top matching sequences including the correct sequence and 0.6% of the alleles resulted in a wrong sequence assignments due to the failure of two out of the four cleavage reactions. Reproducibility of the method on 3 consecutive days in a random subset of 23 samples showed 99.1% correct identification results. Overall 98.1% of the repeated typing events were reproducible. This reflects the stability of the molecular typing approach by MALDI-TOF MS.

The application of comparative sequence analysis by MALDI-TOF MS to the genotypic classification of Salmonella strains will be described in Chapter 20.

Additional challenges are imposed on a typing scheme and technology if the investigation of an outbreak entails a population or a mixture of quasispecies. Genetic heterogeneity is a hallmark of e.g. hepatitis C (HCV). Consensus sequencing of a variety of genomic regions has been frequently used to resolve transmission events. If, however the prevalent genotype in the community is identical to the index cases or transmission has occurred in a distant past, consensus data are not adequate to infer transmission relationships (41). Alternative approaches like a combination of cloning and the analysis of multiple single clones per population sample by end point limiting dilution PCR, even though very accurate, are laborious and not suitable for large-scale outbreak investigations or population level surveillance studies. Base specific cleavage (MassCLEAVE™) and MALDI-TOF MS analysis (iSEQ™) of mixed populations on the other hand has shown to be a very promising tool for quick and robust quasispecies analysis (42). MALDI-TOF MS has the unique ability to analyse mixtures of quasispecies without separation of the components. Mixtures are reflected as intensity changes and the appearance/disappearance of peaks within the peak pattern of the marker region of interest. A phylogenetic analysis example of an outbreak scenario is shown in Figure 19.8. Clustering of sequence data from dideoxy-sequencing of 50–100 individual clones per sample is compared with clustering of the mass spectrometric fingerprints of the mixed populations. Clustered together and circled are two groups of acute cases and the respective chronic source. Sequencing of about 50–100 clones per sample and the analysis of the whole population by MALDI-TOF MS presents equal results. The difference is that the MALDI-TOF MS approach reduces the outbreak investigation time more than 10-fold, is reproducible and provides direct real-time genotype/subtype data. This represents a new approach to measure viral genetic relatedness in an automated and efficient manner.

19.4   Conclusion

MALDI-TOF MS has successfully been applied to solve a wide range of biological problems through the analysis of proteins, peptides, and nucleic acids. SNP and reference sequence based MALDI-TOF MS typing are generic approaches, which facilitate the identification of any microbial and viral taxa with a broad application across the fields of microbiology and epidemiology. With the appropriate marker sets in place, a wide range of isolates to the genus, species or subspecies level are typeable and populations can be monitored. The combination of suitable biochemical assay formats with MALDI-TOF MS provides ease of use, accuracy, reproducibility, automation, high throughput and data portability. These features are important for the recognition of global microbial and viral outbreak situations, monitoring of their spread and trends in virulence or antibiotic resistance.

References

1. Karas M, Bachmann D, Hillenkamp F. Influence of the wavelength in high-irradiance ultra-violet laser desorption mass spectrometry of organic molecules. Anal. Chem. 1985; 57(14): 2935–2939.
2. Karas M, Hillenkamp F. Laser desorption ionization of proteins with molecular masses exceeding 10,000 daltons. Anal. Chem. 1988; 60(20): 2299–2301.
3. Fenn JB, Mann M, Meng CK, Wong SF, Whitehouse CM. Electrospray ionization for mass spectrometry of large biomolecules. Science 1989; 246(4926): 64–71.
4. Nordhoff E, Ingendoh A, Cramer R, Overberg A, Stahl B, Karas M, Hillenkamp F, Crain, PF. Matrix-assisted laser desorption/ionization mass spectrometry of nucleic acids with wavelengths in the ultraviolet and infrared. Rapid Commun. Mass Spectrom. 1992; 6(12): 771–776.
5. Nordhoff E, Cramer R, Karas M, Hillenkamp F, Kirpekar F, Kristiansen K, Roepstorff, P. Ion stability of nucleic acids in infrared matrix-assisted laser desorption/ionization mass spectrometry. Nucleic Acids Res. 1993; 21(15): 3347–3357.
6. Tang W, Zhu L, Smith LM. Controlling DNA fragmentation in MALDI-MS by chemical modification. Anal. Chem 1997; 69(3): 302–312.
7. Gut IG, Beck S. A procedure for selective DNA alkylation and detection by mass spectrometry. Nucleic Acids Res. 1995; 23(8): 1367–1373.
8. Ding C, Cantor CR. A high-throughput gene expression analysis technique using competitive PCR and matrix-assisted laser desorption ionization time-of-flight MS. Proc. Natl Acad. Sci. USA 2003; 100(6): 3059–3064.
9. Ragoussis J, Elvidge GP, Kaur K, Colella S. Matrix-assisted laser desorption/ionisation, time-of-flight mass spectrometry in genomics research. PLoS Genet. 2006; 2(7): e100.
10. Ehrich M, Nelson MR, Stanssens P, Zabeau M, Liloglou T, Xinarianos G, Cantor, CR, Field, JK, Van den Boom D. Quantitative high-throughput analysis of DNA methylation patterns by base-specific cleavage and mass spectrometry. Proc. Natl Acad. Sci. USA 2005; 102(44): 15785–15790.
11. van den Boom D, Ehrich M. Mass spectrometric analysis of cytosine methylation by base-specific cleavage and primer extension methods. Methods Mol. Biol. 2009; 507: 207–227.
12. Vanaja DK, Ehrich M, Van den Boom D, Cheville JC, Karnes RJ, Tindall DJ, Cantor, CR, Young, CY. Hypermethylation of Genes for Diagnosis and Risk Stratification of Prostate Cancer. Cancer Invest 2009:1.
13. Hartmer R, Storm N, Boecker S, Rodi CP, Hillenkamp F, Jurinke C, Van den Boom D. RNase T1 mediated base-specific cleavage and MALDI-TOF MS for high-throughput comparative sequence analysis. Nucleic Acids Res. 2003; 31(9): e47.
14. Stanssens P, Zabeau M, Meersseman G, Remes G, Gansemans Y, Storm N, Hartmer R, Honisch C, Rodi CP, Bocker S, Van den Boom D. High-throughput MALDI-TOF discovery of genomic sequence polymorphisms. Genome Res. 2004; 14(1): 126–133.
15. Hahner S, Ludemann HC, Kirpekar F, Nordhoff E, Roepstorff P, Galla HJ, Hillenkamp F. Matrix-assisted laser desorption/ionization mass spectrometry (MALDI) of endonuclease digests of RNA. Nucleic Acids Res. 1997; 25(10): 1957–1964.

16. Ehrich M, Bocker S, van den Boom D. Multiplexed discovery of sequence polymorphisms using base-specific cleavage and MALDI-TOF MS. Nucleic Acids Res. 2005; 33(4): e38.
17. Bocker S. Simulating multiplexed SNP discovery rates using base-specific cleavage and mass spectrometry. Bioinformatics 2007; 23(2): e5–e12.
18. Nesvizhskii AI, Vitek O, Aebersold R. Analysis and validation of proteomic data generated by tandem mass spectrometry. Nat. Methods 2007; 4(10): 787–797.
19. Lefmann M, Honisch C, Bocker S, Storm N, von Wintzingerode F, Schlotelburg C, Moter A, Van den Boom D, Gobel UB. Novel mass spectrometry-based tool for genotypic identification of mycobacteria. J. Clin. Microbiol.2004; 42(1): 339–346.
20. Bocker S. SNP and mutation discovery using base-specific cleavage and MALDI-TOF mass spectrometry. Bioinformatics 2003; 19 (Suppl. 1): i44–i53.
21. Vanlaere E, Sergeant K, Dawyndt P, Kallow W, Erhard M, Sutton H, Dare D, Devreese B, Samyn B, Vandamme P. Matrix-assisted laser desorption ionisation-time-of of-flight mass spectrometry of intact cells allows rapid identification of Burkholderia cepacia complex. J. Microbiol. Methods 2008; 75(2): 279–286.
22. Mellmann A, Cloud J, Maier T, Keckevoet U, Ramminger I, Iwen P, Dunn J, Hall G, Wilson D, Lasala P, Kostrzewa M, Harmsen D. Evaluation of matrix-assisted laser desorption ionization-time-of-flight mass spectrometry in comparison to 16S rRNA gene sequencing for species identification of nonfermenting bacteria. J. Clin. Microbiol. 2008; 46(6): 1946–1954.
23. Keys CJ, Dare DJ, Sutton H, Wells G, Lunt M, McKenna T, McDowall M, Shah HN. Compilation of a MALDI-TOF mass spectral database for the rapid screening and characterisation of bacteria implicated in human infectious diseases. Infect. Genet. Evol. 2004; 4(3): 221–242.
24. Schwartz DC, Cantor CR. Separation of yeast chromosome-sized DNAs by pulsed field gradient gel electrophoresis. Cell 1984; 37(1): 67–75.
25. Garaizar J, Rementeria A, Porwollik S. DNA microarray technology: a new tool for the epidemiological typing of bacterial pathogens? FEMS Immunol. Med. Microbiol. 2006; 47(2): 178–189.
26. Mellmann A, Cloud JL, Andrees S, Blackwood K, Carroll KC, Kabani A, Roth A, Harmsen D. Evaluation of RIDOM, MicroSeq, and Genbank services in the molecular identification of Nocardia species. Int. J. Med. Microbiol. 2003; 293(5): 359–370.
27. Woo PC, Lau SK, Teng JL, Tse H, Yuen KY. Then and now: use of 16S rDNA gene sequencing for bacterial identification and discovery of novel bacteria in clinical microbiology laboratories. Clin. Microbiol. Infect. 2008; 14(10): 908–934.
28. Fournier PE, Drancourt M, Raoult D. Bacterial genome sequencing and its use in infectious diseases. Lancet Infect. Dis. 2007; 7(11): 711–723.
29. Gutacker MM, Mathema B, Soini H, Shashkina E, Kreiswirth BN, Graviss EA, Musser JM. Single-nucleotide polymorphism-based population genetic analysis of Mycobacterium tuberculosis strains from 4 geographic sites. J. Infect. Dis. 2006; 193(1): 121–128.
30. Soderlund-Strand A, Dillner J, Carlson J. High-throughput genotyping of oncogenic human papilloma viruses with MALDI-TOF mass spectrometry. Clin. Chem. 2008; 54(1): 86–92.
31. Yang H, Yang K, Khafagi A, Tang Y, Carey TE, Opipari AW, LIeberman R, Oeth PA, Lancaster W, Klinger HP, Kaseb AO, Metwall A, Khaled H, Kurnit DM. Sensitive detection of human papillomavirus in cervical, head/neck, and schistosomiasis-associated bladder malignancies. Proc. Natl Acad. Sci. USA 2005; 102(21): 7683–7688.
32. Honisch C, Raghunathan A, Cantor CR, Palsson BO, van den Boom D. High-throughput mutation detection underlying adaptive evolution of *Escherichia coli*-K12. Genome Res. 2004; 14(12): 2495–2502.
33. Herring CD, Raghunathan A, Honisch C, Patel T, Applebee MK, Joyce AR, Albert TJ, Blattner FR, Van den Boom D, Cantor CR, Palsson BO. Comparative genome sequencing of *Escherichia coli* allows observation of bacterial evolution on a laboratory timescale. Nat. Genet. 2006; 38(12): 1406–1412.
34. Applebee MK, Herrgard MJ, Palsson BO. Impact of individual mutations on increased fitness in adaptively evolved strains of *Escherichia coli*. J. Bacteriol. 2008; 190(14): 5087–5094.
35. Lenski RE, Rose MR, Simpson SC, Tadler SC. Long-term experimental evolution in *Escherichia coli*. I. Adaptation and divergence during 2,000 generations. Am. Naturalist 1991; 138: 1315–1341.

36. Hegreness M, Shoresh N, Hartl D, Kishony R. An equivalence principle for the incorporation of favorable mutations in asexual populations. Science 2006; 311(5767): 1615–1617.
37. Imhof M, Schlotterer C. Fitness effects of advantageous mutations in evolving *Escherichia coli* populations. Proc. Natl Acad. Sci. USA 2001; 98(3): 1113–1117.
38. von Wintzingerode F, Bocker S, Schlotelburg C, Chiu NH, Storm N, Jurinke C, Cantor CR, Gobel UB, Van den Boom D. Base-specific fragmentation of amplified 16S rRNA genes analyzed by mass spectrometry: a tool for rapid bacterial identification. Proc. Natl Acad. Sci. USA 2002; 99(10): 7039–7044.
39. Honisch C, Chen Y, Mortimer C, Arnold C, Schmidt O, van den Boom D, Cantor CR, Shah HN, Gharbia SE. Automated comparative sequence analysis by base-specific cleavage and mass spectrometry for nucleic acid-based microbial typing. Proc. Natl Acad. Sci. USA 2007; 104(25): 10649–10654.
40. Maiden MC. Multilocus sequence typing of bacteria. Annu. Rev. Microbiol. 2006; 60: 561–588.
41. Bracho MA, Gosalbes MJ, Blasco D, Moya A, Gonzalez-Candelas F. Molecular epidemiology of a hepatitis C virus outbreak in a hemodialysis unit. J. Clin. Microbiol. 2005; 43(6): 2750–2755.
42. Ganova-Raeva LM, Ramachandran S, Lin Y, Dimitrova Z, Campo D, Xia GL, Khudyakov Y. Mass fingerprint analysis of hepatitis C virus quasispecies complexity for outbreak investigations. Poster presentation at Epidemics: First International Conference on Infectious Disease Dynamics, Asilomar, CA, USA, 2008.

20

Transfer of a Traditional Serotyping System (Kauffmann–White) onto a MALDI-TOF-MS Platform for the Rapid Typing of Salmonella Isolates

Chloe Bishop, Catherine Arnold and Saheer E. Gharbia
Department for Bioanalysis and Horizon Technologies, Health Protection Agency Centre for Infections, London, UK

20.1   Introduction

The characterization of pathogenic microorganisms and monitoring of their global spread demands schemes that are easy to standardize and use, portable, high-throughput and automated. In a routine clinical laboratory, characterization of isolates often still uses phenotypes, such as biochemical pathways, antibiotic resistance, antigens presented on the cell's surface and fatty acid patterns. Such methods rely on gene expression, which can be affected by growth conditions, growth phase, and the frequency of occurrence of spontaneous mutations, so these tests all have a tendency to vary.

Typing schemes used are different between organisms, not only because of the uniqueness of genetic and phenotypic traits seen in some taxa, but also because of differences in ecological distribution and diversity of strains. Strains within some species are much more variable than others and require a different approach than, for example, clonal strains. There is a demand for harmonized, molecular approaches for microbial typing. Databases of microbial genetic information are rapidly expanding and the technologies are developing continuously. Unlike phenotypic methods, DNA sequence-based approaches give the flexibility to target selective genes marking for the organism. The use of comparative sequencing by matrix assisted laser desorption/ionization time-of-flight mass spectrometry

Mass Spectrometry for Microbial Proteomics   Edited by Haroun N. Shah and Saheer E. Gharbia
© 2010 John Wiley & Sons, Ltd

(MALDI-TOF-MS) for automated high-throughput microbial DNA sequence analysis is a way towards the ideal (rapid, standardized and high-throughput).

In this chapter, *Salmonella enterica* is used to illustrate the possible application of MALDI-TOF-MS for microbial characterization. Along with *Escherichia coli*, *S. enterica* probably represents the most commonly used laboratory organism for biochemical and genetic studies; however, it remains a major threat to human and animal health in both the developing and developed world. Salmonellosis is one of the most common and widely distributed diseases of humans and animals and constitutes a major public health burden (134). Isolations of *Salmonella* causing gastroenteritis in humans have increased in recent years in developed countries, primarily because modern methods of animal husbandry, food preparation, and distribution encourage the spread of *Salmonella*. The health services in England and Wales report approximately 15 000 annual cases of salmonellosis (45). There are millions of human cases worldwide every year and the disease results in thousands of deaths but because many milder cases are not diagnosed or reported, the actual number of infections may be very much higher (134).

The current, widely accepted, method targets variation of *Salmonella* antigens to separate them into over 2500 serotypes. This allows discrimination between isolates for diagnosis and infections to be tracked. New techniques for the molecular typing of isolates have been described, usually a combination of phenotypic and genotypic approaches. *Salmonella* represents a challenging case study to test a rapid and universal scheme. This chapter describes specifically the transfer of the serotyping scheme for *Salmonella* analysis from phenotypic detection to a rapid, molecular-based characterization.

The following pages introduce the pathogen and its underlying genetics, the nature of its antigens and current typing methods. To illustrate how serotypes can be rapidly determined from mass spectra of antigen genes, the sequence variation found in salmonella antigens will be illustrated.

The use of MALDI-TOF to detect known sequence motifs and identify novel sequences will then be described. Finally, how the typing approach using re-sequencing can be transferred to other gene targets or other organisms will be discussed.

20.2   Salmonella, the Pathogen

20.2.1   Biology

Salmonella was discovered in pigs suffering from hog cholera in 1885, by pathologist Theobald Smith, and Daniel Salmon termed it *Hog-cholerabacillus*, mistakenly thinking it was the cause of swine plague (108).

Salmonella species are facultative anaerobic rods, which measure about 2 μm by 0.5 μm. They are Gram-negative by staining, indicating only a thin layer of peptidoglycan between the cytoplasmic and outer membrane compared with Gram-positive organisms. The outer membrane is covered with flagella in a peritrichous arrangement; these organelles enable cells to move towards nutrients and away from unfavourable conditions (118,130). As with other Gram-negative bacilli, the cell envelope of salmonellae contains a complex lipopolysaccharide (LPS) structure that is released on lysis of the cell. Important in determining virulence of the strain, LPS may function as an endotoxin, which can activate the immune system and induce fever. Circulating endotoxin may be responsible in part for septic shock that can occur in systemic infections (86).

Salmonella is a food- and water-borne pathogen that can be easily disseminated in a population. The primary reservoir of the species *S. enterica* is the intestinal tracts of livestock, poultry, and reptiles. Contaminated water, undercooked poultry and eggs are the most common sources of human infections, typically resulting in enterocolitis or food poisoning accompanied by nausea, vomiting, and watery or bloody diarrhoea. Although salmonellosis can be controlled by antibiotics, drug resistant and hypervirulent strains are difficult to combat.

20.2.2 Pathogenesis

Pathogenicity of *Salmonella* strains depends on a combination of virulence factors produced by the organism. *Salmonella* infection may be restricted to the gut, or may lead to systemic infection. In infection restricted to the gut, the expression of genes in *Salmonella* pathogenicity island (SPI)-1 is upregulated, inducing flagellin release. Flagellin is translocated to the basolateral surface of host epithelial cells. It is recognized by the host immune system and is bound to a toll-like receptor (TLR), inducing interleukin (IL)-8 secretion. IL-8 guides the influx of neutrophils and pathogen-elicited epithelial chemoattractants (PEECs) direct the transmigration of neutrophils into the gut lumen. The influx of neutrophils leads to epithelial tissue injury and diarrhoea which helps clear the infection from the host and prevents systemic spread of *Salmonella*.

In systemic infection, SPI-1 expression is upregulated and flagellin released. Virulence antigen expression is down-regulated, further promoting effector protein secretion and flagellin release. *Salmonella* cells traverse the epithelium. IL-6 secretion is induced and the invading bacteria are taken up by macrophages. Infected macrophages die and IL-1 is released and a pro-inflammatory response is induced. Invading bacteria are disseminated throughout the body, possibly within CD18+ cells (124).

20.2.3 Clinical Disease

Salmonellosis includes several syndromes: gastroenteritis, enteric fevers, septicaemia, focal infections, and an asymptomatic carrier state. Particular serotypes are often associated with a particular syndrome (18,64). Host-adapted serotypes produce enteric fever: *S.* Typhi, *S.* Paratyphi, and *S.* Sendai in man; *S.* Abortusovis in sheep; *S.* Gallinarum in fowl and *S.* Typhi-suis in swine. In man, *S.* Choleraesuis produces septicaemia or focal infections; *S.* Typhimurium and *S.* Enteritidis produce gastroenteritis. Depending on the strain, the typical symptoms are mild fever and diarrhoea which normally begin 6–48 h after ingestion of contaminated water or food. In healthy adults the disease is usually self-limiting but it is more serious in the young, the old, and those with underlying medical conditions. *S.* Typhimurium causes a systemic disease in mice as does typhoid in humans, thus *S.* Typhimurium has been widely used as a model for typhoid. Typhoid fever is associated with more severe symptoms in the digestive system in the second phase of the illness as it penetrates to the small intestine (52). It can be life threatening without antibiotic treatment. Of those infected with *S.* Typhi, 3% become chronic carriers, compared with 0.6% of those infected with a nontyphoidal strain (134). An important property of *S.*

Typhi is its ability to survive in the environment, water and food, sometimes for several months. Paratyphoid fever is a similar but generally milder disease.

20.3   Complex Genetic Structure and the Need to Subtype this Genus

20.3.1   Phylogeny

The genus *Salmonella* is part of the Enterobacteriaceae (nonsporulating, Gram-negative rods) which includes many clinically important bacteria, usually associated with intestinal infections. Sequences of 16S rDNA are widely used as evolutionary chronometers in phylogenetic studies as well as for species identification (16,78,131). The 16S tree fits broadly with taxonomic classification of bacterial species based on alternative phylogenetic markers but due to the high level of sequence conservation between the Enterobacteriaceae (>98.5%) (66), phylogenetic trees based on 16S rDNA cannot be used for the separation of many of the species (35).

It has been estimated that the genus *Salmonella* diverged from *E. coli* 100 million–160 million years ago (94), however, they remain close relatives so parallels can be drawn between them. *Salmonella* has acquired a number of genes and abilities since its divergence from *E. coli* (18,42,101), contributing to the evolution of subspecies, encoding new pathogenic abilities, allowing *Salmonella* to colonize new hosts and niches (6,10,65,126).

Organisms in the genus *Salmonella* are so genetically similar that they are now considered as belonging to one of two species; *Salmonella enterica* or *Salmonella bongori*. Subdivision of the genus has been made to at least seven genetically distinct subspecies and further to serotypes (99). There are over 2500 serotypes of *Salmonella* and the disease produced depends to some extent on which strain is causing the infection. Serotype recognition is therefore important in epidemiology; often an outbreak or epidemic caused by strains of one serotype can be traced to a common source. The serotypes *Salmonella* Typhimurium, and *S.* Enteritidis are responsible for the majority of the cases of human *Salmonella* infection in the UK (46).

20.3.2   Virulence and Gene Transfer

Salmonella infections in humans vary with the infectious dose, the nature of the contaminated food, the susceptibility of the host, the serotype and the strain. The Virulence (Vi) antigen is a capsular polysaccharide antigen that occurs in only three *Salmonella* serotypes: *S.* Typhi, *S.* Paratyphi C, and *S.* Dublin. Strains of these three serovars may or may not have the Vi antigen; those with the Vi antigen are less prone to lysis by nonimmune serum (77).

A variety of virulence factors is required for the *Salmonella* to be fully pathogenic and some factors are host-specific. Virulence factors include a complete LPS coat, the ability to invade and replicate in cells and possibly the production of toxin(s) (95). These factors can be encoded by genes present on the chromosome, bacteriophages, pathogenicity islands or on plasmids (33,65,82). Plasmids are typically found in serotypes associated with infections in humans and farm animals. Genes on high molecular weight plasmids can encode the ability to survive within macrophages or antibiotic resistance, however the biological functions of low molecular weight plasmids are largely unknown (36,107). The variability of low molecular weight plasmids has been applied to the molecular typing of *Salmonella* isolates (105,123). Bacteriophages typically encode toxins and proteins that

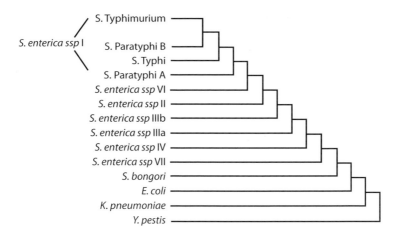

Figure 20.1 *Lateral gene transfer in* Salmonella. *The phylogenetic tree was based on sequence comparisons for the* non-Salmonella *species and comparative genomic hybridization of strains to a* S. Typhimurium *LT2 microarray (100). Maximum parsimony analysis and 1000 bootstraps were performed in the construction of the tree. The numbers at the nodes refer to the number of genes acquired and retained by subsequent subspecies*

alter the antigenicity of the cells they infect (11). Pathogenicity islands are large, unstable chromosomal segments that are absent from related nonpathogenic organisms (38,70). *Salmonella* has acquired two pathogenicity islands, probably by horizontal gene transfer (48,112,126). These gene transfers represent important stages in *Salmonella* evolution (Figure 20.1). The acquisition of SPI-1 marks the divergence of *Salmonella* from *E. coli* to bring about *S. bongori*, able to invade host epithelial cells; the acquisition of SPI-2 brought about a new species – *S. enterica*, able to survive within macrophages (41).

Within related species, lineages and points of divergence have been inferred from genomic DNA hybridization (14) and DNA digestion (76), multi-locus enzyme electrophoresis (MLEE) (13) and by comparing amino acid or DNA sequences of housekeeping genes (40,62,125). DNA-DNA hybridization of fragments produced by shearing genomic DNA has shown *Salmonella* strains to be 85–100% related to one another, while only 45–55% related to *E. coli* and *Citrobacter* (19). Horizontal gene transfer and recombination between species is relatively uncommon (17) but more likely between similar species, such as *S. enterica* and *E. coli*, because they are susceptible to the same phages, plasmids and transposons (85). Genetic material conferring antibiotic resistance is more readily transmissible (9) and the spread of antibiotic resistance in *Salmonella* is of increasing concern; an additional test for also determining the antibiotic resistance profile of the strain would be key to subtyping.

20.3.3 Necessity to Subtype

The disease produced by *Salmonella* infection depends to some extent on which strain is causing the infection. Recognition of the over 2500 different serotypes is important in epidemiology; often a common source can be linked to an outbreak or epidemic caused by strains of one serotype. Salmonella is a very clonal genus and it is often necessary to type beyond the serotype level to distinguish between outbreak strains. Outbreaks can arise due to genetic change in the strain, such as antigenic variation but are also associated with a food contamination event or production process. Strains may acquire genetic changes that increase their fitness for a new niche, leading to a rise in their prevalence. Genome sampling methods may allow the detection of horizontal gene transfer and thereby help predict and detect the emergence of new types.

Pulsed-field gel electrophoresis (PFGE) (90,105) is currently the benchmark for molecular subtyping of *Salmonella*, however it is best used in combination with plasmid profiling and ribotyping for strain discrimination (133). Other approaches include fluorescent amplified fragment length polymorphism (FAFLP) (113) and MLEE (7) which sample genomic DNA and provide a view of genetic diversity between strains and partially group some serotypes but on the whole do not group or identify serotypes. Multi-locus sequence typing (MLST) has been used to discriminate between *Salmonella* strains by sampling variation in a set of housekeeping genes which precludes antigen encoding genes (67). As antigenic analysis has formed the basis of *Salmonella* typing for 80 years, the genes encoding antigens are a logical target for molecular typing of *Salmonella*.

20.4   Antigenic Analysis – the Traditional Kauffmann–White Scheme and Its Future

20.4.1   Serotyping

Driven by selective pressures from the host immune system (37), the rate of divergence of proteins expressed on the cell surface is much greater than those located internally (132), making antigenic variation useful for identifying and differentiating between strains. *Salmonella* antigen expression determines strain pathogenicity and therefore variation of these antigens has formed the basis for *Salmonella* serotyping.

The Kauffmann–White scheme, first published in 1929, divides *Salmonella* into more than 2500 serotypes according to their antigenic formulae. Routine clinical laboratories classify *Salmonella* into serotypes according to their particular combination of antigens by observing agglutination with specific antisera (99). *Salmonella* express flagella (phase 1 and phase 2) and somatic (O) antigens. Within the 2500 serotypes, more than 100 distinct flagella antigens and 46 O antigen groups are recognized by agglutination methods.

Antiserum production involves inoculating rabbits with an antigen preparation, typically several times at intervals over a 2-week period (53). Six to eight days after the last injection blood is collected from which serum is prepared. O and H antigens are separately prepared (formalized) from bacterial solutions, and then mixed with specific antiserum. Somatic (granular) and flagellar (floccular) agglutination must be distinguished and in some cases, the specificity of the agglutination reaction must be confirmed by titration. As most *Salmonella* are biphasic, it is necessary to determine the specificity of both flagellin phases for the complete serotype. The isolate is applied to a semi-solid medium containing antibody against the expressed flagellar antigen. A binding reaction between the known flagellae and antibody inhibits cell motility, so any motile bacteria are known to have switched phase expression and the specificity of the alternate antigen can be determined by serology (136).

20.4.2 Flagellar Antigens

Expressed on the cell surface, flagella are responsible for cell motility, virulence and antigenicity (81). Most serotypes exhibit biphasic flagellar antigen expression by alternately expressing two genes, *fliC* (phase 1) and *fljB* (phase 2) which encode flagellins of different antigenicity. *Salmonella* serotyping methods recognize 70 distinct phase 1 flagellar antigenic factors and 50 phase 2 flagellar antigenic factors although the latter are not always present. Single factors (e.g. d or i) define some antigens; others are defined by several factors, separated by commas in the formula (e.g. l,v or e,n,x).

Most cells have twenty or so flagellae, distributed all over the cell. Type III secretion systems are used by *Salmonella* to inject toxic molecules (mostly proteins) into host cells to invade or kill (3). Genetic similarities between the systems for type III secretion and flagella synthesis suggest the latter arose from secretory systems, or at least that the two have evolved from a common ancestral system (32).

20.4.2.1 Structure and Function of Flagella

Structure. The flagellum filament is a hollow, rigid cylinder whose wall consists of flagellin proteins. The ends of the protein are conserved and responsible for the hairpin shape of the subunit while variation in the central region generates the antigenic diversity. Twenty thousand or so flagellin proteins polymerize to make up the flagellum filament (80). Flagella synthesis occurs by the export of flagellin proteins up through the base of the growing hollow filament; the cap structure positions the proteins at the distal end of the filament (2). The tertiary structure of a filament can be altered by environmental conditions, but also by amino acid substitution (60).

Function. A rotary motor turns the flagellum, giving the bacterial cell motility. Motile bacteria can respond to environmental stimuli accordingly, such as movement towards nutrients or away from waste products (31). In *Salmonella*, the flagella may also have a role in moving to epithelial cells to invade (97).

Subunit Structure. The *Salmonella* flagellin subunit forms a hairpin structure; the ends of the protein are highly conserved since these regions are responsible for the flagellin shape (Figure 20.2). Terminal regions of flagellin are essential for its polymerization and excretion (44,49). The loop of the hairpin, the central part of the flagellin subunit, is the most exposed part on the filament surface; amino acid substitutions in this region generate antigenic diversity (58).

20.4.3 Flagellar Variation

20.4.3.1 Gene Structure

Parish first showed that the central region of the *Salmonella* flagellin protein is responsible for antigenic specificities (96). Amino acid substitutions across a 300 bp region are responsible for the antigenic variation assayed in serotyping (22,64,116), and one or a few amino acid replacements are sufficient to alter flagellar antigenic character (44,91). Flagellins have diversified by point mutations, horizontal gene transfer and recombination leading to new antigenic types (73).

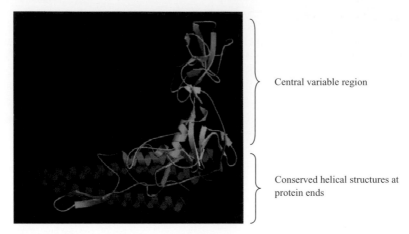

Central variable region

Conserved helical structures at protein ends

Figure 20.2 *Secondary structures within flagellin type i exhibited by S. Typhimurium. This image was created by Protein Data Bank Viewer based on the crystal structure (109). The variable central region is shown in green and the conserved protein ends are helices shown as dark blue and red*

Nucleotide sequence encodes antigenic factors (57,74,84,114). Serological analysis of mutant genes determined that amino acids 229 and 230 were at the centre of dominant surface epitope of flagellin d because their replacement eliminated most of the reactivity with anti-d antisera (44). However, it was not determined how much reactivity was lost by replacement of other amino acids in the sequence compared with these 'central' amino acids. Amino acids that are involved in epitope structure are not necessarily adjacent in protein secondary structure.

20.4.3.2 *Phase Variation*

Flagellin is a virulence factor that is recognized by the innate immune system in organisms as diverse as flies, plants and mammals (43,137). Phase variation in *Salmonella* is a unique mechanism for evasion of host defence systems, involving the switching between expression of serologically distinct flagellin proteins; phase 1 (which is encoded by the *fliC* gene) and phase 2 (encoded by the *fljB* gene). The shift in phase occurs every 10^{-3}–10^{-5} generations. *S. bongori* and *S. enterica* subspecies IV and IIIa are monophasic; each serotype expressing just one type of flagellin. *S. enterica* subspecies IIIb, II, VI and I are typically, but not exclusively, biphasic. A few serotypes express a third distinct flagellin, encoded by plasmid-encoded *flpA* (116). Serotypes *S.* Gallinarum and *S.* Pullorum are nonmotile, expressing no flagella.

Mechanism.   The genes *fliC* and *fljB* are at separate locations on the chromosome. Within the *S.* Typhimurium LT2 chromosome *fliC* is located at 42.5 Cs (2047658 bp) and *fljB* at 60.1 Cs (2913230 bp). *fljB* is part of an operon containing a recombinase (encoded by *hin*) and a *fliC* repressor (39; Figure 20.3).

Phase variation occurs by a recombination event involving a revertible element called the H segment, a 993 bp region containing a promoter (115). Either *fliC* or *fljB* are expressed depending on the orientation of the element because in one orientation the

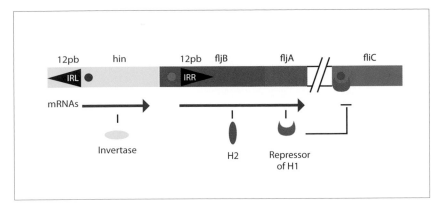

Figure 20.3   *Mechanism for phase variation. Antigenically distinct flagellins are alternately expressed depending on the orientation of hin. fliC is repressed by the product of fljA when fljAB are transcribed*

promoter becomes inactive as it is hundreds of base pairs upstream and inverted. *fljA* is co-transcribed with *fljB* resulting in exhibition of a phase 2 phenotype. When *fljB* is inactive, *fliC* is de-repressed and a phase 1 phenotype results. Homologies between the invertase genes *hin* in *Salmonella*, *cin* in bacteriophage P1 and *gin* in bacteriophage Mu suggest the phase variation mechanism in *Salmonella* originated from a phage (120). *fljB* is postulated to be a duplication of *fliC* (121).

20.4.4   Somatic Antigens

20.4.4.1   *Structural Variation of the O-Antigen*

Lipopolysaccharides are expressed on the bacterial cell surface. They are hydrophilic and enable the bacteria to form stable, homogeneous suspensions in saline solution. These consist of three parts: the innermost lipid A, a core oligosaccharide and the outermost linear polymer of repeating polysaccharide units, the somatic (O) antigen. Between three and six sugars make up an O-unit; 1–40 repeats of the unit make up an O-antigen. This antigen is particularly complex; the nature of its epitopes are still poorly understood.

The O-antigen is subject to diversifying selection by the host immune system and is a receptor for bacteriophages (25). The pattern of sugars is highly variable and is used to divide *S. enterica* into 46 serogroups (99), each group possessing a unique O-antigen structure which can often be correlated with the mode of pathogenesis (34). *Salmonella* strains from major groups A, B or D have similar O-antigen structures and are often pathogenic.

The fact that there are very few O-antigens common to *E. coli* and *Salmonella* shows that there has been extreme turnover of O-antigens since species divergence. The existing variation in O-antigens is such that it cannot have occurred by recent mutations (102). The *rfb* gene cluster has evolved by acquiring clusters of genes by gene transfer from other species (68,135). The level of genetic mutation in the population is maintained by selection.

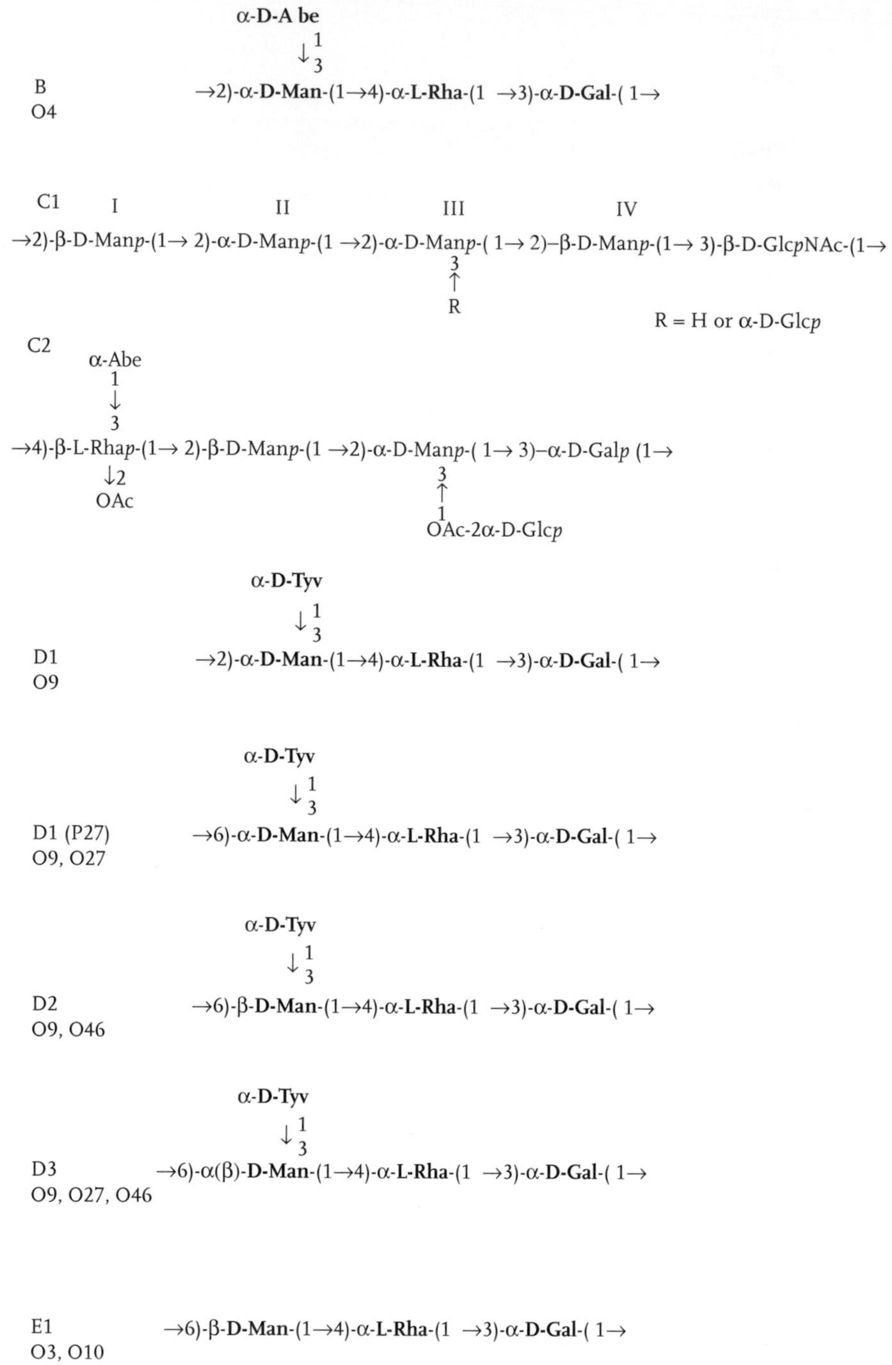

α-**D-A be**
↓ 1
　 3

B
O4　　　　　　　→2)-α-**D-Man**-(1→4)-α-**L-Rha**-(1 →3)-α-**D-Gal**-(1→

C1　　I　　　　　　　II　　　　　　　III　　　　　　　IV

→2)-β-D-Man*p*-(1→ 2)-α-D-Man*p*-(1 →2)-α-D-Man*p*-(1→ 2)–β-D-Man*p*-(1→ 3)-β-D-Glc*p*NAc-(1→
　　　　　　　　　　　　　　　　　　　　　　　　　3
　　　　　　　　　　　　　　　　　　　　　　　　　↑
　　　　　　　　　　　　　　　　　　　　　　　　　R

　　　　　　　　　　　　　　　　　　　　　R = H or α-D-Glc*p*

C2
　　α-Abe
　　1
　　↓
　　3
→4)-β-L-Rha*p*-(1→ 2)-β-D-Man*p*-(1 →2)-α-D-Man*p*-(1→ 3)–α-D-Gal*p* (1→
↓2　　　　　　　　　　　　　　　　　　　　3
OAc　　　　　　　　　　　　　　　　　　　↑
　　　　　　　　　　　　　　　　　　　　1
　　　　　　　　　　　　　　　　　OAc-2α-D-Glc*p*

α-**D-Tyv**
↓ 1
　 3

D1
O9　　　　　　　→2)-α-**D-Man**-(1→4)-α-**L-Rha**-(1 →3)-α-**D-Gal**-(1→

α-**D-Tyv**
↓ 1
　 3

D1 (P27)
O9, O27　　　　　→6)-α-**D-Man**-(1→4)-α-**L-Rha**-(1 →3)-α-**D-Gal**-(1→

α-**D-Tyv**
↓ 1
　 3

D2
O9, O46　　　　　→6)-β-**D-Man**-(1→4)-α-**L-Rha**-(1 →3)-α-**D-Gal**-(1→

α-**D-Tyv**
↓ 1
　 3

D3
O9, O27, O46　　→6)-α(β)-**D-Man**-(1→4)-α-**L-Rha**-(1 →3)-α-**D-Gal**-(1→

E1　　　　　　　→6)-β-**D-Man**-(1→4)-α-**L-Rha**-(1 →3)-α-**D-Gal**-(1→
O3, O10

Figure 20.4 *Structures of O-antigen repeat units synthesized by various* Salmonella *species (20). Sugar residues are abbreviated: Abe (abequose), Gal (galactose), Man (mannose), Rha (rhamnose), Tyv (tyvelose), GlcNAc (N-acetylglucosamine). Linkages are shown in parentheses and can be alpha or beta forms*

Table 20.1   *Major somatic antigen groups in Salmonella. Each O-group is recognized by characteristic antigenic factor(s). As well as the variation in the sugars that constitute the backbone, different linkages between the sugars (α and β) can determine antigenic specificity*

Antigen group	Characteristic factors	Backbone	Side chain
A	2	Mannose–rhamnose–galactose	Paratose
B	4	Mannose–rhamnose–galactose	Abequose
C1	6,7	Mannose–mannose–mannose–mannose–GlcpNA	H or α-D-Glcp
C2	8	Rhamnose–mannose–mannose–galactose	Abequose
D1	9	α-Mannose–rhamnose–galactose	Tyvelose
D2	9,46	β-Mannose–rhamnose–galactose	Tyvelose
D3	9,12,27,46	α- or β-Mannose–rhamnose–galactose	Tyvelose
E1	3,10	β-Mannose–rhamnose–galactose	

Side Chains and Linkages.   The extensive range of sugars found in O-antigens allows for numerous combinations of donor sugar, acceptor sugar and acceptor carbon atom for the glycosidic linkages. In turn, this provides for a very large number of linkage specificities (110). Different types of linkages between O-units can result in serological variation and different side chains confer other or additional O epitopes (Figure 20.4). Further antigenic heterogeneity is associated with nonessential modifications involving acetylation or glucosylation of specific residues in the O-units (72). The glucosylation of D-galactose residues which determines factor O:12 in *S.* Typhimurium is known to be incomplete and random and to involve only chains longer than six repeat units (47).

The O-antigens of groups A, B and D possess the same pattern of sugars in their backbone, but have different side chains (Table 20.1). O-antigens C2 and B consist of the same sugars, but in a different structure. C1 antigenic determinants are variable (93). Group specific transferases may be responsible for antigenic differences and this rearrangement of sugars forms the basis of antigenic variation.

20.4.4.2   rfb *Gene Cluster*

In *Salmonella*, O-antigen synthesis and assembly is encoded by *rfb* genes (Table 20.2) in a cluster that typically contains 12 open reading frames, and ranges in size between serotypes, from approximately 8 to 23 kbp (Table 20.3). The variation of the somatic antigen is due to the presence or absence and the arrangement of genes rather than due to individual gene sequence variation. Reeves and co-workers have shown that *rfb* gene clusters sequenced from different strains of *Salmonella* and *E. coli* are highly divergent (103).

Some genes for somatic antigen synthesis may be outside the *rfb* gene cluster or even plasmid encoded. Members of antigenic Group 54 often possess other O-antigens due to the acquisition of a 7.5 kb plasmid on which O-factor O:54 is encoded (61). Serotypes of Group 54 may lose this plasmid, resulting in loss of the factor.

Transposable elements seen in *rfb* gene clusters are often found adjacent to sites where there is evidence of recombination. Strains belonging to serogroups B and D2 have transposable elements next to the *wzy* gene, which have been associated with the evolution of

Table 20.2 Rfb genes of Salmonella and their functions. The bacterial polysaccharide gene nomenclature (BPGN) was proposed when all 26 available rfb genetic symbols were used(104)

BPGN	Genetic symbol	Function
abe	rfbJ	CDP-Abequose synthesis
ddhA	rfbF	CDP-Abequose synthesis
ddhB	rfbG	CDP-Abequose synthesis
ddhC	rfbH	Rhamnose transferase. dehydrase
ddhD	rfbI	Reductase
fcl		GDP-L-fucose synthetase
galF		Glucose-1-phosphate uridylyltransferase modification
gmD		GDP-mannose-4,6-dehyratase
gmm		GDP-mannose mannosyl hydrolase
manB	rfbL	Phosphomannomutase structural protein
manC	rfbM	GDP-mannose synthesis
rmlA	rfbA	Glucose-1-phosphate thymidylyltransferase
rmlB	rfbB	dTDP-D-glucose 4,6-dehydratase
rmlC	rfbC	dTDP-4-keto-6-deoxy-D-glucose 3,5-epimerase
rmlD	rfbD	dTDP-4-keto-L-rhamnose reductase
wbaL		O-acetyltransferase
wbaN	rfbN	Rhamnose transferase
wbaO	rfbO	Mannose transferase [catalyses β(1–4) linkage]
wbaP	rfbP	Galactose-1-P transferase. May also have a role in O-antigen transport
wbaQ		Rhamnosyltransferase
wbaR		Abequosyltransferase
wbaU	rfbU	Mannose transferase [catalyses α(1–4) linkage]
wbaV	rfbV	Abequose transferase
wbaW		Mannosyltransferase II
wbaZ		Mannosyltransferase I
wbyL		Glycosyltransferase
wzx	rfbX	O-unit translocation across the cytoplasmic membrane
wzy/ wbaA		O-antigen polymerase
wzz		Chain length determinant
	rfbE	Involved in conversion of CDP-paratose to CDP-tyvelose

Table 20.3 The known genetic variation in rfb gene cluster between groups

O-group	rfb Gene cluster structure
A	rmlB-rmlD-rmlA-rmlC-ddhD-ddhA-ddhB-ddhC-prt-tyv-wzx-wbaV-wbaU-wbaN-manC-manB-wbaP
B	rmlB-rmlD-rmlA-rmlC-ddhD-ddhB-ddhC-abe-wzx-wbaV-wbaU-wbaN-manC-manB-wbaP
C1 C4	wbaA-wbaB-wbaC-wbaD-manC-manB-wzx
C2 C3	rmlB-rmlD-rmlA-rmlC-ddhD-ddhA-ddhB-ddhC-abe-wzx-wbaR-wbaL-wbaQ-wzy-wbaW-wbaZ-manC-manB-wbaP
D1	rmlB-rmlD-rmlA-rmlC-abeI-abeF-abeG-abeH-tyvS-tyvE-wzx-wbaV-wbaU-wbaN-manC-manB-wbaP
D2	rmlB-rmlD-rmlA-rmlC-abeI-abeF-abeG-abeH-tyvS-tyvE-wzx-wbaV-H-wzy-O-N-manC-manB-wbaP
E	rmlB-rmlD-rmlA-rmlC-wzx-wzy-wbaO-wbaN-manC-manB
H	wbaC-wbaD-manC-manB-wzy-wzx-IS3 remnant
I	wzx-wbyL-wzy-galactosyl transferase-galE-gmd-fcl-gmm-manC-wbyL-manB
K	wbaC-wbaD-manC-manB-wzy-wzx

these strains via transposon-mediated recombination (127). Of the *rfb* genes studied thus far, *wzy*, which encodes O-antigen polymerase, is reported to be the most variable. Different *wzy* alleles confer different linkages between the O-units and therefore different O-factors. Sequences in related serotypes can have no detectable similarity (20,29,71). Some serotypes have been found to possess more than one functional *wzy* gene (87).

The *rfb* gene cluster is of keen interest for epidemiological and evolutionary studies as well as being a potential target for molecular typing assays. There is DNA sequence variation among *rfb* genes but, unlike the flagella antigens, where DNA polymorphisms determine antigenic variation the presence and arrangement of genes determine somatic antigen variation. It is not known precisely how the antigenic factors correspond to the sugar-linkage configuration of the somatic antigen. It is complex and multiple sugars that are likely to be involved (92). However, some proposals have been made towards what forms the specific epitopes (20). Nucleotide analysis of *rfb* genes is not expected to reveal epitopes, but sequence differences between serotypes do exist. The level of variation and the stability of these sequences as markers for O-groups and O-factors remain unknown.

DNA sequence-based methods offer many advantages over traditional serotyping and gel-based methods. Traditional serotyping works by agglutination with antisera. These methods are stable, reproducible and have high typeability but the drawbacks are the intensity of labour, an extended turnaround time if the strain needs to be forced to switch phases, and difficulties maintaining good stocks of antisera. The need for a robust single molecular technology to discriminate different serotypes is clear. PFGE is a molecular method used routinely to subtype *Salmonella* (90). Other methods have also been applied, including FAFLP (23,69,113), MLEE (8,12,74) and MLST (26,67,119). These approaches provide a view of genetic diversity between strains and partially group some serotypes, but do not group or identify all serotypes. Faster, more reproducible and more ethical ways to type this complex genus need to be indentified.

20.5   Sequence-Based Methods to Determine Serotypes

The genes encoding antigens are often specifically targeted for molecular typing of *Salmonella*; however, the need for a robust single molecular technology for this is clear. In 1993, Luk *et al.* (79) published a length heterogeneity PCR (LH-PCR)-based method that targeted genes only associated with particular O-antigens (A, B, C2 and D), while a more recent study by Fitzgerald *et al.* (29) developed a serotype-specific PCR assay targeting a single O serotype (O:6,14). Several studies have used a molecular approach to discriminate between particular flagellar serotypes (21,50,63). *FliC* fragment restriction patterns using a dual enzyme combination allowed differentiation of flagellar types b, i, d, j, l,v, and z10 but not r and e,h or [f],g,m, [p], g,p, and g,m,s. (10). Hong used restriction fragment patterns of *fliC* and *fljB* for serotyping of poultry *Salmonella* but could not distinguish *S.* Enteritidis from *S.* Gallinarum and *S.* Dublin (11). Design of a multiplex-polymerase chain reaction (multiplex-PCR) to identify 1,2, 1,5, 1,6, 1,7, 1,w, e,n,x and e,n,z15 second-phase antigens has been reported (5). Peters and Threlfall reported *fliC* restriction fragment length polymorphism (RFLP) profiles were not specific enough to differentiate between certain serotypes (98). To date no studies have attempted a universal molecular serotyping approach. Relevant publicly available sequence data are incomplete, and so is information about epitopes in specific serotypes, therefore approaches are currently being explored to

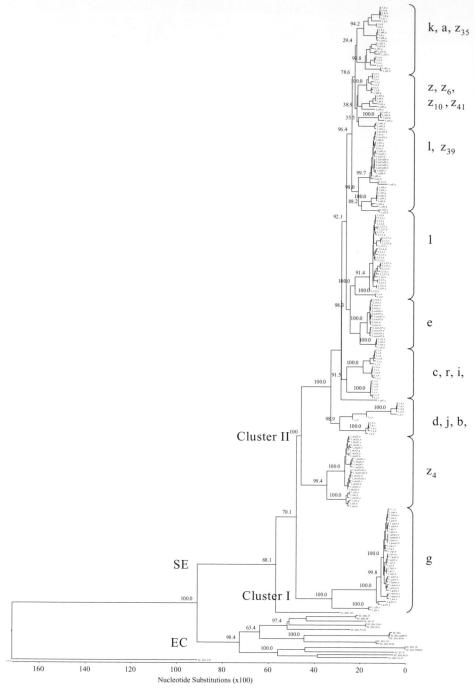

Figure 20.5 *Phylogenetic tree of* Salmonella *flagellin* full gene amino acid sequences, *inferred from DNA sequences determined experimentally and from databases. The tree features all unique complete sequences from* Salmonella *loci fliC and fljB and* E. coli *fliC included in this study. Bootstrap analyses were performed 1000 times; results are shown at the major nodes. SE (S. enterica) and EC (E. coli) are separated by a bootstrap value of 100*

characterize the expressed antigen or the encoding genes as an alternative to traditional serotyping. For *fliC*, evidence from antibody binding studies suggests that sequences of ~300 nucleotides of the central variable region of flagellin correlate with serotype (44) and differences in amino acid sequence can be associated with differences in antigenic specificity. Comparative sequencing has distinguished some *salmonella* serotypes or biotypes (30,88,89,117).

20.5.1   Flagellin Sequences Correspond Directly to *Salmonella* Serotype

fliC genes of *Salmonella* have evolved a mosaic structure made up of recombined segments (73) and sequences in the central region of flagellin gene correspond to serotype (55). Following comprehensive studies, public sequence databases now contain full length gene sequences of all the 43 *E. coli* flagellin types (129) and around 95% of the repertoire of *Salmonella* flagellin gene sequences (88,89). Generally, complete flagellin gene sequences cluster by the antigen they encode. *E. coli fliC* sequences form two groups; *Salmonella* sequences form three main clusters revealing the relatedness of serologically distinct antigens. Cluster analysis group sequences encoding common factors (such as characters e, g, or 1) and demonstrates the relatedness of other antigens (such as r and i, or z_{35} and a). For *Salmonella*, some antigens do not cluster as predicted by serology, and sequences encoding antigen z_{29} do not fall within any of the three main clusters.

20.5.2   Specific SNPs

Antigens have been characterized and sero-specific motifs identified by analysis of multiple sequence alignments (88,89). *Salmonella* sequences fall into two groups, labelled Cluster I and Cluster II in Figure 20.5. Cluster I includes all sequences encoding phase 1 flagella antigens exhibiting factors g or m (the G-group) and antigen z_{29}. Cluster II includes all other *Salmonella* antigens, phases 1 and 2, here termed Non-G. Sero-specific polymorphisms can be identified within the central variable region where consensus sequences of Cluster I and Cluster II diverges, between amino acid positions 175–407.

Sequences of the G-group flagellins are highly homologous, with certain sequences encoding different antigenic factors varying by single base changes. Notwithstanding possible post-translational modifications encoded outside flagellin loci, such polymorphisms must be responsible for changes in epitope specificity. Pairwise comparison of *fliC* sequences within the G-group can pinpoint epitopes, for example: *S.* Enteritidis (g,m) and *S.* Antarctica (g,z63) differ by a duplication of codons; *S.* Enteritidis (g,m) and *S.* Blegdam (g,m,q) differ by a single substitution; *S.* Dublin (g,p) and *S.* Rostock (g,p,u) differ by a single amino acid substitution.

Cluster II, here termed the Non-G-group, includes all *fljB* antigens, and all *fliC* antigens except z_4 and those falling into the G-group. Cluster II contains the majority of *Salmonella* sequences and lower levels (80.3%) of amino acid homology were observed. The longer branch lengths indicate much more variation within this group of sequences than the G-group. There are no common factors linking this group of antigens though some antigens occur together in the tree, indicating a possible relationship between factors b and d; c and e,h; i and r and factors k, a and z_{35}. Sequence conservation within Non-G alleles is demonstrated by 97.8–99.1% homology, while 80.35% homology can be measured within the complex. The high level of variability between alleles in this group does not allow the

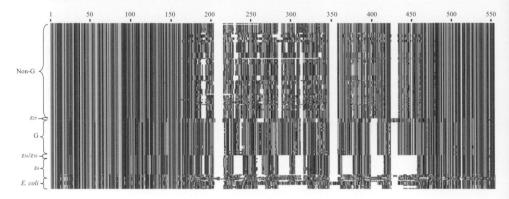

Figure 20.6   *Alignment of flagellin amino acid sequences. BioEdit was used to generate this image where amino acids are represented by colours, assigned arbitrarily. Alignment of sequences revealed conserved ends and variable regions in the gene centre. Large differences in Non-G sequences were apparent, compared with SNPs in the G-group. Two conserved region sequences could be recognized, one sequence specific to fliC G-group antigens and the other to Non-G (fliC and fljB) antigens*

association of specific amino acids to epitope formation that is possible with the G-group sequences.

20.5.3   Subtyping by Antigen Sequence

By characterizing the genetic variation underlying antigenic specificity, levels of variation within and between serotypes can be determined and genetic variation that is not expressed by serotyping can be detected (Figure 20.6). The increased resolution from the DNA sequence as opposed to serology allows discrimination within some serotypes (i.e. subtyping) and provides the potential to dispense with the need for multiple methods for strain discrimination in these cases. Furthermore, analysis of multiple representatives revealed that some antigens were coded for by multiple sequences. Because of this, there is potential for DNA sequence-based typing methods to provide higher levels of discrimination than traditional serotyping methods that do not recognize these as distinct antigens.

Two nucleotide sequences encode *fliC*-i in *S.* Typhimurium and three sequences encode *fliC*-g,m in *S.* Enteritidis but amino acid sequences are conserved in each. *S.* Typhimurium and *S.* Enteritidis are the most commonly occurring serotypes in England and Wales, together constituting 72% of the >11 000 isolates received by a reference laboratory in 2005 (45). Further discrimination of these serotypes would therefore be very advantageous for the purposes of outbreak investigation.

20.5.4   Variation of the *Rfb* Genes

Differences between species and strains exist in all parts of the *Salmonella* LPS, but most variation occurs in the O-polysaccharide, the somatic antigen. Somatic antigens differ in their sugar composition and in the linkages between sugars and between O-units. Variation in the somatic antigen can affect the pathogenicity of the organism, for example by avoiding specific recognition by immunoglobulins and bacteriophages. In *S. enterica*, somatic antigen types are subdivided into 46 groups according to characteristic antigenic factors they exhibit. Even so, extensive genetic variation has been found within some groups. O-antigen gene clusters are not subject to extensive recombination between strains (135) but some movement between subspecies I and II has been traced (127).

The somatic antigen structure is more complex than the flagellin antigen and the *rfb* gene cluster is up to 23 kb in size. Furthermore, it is subject to post-translational modifications that can affect the antigenic specificity. In addition, antigenic factors can be encoded by extra-chromosomal DNA. The gene sequence does not determine the O-type, but, depending on previous horizontal gene transfer events, analysis may show O-antigen markers to be present. Here, a re-sequencing approach taken to study selected somatic antigen synthesis genes is described. These sequences were analysed for differences by alignment and clustering. Potential sero-specific markers to determine whether molecular typing of the somatic antigen is possible are identified.

In the *rfb* gene clusters of salmonellae, variable sero-group specific genes are often flanked by homologous pathway genes or common genes. These flanking genes are thought to play a role in mediating the exchange of the central genes (15,56,75,128). Pathway genes are generally homologous throughout all species, while transferase and processing genes, which are responsible for a wide range of linkages among different O-antigens, are very heterogeneous.

Due to the size and complexity of *rfb* gene clusters, nucleotide sequence data are very limited. The *rfb* gene cluster structures have been determined for 10 *Salmonella* serogroups, comprising A, B, C1, C2, D1, D2, D3, E1, O54 and O35. Other studies have examined sequence differences of individual genes or pathways in multiple strains, specifically, genes for mannose synthesis (*man*), rhamnose synthesis (*rml*), and O-antigen polymerase (*wzy*).

Rhamnose, mannose and galactose are backbone components in the majority of the predominant O-antigens (groups A, B, C2, D1, D2, D3 and E1). Group C1 differs in that it has a pure mannose backbone. The *rml* gene set includes genes with two different evolutionary histories and there has probably been O-antigen transfer between subspecies. *rmlB*, *rmlD* and most of *rmlA* are subspecies specific, while part of *rmlA* and all of *rmlC* is O-antigen specific.

In *Salmonella*, there are two pathways for the synthesis of mannose as it is also a constituent of the colanic acid (CA) antigen; the *manB* and *manC* genes for CA are distinct to *manB* and *manC* for the O-antigen. *manB* and *manC* within the *rfb* cluster of some serotypes demonstrate sequence homology to CA genes, while other serotypes are chimeric (54). Some isolates have been found to have two copies of *manB* (4).

Cluster differences and *rfb* gene sequence differences have previously been used for typing a limited range of O-antigens. O-antigens in group A have a unique paratose sugar side chain; groups B and C2 have an abequose side chain, and group D has a tyvelose side chain. Sequence differences in genes encoding these side chains, or their transfer, have been applied in a PCR assay for the discrimination of these groups (79). The concept of other assays based on *rfb* genes has also been exclusive to one or two O-antigens (28,29).

20.6   Transferring to a MALDI Platform for Rapid Analysis

The previous section described specific sequence differences in antigen genes and their potential use as serotype markers. In this section, transfer of these sequences to a platform allowing rapid analysis will be discussed. There are a number of chemistries and DNA detection technologies available for DNA sequencing, although each ultimately results in nucleotide sequence data, the different methods suit different experimental situations as described here.

20.6.1   Different Methods Available

Certainly the most commonly used method is dye-terminator sequencing, which was developed by Sanger (111). It uses a mixture of deoxynucleotides (dNTPs) and lower proportion fluorescently labelled dideoxynucleotides (ddNTPs). Each nucleotide is labelled with a different fluorescent dye. A sequencing primer is bound to a single-stranded DNA template and extended by polymerase by rounds of thermocycling. As the template is copied, both dNTPs and ddNTPs are available for polymerization to the growing strand. Lacking the 3′OH group required for a subsequent nucleotide addition, ddNTPs in the chain cannot be extended, causing chain termination. The result of the sequencing reaction is a series of fragments ending in colour-coded fluorescently labelled nucleotides. The fragments are separated by gel electrophoresis to a single nucleotide resolution. The different sized fragments are read as they pass a laser allowing the sequence of the template DNA to be acquired. This method is ideal for *de novo* sequencing, requiring knowledge of upstream sequence data for design of sequencing primers.

A second method is Pyrosequencing™, developed in 1996 (106). The technology is based on sequencing by synthesis. Four nucleotides are added step-wise to a primer-template mix. Target regions of up to 100 base pairs can be sequenced by the Pyrosequencing™ method (83). Incorporation of a nucleotide i.e. extension of the DNA strand, leads to an enzymatic reaction resulting in a light flash. Each nucleotide incorporation event is accompanied by release of inorganic pyrophosphate (PPi) in a quantity equimolar to the number of incorporated nucleotides. Release of PPi triggers the adenosine triphosphate (ATP) sulfurylase reaction resulting in a quantitative conversion of PPi to ATP. ATP is readily sensed by firefly luciferase producing light, which is proportional to the amount of DNA and number of incorporated nucleotides. Nucleotides, including unincorporated dNTP and the generated ATP are degraded by apyrase allowing repetitive addition of dNTP to the solution (1). This chemistry has been applied to different platforms and is now typically used for large scale projects, such as microbial genome sequencing.

DNA sequencing using MALDI-TOF-MS instruments is a different approach. This is fully described in the previous chapter; the principle is to generate a spectrum of cleaved DNA fragments from a sample amplicon and compare this with virtual spectra from a reference sequence. Differences in the spectra reflect sequence variations (24,59). Briefly, from sequence alignments of publicly available data, primers are designed to amplify fragments of the target gene. A 300–1200 bp region of interest is amplified by two PCRs. By using tagged primers, one reaction incorporates a T7-promoter tag into the forward strand, and the second reaction incorporates a T7-promoter tag into the reverse strand. Each PCR product is split into two parts for transcription using modified C and U nucleotides. Treatment by RNase A cleaves the products at C or U due to the protective modification. Nucleotide fragments in the four base-specific cleavage reactions are measured by MALDI-TOF-MS. The strength of this method lies in re-sequencing; *de novo* sequencing is not possible. Suitable applications would include comparative sequence analysis of a set of characterized gene targets, for example multi-locus sequence typing (MLST) studies are ideal for this platform. The MLST scheme for microbial typing of *Neisseria meningitidis* has been transferred to the MALDI-TOF platform (51) in a study demonstrating the concordance with dideoxy sequencing and inter-laboratory reproducibility. Re-sequencing of conserved genes is not the limit of this technology; it can be used in assays where amplicons are expected to vary considerably, for example, *Salmonella* antigens.

20.6.2   MALDI-TOF Data Analysis

Re-sequencing depends on the availability of a reference sequence database. These may be assembled from publicly available data or generated by traditional sequencing methods. Each nucleotide difference between the reference sequence and mutant would be reflected by a series of peak shifts. Algorithms within the SNP Discovery software are used to calculate the sample sequence by collective peak variations, described in the previous chapter. In cases where the algorithm is not able to interpret all sample spectra, dependant on the number and distribution of nucleotide differences between the sample and reference sequence, dideoxy sequencing can be used to resolve the sequence.

Re-sequencing by MALDI-TOF-MS ultimately produces the same format of data as more traditional methods and can therefore be analysed in the same ways i.e. cluster analysis and sequence alignments. Cluster analysis can be used to determine levels of sequence variation and the relationship between sequence type and phenotype (such as serotype). Multiple sequence alignments can be used to identify polymorphisms between strains, for example sero-specific motifs.

20.6.3   *Salmonella* Molecular Serotyping as a Case Study

Salmonella antigen sequence data were transferred to the MALDI-TOF-MS platform. For this, MS and SNP discovery analysis were performed on a Sequenom system; specifically, *manB*, *manC*, *fliD*, *gapA*, while *fimA*, *rmlA* and *wbaV* amplicon data were collected.

With respect to the transfer of *Salmonella* serotyping to a molecular platform, there are two challenges. The first, to identify flagellin types from their sequence. As discussed earlier, flagellin genes were characterized and sero-specific motifs identified. Sequence data showed that distinct antigens may be encoded by genes that differ by a single nucleotide, while some antigens that are recognized as the same by agglutination may be encoded by diverse DNA sequences. The second challenge is to successfully identify novel sequences by recognizing and correctly characterizing new polymorphisms. Reports on sero-specific motifs in somatic antigen genes are limited and, for some genes, only a few reference sequences are available.

20.6.4   Gene Selection

The re-sequencing approach was used to detect and characterize DNA sequence variation in selected genes. Genes were selected on their predicted discriminating power; speciation, typing, sub-typing, etc. From the rfb region, genes *manB*, *manC*, *rmlA*, and *wbaV*, and *fliC* encoding the phase 1 flagellin can be used for molecular serotyping. As a control, a 656 bp amplicon of housekeeping gene *gapA*, encoding glyceraldehyde 3-phosphate dehydrogenase, was re-sequenced by the same method. The power of this amplicon to measure

Table 20.4 *Overview of results for each amplicon. Primary findings of the re-sequencing study are displayed. The number of sequence types (alleles) in subspecies I and the total number of sequence types are shown*

Amplicon	manB	manC	rmlA	wbaV	fimA	fliD	gapA	fliC
Amplicon size (bp)	415	282	458	347	303	625	656	1147
Total number of sequences	91	79	79	66	94	76	101	125
Number of subspecies 1 sequences	86	78	78	66	86	74	91	113
Number of non-subspecies 1 sequences	4	0	0	0	7	1	9	12
Number of alleles (subspecies I/total)	34/37	18/18	15/15	08/08	22/35	19/20	27/38	
Similarity of subspecies I sequences (%)	96.78	97.94	98.34	99.48	97.6	98.1	98.8	
Number of polymorphic sites per 100 bp	13.7	7.44	4.58	2.01	11.88	6.4	5.94	

overall genetic relatedness of strains was compared with strain relatedness as previously determined by other methods.

20.6.5 Results Overview

The study on *Salmonella* antigen genes generated sequences for a set of eight amplicons (Table 20.4). Amplicons were not generated for all targets; the success rate for amplification ranged from 53% for *wbaV* to 100% for *fliC*. Lack of amplification reflects the paucity of data available for assay design, for example on two sequences for *wbaV*, while the success rate for amplification of *fliC* can be attributed to the comprehensive data set used to design the assay for this amplicon.

All loci were polymorphic for DNA sequences determined for subspecies I strains. The most diverse strains (strains from subspecies other than subspecies I) did not produce amplicons for *rmlA*, *wbaV* or *manC*; primers designed for one subspecies may not be complementary to other subspecies.

The sequencing of selected genes by MALDI-TOF-MS revealed a number of things about the strains, the genes and the technology used in the study, as well as data relating to the *Salmonella* population structure. The correlation between sequence-type and serotype for each amplicon is discussed below.

20.6.6 Clustering and Sequence Variation of Amplicons

Strains were clustered by sequences of each amplicon in the study, revealing the varying discriminatory power of different amplicons and the differences in correlation to phenotype.

Of the amplicons included, *manB*, *rmlA* and *fliC* relate most strongly to serotype (see below; Figures 20.7–20.9). The value of clustering concatenated sequences is discussed below.

20.6.6.1 manB

manB DNA sequences from *S. enterica* subspecies I (*n* = 61) exhibit a 96.78% homology. The most similar non-*Salmonella* sequence as found by BLAST analysis of public databases is *E. coli* O66, which is used to root the UPGMA tree (Figure 20.7). There is a correlation between *manB* sequence and phenotype but it is not definitive. Sequences are variable for all O-groups and all serotypes. The sequences fell into three broad clusters, each containing serotypes of two or more O-groups. Cluster 1 contained the widest range of O-groups. All six C1 serotypes fell into this cluster, but there were still five sequence-types among them. Cluster 2 contained strains of serotype *S.* Muenchen, identical for *manB*. Two *S.* Muenchen strains cluster separately though – this was seen in the analyses of other amplicons. Cluster 3 contained all the *S.* Typhimurium strains, identical for *manB* barring two strains. All *S.* Heidelberg strains except one fell into cluster 3. Outliers were one strain belonging to subspecies II and two strains belonging to subspecies IIIa; the latter two both exhibited antigen O62 and had identical *manB* sequences.

In three instances *manB* sequences were identical across a range of serotypes e.g. some strains belonging to groups D, F and B share a sequence; strains of groups D and B share a sequence; C2 and R share a sequence. Differences in O-factors within O-groups do not correlate with the clustering of sequences.

20.6.6.2 rmlA

rmlA DNA sequences from subspecies I strains exhibit a 98.34% homology. Sequences in the other *S. enterica* subspecies and *S. bongori* have not been determined. BLAST analysis reveals a sequence from *E. coli* to be the most similar to the *rmlA* amplicon beyond *Salmonella*. Cluster analysis confirms a strong correlation between *rmlA* sequence and O-group (Figure 20.8). There are three main clusters and four predominant sequences in *rmlA*. All group C2 strains fall into cluster 1, having identical sequences except for one strain of *S.* Newport which differs at one nucleotide, and *S.* Agona, which falls into cluster 3. Almost 89% (39/44) of the group B strains fall into cluster 2 with 84% (37/44) of these having identical sequences. The third cluster of *rmlA* sequences is much more diverse than clusters 1 and 2. *rmlA* sequences vary most among group D strains. Although O-antigen C1 does not contain a rhamnose sugar, an *rmlA* amplicon has been identified in a strain of *S.* Decatur and a strain of *S.* Choleraesuis. C1 sequences are identical to each other and most similar to O-group B strain *S.* Agona.

20.6.6.3 *Clustering by Amplicons Reflects Strain Phylogenies*

Clustering patterns by sequences produced by re-sequencing are broadly reflective of other profiling methods. Previous studies using these strain sets include MLEE (7,13) and FAFLP (113), which did not always concur on strain groupings, as was found here with these seven amplicons (Figure 20.10).

Four sets of strains, belonging to four serotypes, are identical for every amplicon.

Some strains did not cluster as expected from their serotype. The separation of these strains from others in their serotype confirms previous studies using MLEE and FAFLP.

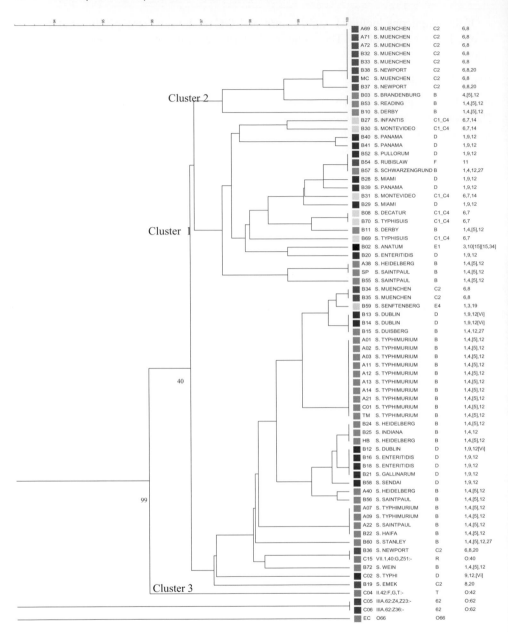

Figure 20.7 *Clustering of manB amplicons. Cluster analysis is performed using UPGMA algorithm and 1000 bootstrap simulations. Bootstrap results are shown at the major nodes. TM represents 10 S. Typhimurium strains, HB represents six S. Heidelberg strains, MC represents four S. Muenchen strains and SP represents four S. Saintpaul strains. The sequence of E. coli, serotype O66, strain DQ069297 roots the tree*

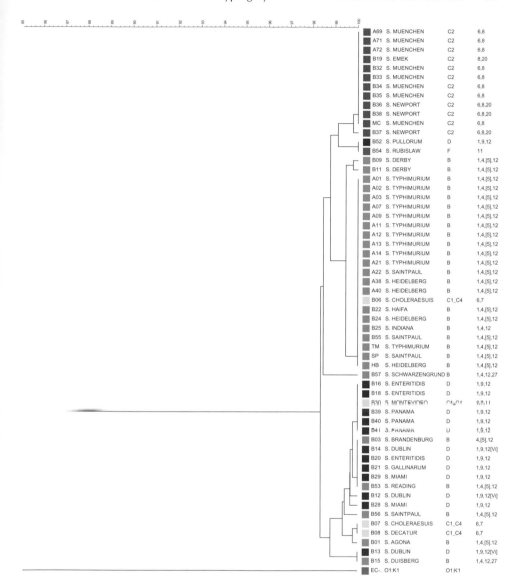

Figure 20.8 *Clustering of rmlA amplicons. Cluster analysis is performed using UPGMA algorithm and 1000 bootstrap simulations. Bootstrap results are shown at the major nodes. TM represents 10 S. Typhimurium strains, HB represents six S. Heidelberg strains, MC represents four S. Muenchen strains and SP represents four S. Saintpaul strains. The sequence of E. coli serotype K1 roots the tree*

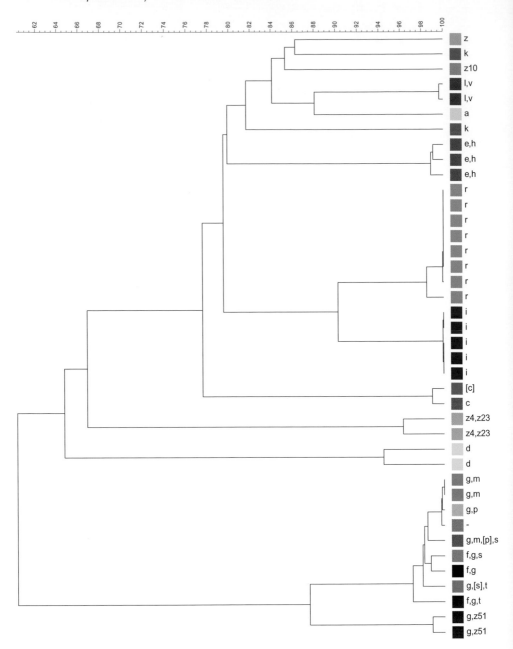

Figure 20.9   *Clustering of fliC amplicons. Cluster analysis is performed using UPGMA algorithm. Representative sequences for 21 different flagellin serotypes are included*

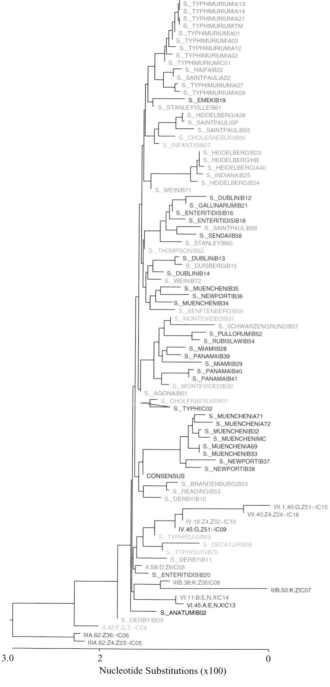

Figure 20.10 *Tree generated from alignment of DNA sequences of seven amplicons as concatenated sequences. Sequences are labelled by the serotype and strain identifier*

S. Enteritidis strain B18 had identical sequence to *S*. Muenchen strains at *manC*, *gapA*, *wbaV*, *fimA*, and similar at *fliD*. However, clustering of concatenated sequences revealed a similarity to *S*. Sendai strain B58, which was congruent with MLEE profiling. B13 *S*. Dublin and B15 *S*. Duisburg were identical at *fimA*, *wbaV*, *rmlA*, *manC*, *manB*, and both failed to amplify *fliD*, *gapA*, confirming the similar profiles from MLEE analysis.

Serotypes of the same O-group often shared sequences. *S*. Typhimurium strains clustered together, though MLEE analysis showed closer relationships between *S*. Heidelberg and *S*. Typhimurium than the sequence data generated here indicated.

Several apparently unrelated strains clustered together. *S*. Heidelberg strain A38 was observed clustering more closely with strains of *S*. Saintpaul than with other *S*. Heidelberg strains for *manB*, *manC*, *gapA* and *fimA* analyses. This does not fit with the monophyletic background of the serotype, or FAFLP analysis where strain A38 grouped with the other *S*. Heidelberg strains (113). Neither does it fit with MLEE, where it branched alone but among the other *S*. Heidelberg strains. B52 and B54 showed 100% identity at *rmlA*, *manC*, *manB*, *fliD*, *gapA* (sequence similarities were not determined for *fimA* or *wbaV*). However, B52 and B54 MLEE profiles were very different. Strains A38 and B52 require further investigation.

20.6.6.4   Congruence of Trees

Clustering of the sequences of *manB*, *manC* and *fliD* produced many of the same groupings. Thirteen clusters of matching O-groups were identified from the *manB* cluster analysis; *manC* clustering supported eight of these clusters. Many of the same clusters appear in the tree for amplicon *fliD*, although there is less variation within serotypes. *S*. Muenchen strains B34 and B35, which are separated from the other *S*. Muenchen strains by their *manB* sequence, have similar *fliD* sequences.

20.7   Conclusions and Summary

To address the identification of O-groups, gene sequences of the *rfb* cluster were studied for sero-specific motifs. Having established the sequence-types of strains, re-sequencing was used to determine the serotype of unknown isolates from marker sequence(s). This amplicon set has potential use in molecular serotyping of the O-antigen but the stability of these sequences in typical populations would need assessment. The separation of antigens exhibiting different factors shows the link between DNA sequence and serotype/O-group. Sero-specific SNPs have been identified where sequences from multiple representatives had been determined. The re-sequencing provided a high resolution and separated sequence types with only a SNP. This can not be readily achieved using hybridization-based methods. Assays targeting fragments of single genes for the identification of the O-group are subject to false negatives when sequence differences occur in primer or probe regions (27). An approach that detects multiple markers is more robust. Furthermore, the re-sequencing method described here would be applicable to the full range of O-antigens and it can distinguish between some serotypes.

Clustering of concatenated sequences of seven amplicons, including *rfb* genes, correlated well with previous phylogenetic analyses. This confirmed an essentially stable association between serotype and genetic relatedness. Despite this, there is evidence of past horizontal gene transfer events blurring this correlation (73). Considering this, new technologies should be exploited to detect genomic evolution by scanning for mobility, addition and loss of mobile genetic elements such as transposons, plasmids and phage which are often associated with the acquisition of virulence factors.

As well as impacting upon *Salmonella* typing methods, characterization of *Salmonella* flagellin can provide further understanding of epitope formation and the evolution of flagellin sequences.

By assessing the variation in these genes within populations, within serogroups and within serotypes, relationships between somatic antigen variation and genetic relatedness can be determined. Previously, MLEE and FAFLP broadly separated strains by their subspecies and serotype, showing that these markers are typically indicative of overall genetic relatedness. Examining the DNA variation that underlies antigenic variation, i.e. a higher resolution, the extent of this correlation can be determined.

manB, *manC*, *rmlA* and *wbaV* are all encoded in the *rfb* gene cluster, which carries most of the genes that determine O-group specificity. Sequences of these genes were anticipated to cluster by O-group or O-factors, which they largely did. O-groups A, B, D and E1 are closely related but sequence variation was still detected between and within serotypes, showing the accumulation of mutations since serotypes diverged. Shared sequences among multiple O-types, indicated either that the gene has been recently transferred or that the strains are related.

Similar clustering results would be predicted from adjacent genes, such as *manB* and *manC*, but in some O-groups *manB* genes originate from the colanic acid pathway (54). The similar clustering of strains by *manB* and *manC* reflects their linkage. Within O-groups B, C2 and D *rmlA* sequences were conserved, showing this gene is stable within serotypes and has been vertically inherited. Amplicons matching the *rmlA* sequences were produced from strains that could not be predicted from their O-antigen structure to have the gene. For example, C1 antigen does not contain a rhamnose sugar, yet *rmlA* amplicons were detected in two out of 10 C1 strains in the study. This may indicate a nonfunctional gene or remnant of *rmlA* in C1 strains, or an error in the typing of these strains. It would be valuable to re-serotype the strains used in this study. Gene clusters with remnants of O-antigen polymerase (*wzy*) have been documented in strains of other groups (127). There is paucity in data for group C1 probably due to its lack of homology (135).

The invariance of *wbaV* sequences may be explained by differences within the primer-binding region, so that sequences were not amplified from the variable strains. Despite their reduced range, *wbaV* amplicons were detected in C1 and C2 strains that have not previously been shown to contain this gene. This may indicate sequence homology with other genes, possibly other transferases in the cluster or elsewhere in the chromosome.

FliD sequences gave a good correlation with serotype and similar clustering patterns to *manB*, *manC* and *rmlA*. This fits with the proximity of the genes on the chromosome – for example in *S.* Typhimurium *fliC* and *fliD* are positioned at 42.5 centisomes and the *rfb* genes are at 45 centisomes. *fljB* is at 60.1 centisomes so is less likely to be transferred with the other antigen genes. *fliD* has a role in the assembly of flagellin subunits in the flagellar filament, however sequences were found not to have a direct relationship to flagellin sequence as observed in *Clostridium* sp. (122) The amino acid sequence was invariable across serotypes, and there is only one cap protein for assembly of the differing specificities of *fliC* and *fljB* subunits.

FimA encodes the subunit of fimbriae and it was expected to enhance the differentiation between similar strains (12). This gene has been found to harbour small deletions which may act as a mechanism for rapid diversification of surface molecules (4). As results have shown, there was sufficient variation within the *rfb* genes to separate O-groups. Overall, *fimA* did not further discriminate groups. A clear area for further work would be to include more fimbrial genes and investigate how sequences correlate with phage type.

20.7.1   Closing Remarks

Historically, combinations of typing methods have been used to detect serological and genomic variation. Serology identifies differences in the most variable part of the *Salmonella* genome, so is useful for discriminating between many similar strains. Serology has been incredibly successful in its application to *Salmonella* because it is low-tech, reflects immune pressure and serotype does not switch as often as previously thought – maybe less often than changes in virulence factors. However, it misses important genetic changes that affect pathogenicity, which is why additional molecular methods such as PFGE and plasmid profiling are needed. Extra-chromosomal DNA in *Salmonella* is where much of the variability lies and can confer medically important phenotypic traits such as antibiotic resistance and flagellin type. The acquisition of virulence factors and antibiotic resistance genes leads to the emergence of new strains and outbreaks, so these are the genetic differences that should be the targets of a clinically relevant typing method. A scheme for *Salmonella* would ideally include combinations of clinically significant genetic markers with molecular serotype for cross-referencing to previous strains. This is almost equivalent to the multiple methods commonly required to differentiate strains but by re-sequencing these can potentially be determined on one molecular platform producing unequivocal and portable sequence data.

References

1 Agah A, Aghajan M, Mashayekhi F, Amini S, Davis RW, Plummer JD, Ronaghi M, and Griffin PB: A multi-enzyme model for Pyrosequencing. Nucleic Acids Res *32*: e166, 2004.
2 Aizawa SI: Flagellar assembly in *Salmonella* typhimurium. Mol Microbiol *19*: 1–5, 1996.
3 Aizawa SI: Bacterial flagella and type III secretion systems. FEMS Microbiol Lett *202*: 157–164, 2001.
4 Alcaine SD, Soyer Y, Warnick LD, Su WL, Sukhnanand S, Richards J, Fortes ED, McDonough P, Root TP, Dumas NB, Grohn Y, and Wiedmann M: Multilocus sequence typing supports the hypothesis that cow- and human-associated *Salmonella* isolates represent distinct and overlapping populations. Appl Environ Microbiol *72*: 7575–7585, 2006.
5 Alvarez J, Sota M, Vvanco A, Perales I, Cisterna R, Rementeria R, and Garaizar J: Development of a multiplex PCR technique for detection and epidemiological typing of *Salmonella* in human clinical samples. J Clin Microbiol *42*: 1734–1738, 2004.
6 Baumler AJ: The record of horizontal gene transfer in *Salmonella*. Trends Microbiol *5*: 318–322, 1997.
7 Beltran P, Musser JM, Helmuth R, Farmer JJ, III, Frerichs WM, Wachsmuth IK, Ferris K, McWhorter AC, Wells JG, and Cravioto A: Toward a population genetic analysis of *Salmonella*: genetic diversity and relationships among strains of serotypes S. choleraesuis, S. derby, S. dublin, S. enteritidis, S. heidelberg, S. infantis, S. newport, and S. typhimurium. Proc Natl Acad Sci USA *85*: 7753–7757, 1988.

8 Beltran P, Plock SA, Smith NH, Whittam TS, Old DC, and Selander RK: Reference collection of strains of the *Salmonella* typhimurium complex from natural populations. J Gen Microbiol *137* (Pt 3): 601–606, 1990.

9 Blake DP, Hillman K, Fenlon DR, and Low JC: Transfer of antibiotic resistance between commensal and pathogenic members of the Enterobacteriaceae under ileal conditions. J Appl Microbiol *95*: 428–436, 2003.

10 Blanc-Potard AB and Lafay B: MgtC as a horizontally-acquired virulence factor of intracellular bacterial pathogens: evidence from molecular phylogeny and comparative genomics. J Mol Evol *57*: 479–486, 2003.

11 Boyd EF and Brussow H: Common themes among bacteriophage-encoded virulence factors and diversity among the bacteriophages involved. Trends Microbiol *10*: 521–529, 2002.

12 Boyd EF and Hartl DL: Analysis of the type 1 pilin gene cluster *fim* in *Salmonella*: its distinct evolutionary histories in the 5' and 3' regions. J Bacteriol *181*: 1301–1308, 1999.

13 Boyd EF, Wang FS, Whittam TS, and Selander RK: Molecular genetic relationships of the salmonellae. Appl Environ Microbiol *62*: 804–808, 1996.

14 Brenner DJ, Fanning GR, Johnson KE, Citarella RV, and Falkow S: Polynucleotide sequence relationships among members of Enterobacteriaceae. J Bacteriol *98*: 637–650, 1969.

15 Brown PK, Romana LK, and Reeves PR: Molecular analysis of the *rfb* gene cluster of *Salmonella* serovar muenchen (strain M67): the genetic basis of the polymorphism between groups C2 and B. Mol Microbiol *6*: 1385–1394, 1992.

16 Chang HR, Loo LH, Jeyaseelan K, Earnest L, and Stackebrandt E: Phylogenetic relationships of *Salmonella* typhi and *Salmonella* typhimurium based on 16S rRNA sequence analysis. Int J Syst Bacteriol *47*: 1253–1254, 1997.

17 Choi IG and Kim SH: Global extent of horizontal gene transfer. Proc Natl Acad Sci USA *104*: 4489–4494, 2007.

18 Conner CP, Heithoff DM, Julio SM, Sinsheimer RL, and Mahan MJ: Differential patterns of acquired virulence genes distinguish *Salmonella* strains. Proc Natl Acad Sci USA *95*: 4641–4645, 1998.

19 Crosa JH, Brenner DJ, Ewing WH, and Falkow S: Molecular relationships among the *Salmonelleae*. J Bacteriol *115*: 307–315, 1973.

20 Curd H, Liu D, and Reeves PR: Relationships among the O-antigen gene clusters of *Salmonella enterica* groups B, D1, D2, and D3. J Bacteriol *180*: 1002–1007, 1998.

21 Dauga C, Zabrovskaia A, and Grimont PA: Restriction fragment length polymorphism analysis of some flagellin genes of *Salmonella enterica*. J Clin Microbiol *36*: 2835–2843, 1998.

22 de Vries N, Zwaagstra KA, Huis, V, van Knapen F, van Zijderveld FG, and Kusters JG: Production of monoclonal antibodies specific for the i and 1,2 flagellar antigens of *Salmonella* typhimurium and characterization of their respective epitopes. Appl Environ Microbiol *64*: 5033–5038, 1998.

23 Desai M, Threlfall EJ, and Stanley J: Fluorescent amplified-fragment length polymorphism subtyping of the *Salmonella enterica* serovar enteritidis phage type 4 clone complex. J Clin Microbiol *39*: 201–206, 2001.

24 Ehrich M, Bocker S, and van den BD: Multiplexed discovery of sequence polymorphisms using base-specific cleavage and MALDI-TOF MS. Nucleic Acids Res *33*: e38, 2005.

25 Erridge C, Bennett-Guerrero E, and Poxton IR: Structure and function of lipopolysaccharides. Microbes Infect *4*: 837–851, 2002.

26 Fakhr MK, Nolan LK, and Logue CM: Multilocus sequence typing lacks the discriminatory ability of pulsed-field gel electrophoresis for typing *Salmonella enterica* serovar Typhimurium. J Clin Microbiol *43*: 2215–2219, 2005.

27 Fitzgerald C, Collins M, van Duyne S, Mikoleit M, Brown T, and Fields P: Multiplex, bead-based suspension array for the molecular determination of common *Salmonella* serogroups. J Clin Microbiol E.pub ahead of print, 2007.

28 Fitzgerald C, Gheesling L, Collins M, and Fields PI: Sequence analysis of the *rfb* loci encoding biosynthesis of the *S. enterica* O17 and O18 antigens: serogroup-specific identification by PCR. Appl Environ Microbiol *72*: 7949–7953, 2006.

29 Fitzgerald C, Sherwood R, Gheesling LL, Brenner FW, and Fields PI: Molecular analysis of the *rfb* O antigen gene cluster of *Salmonella enterica* serogroup O:6,14 and development of a serogroup-specific PCR assay. Appl Environ Microbiol *69*: 6099–6105, 2003.

30 Frankel G, Newton SM, Schoolnik GK, and Stocker BA: Unique sequences in region VI of the flagellin gene of *Salmonella typhi*. Mol Microbiol *3*: 1379–1383, 1989.

31 Fraser GM and Hughes C: Swarming motility. Curr Opin Microbiol *2*: 630–635, 1999.

32 Gophna U, Ron EZ, and Graur D: Bacterial type III secretion systems are ancient and evolved by multiple horizontal-transfer events. Gene *312*: 151–163, 2003.

33 Groisman EA and Ochman H: How *Salmonella* became a pathogen. Trends Microbiol *5*: 343–349, 1997.

34 Guard-Petter J, Keller LH, Rahman MM, Carlson RW, and Silvers S: A novel relationship between O-antigen variation, matrix formation, and invasiveness of *Salmonella* enteritidis. Epidemiol Infect *117*: 219–231, 1996.

35 Guerra B, Landeras E, Gonzalez-Hevia MA, and Mendoza MC: A three-way ribotyping scheme for *Salmonella* serotype Typhimurium and its usefulness for phylogenetic and epidemiological purposes. J Med Microbiol *46*: 307–313, 1997.

36 Guiney DG, Fang FC, Krause M, Libby S, Buchmeier NA, and Fierer J: Biology and clinical significance of virulence plasmids in *Salmonella* serovars. Clin Infect Dis *21* (Suppl. 2): S146–S151, 1995.

37 Gupta S and Maiden MC: Exploring the evolution of diversity in pathogen populations. Trends Microbiol *9*: 181–185, 2001.

38 Hacker J and Kaper JB: Pathogenicity islands and the evolution of microbes. Annu Rev Microbiol *54*: 641–679, 2000.

39 Hanafusa T, Saito K, Tominaga A, and Enomoto M: Nucleotide sequence and regulated expression of the *Salmonella fljA* gene encoding a repressor of the phase 1 flagellin gene. Mol Gen Genet *236*: 260–266, 1993.

40 Hanage WP, Fraser C, and Spratt BG: Sequences, sequence clusters and bacterial species. Philos Trans R Soc Lond, Ser B *361*: 1917–1927, 2006.

41 Hansen-Wester I and Hensel M: *Salmonella* pathogenicity islands encoding type III secretion systems. Microbes Infect *3*: 549–559, 2001.

42 Hansen-Wester I and Hensel M: Genome-based identification of chromosomal regions specific for *Salmonella* spp. Infect Immun *70*: 2351–2360, 2002.

43 Hayashi F, Smith KD, Ozinsky A, Hawn TR, Yi EC, Goodlett DR, Eng JK, Akira S, Underhill DM, and Aderem A: The innate immune response to bacterial flagellin is mediated by Toll-like receptor 5. Nature *410*: 1099–1103, 2001.

44 He XS, Rivkina M, Stocker BA, and Robinson WS: Hypervariable region IV of *Salmonella* gene *fliCd* encodes a dominant surface epitope and a stabilizing factor for functional flagella. J Bacteriol *176*: 2406–2414, 1994.

45 Health Protection Agency. *Salmonella* in humans (excluding *S.* Typhi & *S.* Paratyphi) reported to the Health Protection Agency Centre for Infections, England and Wales, 1981–2006. http://www.hpa.org.uk/infections/topics_az/salmonella/data_human.htm. 2007.

46 Health Protection Agency. Salmonella in humans (excluding S. typhi; S. Paratyphi). Faecal and lower gastrointestinal tract isolates reported to the Health Protection Agency Centre for Infections. England and Wales, 1990–2008. http://www.hpa.org.uk/infections/topics_az/salmonella/data_human.htm. 2009.

47 Helander IM, Moran AP, and Makela PH: Separation of two lipopolysaccharide populations with different contents of O-antigen factor 122 in *Salmonella enterica* serovar typhimurium. Mol Microbiol *6*: 2857–2862, 1992.

48 Hensel M, Shea JE, Baumler AJ, Gleeson C, Blattner F, and Holden DW: Analysis of the boundaries of *Salmonella* pathogenicity island 2 and the corresponding chromosomal region of *Escherichia coli* K-12. J Bacteriol *179*: 1105–1111, 1997.

49 Homma M, Fujita H, Yamaguchi S, and Iino T: Regions of *Salmonella* typhimurium flagellin essential for its polymerization and excretion. J Bacteriol *169*: 291–296, 1987.

50 Hong Y, Liu T, Hofacre C, Maier M, White DG, Ayers S, Wang L, and Maurer JJ: A restriction fragment length polymorphism-based polymerase chain reaction as an alternative to serotyping for identifying *Salmonella* serotypes. Avian Dis *47*: 387–395, 2003.

51 Honisch C, Chen Y, Mortimer C, Arnold C, Schmidt O, van den BD, Cantor CR, Shah HN, and Gharbia SE: Automated comparative sequence analysis by base-specific cleavage and mass spectrometry for nucleic acid-based microbial typing. Proc Natl Acad Sci USA *104*: 10649–10654, 2007.

52 House D, Bishop A, Parry C, Dougan G, and Wain J: Typhoid fever: pathogenesis and disease. Curr Opin Infect Dis *14*: 573–578, 2001.

53 Ibrahim GF, Fleet GH, Lyons MJ, and Walker RA: Production of potent *Salmonella* H antisera by immunization with polymeric flagellins. J Clin Microbiol *22*: 347–351, 1985.

54 Jensen SO and Reeves PR: Molecular evolution of the GDP-mannose pathway genes (*manB* and *manC*) in *Salmonella enterica*. Microbiology *147*: 599–610, 2001.

55 Ji WS, Hu JL, Qiu JW, Pan BR, Peng DR, Shi BL, Zhou SJ, Wu KC, and Fan DM: Relationship between genotype and phenotype of flagellin C in *Salmonella*. World J Gastroenterol *7*: 864–867, 2001.

56 Jiang XM, Neal B, Santiago F, Lee SJ, Romana LK, and Reeves PR: Structure and sequence of the *rfb* (O antigen) gene cluster of *Salmonella* serovar typhimurium (strain LT2). Mol Microbiol *5*: 695–713, 1991.

57 Joys TM: The covalent structure of the phase-1 flagellar filament protein of *Salmonella* typhimurium and its comparison with other flagellins. J Biol Chem *260*: 15758–15761, 1985.

58 Joys TM and Schodel F: Epitope mapping of the d flagellar antigen of *Salmonella* muenchen. Infect Immun *59*: 3330–3332, 1991.

59 Jurinke C, van den BD, Cantor CR, and Koster H: The use of MassARRAY technology for high throughput genotyping. Adv Biochem Eng Biotechnol *77*: 57–74, 2002.

60 Kanto S, Okino H, Aizawa S, and Yamaguchi S: Amino acids responsible for flagellar shape are distributed in terminal regions of flagellin. J Mol Biol *219*: 471–480, 1991.

61 Keenleyside WJ, Perry M, Maclean L, Poppe C, and Whitfield C: A plasmid-encoded rfbO:54 gene cluster is required for biosynthesis of the O:54 antigen in *Salmonella enterica* serovar Borreze. Mol Microbiol *11*: 437–448, 1994.

62 Kidgell C, Reichard U, Wain J, Linz B, Torpdahl M, Dougan G, and Achtman M: *Salmonella* typhi, the causative agent of typhoid fever, is approximately 50,000 years old. Infect Genet Evol *2*: 39–45, 2002.

63 Kilger G and Grimont PA: Differentiation of *Salmonella* phase 1 flagellar antigen types by restriction of the amplified *fliC* gene. J Clin Microbiol *31*: 1108–1110, 1993.

64 Kingsley RA and Baumler AJ: Host adaptation and the emergence of infectious disease: the *Salmonella* paradigm. Mol Microbiol *36*: 1006–1014, 2000.

65 Kingsley RA and Baumler AJ: Pathogenicity islands and host adaptation of *Salmonella* serovars. Curr Top Microbiol Immunol *264*: 67–87, 2002.

66 Konstantinidis KT, Ramette A, and Tiedje JM: Towards a more robust assessment of intraspecies diversity using fewer genetic markers. Appl Environ Microbiol 2006.

67 Kotetishvili M, Stine OC, Kreger A, Morris JG, Jr, and Sulakvelidze A: Multilocus sequence typing for characterization of clinical and environmental *Salmonella* strains. J Clin Microbiol *40*: 1626–1635, 2002.

68 Lan R and Reeves PR: Gene transfer is a major factor in bacterial evolution. Mol Biol Evol *13*: 47–55, 1996.

69 Lawson AJ, Stanley J, Threlfall EJ, and Desai M: Fluorescent amplified fragment length polymorphism subtyping of multiresistant *Salmonella enterica* serovar Typhimurium DT104. J Clin Microbiol *42*: 4843–4845, 2004.

70 Lee CA: Pathogenicity islands and the evolution of bacterial pathogens. Infect Agents Dis *5*: 1–7, 1996.

71 Lee SJ, Romana LK, and Reeves PR: Sequence and structural analysis of the *rfb* (O antigen) gene cluster from a group C1 *Salmonella enterica* strain. J Gen Microbiol *138* (Pt 9): 1843–1855, 1992.

72 Lerouge I and Vanderleyden J: O-antigen structural variation: mechanisms and possible roles in animal/plant-microbe interactions. FEMS Microbiol Rev *26*: 17–47, 2002.

73 Li J, Nelson K, McWhorter AC, Whittam TS, and Selander RK: Recombinational basis of serovar diversity in *Salmonella enterica*. Proc Natl Acad Sci USA *91*: 2552–2556, 1994.

74 Li J, Smith NH, Nelson K, Crichton PB, Old DC, Whittam TS, and Selander RK: Evolutionary origin and radiation of the avian-adapted non-motile salmonellae. J Med Microbiol *38*: 129–139, 1993.

75 Liu D, Verma NK, Romana LK, and Reeves PR: Relationships among the *rfb* regions of *Salmonella* serovars A, B, and D. J Bacteriol *173*: 4814–4819, 1991.

76 Liu SL, Hessel A, and Sanderson KE: Genomic mapping with I-Ceu I, an intron-encoded endonuclease specific for genes for ribosomal RNA, in *Salmonella* spp., *Escherichia coli*, and other bacteria. Proc Natl Acad Sci USA *90*: 6874–6878, 1993.

77 Looney RJ and Steigbigel RT: Role of the Vi antigen of *Salmonella* typhi in resistance to host defense in vitro. J Lab Clin Med *108*: 506–516, 1986.

78 Ludwig W and Schleifer KH: Bacterial phylogeny based on 16S and 23S rRNA sequence analysis. FEMS Microbiol Rev *15*: 155–173, 1994.

79 Luk JM, Kongmuang U, Reeves PR, and Lindberg AA: Selective amplification of abequose and paratose synthase genes (*rfb*) by polymerase chain reaction for identification of *Salmonella* major serogroups (A, B, C2, and D). J Clin Microbiol *31*: 2118–2123, 1993.

80 Macnab RM: Genetics and biogenesis of bacterial flagella. Annu Rev Genet *26*: 131–158, 1992.

81 Macnab RM: Flagella and motility. In: *Eschericia coli* and *Salmonella*: Cellular and Molecular Biology (Neidhardt FC, Curtiss R, Ingraham JL, and Low KB eds). Washington DC, ASM Press, 1996, pp 123–145.

82 Marcus SL, Brumell JH, Pfeifer CG, and Finlay BB: *Salmonella* pathogenicity islands: big virulence in small packages. Microbes Infect *2*: 145–156, 2000.

83 Mashayekhi F and Ronaghi M: Analysis of read length limiting factors in Pyrosequencing chemistry. Anal Biochem *363*: 275–287, 2007.

84 Masten BJ and Joys TM: Molecular analyses of the *Salmonella* g … flagellar antigen complex. J Bacteriol *176*: 2771, 1994.

85 Matic I, Taddei F, and Radman M: Genetic barriers among bacteria. Trends Microbiol *4*: 69–72, 1996.

86 Mayeux PR: Pathobiology of lipopolysaccharide. J Toxicol Environ Health *51*: 415–435, 1997.

87 McConnell MR, Oakes KR, Patrick AN, and Mills DM: Two functional O-polysaccharide polymerase *wzy* (*rfc*) genes are present in the rfb gene cluster of Group E1 *Salmonella enterica* serovar Anatum. FEMS Microbiol Lett *199*: 235–240, 2001.

88 McQuiston JR, Parrenas R, Ortiz-Rivera M, Gheesling L, Brenner F, and Fields PI: Sequencing and comparative analysis of flagellin genes *fliC*, *fljB*, and *flpA* from *Salmonella*. J Clin Microbiol *42*: 1923–1932, 2004.

89 Mortimer CK, Peters TM, Gharbia SE, Logan JM, and Arnold C: Towards the development of a DNA-sequence based approach to serotyping of *Salmonella enterica*. BMC Microbiol *4*: 31, 2004.

90 Murase T, Okitsu T, Suzuki R, Morozumi H, Matsushima A, Nakamura A, and Yamai S: Evaluation of DNA fingerprinting by PFGE as an epidemiologic tool for *Salmonella* infections. Microbiol Immunol *39*: 673–676, 1995.

91 Newton SM, Wasley RD, Wilson A, Rosenberg LT, Miller JF, and Stocker BA: Segment IV of a *Salmonella* flagellin gene specifies flagellar antigen epitopes. Mol Microbiol *5*: 419–425, 1991.

92 Nghiem HO, Himmelspach K, and Mayer H: Immunochemical and structural analysis of the O polysaccharides of *Salmonella* zuerich [1,9,27,(46)]. J Bacteriol *174*: 1904–1910, 1992.

93 Nnalue NA and Lindberg AA: O-antigenic determinants in *Salmonella* species of serogroup C1 are expressed in distinct immunochemical populations of chains. Microbiology *143* (Pt 2): 653–662, 1997.

94 Ochman H and Wilson AC: Evolution in bacteria: evidence for a universal substitution rate in cellular genomes. J Mol Evol *26*: 74–86, 1987.

95 Ohl ME and Miller SI: *Salmonella*: a model for bacterial pathogenesis. Annu Rev Med *52*: 259–274, 2001.

96 Parish CR, Wistar R, and Ada G: Cleavage of bacterial flagellin with cyanogen bromide: antigenic properties of the protein fragments. Biochem J *113*: 501–506, 1969.

97 Parker CT and Guard-Petter J: Contribution of flagella and invasion proteins to pathogenesis of *Salmonella enterica* serovar enteritidis in chicks. FEMS Microbiol Lett *204*: 287–291, 2001.

98 Peters TM and Threlfall EJ: Single-enzyme amplified fragment length polymorphism and its applicability for *Salmonella* epidemiology. Syst Appl Microbiol *24*: 400–404, 2001.

99 Popoff MY: Antigenic Formulas of *Salmonella* serovars. Paris, Institute Pasteur, 2001.

100 Porwollik S and McClelland M: Lateral gene transfer in *Salmonella*. Microbes Infect *5*: 977–989, 2003.

101 Porwollik S, Wong RM, and McClelland M: Evolutionary genomics of *Salmonella*: gene acquisitions revealed by microarray analysis. Proc Natl Acad Sci USA *99*: 8956–8961, 2002.

102 Reeves P: Evolution of *Salmonella* O antigen variation by interspecific gene transfer on a large scale. Trends Genet *9*: 17–22, 1993.

103 Reeves PR: Variation in O-antigens, niche-specific selection and bacterial populations. FEMS Microbiol Lett *79*: 509–516, 1992.

104 Reeves PR, Hobbs M, Valvano MA, Skurnik M, Whitfield C, Coplin D, Kido N, Klena J, Maskell D, Raetz CR, and Rick PD: Bacterial polysaccharide synthesis and gene nomenclature. Trends Microbiol *4*: 495–503, 1996.

105 Ridley AM, Threlfall EJ, and Rowe B: Genotypic characterization of *Salmonella* enteritidis phage types by plasmid analysis, ribotyping, and pulsed-field gel electrophoresis. J Clin Microbiol *36*: 2314–2321, 1998.

106 Ronaghi M, Karamohamed S, Pettersson B, Uhlen M, and Nyren P: Real-time DNA sequencing using detection of pyrophosphate release. Anal Biochem *242*: 84–89, 1996.

107 Rychlik I, Gregorova D, and Hradecka H: Distribution and function of plasmids in *Salmonella enterica*. Vet Microbiol *112*: 1–10, 2006.

108 Salmon D and Smith T: The bacterium of swine-plague. Am Microscop J *7*: 204–205, 1886.

109 Samatey FA, Imada K, Nagashima S, Vonderviszt F, Kumasaka T, Yamamoto M, and Namba K: Structure of the bacterial flagellar protofilament and implications for a switch for supercoiling. Nature *410*: 331–337, 2001.

110 Samuel G and Reeves P: Biosynthesis of O-antigens: genes and pathways involved in nucleotide sugar precursor synthesis and O-antigen assembly. Carbohydr Res *338*: 2503–2519, 2003.

111 Sanger F, Nicklen S, and Coulson AR: DNA sequencing with chain-terminating inhibitors. Proc Natl Acad Sci USA *74*: 5463–5467, 1977.

112 Schmidt H and Hensel M: Pathogenicity islands in bacterial pathogenesis. Clin Microbiol Rev *17*: 14–56, 2004.

113 Scott F, Threlfall J, and Arnold C: Genetic structure of *Salmonella* revealed by fragment analysis. Int J Syst Evol Microbiol *52*: 1701–1713, 2002.

114 Selander RK, Smith NH, Li J, Beltran P, Ferris KE, Kopecko DJ, and Rubin FA: Molecular evolutionary genetics of the cattle-adapted serovar *Salmonella* dublin. J Bacteriol *174*: 3587–3592, 1992.

115 Silverman M, Zieg J, Hilmen M, and Simon M: Phase variation in *Salmonella*: genetic analysis of a recombinational switch. Proc Natl Acad Sci USA *76*: 391–395, 1979.

116 Smith NH and Selander RK: Molecular genetic basis for complex flagellar antigen expression in a triphasic serovar of *Salmonella*. Proc Natl Acad Sci USA *88*: 956–960, 1991.

117 Sonne-Hansen J and Jenabian SM: Molecular serotyping of Salmonella: identification of the phase 1 H antigen based on partial sequencing of the *fliC* gene. APMIS *113*: 340–348, 2005.

118 Stock JB, Surette MG, McCleary WR, and Stock AM: Signal transduction in bacterial chemotaxis. J Biol Chem *267*: 19753–19756, 1992.

119 Sukhnanand S, Alcaine S, Warnick LD, Su WL, Hof J, Craver MP, McDonough P, Boor KJ, and Wiedmann M: DNA sequence-based subtyping and evolutionary analysis of selected *Salmonella enterica* serotypes. J Clin Microbiol *43*: 3688–3698, 2005.

120 Szekely E and Simon M: Homology between the invertible deoxyribonucleic acid sequence that controls flagellar-phase variation in Salmonella sp. and deoxyribonucleic acid sequences in other organisms. J Bacteriol *148*: 829–836, 1981.

121 Szekely E and Simon M: DNA sequence adjacent to flagellar genes and evolution of flagellar-phase variation. J Bacteriol *155*: 74–81, 1983.

122 Tasteyre A, Karjalainen T, Avesani V, Delmee M, Collignon A, Bourlioux P, and Barc MC: Phenotypic and genotypic diversity of the flagellin gene (*fliC*) among *Clostridium difficile* isolates from different serogroups. J Clin Microbiol *38*: 3179–3186, 2000.

123 Threlfall EJ, Hampton MD, Chart H, and Rowe B: Use of plasmid profile typing for surveillance of *Salmonella* enteritidis phage type 4 from humans, poultry and eggs. Epidemiol Infect *112*: 25–31, 1994.

124 Vazquez-Torres A, Jones-Carson J, Baumler AJ, Falkow S, Valdivia R, Brown W, Le M, Berggren R, Parks WT, and Fang FC: Extraintestinal dissemination of *Salmonella* by CD18-expressing phagocytes. Nature *401*: 804–808, 1999.

125 Vernikos GS, Thomson NR, and Parkhill J: Genetic flux over time in the *Salmonella* lineage. Genome Biol *8*: R100, 2007.

126 Wain J, House D, Pickard D, Dougan G, and Frankel G: Acquisition of virulence-associated factors by the enteric pathogens *Escherichia coli* and *Salmonella enterica*. Philos Trans R Soc Lond, Ser B *356*: 1027–1034, 2001.

127 Wang L, Andrianopoulos K, Liu D, Popoff MY, and Reeves PR: Extensive variation in the O-antigen gene cluster within one *Salmonella enterica* serogroup reveals an unexpected complex history. J Bacteriol *184*: 1669–1677, 2002.

128 Wang L, Romana LK, and Reeves PR: Molecular analysis of a *Salmonella enterica* group E1 *rfb* gene cluster: O antigen and the genetic basis of the major polymorphism. Genetics *130*: 429–443, 1992.

129 Wang L, Rothemund D, Curd H, and Reeves PR: Species-wide variation in the *Escherichia coli* flagellin (H-antigen) gene. J Bacteriol *185*: 2936–2943, 2003.

130 Wang Q, Suzuki A, Mariconda S, Porwollik S, and Harshey RM: Sensing wetness: a new role for the bacterial flagellum. EMBO J *24*: 2034–2042, 2005.

131 Weisburg WG, Barns SM, Pelletier DA, and Lane DJ: 16S ribosomal DNA amplification for phylogenetic study. J Bacteriol *173*: 697–703, 1991.

132 Whittam TS and Bumbaugh AC: Inferences from whole-genome sequences of bacterial pathogens. Curr Opin Genet Dev *12*: 719–725, 2002.

133 Winokur PL: Molecular epidemiological techniques for *Salmonella* strain discrimination. Front Biosci *8*: c14–c24, 2003.

134 World Health Organisation. *Salmonella*. http://www.who.int/topics/salmonella/en/. 2007.

135 Xiang SH, Haase AM, and Reeves PR: Variation of the *rfb* gene clusters in *Salmonella enterica*. J Bacteriol *175*: 4877–4884, 1993.

136 Yu LSL and Fung DYC: Use if selective motility enrichment to isolate foodborne pathogens: A review. J Rapid Methods Automat Microbiol *2*: 167, 1993.

137 Zeng H, Carlson AQ, Guo Y, Yu Y, Collier-Hyams LS, Madara JL, Gewirtz AT, and Neish AS: Flagellin is the major proinflammatory determinant of enteropathogenic *Salmonella*. J Immunol *171*: 3668–3674, 2003.

Index

Note: page numbers in italics refer to figures and tables

Mass Spectrometry for Microbial Proteomics   Edited by Haroun N. Shah and Saheer E. Gharbia
© 2010 John Wiley & Sons, Ltd